ENGINEERING MATHEMATICS-II

L. Mary Florida

M. Felix Nes Mabel

T. Berjin Magizha

S. Asha Alice

Department of Mathematics,

St. Xavier's Catholic College of Engineering,

Nagercoil.

Published by

BONFRING®
Intellectual Integrity

ENGINEERING MATHEMATICS–II

First Edition : 2016
New Revised Edition : 2017

9 789386 176417 >

ISBN 978-93-86176-41-7

₹ 599

Authors

L. Mary Florida

M. Felix Nes Mabel

T. Berjin Magizha

S. Asha Alice

Bonfring

309, 2nd Floor, 5th Street Extension, Gandhipuram,

Coimbatore-641 012, Tamilnadu, India.

E-mail: info@bonfring.org

Website: www.bonfring.org

Phone: 0422 4213231

Authors Wish to Dedicate

This Book

To their Parents

KALASALINGAM UNIVERSITY
(Kalasalingam Academy of Research and Education)
(Under Sec.3 of UGC Act 1956)
Anand Nagar, Krishnankoil-626 126, Srivilliputtur (via), Tamil Nadu, India.
Phone : 04563-289042, Fax : 04563-289322
www.kalasalingam.ac.in email : info@kalasalingam.ac.in
(Accredited by NAAC with B Grade with a CGPA of 2.81)

"Kalvivallal"
T. Kalasalingam
Chairman

'Ilayavallal'
K. Sridharan
Chancellor

Dr. S. Arumugam, M.Sc., M.Phil., Ph.D.,
Senior Professor (Research) & Director (n-CARDMATH)
E-mail: s.arumugam.klu@gmail.com

Off : +91 04563 289303
Mobile : 0 98940 27224

FOREWORD

The great Indian Mathematician Srinivasa Ramanujan has stated that an equation means nothing to me unless it expresses a thought of God. It is well known that Mathematics plays an important role in building up modern civilization and it is the language of science and technology. Hence science becomes exact only when Mathematics is employed.

I am delighted to write this foreword to this book on Engineering Mathematics which caters to the requirements of the Anna University syllabus. The development of the material is very lucid and simple. The material that has been included in this text book is comprehensive and the students of all engineering streams will find this text book useful. I congratulate the authors for bringing out such a nice book and I wish them success.

Dr. S. ARUMUGAM
Director(n-CARDMATH)

Preface

This book will provide a foundation in mathematical principles, which will enable students to solve mathematical, scientific and associated engineering principles. In addition, the material will provide engineering applications and mathematical principles necessary for advancement onto a range of Incorporated Engineer degree profiles. It is widely recognized that a student's ability to use mathematics is a key element in determining subsequent success.

In Engineering Mathematics-II, theory is introduced in each chapter by a simple outline of essential definitions, formulae, theorems and procedures. The theory is kept to a minimum, for problems solving is extensively used to establish and exemplify the theory. It is intended that readers will gain real understanding through seeing problems solved and then through solving similar problems themselves.

For clarity, the text is divided into five topic areas, these being: Vector Calculus, Ordinary Differential Equations, Laplace Transform, Analytic Functions and Complex Integration.

Acknowledgments

We would like to acknowledge our sincere thanks to our learned professor and authors of a number of excellent books on the subject, and inquisitive students who motivated us to understand the subject better and present it in an engaging manner.

Several people have contributed directly and indirectly to this project. We express our thanks to our Correspondent Rev.Dr.Fr.M.Maria William, Principal Dr.S.Joseph Sekhar, Bursar Rev. Fr.P.Benziger and the management of St. Xavier's Catholic College of Engineering who gave an excellent opportunity to develop this project in a wonderful environment.

We thank our department members for their encouragement and support at every stage of the preparation of the book. We conclude the acknowledgements by recording our grateful thanks to the Bonfring publishers and their dedicated team and Editorial Manager for their encouraging support, guidance, patience and prompt response to our queries.

Request to Readers

We request the readers to send their valuable suggestions for any modification on addition by way of text, illustration or case. Although a sincere attempt has been made to keep the book free of inaccuracies, yet it is possible for some errors to have crept in a book of this volume. We shall appreciate if errors, if any, are brought to our notice by users of the book. Also, we shall look forward to comments, observations and suggestions for improving the contents of book from the teachers, students and readers. The Communication to authors could be mailed to **floridanxn@gmail.com.**

L. Mary Florida

M. Felix Nes Mabel

T. Berjin Magizha

S. Asha Alice

MA8251 ENGINEERING MATHEMATICS-II

UNIT I MATRICES

Eigen values and Eigenvectors of a real matrix – Characteristic equation – Properties of Eigen values and Eigenvectors – Cayley-Hamilton theorem – Diagonalization of matrices – Reduction of a quadratic form to canonical form by orthogonal transformation – Nature of quadratic forms.

UNIT II VECTOR CALCULUS

Gradient and directional derivative – Divergence and curl - Vector identities – Irrotational and Solenoidal vector fields – Line integral over a plane curve – Surface integral - Area of a curved surface - Volume integral - Green's, Gauss divergence and Stoke's theorems – Verification and application in evaluating line, surface and volume integrals.

UNIT III ANALYTIC FUNCTIONS

Analytic functions – Necessary and sufficient conditions for analyticity in Cartesian and polar coordinates - Properties – Harmonic conjugates – Construction of analytic function – Conformal mapping – Mapping by functions $w = z + c, cz, \frac{1}{z}, z^2$ *- Bilinear transformation.*

UNIT IV COMPLEX INTEGRATION

Line integral - Cauchy's integral theorem – Cauchy's integral formula – Taylor's and Laurent's series – Singularities – Residues – Residue theorem – Application of residue theorem for evaluation of real integrals – Use of circular contour and semicircular contour.

UNIT V LAPLACE TRANSFORMS

Existence conditions – Transforms of elementary functions – Transform of unit step function and unit impulse function – Basic properties – Shifting theorems -Transforms of derivatives and integrals – Initial and final value theorems – Inverse transforms – Convolution theorem – Transform of periodic functions – Application to solution of linear second order ordinary differential equations with constant coefficients.

Chapter	Contents	Page No

Chapter	Contents	Page No
1	**Matrices**	**1**
	1.1. Introduction	1
	1.2. Eigen values and Eigen vectors of a real matrix	2
	1.3. Properties of Eigen values and Eigen vectors	4
	1.4. Cayley - Hamilton Theorem	30
	1.5. Diagonalization of matrices	44
	1.6. Reduction of a quadratic form to canonical form by orthogonal transformation	46
	1.7. Nature of Quadratic forms	47
	1.8. Short Answer Questions	67
2	**Vector Calculus**	**75**
	2.1. Introduction	75
	2.2. Vector Identities	77
	2.3. Vector Differential Operator **∇**	78
	2.4. Gradient of a Scalar Point Function	78
	2.5. The Divergence and Curl of a Vector	92
	2.6. Line Integrals	113
	2.7. Surface Integrals	123
	2.8. Volume Integrals	130
	2.9. Green's Theorem in a Plane	133
	2.10. Stoke's Theorem	151
	2.11. Gauss Divergence Theorem	163
	2.12. Short Answer Questions	176
3	**Analytic Functions**	**188**
	3.1. Introduction	188
	3.2. Preliminaries	188
	3.3. Analytic Functions	191
	3.4. Properties of Analytic Functions	203
	3.5. Construction of Analytic Functions	215
	3.6. Conformal Mapping	230
	3.7. Some Standard Transformations	231
	3.8. Special Conformal Transformation	231

	3.9. Bilinear Transformation	243
	3.10. Short Answer Questions	255
4	**Complex Integration**	**263**
	4.1. Introduction	263
	4.2. Cauchy's Integral Theorem OR Cauchy's Fundamental Theorem	265
	4.3. Cauchy's Integral Formula	267
	4.4. Taylor's and Laurent's Series	278
	4.5. Singularities	295
	4.6. Residues	298
	4.7. Evaluation of Real Definite Integrals by Contour Integration	313
	4.8. Short Answer Questions	349
5	**Laplace Transforms**	**358**
	5.1. Introduction	358
	5.2. Laplace Transforms	358
	5.3. Existence Conditions	358
	5.4. Transforms of Elementary Functions	359
	5.5. Basic Properties	362
	5.6. Shifting Theorems	371
	5.7. Laplace Transforms of Derivatives and Integrals	377
	5.8. Laplace Transforms of Unit Step and Impulse Functions	383
	5.9. Laplace Transform of Periodic Functions	384
	5.10. Initial and Final Value Theorems	392
	5.11. Inverse Laplace Transforms	395
	5.12. Convolution Theorem	420
	5.13. Solutions of Ordinary Differential Equations	427
	5.14. Short Answer Questions	440
	Anna University Questions	
	Bibliography	

CHAPTER 1

MATRICES

1.1. Introduction

A Matrices is a two dimensional arrangement of numbers in row and column enclosed by a pair of square brackets or can say matrices are nothing but the rectangular arrangement of numbers, expression, symbols which are arranged in columns and rows. Matrices find many applications in scientific field and apply to practical real life problem. Matrices can be solved physical related application and one applied in the study of electrical circuits, quantum mechanics and optics, with the help of matrices, calculation of battery power outputs, resistor conversion of electrical energy into another useful energy, these matrices play a role in calculation, with the help of matrices problem related to Kirchhoff law of voltage and current can be easily solved. Now a day's matrices are used in the ranking of web pages in the Google search. The most important usages of matrices in computer side application are encryption of message codes with the help of encryptions only, internal function are working and even could work with transmission of sensitive & private data. Matrices are also used in robotics & automation in terms of base elements for the robot movements. The movements of the robots are programmed with the calculation of matrices 'row & column "Controlling of matrices are done by calculation of matrices. Mathematics puzzles, games, government and military organization websites financial information like credit card number and bank account, information security, all related encode, decode, theory a secret message in which matrices play a very important role.

Basic Definitions

Linear combination: If $X_1, X_2, ..., X_n$ are n vector and if $a_1, a_2, ..., a_n$ are numbers, then $a_1X_1 + a_2X_2 + \cdots + a_nX_n$ is called a linear combination of the vectors $X_1, X_2, ..., X_n$.

Linearly independent vectors: The vectors $X_1, X_2, ..., X_n$ are said to be linearly independent of $a_1X_1 + a_2X_2 + \cdots + a_nX_n = 0$ such that $a_1 = 0, a_2 = 0, ..., a_n = 0$.

Linearly dependent vectors: The vectors $X_1, X_2, ..., X_n$ are said to be linearly dependent if there exist $a_i \neq 0$ such that $a_1X_1 + a_2X_2 + \cdots + a_nX_n = 0$.

Orthogonal vector: Two column matrices X_1 and X_2 are said to *orthogonal* if $X_1^T X_2 = 0$ or $X_2^T X_1 = 0$.

Symmetric matrix: A real square matrix A is said to be *symmetric* if $A^T = A$.

1.2. Eigen values and Eigen vectors of a real matrix

Eigen values and eigen vectors are important concepts in linear algebra.

They are derived from the German word 'eigen' which means proper or characteristic.

Eigen values and Eigen vectors: Let $A = (a_{ij})$ be a square matrix of order n. If there exists a non-zero column vector X and a scalar λ such that $AX = \lambda X$, then λ is called the *eigen value* of A and X is called the *eigen vector* corresponding to λ.

Characteristic Equation

If X is the eigen vector corresponding to an eigen value λ of A, then $AX = \lambda X$.

$\therefore AX - \lambda IX = 0$

$\Rightarrow (A - \lambda I)X = 0$ where I is the unit matrix of order n.

The matrix equation $(A - \lambda I)X = 0$ gives rise to a system of equations.

$$\left.\begin{aligned} (a_{11} - \lambda)x_1 + a_{12}x_2 + \dots + a_{1n}x_n &= 0 \\ a_{21}x_1 + (a_{22} - \lambda)x_2 + \dots + a_{2n}x_n &= 0 \\ \dots \quad \dots \quad \dots \quad \dots & \\ \dots \quad \dots \quad \dots \quad \dots & \\ a_{n1}x_1 + a_{n2}x_2 + \dots + (a_{nn} - \lambda)x_n &= 0 \end{aligned}\right\} \quad \dots (1)$$

For non-zero vector $= \begin{bmatrix} x_1 \\ x_2 \\ \vdots \\ x_n \end{bmatrix}$, the solution of the above system should be a non-trivial solution.

The condition for the system (1) to have a non-trivial solution is

$$\begin{vmatrix} a_{11} - \lambda & a_{12} & \dots & a_{1n} \\ a_{21} & a_{22} - \lambda & \dots & a_{2n} \\ \dots & \dots & \dots & \dots \\ a_{n1} & a_{n2} & \dots & a_{nn} - \lambda \end{vmatrix} = 0$$

ie. $|A - \lambda I| = 0$.

The equation $|A - \lambda I| = 0$ is called the characteristic equation of the matrix A. The determinant $|A - \lambda I|$ is a polynomial in λ of degree n and it is called the *characteristic polynomial* of A.

Solving the characteristic equation $|A - \lambda I| = 0$ we get n values of λ and these n roots are called the eigen values or latent roots or characteristic values of λ.

Corresponding to each value of λ, the equation $(A - \lambda I)X = 0$ gives a non-zero solution vector X. The vector X is called the eigen vector or latent vector or characteristic vector of A corresponding to the eigen value λ.

Method for finding Characteristic equation

Case (i) Let $A = \begin{pmatrix} a_{11} & a_{12} \\ a_{21} & a_{22} \end{pmatrix}$ be a square matrix of order 2, then the characteristic equation of A is, $|A - \lambda I| = 0$.

$$\Rightarrow \begin{vmatrix} a_{11} - \lambda & a_{12} \\ a_{21} & a_{22} - \lambda \end{vmatrix} = 0$$

$$\Rightarrow (a_{11} - \lambda)(a_{22} - \lambda) - a_{21}a_{12} = 0$$

$$\Rightarrow a_{11}a_{22} - (a_{11} + a_{22})\lambda + \lambda^2 - a_{21}a_{12} = 0$$

$$\Rightarrow \lambda^2 - (a_{11} + a_{22})\lambda + (a_{11}a_{22} - a_{21}a_{12}) = 0$$

$$\Rightarrow \lambda^2 - S_1\lambda + S_2 = 0$$

where $S_1 = a_{11} + a_{22}$ = sum of the main diagonal elements of A

$S_2 = a_{11}a_{22} - a_{21}a_{12} = |A|$.

Case (ii) If A is a square matrix of order 3, then its characteristic equation can be simply written as $\lambda^3 - S_1\lambda^2 + S_2\lambda - S_3 = 0$

where S_1 = sum of the principal diagonal elements of A

S_2 = sum of the minors of principal diagonal elements

$S_3 = |A|$.

Method to find the Eigen vector

The eigenvector X is the solution of the system, of equations $(A - \lambda I)X = 0$ where λ is the eigen value of A.

If $A = (a_{ij})_{3X3}$ matrix, $(A - \lambda I)X = 0$ gives the system of equations

$$a_1x + b_1y + c_1z = 0$$

$$a_2x + b_2y + c_2z = 0$$

$$a_3x + b_3y + c_3z = 0 \text{ where } X = (x\ y\ z)^T.$$

The solution of first two equations by cross multiplication rule is given by

$$\frac{x}{\begin{vmatrix} b_1 & c_1 \\ b_2 & c_2 \end{vmatrix}} = \frac{y}{\begin{vmatrix} c_1 & a_1 \\ c_2 & a_2 \end{vmatrix}} = \frac{z}{\begin{vmatrix} a_1 & b_1 \\ a_2 & b_2 \end{vmatrix}}$$

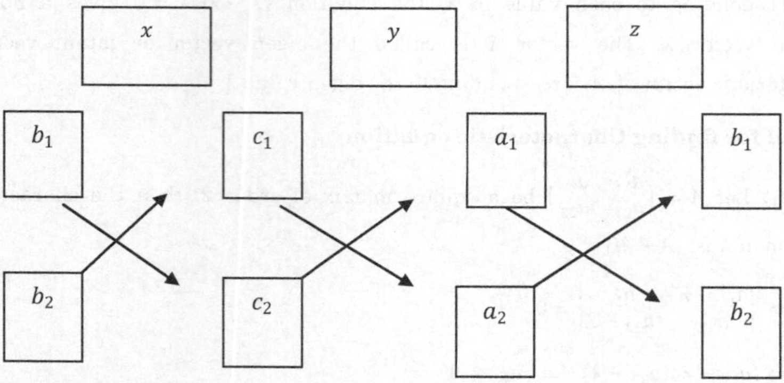

ie. $x = b_1c_2 - b_2c_1$; $y = c_1a_2 - c_2a_1$ and $z = a_1b_2 - a_2b_1$.

1.3. Properties of Eigen values and Eigen vectors

Property 1:

A square matrix A and it transpose A^T have the same eigenvalues.

Proof: Let $A = \begin{bmatrix} a_{11} & a_{12} & \dots & a_{1n} \\ a_{21} & a_{22} & \dots & a_{2n} \\ \dots & \dots & \dots & \dots \\ a_{n1} & a_{n2} & \dots & a_{nn} \end{bmatrix}$

The characteristic polynomial of A is

$$|A - \lambda I| = \begin{vmatrix} a_{11} - \lambda & a_{12} & \dots & a_{1n} \\ a_{21} & a_{22} - \lambda & \dots & a_{2n} \\ \dots & \dots & \dots & \dots \\ a_{n1} & a_{n2} & \dots & a_{nn} - \lambda \end{vmatrix} \quad \dots (1)$$

The characteristic polynomial of A^T is

$$| A^T - \lambda I | = \begin{vmatrix} a_{11} - \lambda & a_{21} & \dots & a_{n1} \\ a_{12} & a_{22} - \lambda & \dots & a_{n2} \\ \dots & \dots & \dots & \dots \\ a_{1n} & a_{2n} & \dots & a_{nn} - \lambda \end{vmatrix} \quad \dots (2)$$

Determinant (2) can be obtained by changing rows into columns of determinant (1).

$\therefore$ $|A - \lambda I| = |A^T - \lambda I|$.

$\therefore$ The characteristic equations of A and A^T are same.

$\therefore$ The eigenvalues of A and A^T are the same.

Property 2:

The sum of the eigenvalues of a matrix A is the trace of A.

(or)

The sum of the eigenvalues of a matrix A is equal to the sum of the principal diagonal elements of A.

Proof: First we verify the result for a third order matrix.

Let $A = \begin{bmatrix} a_{11} & a_{12} & a_{13} \\ a_{21} & a_{22} & a_{23} \\ a_{31} & a_{32} & a_{33} \end{bmatrix}$ be a square matrix of order 3

The characteristic equation of A is $|A - \lambda I| = 0$.

$$\Rightarrow \begin{vmatrix} a_{11} - \lambda & a_{12} & a_{13} \\ a_{21} & a_{22} - \lambda & a_{23} \\ a_{31} & a_{32} & a_{33} - \lambda \end{vmatrix} = 0$$

$$\Rightarrow -\lambda^3 + \lambda^2(a_{11} + a_{22} + a_{33}) + \cdots = 0 \quad \dots (1)$$

But we know that $|A - \lambda I| = (-1)^3 (\lambda - \lambda_1)(\lambda - \lambda_2)(\lambda - \lambda_3)$

$$= -\lambda^3 + \lambda^2 (\lambda_1 + \lambda_2 + \lambda_3) \quad \dots (2)$$

Comparing equations (1) & (2), we get

$$\lambda_1 + \lambda_2 + \lambda_3 = a_{11} + a_{22} + a_{33}$$

In general the characteristic equation of nth order is

$\lambda^n - s_1\lambda^{n-1} + s_2\lambda^{n-2} - \ldots . s_n = 0$, where S_r is the sum of the rth order minors of A.

$$\lambda_1 + \lambda_2 + \cdots + \lambda_n = a_{11} + a_{22} + a_{33} + \cdots + a_{nn}$$

Hence proved.

Property 3:

The product of the eigenvalues of a square matrix A is equal to the determinant of A.

Proof: Let $\lambda_1, \lambda_2, \ldots \lambda_n$ be the eigen values of A.

The characteristic polynomial of A is

$$|A - \lambda I| = (-1)^n (\lambda - \lambda_1)(\lambda - \lambda_2) \ldots \ldots (\lambda - \lambda_n)$$

Put $\lambda = 0$, we get $|A| = (-1)^n (-1)^n \lambda_1, \lambda_2, \ldots \lambda_n$

$$\Rightarrow \quad |A| = \lambda_1, \lambda_2, \ldots, \lambda_n$$

Hence the proof.

Property 4:

The eigenvalues of a triangular matrix are the diagonal elements of the matrix.

Proof: Let $A = \begin{bmatrix} a_{11} & a_{12} & . & \ldots & a_{1n} \\ 0 & a_{22} & . & \ldots & a_{2n} \\ \ldots & \ldots & \ldots & \ldots & \ldots \\ 0 & 0 & . & \ldots & a_{nn} \end{bmatrix}$ be a triangular matrix of order n.

The characteristic equation of A is $|A - \lambda I| = 0$

$$\Rightarrow \begin{vmatrix} a_{11} - \lambda & a_{12} & \ldots & a_{1n} \\ 0 & a_{22} - \lambda & \ldots & a_{2n} \\ \ldots & \ldots & \ldots & \ldots \\ 0 & 0 & \ldots & a_{nn} - \lambda \end{vmatrix} = 0$$

$$\Rightarrow \quad (a_{11} - \lambda)(a_{22} - \lambda) \ldots . (a_{nn} - \lambda) = 0.$$

$$\Rightarrow \quad \lambda = a_{11}, a_{22}, \ldots \ldots a_{nn}.$$

Hence, the eigenvalues of A are $a_{11}, a_{22}, \ldots, a_{nn}$ which are the diagonal elements of the matrix.

Property 5:

If $\lambda_1,\ \lambda_2, \dots \lambda_n$ are the eigenvalues of a square matrix A of order n, then

(i) $k\lambda_1,\ k\lambda_2,\ \dots\ k\lambda_n$ are the eigenvalues of kA where k is a non-zero scalar.

(ii) $\frac{1}{\lambda_1}, \frac{1}{\lambda_2}\ \dots\dots\ \frac{1}{\lambda_n}$ are the eigenvalues of A^{-1} provided $\lambda_i\ \neq\ 0$, for $i\ =\ 1, 2, \dots, .n.$

(iii) $\lambda_1^m, \lambda_2^m \dots\dots \lambda_n^m$ are the eigenvalues of A^m, where m is a positive integer.

(iv) $\lambda_1 \pm k,\ \lambda_2 \pm k, \dots \lambda_n \pm k$ are the eigenvales of $A\ \pm K\,I$.

Proof: Let A be a square matrix of order n.

Let $\lambda_1,\ \lambda_2, \dots \lambda_n$ are the eigenvalues of A.

(i) Let λ_i be an eigenvalue of A and X_i be the corresponding eigenvector.

Then, by definition, $AX_i\ =\ \lambda_i X_i$ ... (1)

Multiplying both sides of (1) by k, we get

$$kAX_i\ =\ k\lambda_i X_i \qquad \dots (2)$$

From (2), we see that $k\lambda_i$ is an eigenvalue of kA and the corresponding eigenvector is the same as that of λ_i , namely X_i.

(ii) Pre multiplying both sides of (1) by A^{-1}, we get

$$A^{-1}(AX_i) = A^{-1}(\lambda_i X_i)$$

$$\Rightarrow \quad X_i = \lambda_i (A^{-1} X_i)$$

$$\Rightarrow \quad A^{-1} X_i = \frac{1}{\lambda_i}(X_i) \qquad \dots (3)$$

From (3) we see that $\frac{1}{\lambda_i}$ is an eigenvalue of A^{-1} with the corresponding eigenvector equal to X_i, which is the same of λ_i .

(iii) Pre multiplying both sides of (1) by A, we get

$$A^2 X_i = A(AX_i)$$

$= A(\lambda_i X_i)$

$= \lambda_i (A X_i)$

$= \lambda_i^2 X_i \because [AX_i = \lambda_i X_i]$

Similarly $A^3 X_i = \lambda_i^3 X_i$ and so on.

In general $A^m X_i = \lambda_i^m X_i$ ... (4)

From (4), we see that $\lambda_i^m [\lambda_1^m, \lambda_2^m \ldots\ldots \lambda_n^m]$ is an eigenvalue of A^m with the corresponding eigenvectors equal to X_i, which is the same for λ_i.

(iv) Let λ be an eigenvalue of the matrix A.

$\therefore \quad AX = \lambda X$ ... (5)

Adding kIX on both sides of equation (5),

$$AX + kIX = \lambda X + kIX$$

$$(A + kI) X = \lambda X + kX$$

$$= (\lambda + k)X$$

Similarly, $(A - kI)X = (\lambda - k)X$

In general $(A \pm kI)X = (\lambda \pm kI)X$

Hence $\lambda \pm Ik$ is an eigenvalue of $(A \pm kI)$.

Property 6:

The eigenvalues of a real symmetric matrix are all real.

Proof: Let λ be an eigenvalue of the real symmetric matric A and X be the corresponding eigenvector.

Then $AX = \lambda X$ ... (1)

Taking complex conjugate on both sides in (1), we get

$$\bar{A}\,\bar{X} = \bar{\lambda}\bar{X}$$

$$\Rightarrow \quad A\bar{X} = \bar{\lambda}\bar{X} \quad [\text{since } \bar{A} = A] \quad \text{... (2)}$$

Taking Transpose in equation (2), we get $(A\bar{X})^T = (\bar{\lambda}\bar{X})^T$

$$\Rightarrow \quad \bar{X}^T A^T = \bar{\lambda}\bar{X}^T$$

$$\Rightarrow \quad \bar{X}^T A = \bar{\lambda}\bar{X}^T \quad [A = A^T] \quad \text{... (3)}$$

Post multiplying both sides of equation (3) by X, we get

$$\bar{X}^T AX = \bar{\lambda}\bar{X}^T X$$

$$\Rightarrow \quad \bar{X}^T \lambda X = \bar{\lambda}\bar{X}^T X \quad [\text{since } AX = \lambda X]$$

$$\Rightarrow \quad \lambda\bar{X}^T X = \bar{\lambda}\bar{X}^T X$$

$$\Rightarrow \quad (\lambda - \bar{\lambda})\bar{X}^T X = 0$$

Since $\bar{X}^T X \neq 0$, so $(\lambda - \bar{\lambda}) = 0$.

$\therefore \lambda = \bar{\lambda} \quad \Rightarrow \quad \lambda$ is real.

Hence all the eigenvalues of a real symmetric matrix are real.

Property 7:

The eigenvectors corresponding to district eigenvalues of a real symmetric matrix are orthogonal.

Proof: Let X_1, X_2 be eigenvectors corresponding to two distinct eigenvalues λ_1, λ_2.

Recall that the eigenvalues of a real symmetric matrix are real.

$\therefore \lambda_1, \lambda_2$ are real numbers.

By definition, $AX_1 = \lambda_i X_1$ ... (1)

$$AX_2 = \lambda_2 X_2 \quad \text{... (2)}$$

Pre-multiplying equation (1) by $X_2{}^T$, we get

$$X_2{}^T AX_1 = \lambda_1 X_2{}^T X_1 \quad \text{... (3)}$$

Pre-multiplying equation (2) by $X_1{}^T$, we get

$$X_1{}^T A X_2 = \lambda_2 X_1{}^T X_2 \qquad \text{... (4)}$$

Taking transpose on both sides of equation (3), we get

$$(X_2{}^T A X_1)^T = (\lambda_1 X_2{}^T X_1)^T.$$

$$\Rightarrow \qquad X_1{}^T A^T X_2 = \lambda_1 X_1{}^T X_2$$

$$\Rightarrow \qquad X_1{}^T A X_2 = \lambda_1 X_1{}^T X_2 \quad [\ = A^T \text{ (ie) } A \text{ is symmetric}] \qquad \text{... (5)}$$

From (4) & (5), we get $\lambda_1 X_1{}^T X_2 = \lambda_2 X_1{}^T X_2$

$$\Rightarrow \qquad (\lambda_1 - \lambda_2) X_1{}^T X_2 = 0$$

Since $\lambda_1 \neq \lambda_2$, $X_1{}^T X_2 = 0$

$\therefore$ X_1 and X_2 are orthogonal.

Property 8:

The eigenvectors corresponding to distinct eigenvalues of a matrix are linearly independent.

Proof: Let $\lambda_1, \lambda_2, \ldots \lambda_m \ (m \leq n)$ be the distinct eigenvalues of a square matrix of order n.

Let $X_1, X_2, \ldots, X_m \ (m \leq n)$ be their corresponding eigenvectors.

We have to prove $\sum_{i=1}^{m} \alpha_i X_i = 0$ implies $\alpha_i = 0, \ i = 1, 2, \ldots, m$

Multiplying $\sum_{i=1}^{m} \alpha_i X_i = 0$ by $(A - \lambda_1 I)$, we see

$$\begin{aligned}(A - \lambda_1 I)\alpha_1 X_1 &= \alpha_1 (A X_1 - \lambda_1 X_1) \\ &= \alpha_1 (0) \\ &= 0\end{aligned}$$

When $\sum_{i=1}^{m} \alpha_i X_i = 0$ is multiplied by

$(A - \lambda_1 I)(A - \lambda_2 I) \ldots (A - \lambda_{i-1} I)(A - \lambda_{i+1} I) \ldots (A - \lambda_m I)$, we get

$$\alpha_i (\lambda_i - \lambda_1)(\lambda_i - \lambda_2) \ldots (\lambda_i - \lambda_{i-1} I)(\lambda_i - \lambda_{i+1} I) \ldots (\lambda_i - \lambda_m) = 0$$

Since λ_i are distinct, $\alpha_i = 0$.

Since i is arbitrary, each $\alpha_i = 0, i = 1,2,\dots m$.

$\sum_{i=1}^{m} \alpha_i X_i = 0$ implies $\alpha_1 = \alpha_2 = \cdots = \alpha_m = 0$.

Hence $X_1, X_2, \dots, X_m$ are linearly independent.

Example 1 : Find the eigenvalues and eigenvectors of $\begin{bmatrix} 1 & 1 \\ 3 & -1 \end{bmatrix}$.

Solution : Let $A = \begin{bmatrix} 1 & 1 \\ 3 & -1 \end{bmatrix}$.

The characteristic equation of A is $\lambda^2 - S_1\lambda + S_2 = 0$.

where $S_1 =$ sum of the diagonal elements of A.

$$= 1 - 1 = 0$$

$$S_2 = |A| = \begin{vmatrix} 1 & 1 \\ 3 & -1 \end{vmatrix} = -1 - 3 = -4$$

Hence, the characteristic equation of A is $\lambda^2 - 4 = 0$.

By factorizing,

$$(\lambda - 2)(\lambda + 2) = 0$$

$\therefore$ The eigenvalues of A are $\lambda = -2, 2$.

To find eigenvectors:

Let $X = \begin{bmatrix} x_1 \\ x_2 \end{bmatrix}$ be the eigenvector of A corresponding to the eigen value λ.

Now, $(A - \lambda I)X = 0 \Rightarrow \begin{bmatrix} 1-\lambda & 1 \\ 3 & -1-\lambda \end{bmatrix} \begin{bmatrix} x_1 \\ x_2 \end{bmatrix} = \begin{bmatrix} 0 \\ 0 \end{bmatrix}$

$$\Rightarrow \quad (1-\lambda)x_1 + x_2 = 0$$

$$3x_1 + (-1-\lambda)x_2 = 0 \qquad \dots \qquad (1)$$

Case (i) : When $\lambda = -2$, equation (1) becomes,

$$3x_1 + x_2 = 0$$

$$3x_1 + x_2 = 0$$

Here we get only one equation $3x_1 + x_2 = 0$.

$$\Rightarrow \qquad 3x_1 = -x_2$$

$$\Rightarrow \qquad \frac{x_1}{-1} = \frac{x_2}{3}$$

$\therefore$ The eigenvector corresponding to $\lambda = -2$ is $X_1 = \begin{bmatrix} -1 \\ 3 \end{bmatrix} = \begin{bmatrix} 1 \\ -3 \end{bmatrix}$.

Case (i) : When $\lambda = 2$, equation (1) becomes,

$$-x_1 + x_2 = 0$$

$$3x_1 - 3x_2 = 0$$

Here we get only one equation $x_1 - x_2 = 0$.

$$\Rightarrow \qquad x_1 = x_2$$

$$\Rightarrow \qquad \frac{x_1}{1} = \frac{x_2}{1}$$

$\therefore$ The eigenvector corresponding to $\lambda = 2$ is $X_2 = \begin{bmatrix} 1 \\ 1 \end{bmatrix}$.

Example 2 : Find the eigenvalues and eigenvectors of $\begin{bmatrix} 2 & 2 & 1 \\ 1 & 3 & 1 \\ 1 & 2 & 2 \end{bmatrix}$. ***[AU, Nov/Dec 2015]***

Solution : Let $A = \begin{bmatrix} 2 & 2 & 1 \\ 1 & 3 & 1 \\ 1 & 2 & 2 \end{bmatrix}$.

To find eigen values:

The characteristic equation of A is $|A - \lambda I| = 0$.

$$\Rightarrow \qquad \lambda^3 - S_1\lambda^2 + S_2\lambda - S_3 = 0$$

Where $S_1 =$ sum of the diagonal elements of A.

$$= 2 + 3 + 2 = 7$$

$S_2 =$ sum of minors of the diagonal elements of A.

$$= \begin{vmatrix} 3 & 1 \\ 2 & 2 \end{vmatrix} + \begin{vmatrix} 2 & 1 \\ 1 & 2 \end{vmatrix} + \begin{vmatrix} 2 & 2 \\ 1 & 3 \end{vmatrix}$$

$$= 6 - 2 + 4 - 1 + 6 - 2 = 11$$

$$S_3 = |A|$$

$$= \begin{vmatrix} 2 & 2 & 1 \\ 1 & 3 & 1 \\ 1 & 2 & 2 \end{vmatrix}$$

$$= 2(6 - 2) - 2(2 - 1) + 1(2 - 3) = 8 - 2 - 1 = 5$$

$\therefore$ The characteristic equation is $\lambda^3 - 7\lambda^2 + 11\lambda - 5 = 0$

By synthetic division method,

$$\begin{array}{c|cccc} 1 & 1 & -7 & 11 & -5 \\ & 0 & 1 & 6 & 5 \\ \hline & 1 & -6 & 5 & 0 \end{array}$$

$$(\lambda - 1)(\lambda^2 - 6\lambda + 5) = 0$$

$$(\lambda - 1)(\lambda - 1)(\lambda - 5) = 0$$

$\therefore$ The eigenvalues of A are $\lambda = 1, 1, 5$.

To find eigenvectors:

Let $X = \begin{bmatrix} x_1 \\ x_2 \\ x_3 \end{bmatrix}$ be the eigenvector of A corresponding to the eigen value λ.

Now, $(A - \lambda I)X = 0 \Rightarrow \begin{bmatrix} 2-\lambda & 2 & 1 \\ 1 & 3-\lambda & 1 \\ 1 & 2 & 2-\lambda \end{bmatrix} \begin{bmatrix} x_1 \\ x_2 \\ x_3 \end{bmatrix} = \begin{bmatrix} 0 \\ 0 \\ 0 \end{bmatrix}$

$$\Rightarrow \quad (2-\lambda)x_1 + 2x_2 + x_3 = 0$$

$$x_1 + (3-\lambda)x_2 + x_3 = 0 \quad \ldots (1)$$

$$x_1 + 2x_2 + (2-\lambda)x_3 = 0$$

Case (i) : When $\lambda = 5$, equation (1) becomes,

$$-3x_1 + 2x_2 + x_3 = 0$$

$$x_1 - 2x_2 + x_3 = 0$$

$$x_1 + 2x_2 - 3x_3 = 0$$

Solving the first two equations by the rule of cross multiplication we get,

$$\frac{x_1}{2+2} = \frac{x_2}{1+3} = \frac{x_3}{6-2}$$

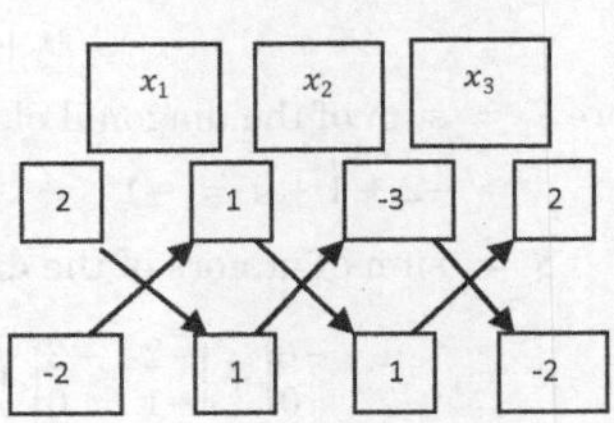

$$\Rightarrow \quad \frac{x_1}{4} = \frac{x_2}{4} = \frac{x_3}{4}$$

$$\Rightarrow \quad \frac{x_1}{1} = \frac{x_2}{1} = \frac{x_3}{1}$$

$\therefore$ The eigenvector corresponding to $\lambda = 5$ is $X_1 = \begin{bmatrix} 1 \\ 1 \\ 1 \end{bmatrix}$.

Case (ii) : When $\lambda = 1$, equation (1) becomes,

$$x_1 + 2x_2 + x_3 = 0$$
$$x_1 + 2x_2 + x_3 = 0$$

$$x_1 + 2x_2 + x_3 = 0$$

Here we get only one equation $x_1 + 2x_2 + x_3 = 0$.

$\therefore$ By assigning arbitrary values for any two variables, we can find an independent eigenvector.

Take $x_1 = 0$, then we get $2x_2 + x_3 = 0$.

$$\Rightarrow \quad 2x_2 = -x_3$$

$$\Rightarrow \quad \frac{x_2}{-1} = \frac{x_3}{2}$$

The eigenvector is $X_2 = \begin{bmatrix} 0 \\ -1 \\ 2 \end{bmatrix} = \begin{bmatrix} 0 \\ 1 \\ -2 \end{bmatrix}$

Also take $x_2 = 0$, then we get $x_1 + x_3 = 0$.

$$\Rightarrow \quad x_1 = -x_3$$

$$\Rightarrow \quad \frac{x_1}{-1} = \frac{x_3}{1}$$

$\therefore$ The eigenvector is $X_3 = \begin{bmatrix} -1 \\ 0 \\ 1 \end{bmatrix} = \begin{bmatrix} 1 \\ 0 \\ -1 \end{bmatrix}$.

Example 3 : Find the eigenvalues and eigenvectors of $\begin{bmatrix} -2 & 2 & -3 \\ 2 & 1 & -6 \\ -1 & -2 & 0 \end{bmatrix}$.

[AU, Nov/Dec 2015]

Solution : Let $A = \begin{bmatrix} -2 & 2 & -3 \\ 2 & 1 & -6 \\ -1 & -2 & 0 \end{bmatrix}$.

To find eigen values:

The characteristic equation of A is $|A - \lambda I| = 0$.

$$\Rightarrow \quad \lambda^3 - S_1\lambda^2 + S_2\lambda - S_3 = 0$$

Where $S_1 =$ sum of the diagonal elements of A.

$= -2 + 1 + 0 = -1$

$S_2 =$ sum of minors of the diagonal elements of A.

$$= \begin{vmatrix} 1 & -6 \\ -2 & 0 \end{vmatrix} + \begin{vmatrix} -2 & -3 \\ -1 & 0 \end{vmatrix} + \begin{vmatrix} -2 & 2 \\ 2 & 1 \end{vmatrix} = 0 - 12 + (0 - 3) + (-2 - 4) = -21$$

$S_3 = |A|$

$$= \begin{vmatrix} -2 & 2 & -3 \\ 2 & 1 & -6 \\ -1 & -2 & 0 \end{vmatrix}$$

$$= -2(0 - 12) - 2(0 - 6) - 3(-4 + 1) = 24 + 12 + 9 = 45$$

$\therefore$ The characteristic equation is $\lambda^3 + \lambda^2 - 21\lambda - 45 = 0$

By synthetic division method,

$$\begin{array}{r|rrrr} -3 & 1 & 1 & -21 & -45 \\ & 0 & -3 & 6 & 45 \\ \hline & 1 & -2 & -15 & 0 \end{array}$$

$$(\lambda + 3)(\lambda^2 - 2\lambda - 15) = 0$$

$$(\lambda + 3)(\lambda + 3)(\lambda - 5) = 0$$

$\therefore$ The eigenvalues of A are $\lambda = -3, -3, 5$.

To find eigenvectors:

Let $X = \begin{bmatrix} x_1 \\ x_2 \\ x_3 \end{bmatrix}$ be the eigenvector of A corresponding to the eigen value λ.

Now, $(A - \lambda I)X = 0 \Rightarrow \begin{bmatrix} -2-\lambda & 2 & -3 \\ 2 & 1-\lambda & -6 \\ -1 & -2 & 0-\lambda \end{bmatrix} \begin{bmatrix} x_1 \\ x_2 \\ x_3 \end{bmatrix} = \begin{bmatrix} 0 \\ 0 \\ 0 \end{bmatrix}$

$$\Rightarrow \quad (-2 - \lambda)x_1 + 2x_2 - 3x_3 = 0$$

$$2x_1 + (1 - \lambda)x_2 - 6x_3 = 0 \qquad \dots \qquad (1)$$

$$-x_1 - 2x_2 - \lambda x_3 = 0$$

Case (i) : When $\lambda = 5$, equation (1) becomes,

$$-7x_1 + 2x_2 - 3x_3 = 0$$

$$2x_1 - 4x_2 - 6x_3 = 0$$

$$-x_1 - 2x_2 - 5x_3 = 0$$

Solving the first two equations by the rule of cross multiplication we get,

$$\frac{x_1}{-12-12} = \frac{x_2}{-6-42} = \frac{x_3}{28-4}$$

$$\Rightarrow \quad \frac{x_1}{-24} = \frac{x_2}{-48} = \frac{x_3}{24}$$

$$\Rightarrow \quad \frac{x_1}{1} = \frac{x_2}{2} = \frac{x_3}{-1}$$

$\therefore$ The eigenvector corresponding to $\lambda = 5$ is $X_1 = \begin{bmatrix} 1 \\ 2 \\ -1 \end{bmatrix}$.

Case (ii) : When $\lambda = -3$, equation (1) becomes,

$$x_1 + 2x_2 - 3x_3 = 0$$

$$2x_1 + 4x_2 - 6x_3 = 0$$
$$-x_1 - 2x_2 + 3x_3 = 0$$

Here we get only one equation $x_1 + 2x_2 - 3x_3 = 0$.

$\therefore$ By assigning arbitrary values for any two variables, we can find an independent eigenvector.

Take $x_1 = 0$, then we get $2x_2 - 3x_3 = 0$.

$$\Rightarrow \quad 2x_2 = 3x_3$$

$$\Rightarrow \quad \frac{x_2}{3} = \frac{x_3}{2}$$

The eigenvector is $X_2 = \begin{bmatrix} 0 \\ 3 \\ 2 \end{bmatrix}$

Also take $x_2 = 0$, then we get $x_1 - 3x_3 = 0$.

$$\Rightarrow \quad x_1 = 3x_3$$

$$\Rightarrow \quad \frac{x_1}{3} = \frac{x_3}{1}$$

$\therefore$ The eigenvector is $X_3 = \begin{bmatrix} 3 \\ 0 \\ 1 \end{bmatrix}$.

Example 4 : Find the eigenvalues and eigenvectors of $\begin{bmatrix} 7 & -2 & 0 \\ -2 & 6 & -2 \\ 0 & -2 & 5 \end{bmatrix}$.

[*AU, Nov/Dec 2014*]

Solution : Let $A = \begin{bmatrix} 7 & -2 & 0 \\ -2 & 6 & -2 \\ 0 & -2 & 5 \end{bmatrix}$.

To find eigen values:

The characteristic equation of A is $A - \lambda I = 0$.

$$\Rightarrow \quad \lambda^3 - S_1\lambda^2 + S_2\lambda - S_3 = 0$$

Where S_1 = sum of the diagonal elements of A.

$= 7 + 6 + 5 = 18$

S_2 = sum of minors of the diagonal elements of A.

$$= \begin{vmatrix} 6 & -2 \\ -2 & 5 \end{vmatrix} + \begin{vmatrix} 7 & 0 \\ 0 & 5 \end{vmatrix} + \begin{vmatrix} 7 & -2 \\ -2 & 6 \end{vmatrix}$$

$= 30 - 4 + 35 - 0 + 42 - 4 = 99$

$S_3 = |A|$

$$= \begin{vmatrix} 7 & -2 & 0 \\ -2 & 6 & -2 \\ 0 & -2 & 5 \end{vmatrix}$$

$$= 7(30-4) + 2(-10-0) + 0 = 182 - 20 = 162$$

$\therefore$ The characteristic equation is $\lambda^3 - 18\lambda^2 + 99\lambda - 162 = 0$

By synthetic division method,

$$(\lambda - 3)(\lambda^2 - 15\lambda + 54) = 0$$

$$(\lambda - 3)(\lambda - 6)(\lambda - 9) = 0$$

3	1	−18	99	−162
	0	3	−45	162
	1	−15	54	0

$\therefore$ The eigenvalues of A are $\lambda = 3,6,9$.

To find eigenvectors:

Let $X = \begin{bmatrix} x_1 \\ x_2 \\ x_3 \end{bmatrix}$ be the eigenvector of A corresponding to the eigenvalue λ.

Now, $(A - \lambda I)X = 0 \Rightarrow \begin{bmatrix} 7-\lambda & -2 & 0 \\ -2 & 6-\lambda & -2 \\ 0 & -2 & 5-\lambda \end{bmatrix} \begin{bmatrix} x_1 \\ x_2 \\ x_3 \end{bmatrix} = \begin{bmatrix} 0 \\ 0 \\ 0 \end{bmatrix}$

$$\Rightarrow \quad (7-\lambda)x_1 - 2x_2 + 0x_3 = 0$$

$$-2x_1 + (6-\lambda)x_2 - 2x_3 = 0 \quad \ldots \quad (1)$$

$$0x_1 - 2x_2 + (5-\lambda)x_3 = 0$$

Case (i) : When $\lambda = 3$, equation (1) becomes,

$$4x_1 - 2x_2 + 0x_3 = 0$$

$$-2x_1 + 3x_2 - 2x_3 = 0$$

$$0x_1 - 2x_2 + 2x_3 = 0$$

Solving the first two equations by the rule of cross multiplication we get,

$$\frac{x_1}{4-0} = \frac{x_2}{0+8} = \frac{x_3}{12-4}$$

$$\Rightarrow \quad \frac{x_1}{4} = \frac{x_2}{8} = \frac{x_3}{8}$$

$$\Rightarrow \quad \frac{x_1}{1} = \frac{x_2}{2} = \frac{x_3}{2}$$

$\therefore$ The eigenvector corresponding to $\lambda = 3$ is $X_1 = \begin{bmatrix} 1 \\ 2 \\ 2 \end{bmatrix}$.

Case (ii) : When $\lambda = 6$, equation (1) becomes,

$$x_1 - 2x_2 + 0x_3 = 0$$

$$-2x_1 + 0x_2 - 2x_3 = 0$$

$$0x_1 - 2x_2 - 1\,x_3 = 0$$

Solving the first two equations by the rule of cross multiplication we get,

$$\frac{x_1}{4-0} = \frac{x_2}{0+2} = \frac{x_3}{0-4}$$

$$\Rightarrow \quad \frac{x_1}{4} = \frac{x_2}{2} = \frac{x_3}{-4}$$

$$\Rightarrow \quad \frac{x_1}{2} = \frac{x_2}{1} = \frac{x_3}{-2}$$

∴ The eigenvector corresponding to $\lambda = 6$ is $X_2 = \begin{bmatrix} 2 \\ 1 \\ -2 \end{bmatrix}$.

Case (iii) : When $\lambda = 9$, equation (1) becomes,

$$-2x_1 - 2x_2 + 0x_3 = 0$$

$$-2x_1 - 3x_2 - 2x_3 = 0$$

$$0x_1 - 2x_2 - 4\,x_3 = 0$$

Solving the first two equations by the rule of cross multiplication we get,

$$\frac{x_1}{4-0} = \frac{x_2}{0-4} = \frac{x_3}{6-4}$$

$$\Rightarrow \quad \frac{x_1}{4} = \frac{x_2}{-4} = \frac{x_3}{2}$$

$$\Rightarrow \quad \frac{x_1}{2} = \frac{x_2}{-2} = \frac{x_3}{1}$$

∴ The eigenvector corresponding to $\lambda = 9$ is $X_3 = \begin{bmatrix} 2 \\ -2 \\ 1 \end{bmatrix}$.

Example 5 : Find the eigenvalues and eigenvectors of $\begin{bmatrix} 2 & 0 & 1 \\ 0 & 2 & 0 \\ 1 & 0 & 2 \end{bmatrix}$.

[*AU, Dec 2015/Jan 2016*]

Solution : Let $A = \begin{bmatrix} 2 & 0 & 1 \\ 0 & 2 & 0 \\ 1 & 0 & 2 \end{bmatrix}$.

To find eigen values:

The characteristic equation of A is $A - \lambda I = 0$.

$$\Rightarrow \quad \lambda^3 - S_1\lambda^2 + S_2\lambda - S_3 = 0$$

Where $S_1 =$ sum of the diagonal elements of A.

$= 2 + 2 + 2 = 6$

$S_2 =$ sum of minors of the diagonal elements of A.

$$= \begin{vmatrix} 2 & 0 \\ 0 & 2 \end{vmatrix} + \begin{vmatrix} 2 & 1 \\ 1 & 2 \end{vmatrix} + \begin{vmatrix} 2 & 0 \\ 0 & 2 \end{vmatrix}$$

$= 4 + 4 - 1 + 4 = 11$

$S_3 = |A|$

$$= \begin{vmatrix} 2 & 0 & 1 \\ 0 & 2 & 0 \\ 1 & 0 & 2 \end{vmatrix}$$

$= 2(4 - 0) + 0 + 1(0 - 2) = 8 - 2 = 6$

$\therefore$ The characteristic equation is $\lambda^3 - 6\lambda^2 + 11\lambda - 6 = 0$

By synthetic division method,

1	1	−6	11	−6
	0	1	−5	6
	1	−5	6	0

$(\lambda - 1)(\lambda^2 - 5\lambda + 6) = 0$

$(\lambda - 1)(\lambda - 2)(\lambda - 3) = 0$

$\therefore$ The eigenvalues of A are $\lambda = 1,2,3$.

To find eigenvectors:

Let $X = \begin{bmatrix} x_1 \\ x_2 \\ x_3 \end{bmatrix}$ be the eigenvector of A corresponding to the eigen value λ.

Now, $(A - \lambda I)X = 0 \Rightarrow \begin{bmatrix} 2-\lambda & 0 & 1 \\ 0 & 2-\lambda & 0 \\ 1 & 0 & 2-\lambda \end{bmatrix} \begin{bmatrix} x_1 \\ x_2 \\ x_3 \end{bmatrix} = \begin{bmatrix} 0 \\ 0 \\ 0 \end{bmatrix}$

$$\Rightarrow \quad \begin{aligned} (2-\lambda)x_1 + 0x_2 + x_3 &= 0 \\ 0x_1 + (2-\lambda)x_2 + 0x_3 &= 0 \\ x_1 + 0x_2 + (2-\lambda)x_3 &= 0 \end{aligned} \qquad \ldots (1)$$

Case (i) : When $\lambda = 1$, equation (1) becomes,

$x_1 + 0x_2 + x_3 = 0$

$$0x_1 + x_2 + 0x_3 = 0$$
$$x_1 + 0x_2 + x_3 = 0$$

Solving the first two equations by the rule of cross multiplication we get,

$$\frac{x_1}{0-1} = \frac{x_2}{0-0} = \frac{x_3}{1-0}$$

$$\Rightarrow \quad \frac{x_1}{-1} = \frac{x_2}{0} = \frac{x_3}{1}$$

$\therefore$ The eigenvector corresponding to $\lambda = 1$ is $X_1 = \begin{bmatrix} -1 \\ 0 \\ 1 \end{bmatrix}$.

Case (ii) : When $\lambda = 2$, equation (1) becomes,

$$0x_1 + 0x_2 + x_3 = 0$$
$$0x_1 + 0x_2 + 0x_3 = 0$$
$$x_1 + 0x_2 + 0x_3 = 0$$

Solving the first and the last equations by the rule of cross multiplication we get,

$$\frac{x_1}{0} = \frac{x_2}{1} = \frac{x_3}{0}$$

$\therefore$ The eigenvector corresponding to $\lambda = 2$ is $X_2 = \begin{bmatrix} 0 \\ 1 \\ 0 \end{bmatrix}$.

Case (iii) : When $\lambda = 3$, equation (1) becomes,

$$-x_1 + 0x_2 + x_3 = 0$$
$$0x_1 - x_2 + 0x_3 = 0$$
$$x_1 + 0x_2 - x_3 = 0$$

Solving the first two equations by the rule of cross multiplication we get,

$$\frac{x_1}{0+1} = \frac{x_2}{0} = \frac{x_3}{1-0}$$

$$\Rightarrow \quad \frac{x_1}{1} = \frac{x_2}{0} = \frac{x_3}{1}$$

$\therefore$ The eigenvector corresponding to $\lambda = 3$ is $X_3 = \begin{bmatrix} 1 \\ 0 \\ 1 \end{bmatrix}$.

Example 6 : Find the eigenvalues and eigenvectors of $\begin{bmatrix} 0 & 1 & 0 \\ 0 & 0 & 1 \\ 1 & -3 & 3 \end{bmatrix}$.

Solution : Let $A = \begin{bmatrix} 0 & 1 & 0 \\ 0 & 0 & 1 \\ 1 & -3 & 3 \end{bmatrix}$.

To find eigenvalues:

The characteristic equation of A is $A - \lambda I = 0$.

$$\Rightarrow \qquad \lambda^3 - S_1\lambda^2 + S_2\lambda - S_3 = 0$$

Where $S_1 =$ sum of the diagonal elements of A.

$$= 0 + 0 + 3 = 3$$

$S_2 =$ sum of minors of the diagonal elements of A.

$$= \begin{vmatrix} 0 & 1 \\ -3 & 3 \end{vmatrix} + \begin{vmatrix} 0 & 0 \\ 1 & 3 \end{vmatrix} + \begin{vmatrix} 0 & 1 \\ 0 & 0 \end{vmatrix}$$

$$= 0 + 3 + 0 + 0 = 3$$

$$S_3 = |A|$$

$$= \begin{vmatrix} 0 & 1 & 0 \\ 0 & 0 & 1 \\ 1 & -3 & 3 \end{vmatrix}$$

$$= 0 - 1(0 - 1) + 0 = 1$$

$\therefore$ The characteristic equation is $\lambda^3 - 3\lambda^2 + 3\lambda - 1 = 0$

By synthetic division method,

$$\begin{array}{c|cccc} 1 & 1 & -3 & 3 & -1 \\ & 0 & 1 & -2 & 1 \\ \hline & 1 & -2 & 1 & 0 \end{array}$$

$$(\lambda - 1)(\lambda^2 - 2\lambda + 1) = 0$$

$$(\lambda - 1)(\lambda - 1)(\lambda - 1) = 0$$

$\therefore$ The eigenvalues of A are $\lambda = 1,1,1$.

To find eigenvectors:

Let $X = \begin{bmatrix} x_1 \\ x_2 \\ x_3 \end{bmatrix}$ be the eigenvector of A corresponding to the eigenvalue λ.

Now, $(A - \lambda I)X = 0 \Rightarrow \begin{bmatrix} -\lambda & 1 & 0 \\ 0 & -\lambda & 1 \\ 1 & -3 & 3-\lambda \end{bmatrix} \begin{bmatrix} x_1 \\ x_2 \\ x_3 \end{bmatrix} = \begin{bmatrix} 0 \\ 0 \\ 0 \end{bmatrix}$

$$\Rightarrow \qquad -\lambda\, x_1 + x_2 + 0x_3 = 0$$

$$0x_1 - \lambda\, x_2 + x_3 = 0 \qquad \ldots (1)$$

$$x_1 - 3x_2 + (3 - \lambda)\, x_3 = 0$$

When $\lambda = 1$, equation (1) becomes,

$$-x_1 + x_2 + 0x_3 = 0$$

$$0x_1 - x_2 + x_3 = 0$$

$$x_1 - 3x_2 + 2x_3 = 0$$

Solving the first two equations by the rule of cross multiplication we get,

$$\frac{x_1}{1-0} = \frac{x_2}{0+1} = \frac{x_3}{1-0}$$

$$\Rightarrow \quad \frac{x_1}{1} = \frac{x_2}{1} = \frac{x_3}{1}$$

$\therefore$ The eigenvector corresponding to $\lambda = 1$ is $X_1 = \begin{bmatrix} 1 \\ 1 \\ 1 \end{bmatrix}$.

Hence, there is only one eigenvector corresponding to the repeated eigen value $\lambda = 1$.

Thus, $X_1 = X_2 = X_3 = \begin{bmatrix} 1 \\ 1 \\ 1 \end{bmatrix}$. All eigenvectors are dependent.

Example 7 : Find the eigenvalues and eigenvectors of $\begin{bmatrix} 11 & -4 & -7 \\ 7 & -2 & -5 \\ 10 & -4 & -6 \end{bmatrix}$.

[AU, Nov/Dec 2016]

Solution : Let $A = \begin{bmatrix} 11 & -4 & -7 \\ 7 & -2 & -5 \\ 10 & -4 & -6 \end{bmatrix}$.

To find eigen values:

The characteristic equation of A is $|A - \lambda I| = 0$.

$$\Rightarrow \quad \lambda^3 - S_1\lambda^2 + S_2\lambda - S_3 = 0$$

Where $S_1 =$ sum of the diagonal elements of A.

$= 11 - 2 - 6 = 3$

$S_2 =$ sum of minors of the diagonal elements of $|A|$.

$$= \begin{vmatrix} -2 & -5 \\ -4 & -6 \end{vmatrix} + \begin{vmatrix} 11 & -7 \\ 10 & -6 \end{vmatrix} + \begin{vmatrix} 11 & -4 \\ 7 & -2 \end{vmatrix}$$

$$= (12 - 20) + (-66 + 70) + (-22 + 28) = -8 + 4 + 6 = 2$$

$S_3 = |A|$

$$= \begin{vmatrix} 11 & -4 & -7 \\ 7 & -2 & -5 \\ 10 & -4 & -6 \end{vmatrix}$$

$$= 11(12-20)+4(-42+50)-7(-28+20) = -88+32+56 = 0$$

$\therefore$ The characteristic equation is $\lambda^3 - 3\lambda^2 + 2\lambda = 0$

$$\Rightarrow \quad \lambda(\lambda^2 - 3\lambda + 2) = 0$$

$$\lambda(\lambda-1)(\lambda-2) = 0$$

$\therefore$ The eigenvalues of A are $\lambda = 0, 1, 2$.

To find eigenvectors:

Let $X = \begin{bmatrix} x_1 \\ x_2 \\ x_3 \end{bmatrix}$ be the eigenvector of A corresponding to the eigen value λ.

Now, $(A - \lambda I)X = 0 \Rightarrow \begin{bmatrix} 11-\lambda & -4 & -7 \\ 7 & -2-\lambda & -5 \\ 10 & -4 & -6-\lambda \end{bmatrix} \begin{bmatrix} x_1 \\ x_2 \\ x_3 \end{bmatrix} = \begin{bmatrix} 0 \\ 0 \\ 0 \end{bmatrix}$

$$\Rightarrow \quad \begin{aligned} (11-\lambda)x_1 - 4x_2 - 7x_3 &= 0 \\ 7x_1 + (-2-\lambda)x_2 - 5x_3 &= 0 \\ 10x_1 - 4x_2 + (-6-\lambda)x_3 &= 0 \end{aligned} \qquad \ldots \quad (1)$$

Case (i) : When $\lambda = 0$, equation (1) becomes,

$$\begin{aligned} 11x_1 - 4x_2 - 7x_3 &= 0 \\ 7x_1 - 2x_2 - 5x_3 &= 0 \\ 10x_1 - 4x_2 - 6x_3 &= 0 \end{aligned}$$

Solving the first two equations by the rule of cross multiplication we get,

$$\frac{x_1}{20-14} = \frac{x_2}{-49+55} = \frac{x_3}{-22+28}$$

$$\Rightarrow \quad \frac{x_1}{6} = \frac{x_2}{6} = \frac{x_3}{6}$$

$$\Rightarrow \quad \frac{x_1}{1} = \frac{x_2}{1} = \frac{x_3}{1}$$

$\therefore$ The eigenvector corresponding to $\lambda = 0$ is $X_1 = \begin{bmatrix} 1 \\ 1 \\ 1 \end{bmatrix}$.

Case (ii) : When $\lambda = 1$, equation (1) becomes,

$$\begin{aligned} 10x_1 - 4x_2 - 7x_3 &= 0 \\ 7x_1 - 3x_2 - 5x_3 &= 0 \\ 10x_1 - 4x_2 - 7x_3 &= 0 \end{aligned}$$

Solving the first two equations by the rule of cross multiplication we get,

$$\frac{x_1}{20-21} = \frac{x_2}{-49+50} = \frac{x_3}{-30+28}$$

$$\Rightarrow \qquad \frac{x_1}{-1} = \frac{x_2}{1} = \frac{x_3}{-2}$$

$$\Rightarrow \qquad \frac{x_1}{1} = \frac{x_2}{-1} = \frac{x_3}{2}$$

$\therefore$ The eigenvector corresponding to $\lambda = 1$ is $X_2 = \begin{bmatrix} 1 \\ -1 \\ 2 \end{bmatrix}$.

Case (iii) : When $\lambda = 2$, equation (1) becomes,

$$9x_1 - 4x_2 - 7x_3 = 0$$
$$7x_1 - 4x_2 - 5x_3 = 0$$
$$10\,x_1 - 4x_2 - 8\,x_3 = 0$$

Solving the first two equations by the rule of cross multiplication we get,

$$\frac{x_1}{20-28} = \frac{x_2}{-49+45} = \frac{x_3}{-36+28}$$

$$\Rightarrow \qquad \frac{x_1}{-8} = \frac{x_2}{-4} = \frac{x_3}{-8}$$

$$\Rightarrow \qquad \frac{x_1}{2} = \frac{x_2}{1} = \frac{x_3}{2}$$

$\therefore$ The eigenvector corresponding to $\lambda = 2$ is $X_3 = \begin{bmatrix} 2 \\ 1 \\ 2 \end{bmatrix}$.

EXERCISES

1. Find the eigenvalues and eigenvectors of $\begin{bmatrix} 4 & 1 \\ 3 & 2 \end{bmatrix}$.

2. Find the eigenvalues and eigenvectors of $\begin{bmatrix} 4 & 2 & -2 \\ -5 & 3 & 2 \\ -2 & 4 & 1 \end{bmatrix}$.

3. Find the eigenvalues and eigenvectors of $\begin{bmatrix} 2 & -2 & 3 \\ 1 & 1 & 1 \\ 1 & 3 & -1 \end{bmatrix}$.

4. Find the eigenvalues and eigenvectors of $\begin{bmatrix} 2 & 1 & 1 \\ 2 & 3 & 2 \\ 3 & 3 & 4 \end{bmatrix}$.

5. Find the eigenvalues and eigenvectors of $\begin{bmatrix} 6 & -6 & 5 \\ 14 & -13 & 10 \\ 7 & -6 & 4 \end{bmatrix}$.

6. Find the eigenvalues and eigenvectors of $\begin{bmatrix}1 & 1 & 3\\1 & 5 & 1\\3 & 1 & 1\end{bmatrix}$.

7. Find the eigenvalues and eigenvectors of $\begin{bmatrix}5 & 0 & 1\\0 & -2 & 0\\1 & 0 & 5\end{bmatrix}$.

8. Find the eigenvalues and eigenvectors of $\begin{bmatrix}1 & 2 & 3\\2 & 4 & 6\\3 & 6 & 9\end{bmatrix}$.

ANSWERS

1. 1,5; $\begin{bmatrix}-1\\3\end{bmatrix}, \begin{bmatrix}1\\1\end{bmatrix}$

2. 1,2,5 ; $\begin{bmatrix}2\\1\\4\end{bmatrix}, \begin{bmatrix}1\\1\\2\end{bmatrix}, \begin{bmatrix}0\\1\\1\end{bmatrix}$

3. −2,1,3 ; $\begin{bmatrix}11\\1\\-14\end{bmatrix}, \begin{bmatrix}-1\\1\\1\end{bmatrix}, \begin{bmatrix}1\\1\\1\end{bmatrix}$

4. 1,1,7 ; $\begin{bmatrix}0\\1\\-1\end{bmatrix}, \begin{bmatrix}1\\0\\-1\end{bmatrix}, \begin{bmatrix}1\\2\\3\end{bmatrix}$

5. −1, −1, −1 ; $\begin{bmatrix}0\\5\\6\end{bmatrix}, \begin{bmatrix}-5\\0\\7\end{bmatrix}, \begin{bmatrix}6\\7\\0\end{bmatrix}$

6. −2,3,6 ; $\begin{bmatrix}0\\1\\-1\end{bmatrix}, \begin{bmatrix}1\\-1\\1\end{bmatrix}, \begin{bmatrix}1\\2\\1\end{bmatrix}$

7. −2,4,6 ; $\begin{bmatrix}0\\1\\0\end{bmatrix}, \begin{bmatrix}1\\0\\-1\end{bmatrix}, \begin{bmatrix}1\\0\\1\end{bmatrix}$

8. 14,0,0 ; $\begin{bmatrix}1\\2\\3\end{bmatrix}, \begin{bmatrix}-2\\1\\0\end{bmatrix}, \begin{bmatrix}3\\6\\-5\end{bmatrix}$.

Problems based on properties

Example 1: Find the characteristic polynomial of $\begin{bmatrix}1 & 4\\2 & 3\end{bmatrix}$.

Solution : Let $A = \begin{bmatrix}1 & 4\\2 & 3\end{bmatrix}$.

The characteristic equation of A is $\lambda^2 - S_1\lambda + S_2 = 0$.

Where $S_1 =$ sum of the diagonal elements of A.

$= 1 + 3 = 4$

$S_2 = |A| = \begin{vmatrix}1 & 4\\2 & 3\end{vmatrix} = 3 - 8 = -5$

Hence, the characteristic equation of A is $\lambda^2 - 4\lambda - 5 = 0$.

∴ The characteristic polynomial of A is $\lambda^2 - 4\lambda - 5$.

Example 2: The product of two eigenvalues of the matrix $A = \begin{bmatrix} 6 & -2 & 2 \\ -2 & 3 & -1 \\ 2 & -1 & 3 \end{bmatrix}$ is 16.

Find the third eigenvalue.

Solution : Given $A = \begin{bmatrix} 6 & -2 & 2 \\ -2 & 3 & -1 \\ 2 & -1 & 3 \end{bmatrix}$.

Let $\lambda_1, \lambda_2, \lambda_3$ be the eigenvalues of A.

Given that $\lambda_1 \times \lambda_2 = 16$

We know that, product of the eigenvalues $= |A|$

$$\Rightarrow \lambda_1 \times \lambda_2 \times \lambda_3 = \begin{vmatrix} 6 & -2 & 2 \\ -2 & 3 & -1 \\ 2 & -1 & 3 \end{vmatrix}$$

$$\Rightarrow 16\lambda_3 = 6(9-1) + 2(-6+2) + 2(2-6)$$

$$= 48 - 8 - 8 = 32$$

$$\Rightarrow \lambda_3 = \frac{32}{16} = 2.$$

Hence the third eigenvalue is 2.

Example 3: Find the sum and product of all the eigenvalues of $\begin{bmatrix} 8 & -6 & 2 \\ -6 & 7 & -4 \\ 2 & -4 & 3 \end{bmatrix}$.

[AU, *April/May 2015*]

Solution : Let $A = \begin{bmatrix} 8 & -6 & 2 \\ -6 & 7 & -4 \\ 2 & -4 & 3 \end{bmatrix}$.

We know that, sum of eigenvalues = sum of main diagonal elements of A

$$= 8 + 7 + 3 = 18$$

Also, product of the eigenvalues $= |A|$

$$= \begin{vmatrix} 8 & -6 & 2 \\ -6 & 7 & -4 \\ 2 & -4 & 3 \end{vmatrix}$$

$$= 8(21-6) + 6(-18+8) + 2(24-14)$$

$$= 0.$$

Example 4: If 2 is an eigenvalue of $A = \begin{bmatrix} 2 & -2 & 2 \\ 1 & 1 & 1 \\ 1 & 3 & -1 \end{bmatrix}$ find the other two values.

Solution : Given , eigenvalue $\lambda_1 = 2$ and $A = \begin{bmatrix} 2 & -2 & 2 \\ 1 & 1 & 1 \\ 1 & 3 & -1 \end{bmatrix}$.

Let λ_2 and λ_3 be the other two eigenvalues.

We know that, sum of eigenvalues = sum of main diagonal elements of A

$\Rightarrow \ \lambda_1 + \lambda_2 + \lambda_3 \ = 2 + 1 - 1$

$\Rightarrow \ 2 + \lambda_2 + \lambda_3 = 2$

$\Rightarrow \ \lambda_2 + \lambda_3 = 0$ ... (1)

Also, product of the eigenvalues = $|A|$

$$\Rightarrow \ \lambda_1 \times \lambda_2 \times \lambda_3 \ = \begin{vmatrix} 2 & -2 & 2 \\ 1 & 1 & 1 \\ 1 & 3 & -1 \end{vmatrix}$$

$\Rightarrow \ 2 \times \lambda_2 \times \lambda_3 \ = 2(-1-3) + 2(-1-1) + 2(3-1) = -8 - 4 + 4 = -8$

$\Rightarrow \ \lambda_2 \ \lambda_3 = -4$

Now, $(\lambda_2 - \lambda_3)^2 = (\lambda_2 + \lambda_3)^2 - 4\lambda_2\lambda_3 = 0 - 4(-4) = 16$

$\therefore \ \lambda_2 - \lambda_3 = 4$ ... (2)

Solving (1) and (2) we get, $\lambda_2 = 2$ and $\lambda_3 = -2$.

Hence, the other two eigenvalues are 2 and -2.

Example 5: Find the sum and product of the eigenvalues of a 3×3 matrix A whose characteristic equation is $\lambda^3 - 7\lambda^2 + 36 = 0$. **[AU, *Nov/Dec 2017*]**

Solution: Given that the characteristic equation of A is $\lambda^3 - 7\lambda^2 + 36 = 0$.

We know that, the characteristic equation of A is $\lambda^3 - S_1\lambda^2 + S_2\lambda - S_3 = 0$,

where $S_1 = 7 =$ sum of the diagonal elements of A

and $S_3 \ = -36 = |A|$.

$\therefore$ Sum of the eigenvalues of A = sum of the diagonal elements of $A = 7$

and product of the eigenvalues of $A = |A| = -36$.

Example 6: If 2, 3 are the eigenvalues of $\begin{bmatrix} 2 & 0 & 1 \\ 0 & 2 & 0 \\ b & 0 & 2 \end{bmatrix}$, then find the value of b.

Solution: Let $A = \begin{bmatrix} 2 & 0 & 1 \\ 0 & 2 & 0 \\ b & 0 & 2 \end{bmatrix}$ **[AU, *Nov/Dec 2014*]**

Given $\lambda_1 = 2, \lambda_2 = 3$.

Let λ_3 be the third eigenvalue.

We know that, product of the eigenvalues = $|A|$

$$\Rightarrow \ \lambda_1 \times \lambda_2 \times \lambda_3 \ = \begin{vmatrix} 2 & 0 & 1 \\ 0 & 2 & 0 \\ b & 0 & 2 \end{vmatrix}$$

$\Rightarrow \ 6\lambda_3 = 8 - 2b$

$\Rightarrow \ \lambda_3 = \frac{8-2b}{6}$

Also, sum of eigenvalues = sum of main diagonal elements of A

$$\Rightarrow \quad 2+3+\frac{8-2b}{6}=6$$

$$\Rightarrow \quad \frac{8-2b}{6}=1$$

$$\Rightarrow \quad 8-2b = 6 \qquad \Rightarrow 2b = 2 \quad \Rightarrow b=1.$$

Example 7: Find the Eigenvalues of $3A+2I$, where $A=\begin{bmatrix}5 & 4\\0 & 2\end{bmatrix}$.

[AU, Dec 2015/Jan 2016]

Solution: Given that $A=\begin{bmatrix}5 & 4\\0 & 2\end{bmatrix}$ which is an upper triangular matrix.

$\therefore$ Eigenvalues of $A = 5, 2.$

$\therefore$ Eigenvalues of $3A+2I = 3(5)+2,\ 3(2)+2 = 17, 8.$

Example 8: If λ is the eigenvalue of the matrix A, then prove that λ^2 is the eigenvalue of A^2. ***[AU, January 2014, May/June 2016]***

Solution: Let λ be an eigenvalue and X be the eigenvector of the matrix A.

Then $\quad AX = \lambda X$ ... (1)

$$\therefore \quad A^2X = A\lambda X$$
$$= \lambda(AX)$$
$$= \lambda(\lambda X) \quad \text{[From (1)]}$$
$$= \lambda^2 X$$

$\therefore$ λ^2 is the eigenvalue of A^2.

Example 9: If $2, -1, -3$ are the eigenvalues of the matrix A, then find the eigenvalues of the matrix A^2-2I. [AU, *May/June 2014*]

Solution: The eigenvalues of A are $2, -1, -3$.

$\therefore$ The eigenvalues of $A^2 = 2^2, (-1)^2, (-3)^2 = 4, 1, 9.$

$\therefore$ The eigenvalues of $A^2-2I = 4-2,\ 1-2,\ 9-2 = 2, -1, 7.$

Example 10: If $\lambda(\neq 0)$ is an eigenvalue of a square matrix A, then show that λ^{-1} is an eigenvalue of A^{-1}. [AU, *Nov/Dec 2017*]

Solution: Let λ be an eigenvalue of A and X be the corresponding eigenvector. Then, by definition, $AX = \lambda X$.

Pre multiplying both sides by A^{-1}, we get

$A^{-1}(AX) = A^{-1}(\lambda X)$

$\Rightarrow X = \lambda(A^{-1}X)$

$\Rightarrow A^{-1}X = \frac{1}{\lambda}X$

$\Rightarrow \frac{1}{\lambda}$ is an eigenvalue of A^{-1}

i.e., λ^{-1} is an eigenvalue of A^{-1}.

Example 11: Find the eigenvalues of the matrix A^{-1} if $A = \begin{bmatrix} 1 & 5 \\ 0 & 4 \end{bmatrix}$.

Solution: Given $A = \begin{bmatrix} 1 & 5 \\ 0 & 4 \end{bmatrix}$.

Since it is an upper triangular matrix, the eigenvalues A are 1,4.

$\therefore$ Eigenvalues of $A^{-1} = \frac{1}{1}, \frac{1}{4} = 1, \frac{1}{4}$.

Example 12: Find the eigenvalues of the inverse of the matrix $A = \begin{bmatrix} 2 & 1 & 0 \\ 0 & 3 & 4 \\ 0 & 0 & 4 \end{bmatrix}$.

[AU, *May/June 2014*]

Solution: Given that $A = \begin{bmatrix} 2 & 1 & 0 \\ 0 & 3 & 4 \\ 0 & 0 & 4 \end{bmatrix}$, which is an upper triangular matrix.

$\therefore$ The eigenvalues of A = the diagonal elements of $A = 2, 3, 4$.

We know that, if $\lambda(\neq 0)$ is an eigenvalue of a A, then λ^{-1} is an eigenvalue of A^{-1}.

$\therefore$ The eigenvalues of $A^{-1} = \frac{1}{2}, \frac{1}{3}, \frac{1}{4}$.

Example 13: If eigenvalues of the matrix A of order 3×3 are 2, 3 and 1, then find the eigenvalues of adjoint. ***[AU, January 2014, May/June 2016]***

Solution: If λ is the eigenvalue of A, then $\frac{|A|}{\lambda}$ is the eigenvalue of adjoint of A.

The eigen values of A are 2, 3 and 1.

The eigenvalues of adjoint of A are $\frac{|A|}{2}, \frac{|A|}{3}$ and $\frac{|A|}{1}$.

We know, $|A|$ = Product of eigenvalues of A

$= 2 \times 3 \times 1 = 6$

$\therefore$ The eigenvalues of adjoint of $A = \frac{6}{2}, \frac{6}{3}, \frac{6}{1} = 3, 2, 6$.

EXERCISES

1. Find the sum and product of the eigenvalues of the matrix $\begin{bmatrix} -2 & 2 & -3 \\ 2 & 1 & -6 \\ -1 & -2 & 0 \end{bmatrix}$.
2. Find the eigenvalues of A^3 if $A = \begin{bmatrix} 1 & 2 & 3 \\ 0 & 2 & -7 \\ 0 & 0 & 3 \end{bmatrix}$.
3. Show that the eigenvalues of the real symmetric matrix $A = \begin{bmatrix} -2 & 2 \\ 2 & 1 \end{bmatrix}$ are real.
4. If 1,1,5 are the eigenvalues of $A = \begin{bmatrix} 2 & 2 & 1 \\ 1 & 3 & 1 \\ 1 & 2 & 2 \end{bmatrix}$, find the eigenvalues of $5A$.
5. The eigenvalues of the matrix $A = \begin{bmatrix} 3 & 10 & 5 \\ -2 & -3 & -4 \\ 3 & 5 & 7 \end{bmatrix}$ are equal to 2,2,3 then find the eigen values of A^{-1}.

ANSWERS

1. Sum = −1; Product = 45 2. 1,8,27 3. −3,2 4. 5,5,25

5. $\frac{1}{2}, \frac{1}{2}, \frac{1}{3}$.

1.4. Cayley - Hamilton Theorem

Cayley - Hamilton Theorem

Every square matrix satisfies its own characteristic equation.

Proof: Let $A = [a_{ij}]_{n\times n}$ be a square matrix.

$$\therefore \quad A = \begin{bmatrix} a_{11} & a_{12} & \dots & a_{1n} \\ a_{21} & a_{22} & \dots & a_{2n} \\ \dots & \dots & \dots & \dots \\ a_{n1} & a_{n2} & \dots & a_{nn} \end{bmatrix}$$

Therefore the characteristic equation of A is $|A - \lambda I| = 0$.

$$\Rightarrow \quad \begin{vmatrix} a_{11} - \lambda & a_{12} & \dots & a_{1n} \\ 0 & a_{22} - \lambda & \dots & a_{2n} \\ \dots & \dots & \dots & \dots \\ 0 & 0 & \dots & a_{nn} - \lambda \end{vmatrix} = 0$$

$$\Rightarrow \quad a_0\lambda^n + a_1\lambda^{n-1} + a_2\lambda^{n-2} + \ldots + a_n = 0$$

We have to prove that $a_0A^n + a_1A^{n-1} + a_2A^{n-2} + \ldots + a_nI = 0 \quad \ldots (1)$

Since the elements of $(A-\lambda I)$ are atmost of the first degree in λ, the elements of adj$(A-\lambda I)$ are atmost of degree $(n-1)$ in λ.

Thus adj$(A-\lambda I) = B_0\lambda^{n-1} + B_1\lambda^{n-2} + \ldots + B_{n-1}$,

where $B_0, B_1, \ldots B_{n-1}$ are square matrices of order n, their elements being polynomial in λ.

We know that $(A-\lambda I)$ adj$(A-\lambda I) = |A-\lambda I|I$

$$\Rightarrow \quad (A-\lambda I)(B_0\lambda^{n-1} + B_1\lambda^{n-2} + \ldots + B_{n-1}) = (a_0\lambda^n + a_1\lambda^{n-1} + a_2\lambda^{n-2} + \ldots + a_n)I$$

This is an identity in scalar λ.

Equating the coefficients of like powers of λ on both sides, we get

$$-B_0 = a_0 I$$

$$AB_0 - B_1 = a_1 I$$

$$AB_1 - B_2 = a_2 I$$

$$\ldots \quad \ldots \quad \ldots$$

$$AB_{r-1} - B_r = a_r I$$

$$\ldots \quad \ldots \quad \ldots$$

$$AB_{n-1} = a_n I$$

Pre-multiplying these equations by $A^n, A^{n-1}, A^{n-2}, \ldots, A, I$ respectively and adding, we get $\quad 0 = a_0A^n + a_1A^{n-1} + a_2A^{n-2} + \ldots + a_{n-1}A + a_nI$.

Which proves the theorem.

Note1: Pre multiplying equation (1) by A^{-1}, we have

$$a_0A^{n-1} + a_1A^{n-2} + a_2A^{n-3} + \ldots + a_{n-1}I + a_nA^{-1} = 0$$

$$\Rightarrow \quad A^{-1} = -\frac{1}{a_n}\left[a_0A^{n-1} + a_1A^{n-2} + a_2A^{n-3} + \ldots + a_{n-1}I\right]$$

Thus Cayley-Hamilton theorem gives another method for computing the inverse of a square matrix. Since this method expresses the inverse of a matrix of order n in terms of $(n-1)$ powers of A, it is most suitable for computing inverses of large matrices.

Note2: If m is a positive integer such that $m > n$ then any positive integral power A^m of A is linearly expressible in terms of those of lower degree.

Example 1: Verify Cayley-Hamilton theorem for the matrix $A = \begin{bmatrix} 1 & 2 \\ 2 & -1 \end{bmatrix}$. Hence find A^4.

Solution: The characteristic equation of A is $|A - \lambda I| = 0$.

$$\Rightarrow \qquad \lambda^2 - S_1\lambda + S_2 = 0$$

where S_1 = sum of the diagonal elements of A.

$$= 1 - 1 = 0$$

$$S_2 = |A| = \begin{vmatrix} 1 & 2 \\ 2 & -1 \end{vmatrix} = -1 - 4 = -5.$$

$\therefore$ The characteristic equation is $\lambda^2 - 0\lambda - 5 = 0$

$$\Rightarrow \qquad \lambda^2 - 5 = 0 \qquad \dots (1)$$

By Cayley -Hamilton theorem, $A^2 - 5I = 0$

Now, $A^2 = A.A = \begin{bmatrix} 1 & 2 \\ 2 & -1 \end{bmatrix}\begin{bmatrix} 1 & 2 \\ 2 & -1 \end{bmatrix} = \begin{bmatrix} 5 & 0 \\ 0 & 5 \end{bmatrix}$

$$\therefore A^2 - 5I = \begin{bmatrix} 5 & 0 \\ 0 & 5 \end{bmatrix} - 5\begin{bmatrix} 1 & 0 \\ 0 & 1 \end{bmatrix}$$

$$= \begin{bmatrix} 5 & 0 \\ 0 & 5 \end{bmatrix} - \begin{bmatrix} 5 & 0 \\ 0 & 5 \end{bmatrix}$$

$$= \begin{bmatrix} 0 & 0 \\ 0 & 0 \end{bmatrix}$$

Hence Cayley-Hamilton theorem is verified.

To find A^4:

By Cayley -Hamilton theorem, $A^2 - 5I = 0 \quad \Rightarrow A^2 = 5I$.

Multiply by A^2 on both sides, we get

$$A^4 = A^2\,(5I)$$

$$= \begin{bmatrix} 5 & 0 \\ 0 & 5 \end{bmatrix} \begin{bmatrix} 5 & 0 \\ 0 & 5 \end{bmatrix}$$

$$= \begin{bmatrix} 25 & 0 \\ 0 & 25 \end{bmatrix}$$

Example 2: Using Cayley-Hamilton theorem find A^{-1} and A^4, if $A = \begin{bmatrix} 1 & 2 & -2 \\ -1 & 3 & 0 \\ 0 & -2 & 1 \end{bmatrix}$.

Solution: ***[AU, April/May 2015]***

The characteristic equation of A is $|A - \lambda I| = 0$.

$$\Rightarrow \quad \lambda^3 - S_1\lambda^2 + S_2\lambda - S_3 = 0$$

where S_1 = sum of the diagonal elements of A.

$$= 1 + 3 + 1 = 5$$

S_2 = sum of minors of the diagonal elements of A.

$$= \begin{vmatrix} 3 & 0 \\ -2 & 1 \end{vmatrix} + \begin{vmatrix} 1 & -2 \\ 0 & 1 \end{vmatrix} + \begin{vmatrix} 1 & 2 \\ -1 & 3 \end{vmatrix}$$

$$= (3 - 0) + (1 - 0) + (3 + 2) = 9$$

$$S_3 = |A|$$

$$= \begin{vmatrix} 1 & 2 & -2 \\ -1 & 3 & 0 \\ 0 & -2 & 1 \end{vmatrix}$$

$$= 1(3 - 0) - 2(-1 - 0) - 2(2 - 0) = 3 + 2 - 4 = 1$$

$\therefore$ The characteristic equation is $\lambda^3 - 5\lambda^2 + 9\lambda - 1 = 0$

By Cayley -Hamilton theorem, $A^3 - 5A^2 + 9A - I = 0$ ... (1)

$$\Rightarrow \quad I = A^3 - 5A^2 + 9A$$

$$\Rightarrow \quad A^{-1} = A^2 - 5A + 9I$$

Now, $A^2 = A.A = \begin{bmatrix} 1 & 2 & -2 \\ -1 & 3 & 0 \\ 0 & -2 & 1 \end{bmatrix} \begin{bmatrix} 1 & 2 & -2 \\ -1 & 3 & 0 \\ 0 & -2 & 1 \end{bmatrix} = \begin{bmatrix} -1 & 12 & -4 \\ -4 & 7 & 2 \\ 2 & -8 & 1 \end{bmatrix}$

$$\therefore A^{-1} = A^2 - 5A + 9I$$

$$= \begin{bmatrix} -1 & 12 & -4 \\ -4 & 7 & 2 \\ 2 & -8 & 1 \end{bmatrix} - \begin{bmatrix} 5 & 10 & -10 \\ -5 & 15 & 0 \\ 0 & -10 & 5 \end{bmatrix} + \begin{bmatrix} 9 & 0 & 0 \\ 0 & 9 & 0 \\ 0 & 0 & 9 \end{bmatrix}$$

$$A^{-1} = \begin{bmatrix} 3 & 2 & 6 \\ 1 & 1 & 2 \\ 2 & 2 & 5 \end{bmatrix}.$$

To find A^4:

Multiply equation (1) by A on both sides,

$$A^4 - 5A^3 + 9A^2 - A = 0$$

$$\Rightarrow \quad A^4 = 5A^3 - 9A^2 + A.$$

Now, $A^3 = A^2.A$

$$= \begin{bmatrix} -1 & 12 & -4 \\ -4 & 7 & 2 \\ 2 & -8 & 1 \end{bmatrix} \begin{bmatrix} 1 & 2 & -2 \\ -1 & 3 & 0 \\ 0 & -2 & 1 \end{bmatrix}$$

$$= \begin{bmatrix} 13 & 42 & -2 \\ -9 & 9 & 10 \\ 10 & -22 & -3 \end{bmatrix}.$$

$$\therefore \quad A^4 = 5\begin{bmatrix} 13 & 42 & -2 \\ -9 & 9 & 10 \\ 10 & -22 & -3 \end{bmatrix} - 9\begin{bmatrix} -1 & 12 & -4 \\ -4 & 7 & 2 \\ 2 & -8 & 1 \end{bmatrix} + \begin{bmatrix} 1 & 2 & -2 \\ -1 & 3 & 0 \\ 0 & -2 & 1 \end{bmatrix}$$

$$= \begin{bmatrix} -65 & 210 & -10 \\ -45 & 45 & 50 \\ 50 & -110 & -15 \end{bmatrix} - \begin{bmatrix} -9 & 108 & -36 \\ -36 & 63 & 18 \\ 18 & -72 & 9 \end{bmatrix} + \begin{bmatrix} 1 & 2 & -2 \\ -1 & 3 & 0 \\ 0 & -2 & 1 \end{bmatrix}$$

$$= \begin{bmatrix} -55 & 104 & 24 \\ -10 & -15 & 32 \\ 32 & -40 & -23 \end{bmatrix}.$$

Example 3: Verify Cayley-Hamilton theorem for the matrix $\begin{bmatrix} 1 & 0 & 3 \\ 2 & 1 & -1 \\ 1 & -1 & 1 \end{bmatrix}$ hence find its inverse. ***[AU, May/June 2014, April/May 2015]***

Solution: Let $A = \begin{bmatrix} 1 & 0 & 3 \\ 2 & 1 & -1 \\ 1 & -1 & 1 \end{bmatrix}.$

The characteristic equation of A is $|A - \lambda I| = \mathbf{0}$.

$$\Rightarrow \quad \lambda^3 - S_1\lambda^2 + S_2\lambda - S_3 = 0$$

where S_1 = sum of the diagonal elements of A.

$= 1 + 1 + 1 = 3$

S_2 = sum of minors of the diagonal elements of A.

$$= \begin{vmatrix} 1 & -1 \\ -1 & 1 \end{vmatrix} + \begin{vmatrix} 1 & 3 \\ 1 & 1 \end{vmatrix} + \begin{vmatrix} 1 & 0 \\ 2 & 1 \end{vmatrix}$$

$$= (1+1) + (1-3) + (1-0) = 1$$

$$S_3 = |A|$$

$$= \begin{vmatrix} 1 & 0 & 3 \\ 2 & 1 & -1 \\ 1 & -1 & 1 \end{vmatrix}$$

$$= 1(1-1) - 0(2+1) + 3(-2-1) = 2 + 0 - 9 = -9$$

$\therefore$ The characteristic equation is $\lambda^3 - 3\lambda^2 - \lambda + 9 = 0$

Now we replace λ by A, $\quad A^3 - 3A^2 - A + 9I = 0. \quad \ldots (1)$

Now $A^2 = A.A = \begin{bmatrix} 1 & 0 & 3 \\ 2 & 1 & -1 \\ 1 & -1 & 1 \end{bmatrix}\begin{bmatrix} 1 & 0 & 3 \\ 2 & 1 & -1 \\ 1 & -1 & 1 \end{bmatrix} = \begin{bmatrix} 4 & -3 & 6 \\ 3 & 2 & 4 \\ 0 & -2 & 5 \end{bmatrix}.$

Also $A^3 = A^2.A = \begin{bmatrix} 4 & -3 & 6 \\ 3 & 2 & 4 \\ 0 & -2 & 5 \end{bmatrix}\begin{bmatrix} 1 & 0 & 3 \\ 2 & 1 & -1 \\ 1 & -1 & 1 \end{bmatrix} = \begin{bmatrix} 4 & -9 & 21 \\ 11 & -2 & 11 \\ 1 & -7 & 7 \end{bmatrix}.$

$$\therefore \quad A^3 - 3A^2 - A + 9I = \begin{bmatrix} 4 & -9 & 21 \\ 11 & -2 & 11 \\ 1 & -7 & 7 \end{bmatrix} - 3\begin{bmatrix} 4 & -3 & 6 \\ 3 & 2 & 4 \\ 0 & -2 & 5 \end{bmatrix} - \begin{bmatrix} 1 & 0 & 3 \\ 2 & 1 & -1 \\ 1 & -1 & 1 \end{bmatrix} + 9\begin{bmatrix} 1 & 0 & 0 \\ 0 & 1 & 0 \\ 0 & 0 & 1 \end{bmatrix}$$

$$= \begin{bmatrix} 0 & 0 & 0 \\ 0 & 0 & 0 \\ 0 & 0 & 0 \end{bmatrix}.$$

Hence Cayley-Hamilton theorem is verified.

To find A^{-1}:

Multiply equation (1) by A^{-1} on both sides,

$$A^2 - 3A - I + 9A^{-1} = 0$$

$$\Rightarrow \quad A^{-1} = \frac{-1}{9}[A^2 - 3A - I]$$

$$= \frac{-1}{9}\left\{\begin{bmatrix} 4 & -3 & 6 \\ 3 & 2 & 4 \\ 0 & -2 & 5 \end{bmatrix} - 3\begin{bmatrix} 1 & 0 & 3 \\ 2 & 1 & -1 \\ 1 & -1 & 1 \end{bmatrix} - \begin{bmatrix} 1 & 0 & 0 \\ 0 & 1 & 0 \\ 0 & 0 & 1 \end{bmatrix}\right\}$$

$$= \frac{-1}{9}\begin{bmatrix} 0 & -3 & -3 \\ -3 & -2 & 7 \\ -3 & 1 & 1 \end{bmatrix}.$$

Example 4: If $A = \begin{bmatrix} 3 & -1 & 1 \\ -1 & 5 & -1 \\ 1 & -1 & 3 \end{bmatrix}$, verify Cayley-Hamilton theorem and hence find A^{-1}.

[*AU, April/May 2015*]

Solution: The characteristic equation of A is $|A - \lambda I| = 0$.

$$\Rightarrow \qquad \lambda^3 - S_1\lambda^2 + S_2\lambda - S_3 = 0$$

where S_1 = sum of the diagonal elements of A.

$= 3 + 5 + 3 = 11$

S_2 = sum of minors of the diagonal elements of A.

$$= \begin{vmatrix} 5 & -1 \\ -1 & 3 \end{vmatrix} + \begin{vmatrix} 3 & 1 \\ 1 & 3 \end{vmatrix} + \begin{vmatrix} 3 & -1 \\ -1 & 5 \end{vmatrix}$$

$= (15 - 1) + (9 - 1) + (15 - 1) = 36$

$S_3 = |A|$

$$= \begin{vmatrix} 3 & -1 & 1 \\ -1 & 5 & -1 \\ 1 & -1 & 3 \end{vmatrix}$$

$= 3(15 - 1) + 1(-3 - 1) + 1(1 - 5) = 42 - 4 - 4 = 36$

$\therefore$ The characteristic equation is $\lambda^3 - 11\lambda^2 + 36\lambda - 36 = 0$

Now we replace λ by A, $A^3 - 11A^2 + 36A - 36I = 0$. ... (1)

$$\text{Now,} \quad A^2 = A.A = \begin{bmatrix} 3 & -1 & 1 \\ -1 & 5 & -1 \\ 1 & -1 & 3 \end{bmatrix}\begin{bmatrix} 3 & -1 & 1 \\ -1 & 5 & -1 \\ 1 & -1 & 3 \end{bmatrix} = \begin{bmatrix} 11 & -9 & 7 \\ -9 & 27 & -9 \\ 7 & -9 & 11 \end{bmatrix}.$$

$$\text{Also} \quad A^3 = A^2.A = \begin{bmatrix} 11 & -9 & 7 \\ -9 & 27 & -9 \\ 7 & -9 & 11 \end{bmatrix}\begin{bmatrix} 3 & -1 & 1 \\ -1 & 5 & -1 \\ 1 & -1 & 3 \end{bmatrix} = \begin{bmatrix} 49 & -63 & 41 \\ -63 & 153 & -63 \\ 41 & -63 & 49 \end{bmatrix}.$$

$$\therefore \quad A^3 - 11A^2 + 36A - 36I = \begin{bmatrix} 49 & -63 & 41 \\ -63 & 153 & -63 \\ 41 & -63 & 49 \end{bmatrix} - 11\begin{bmatrix} 11 & -9 & 7 \\ -9 & 27 & -9 \\ 7 & -9 & 11 \end{bmatrix}$$

$$+36\begin{bmatrix} 3 & -1 & 1 \\ -1 & 5 & -1 \\ 1 & -1 & 3 \end{bmatrix} - 36\begin{bmatrix} 1 & 0 & 0 \\ 0 & 1 & 0 \\ 0 & 0 & 1 \end{bmatrix}$$

$$= \begin{bmatrix} 0 & 0 & 0 \\ 0 & 0 & 0 \\ 0 & 0 & 0 \end{bmatrix}.$$

To find A^{-1}:

Multiply equation (1) by A^{-1} on both sides,

$$A^2 - 11A + 36I - 36A^{-1} = 0$$

$$\Rightarrow \quad A^{-1} = \frac{1}{36}[A^2 - 11A + 36I]$$

$$= \frac{1}{36}\left\{\begin{bmatrix} 11 & -9 & 7 \\ -9 & 27 & -9 \\ 7 & -9 & 11 \end{bmatrix} - 11\begin{bmatrix} 3 & -1 & 1 \\ -1 & 5 & -1 \\ 1 & -1 & 3 \end{bmatrix} + 36\begin{bmatrix} 1 & 0 & 0 \\ 0 & 1 & 0 \\ 0 & 0 & 1 \end{bmatrix}\right\}$$

$$= \frac{1}{36}\begin{bmatrix} 14 & 2 & -4 \\ 12 & 8 & 2 \\ -4 & 2 & 14 \end{bmatrix}.$$

Example 5: Verify Cayley-Hamilton theorem for the matrix $A = \begin{bmatrix} 1 & 2 & -2 \\ 2 & 5 & -4 \\ 3 & 7 & -5 \end{bmatrix}$. Hence find A^{-1}. ***[AU, January 2016]***

Solution: The characteristic equation of A is $|A - \lambda I| = 0$.

$$\Rightarrow \quad \lambda^3 - S_1\lambda^2 + S_2\lambda - S_3 = 0.$$

where S_1 = sum of the diagonal elements of A.

$$= 1 + 5 - 5 = 1$$

S_2 = sum of minors of the diagonal elements of A.

$$= \begin{vmatrix} 5 & -4 \\ 7 & -5 \end{vmatrix} + \begin{vmatrix} 1 & -2 \\ 3 & -5 \end{vmatrix} + \begin{vmatrix} 1 & 2 \\ 2 & 5 \end{vmatrix}$$

$$= (-25 + 28) + (-5 + 6) + (5 - 4) = 5$$

$$S_3 = |A|$$

$$= \begin{vmatrix} 1 & 2 & -2 \\ 2 & 5 & -4 \\ 3 & 7 & -5 \end{vmatrix}$$

$$= 1(-25 + 28) - 2(-10 + 12) - 2(14 - 15) = 3 - 4 + 2 = 1$$

$\therefore$ The characteristic equation is $\lambda^3 - \lambda^2 + 5\lambda - 1 = 0$

Now we replace λ by A, $\quad A^3 - A^2 + 5A - I = 0.$ ... (1)

Now, $A^2 = A.A = \begin{bmatrix} 1 & 2 & -2 \\ 2 & 5 & -4 \\ 3 & 7 & -5 \end{bmatrix}\begin{bmatrix} 1 & 2 & -2 \\ 2 & 5 & -4 \\ 3 & 7 & -5 \end{bmatrix} = \begin{bmatrix} -1 & -2 & 0 \\ 0 & 1 & -4 \\ 2 & 6 & -9 \end{bmatrix}$

Also $A^3 = A^2.A = \begin{bmatrix} -1 & -2 & 0 \\ 0 & 1 & -4 \\ 2 & 6 & -9 \end{bmatrix}\begin{bmatrix} 1 & 2 & -2 \\ 2 & 5 & -4 \\ 3 & 7 & -5 \end{bmatrix} = \begin{bmatrix} -5 & -12 & 10 \\ -10 & -23 & 16 \\ -13 & -29 & 17 \end{bmatrix}$

$$\therefore \quad A^3 - A^2 + 5A - I = \begin{bmatrix} -5 & -12 & 10 \\ -10 & -23 & 16 \\ -13 & -29 & 17 \end{bmatrix} - \begin{bmatrix} -1 & -2 & 0 \\ 0 & 1 & -4 \\ 2 & 6 & -9 \end{bmatrix} + 5\begin{bmatrix} 1 & 2 & -2 \\ 2 & 5 & -4 \\ 3 & 7 & -5 \end{bmatrix} - \begin{bmatrix} 1 & 0 & 0 \\ 0 & 1 & 0 \\ 0 & 0 & 1 \end{bmatrix}$$

$$= \begin{bmatrix} 0 & 0 & 0 \\ 0 & 0 & 0 \\ 0 & 0 & 0 \end{bmatrix}.$$

To find A^{-1}:

Multiply equation (1) by A^{-1} on both sides,

$$A^2 - A + 5I - A^{-1} = 0$$

$$\Rightarrow \quad A^{-1} = A^2 - A + 5I$$

$$= \begin{bmatrix} -1 & -2 & 0 \\ 0 & 1 & -4 \\ 2 & 6 & -9 \end{bmatrix} - \begin{bmatrix} 1 & 2 & -2 \\ 2 & 5 & -4 \\ 3 & 7 & -5 \end{bmatrix} + 5\begin{bmatrix} 1 & 0 & 0 \\ 0 & 1 & 0 \\ 0 & 0 & 1 \end{bmatrix}$$

$$= \begin{bmatrix} 3 & -4 & 2 \\ -2 & 1 & 0 \\ -1 & -1 & 1 \end{bmatrix}.$$

Example 6: Using Cayley-Hamilton theorem find A^{-1} where $A = \begin{bmatrix} 7 & 2 & -2 \\ -6 & -1 & 2 \\ 6 & 2 & -1 \end{bmatrix}$.

Solution: ***[AU, Dec 2016]***

The characteristic equation of A is $|A - \lambda I| = 0$.

$$\Rightarrow \quad \lambda^3 - S_1\lambda^2 + S_2\lambda - S_3 = 0.$$

where S_1 = sum of the diagonal elements of A.

$$= 7 - 1 - 1 = 5$$

S_2 = sum of minors of the diagonal elements of A.

$$= \begin{vmatrix} -1 & 2 \\ 2 & -1 \end{vmatrix} + \begin{vmatrix} 7 & -2 \\ 6 & -1 \end{vmatrix} + \begin{vmatrix} 7 & 2 \\ -6 & -1 \end{vmatrix}$$

$$= (1 - 4) + (-7 + 12) + (-7 + 12) = -3 + 5 + 5 = 7$$

$$S_3 = |A|$$

$$= \begin{vmatrix} 7 & 2 & -2 \\ -6 & -1 & 2 \\ 6 & 2 & -1 \end{vmatrix}$$

$$= 7(1-4) - 2(6-12) - 2(-12+6) = -21 + 12 + 12 = 3$$

$\therefore$ The characteristic equation is $\lambda^3 - 5\lambda^2 + 7\lambda - 3 = 0$

By Cayley Hamilton theorem, $A^3 - 5A^2 + 7A - 3I = 0.$ ... (1)

Multiply equation (1) by A^{-1} on both sides,

$$A^2 - 5A + 7I - 3A^{-1} = 0$$

$$\Rightarrow \quad A^{-1} = [A^2 - 5A + 7I]$$

Now, $A^2 = A.A = \begin{bmatrix} 1 & 2 & -2 \\ -1 & 3 & 0 \\ 0 & -2 & 1 \end{bmatrix}\begin{bmatrix} 1 & 2 & -2 \\ -1 & 3 & 0 \\ 0 & -2 & 1 \end{bmatrix} = \begin{bmatrix} -1 & 12 & -4 \\ -4 & 7 & 2 \\ 2 & -8 & 1 \end{bmatrix}$

$$\therefore \quad A^{-1} = A^2 - 5A + 9I$$

$$= \frac{1}{3}\begin{bmatrix} -1 & 12 & -4 \\ -4 & 7 & 2 \\ 2 & -8 & 1 \end{bmatrix} - \begin{bmatrix} 5 & 10 & -10 \\ -5 & 15 & 0 \\ 0 & -10 & 5 \end{bmatrix} + \begin{bmatrix} 9 & 0 & 0 \\ 0 & 9 & 0 \\ 0 & 0 & 9 \end{bmatrix}$$

$$= \frac{1}{3}\begin{bmatrix} 3 & 2 & 6 \\ 1 & 1 & 2 \\ 2 & 2 & 5 \end{bmatrix}.$$

Example 7: Verify Cayley Hamilton theorem for $A = \begin{bmatrix} 1 & 2 & 3 \\ 2 & -1 & 4 \\ 3 & 17 & -1 \end{bmatrix}$. Hence using it find A^{-1}. ***[AU, May/June 2016]***

Solution: The characteristic equation of A is $|A - \lambda I| = \boldsymbol{0}$.

$$\Rightarrow \quad \lambda^3 - S_1\lambda^2 + S_2\lambda - S_3 = 0.$$

where S_1 = sum of the diagonal elements of A.

$$= 1 - 1 - 1 = -1$$

S_2 = sum of minors of the diagonal elements of A.

$$= \begin{vmatrix} -1 & 4 \\ 17 & -1 \end{vmatrix} + \begin{vmatrix} 1 & 3 \\ 3 & -1 \end{vmatrix} + \begin{vmatrix} 1 & 2 \\ 2 & -1 \end{vmatrix}$$

$$= (1-68) + (-1-9) + (-1-4) = -82$$

$$S_3 = |A|$$

$$= \begin{vmatrix} 1 & 2 & 3 \\ 2 & -1 & 4 \\ 3 & 17 & -1 \end{vmatrix}$$

$$= 1(1-68) - 2(-2-12) + 3(34+3) = -67 + 28 + 111 = 72$$

$\therefore$ The characteristic equation is $\lambda^3 + \lambda^2 - 82\lambda - 72 = 0$

Now we replace λ by A, $\quad A^3 + A^2 - 82A - 72I = 0$. ... (1)

$$\text{Now } A^2 = A.A = \begin{bmatrix} 1 & 2 & 3 \\ 2 & -1 & 4 \\ 3 & 17 & -1 \end{bmatrix} \begin{bmatrix} 1 & 2 & 3 \\ 2 & -1 & 4 \\ 3 & 17 & -1 \end{bmatrix} = \begin{bmatrix} 14 & 51 & 8 \\ 12 & 73 & -2 \\ 34 & -28 & 78 \end{bmatrix}.$$

$$\text{Also } A^3 = A^2.A = \begin{bmatrix} 14 & 51 & 8 \\ 12 & 73 & -2 \\ 34 & -28 & 78 \end{bmatrix} \begin{bmatrix} 1 & 2 & 3 \\ 2 & -1 & 4 \\ 3 & 17 & -1 \end{bmatrix} = \begin{bmatrix} 140 & 113 & 238 \\ 152 & -83 & 330 \\ 212 & 1422 & -88 \end{bmatrix}.$$

$$\therefore \quad A^3 + A^2 - 82A - 72I = \begin{bmatrix} 140 & 113 & 238 \\ 152 & -83 & 330 \\ 212 & 1422 & -88 \end{bmatrix} + \begin{bmatrix} 14 & 51 & 8 \\ 12 & 73 & -2 \\ 34 & -28 & 78 \end{bmatrix}$$

$$-82 \begin{bmatrix} 1 & 2 & 3 \\ 2 & -1 & 4 \\ 3 & 17 & -1 \end{bmatrix} - 72 \begin{bmatrix} 1 & 0 & 0 \\ 0 & 1 & 0 \\ 0 & 0 & 1 \end{bmatrix}$$

$$= \begin{bmatrix} 0 & 0 & 0 \\ 0 & 0 & 0 \\ 0 & 0 & 0 \end{bmatrix}.$$

Hence Cayley-Hamilton theorem is verified.

To find A^{-1}:

Multiply equation (1) by A^{-1} on both sides,

$$A^2 + A - 82I - 72IA^{-1} = 0$$

$$\Rightarrow \quad A^{-1} = \frac{1}{72}[A^2 + A - 82I]$$

$$= \frac{1}{72}\left\{ \begin{bmatrix} 14 & 51 & 8 \\ 12 & 73 & -2 \\ 34 & -28 & 78 \end{bmatrix} + \begin{bmatrix} 1 & 2 & 3 \\ 2 & -1 & 4 \\ 3 & 17 & -1 \end{bmatrix} - \begin{bmatrix} 82 & 0 & 0 \\ 0 & 82 & 0 \\ 0 & 0 & 82 \end{bmatrix} \right\}$$

$$\therefore \quad A^{-1} = \frac{1}{72} \begin{bmatrix} -67 & 53 & 11 \\ 14 & -10 & 2 \\ 37 & -11 & -5 \end{bmatrix}.$$

Example 8: Show that $A = \begin{bmatrix} 2 & 1 & 1 \\ 0 & 1 & 0 \\ 1 & 1 & 2 \end{bmatrix}$ satisfies its own characteristic equation and hence find A^{-1}. ***[AU, Nov/Dec 2017]***

Solution: The characteristic equation of A is $|A - \lambda I| = 0$

$$\Rightarrow \quad \lambda^3 - S_1\lambda^2 + S_2\lambda - S_3 = 0.$$

where S_1 = sum of the diagonal elements of A.

$= 2+1+2 = 5$

$S_2 =$ sum of minors of the diagonal elements of A.

$$= \begin{vmatrix} 1 & 0 \\ 1 & 2 \end{vmatrix} + \begin{vmatrix} 2 & 1 \\ 1 & 2 \end{vmatrix} + \begin{vmatrix} 2 & 1 \\ 0 & 1 \end{vmatrix}$$

$= (2-0)+(4-1)+(2-0) = 7$

$S_3 = |A|$

$$= \begin{vmatrix} 2 & 1 & 1 \\ 0 & 1 & 0 \\ 1 & 1 & 2 \end{vmatrix}$$

$= 2(2-0) - 1(0-0) + 1(0-1) = 4-1 = 3$

$\therefore$ The characteristic equation is $\lambda^3 - 5\lambda^2 + 7\lambda - 3 = 0$

Now we replace λ by A, $\quad A^3 - 5A^2 + 7A - 3I = 0.$ ... (1)

$$\text{Now } A^2 = A.A = \begin{bmatrix} 2 & 1 & 1 \\ 0 & 1 & 0 \\ 1 & 1 & 2 \end{bmatrix}\begin{bmatrix} 2 & 1 & 1 \\ 0 & 1 & 0 \\ 1 & 1 & 2 \end{bmatrix} = \begin{bmatrix} 5 & 4 & 4 \\ 0 & 1 & 0 \\ 4 & 4 & 5 \end{bmatrix}.$$

$$\text{Also } A^3 = A^2.A = \begin{bmatrix} 5 & 4 & 4 \\ 0 & 1 & 0 \\ 4 & 4 & 5 \end{bmatrix}\begin{bmatrix} 2 & 1 & 1 \\ 0 & 1 & 0 \\ 1 & 1 & 2 \end{bmatrix} = \begin{bmatrix} 14 & 13 & 13 \\ 0 & 1 & 0 \\ 13 & 13 & 14 \end{bmatrix}.$$

$$\therefore \quad A^3 - 5A^2 + 7A - 3I = \begin{bmatrix} 14 & 13 & 13 \\ 0 & 1 & 0 \\ 13 & 13 & 14 \end{bmatrix} - 5\begin{bmatrix} 5 & 4 & 4 \\ 0 & 1 & 0 \\ 4 & 4 & 5 \end{bmatrix} + 7\begin{bmatrix} 2 & 1 & 1 \\ 0 & 1 & 0 \\ 1 & 1 & 2 \end{bmatrix} - 3\begin{bmatrix} 1 & 0 & 0 \\ 0 & 1 & 0 \\ 0 & 0 & 1 \end{bmatrix}$$

$$= \begin{bmatrix} 0 & 0 & 0 \\ 0 & 0 & 0 \\ 0 & 0 & 0 \end{bmatrix}.$$

Hence A satisfies its own characteristic equation.

To find A^{-1}:

Multiply equation (1) by A^{-1} on both sides,

$$A^2 - 5A + 7I - 3IA^{-1} = 0$$

$$\Rightarrow \quad A^{-1} = \frac{1}{3}[A^2 - 5A + 7I] \quad \frac{1}{3}$$

$$= \frac{1}{3}\left\{\begin{bmatrix} 5 & 4 & 4 \\ 0 & 1 & 0 \\ 4 & 4 & 5 \end{bmatrix} - 5\begin{bmatrix} 2 & 1 & 1 \\ 0 & 1 & 0 \\ 1 & 1 & 2 \end{bmatrix} + \begin{bmatrix} 7 & 0 & 0 \\ 0 & 7 & 0 \\ 0 & 0 & 7 \end{bmatrix}\right\}$$

$$\therefore \quad A^{-1} = \frac{1}{3}\begin{bmatrix} 2 & -1 & -1 \\ 0 & 3 & 0 \\ -1 & -1 & 2 \end{bmatrix}.$$

Example 9: Verify Cayley-Hamilton theorem, find A^4 and A^{-1} for the matrix $A = \begin{bmatrix} 2 & -1 & 2 \\ -1 & 2 & -1 \\ 1 & -1 & 2 \end{bmatrix}$. ***[AU, Nov/Dec 2014, April/May 2015, April/May 2017]***

Solution: The characteristic equation of A is $|A - \lambda I| = \boldsymbol{0}$

$$\Rightarrow \quad \lambda^3 - S_1\lambda^2 + S_2\lambda - S_3 = 0$$

where S_1 = sum of the diagonal elements of A.

$= 2 + 2 + 2 = 6$

S_2 = sum of minors of the diagonal elements of A.

$$= \begin{vmatrix} 2 & -1 \\ -1 & 2 \end{vmatrix} + \begin{vmatrix} 2 & 2 \\ 1 & 2 \end{vmatrix} + \begin{vmatrix} 2 & -1 \\ -1 & 2 \end{vmatrix}$$

$= (4 - 1) + (4 - 2) + (4 - 1) = 8$

$S_3 = |A|$

$$= \begin{vmatrix} 2 & -1 & 2 \\ -1 & 2 & -1 \\ 1 & -1 & 2 \end{vmatrix}$$

$= 2(4 - 1) + 1(-2 + 1) + 2(1 - 2) = 6 - 1 - 2 = 3$

$\therefore$ The characteristic equation is $\lambda^3 - 6\lambda^2 + 8\lambda - 3 = 0$

Now we replace λ by A, $A^3 - 6A^2 + 8A - 3I = 0$. ...(1)

$$\text{Now, } A^2 = A.A = \begin{bmatrix} 2 & -1 & 2 \\ -1 & 2 & -1 \\ 1 & -1 & 2 \end{bmatrix}\begin{bmatrix} 2 & -1 & 2 \\ -1 & 2 & -1 \\ 1 & -1 & 2 \end{bmatrix} = \begin{bmatrix} 7 & -6 & 9 \\ -5 & 6 & -6 \\ 5 & -5 & 7 \end{bmatrix}.$$

$$\text{Also } A^3 = A^2.A = \begin{bmatrix} 7 & -6 & 9 \\ -5 & 6 & -6 \\ 5 & -5 & 7 \end{bmatrix}\begin{bmatrix} 2 & -1 & 2 \\ -1 & 2 & -1 \\ 1 & -1 & 2 \end{bmatrix} = \begin{bmatrix} 29 & -28 & 38 \\ -22 & 23 & -28 \\ 22 & 22 & 29 \end{bmatrix}.$$

$$\therefore \quad A^3 - 6A^2 + 8A - 3I = \begin{bmatrix} 29 & -28 & 38 \\ -22 & 23 & -28 \\ 22 & 22 & 29 \end{bmatrix} - 6\begin{bmatrix} 7 & -6 & 9 \\ -5 & 6 & -6 \\ 5 & -5 & 7 \end{bmatrix}$$

$$+8\begin{bmatrix} 2 & -1 & 2 \\ -1 & 2 & -1 \\ 1 & -1 & 2 \end{bmatrix} - 3\begin{bmatrix} 1 & 0 & 0 \\ 0 & 1 & 0 \\ 0 & 0 & 1 \end{bmatrix}$$

$$= \begin{bmatrix} 0 & 0 & 0 \\ 0 & 0 & 0 \\ 0 & 0 & 0 \end{bmatrix}.$$

To find A^4:

Multiply equation (1) by A on both sides,

$$A^4 - 6A^3 + 8A^2 - 3A = 0$$

$$\Rightarrow \quad A^4 = 6A^3 - 8A^2 + 3A$$

$$= 6[6A^2 - 8A + 3I] - 8A^2 + 3A$$

$$= 28A^2 - 45A + 18I$$

$$= 28\begin{bmatrix} 7 & -6 & 9 \\ -5 & 6 & -6 \\ 5 & -5 & 7 \end{bmatrix} - 45\begin{bmatrix} 2 & -1 & 2 \\ -1 & 2 & -1 \\ 1 & -1 & 2 \end{bmatrix} + 18\begin{bmatrix} 1 & 0 & 0 \\ 0 & 1 & 0 \\ 0 & 0 & 1 \end{bmatrix}$$

$$= \begin{bmatrix} 124 & -123 & 162 \\ -95 & 96 & -123 \\ 95 & -95 & 124 \end{bmatrix}.$$

To find A^{-1}:

Multiply equation (1) by A^{-1} on both sides,

$$A^2 - 6A + 8I - 3A^{-1} = 0$$

$$\therefore \quad A^{-1} = \frac{1}{3}[A^2 - 6A + 8I]$$

$$= \frac{1}{3}\left\{\begin{bmatrix} 7 & -6 & 9 \\ -5 & 6 & -6 \\ 5 & -5 & 7 \end{bmatrix} - 6\begin{bmatrix} 2 & -1 & 2 \\ -1 & 2 & -1 \\ 1 & -1 & 2 \end{bmatrix} + 8\begin{bmatrix} 1 & 0 & 0 \\ 0 & 1 & 0 \\ 0 & 0 & 1 \end{bmatrix}\right\}$$

$$= \frac{1}{3}\begin{bmatrix} 3 & 0 & -3 \\ 1 & 2 & 0 \\ -1 & 1 & 3 \end{bmatrix}$$

EXERCISES

1. Using Cayley-Hamilton theorem compute A^3 for $A = \begin{bmatrix} 1 & 4 \\ 2 & 3 \end{bmatrix}$.

2. Verify Cayley-Hamilton theorem for the matrix $A = \begin{bmatrix} 1 & 0 & 3 \\ 2 & 1 & -1 \\ 1 & -1 & 1 \end{bmatrix}$.

3. Verify Cayley- Hamilton theorem and find its inverse.

   (i) $A = \begin{bmatrix} 7 & 2 & -2 \\ -6 & -1 & 2 \\ 6 & 2 & -1 \end{bmatrix}$ (ii) $A = \begin{bmatrix} 1 & 2 & -2 \\ 2 & 5 & -4 \\ 3 & 7 & -5 \end{bmatrix}$

   (iii) $A = \begin{bmatrix} 1 & 3 & 7 \\ 4 & 2 & 3 \\ 1 & 2 & 1 \end{bmatrix}$ (iv) $A = \begin{bmatrix} 2 & -1 & 1 \\ -1 & 2 & -1 \\ 1 & -1 & 2 \end{bmatrix}$.

4. If $A = \begin{bmatrix} 1 & 2 & 2 \\ 2 & 1 & 2 \\ 2 & 2 & 1 \end{bmatrix}$, then prove that $A^3 - 3A^2 - 9A - 5I = \mathbf{0}$. Hence find A^4.

ANSWERS

1. $\begin{bmatrix} 41 & 84 \\ 42 & 83 \end{bmatrix}$ 3.(i) $A^{-1} = \frac{1}{3}\begin{bmatrix} -3 & -2 & 2 \\ 6 & 5 & -2 \\ -6 & -2 & 5 \end{bmatrix}$ 3.(ii) $A^{-1} = \begin{bmatrix} 3 & -4 & 2 \\ -2 & 1 & 0 \\ -1 & -1 & 1 \end{bmatrix}$

3.(iii). $A^{-1} = \frac{1}{35}\begin{bmatrix} -4 & 11 & -5 \\ -1 & -6 & 25 \\ 6 & 1 & -10 \end{bmatrix}$ 3.(iv) $A^{-1} = \frac{1}{4}\begin{bmatrix} 3 & 1 & -1 \\ 1 & 3 & 1 \\ -1 & 1 & 3 \end{bmatrix}$

4. $A^4 = \begin{bmatrix} 209 & 208 & 208 \\ 208 & 209 & 208 \\ 208 & 208 & 209 \end{bmatrix}$.

1.5. Diagonalization of matrices

Diagonalization of a real matrix can be done in two ways.

(i) Similarity transformation

(ii) Orthogonal transformation.

Similarity transformation

Similar matrices: Two matrices A and B are said to be *similar*, if there exists a non-singular matrix P such that $B = P^{-1}AP$.

Diagonalization: The process of finding a matrix M such that $M^{-1}AM = D$, where D is a diagonal matrix, is called *diagonalization* of the matrix A.

Similarity transformation: The transformation $M^{-1}AM = D$ is called a *similarity transformation*, where A and D are similar matrices.

Theorem

If A is a square matrix with distinct eigen values and M is the matrix whose columns are the eigen vectors of A, then A can be diagonalised by the similarity transformation $M^{-1}AM = D$, where D is the diagonal matrix whose diagonal elements are the eigen values of A.

Note: (i) The similar matrices A and D have the same eigen values.

(ii) When $\lambda_1 = \lambda_2$, if we get two distinct eigen vectors X_1, X_2 we can diagonalise A.

(iii) When $\lambda_1 = \lambda_2$, if we get only one eigen vector, then we cannot diagonalise A.

Calculation of powers of a matrix

From the previous theorem, $D = M^{-1}AM \Rightarrow A = MDM^{-1}$.

Therefore, $A^2 = (MDM^{-1})(MDM^{-1})$

$= MD(M^{-1}M)DM^{-1}$

$= MD^2M^{-1}$

Similarly, $A^3 = MD^3M^{-1}$

Extending , $A^k = MD^kM^{-1}$ where $D^k = \begin{bmatrix} \lambda_1^k & 0 & 0 & \dots & 0 \\ 0 & \lambda_2^k & 0 & \dots & 0 \\ \dots & \dots & \dots & \dots & \dots \\ 0 & 0 & 0 & \dots & \lambda_n^k \end{bmatrix}$.

Working rule to diagonalise a n x n matrix A by similarity transformation:

Step 1: Find the eigenvalues $\lambda_1, \lambda_2, \dots, \lambda_n$

Step 2: Find the linearly independent eigenvectors $X_1, X_2, \dots, X_n$

Step 3: Form the modal matrix $M = [X_1\ X_2\ \dots\ X_n]$

Step 4: Find M^{-1} and AM

Step 5: Calculate $M^{-1}AM = D = \begin{bmatrix} \lambda_1 & 0 & 0 & \dots & 0 \\ 0 & \lambda_2 & 0 & \dots & 0 \\ \dots & \dots & \dots & \dots & \dots \\ 0 & 0 & 0 & \dots & \lambda_n \end{bmatrix}$.

Orthogonal Transformation (Or) Orthogonal Reduction

Orthogonal matrices: A real square matrix A is said to be *orthogonal* if $AA^T = A^TA = I$.

Note: If A is an orthogonal matrix, then $A^{-1} = A^T$.

Normalized vector: If $X = \begin{bmatrix} x_1 \\ x_2 \\ x_3 \end{bmatrix}$ is a column vector, then the *normalized vector* of X is $\frac{X}{\sqrt{x_1^2+x_2^2+x_3^2}}$.

Normalized modal matrix: A *normalized modal matrix* of A is a matrix N, whose columns are the normalized eigen vectors of A such that $N^{-1}AN = D$ where D is a diagonal matrix.

Working rule for Orthogonal reduction of a symmetric matrix:

Step 1: Find the eigenvalues of the square matrix A.

Step 2: Find the eigenvectors corresponding to each eigenvalue.

Step 3: Form the normalized modal matrix N whose columns are normalized eigen vectors.

Step 4: Find N^T and AN.

Step 5: Find the diagonal matrix $D = N^T AN = \begin{bmatrix} \lambda_1 & 0 & 0 & \dots & 0 \\ 0 & \lambda_2 & 0 & \dots & 0 \\ \dots & \dots & \dots & \dots & \dots \\ 0 & 0 & 0 & \dots & \lambda_n \end{bmatrix}$.

1.6. Reduction of a quadratic form to canonical form by orthogonal transformation

Quadratic form : A homogeneous polynomial of second degree in n variables is called a quadratic form. An expression of the form $\sum_{i=1}^{n}\sum_{j=1}^{n} c_{ij}\, x_i x_j$ where $c_{ij} = c_{ji}$ are all real, is called a quadratic form in n variables $x_1, x_2, \dots, x_n$.

In the expression, the coefficient of $x_i x_j = c_{ij} + c_{ji}$.

Define $2a_{ij} = c_{ij} + c_{ji}$ where $a_{ij} = a_{ji}$ and $a_{ii} = c_{ii}$.

Therefore, $Q = \sum_{i=1}^{n}\sum_{j=1}^{n} a_{ij}\, x_i x_j$ where $a_{ij} = a_{ji}$ and hence the matrix $A = [a_{ij}]$ is a symmetric matrix.

Matrix of a quadratic form : The quadratic form Q can be represented as $Q = X^T AX$, where $A = [a_{ij}]$, $X = \begin{bmatrix} x_1 \\ x_2 \\ \vdots \\ x_n \end{bmatrix}$. The symmetric matrix $A = [a_{ij}]$ is called the *matrix of the quadratic form* Q.

Note: To form the symmetric matrix A from the quadratic form Q, we apply the following steps.

(i) a_{ij} = the coefficient of x_i^2 in Q

(ii) $a_{ij} = a_{ji} = \frac{1}{2}$ (the coefficient of $x_i x_j$) in Q.

Linear Transformation : Let $Q = X^T AX$ be a quadratic form and $X = PY$ be a nonsingular linear transformation.

$$Q = X^T AX = (PY)^T A(PY)$$

$$= Y^T P^T APY = Y^T BY$$

The form $Y^T BY$ is called a *linear transformation* of the quadratic form $X^T AX$ under a non singular transformation $X = PY$ and P is called the *matrix of the transformation.*

Further, $B^T = (P^T AP)^T = P^T A^T (P^T)^T + P^T AP = B$.

Hence the matrix B is also symmetric.

Canonical form : A *canonical form* is a quadratic form Q containing only square terms.

Steps for reducing quadratic form to canonical form:

Step 1: Write the matrix A of the quadratic form

Step 2: Find the eigen values and the corresponding eigen vectors of A.

Step 3: Find the normalized model matrix N.

Step 4: Find N^{-1} and AN.

Step 5: Find the diagonal matrix $D = N^{-1}AN$, whose diagonal elements are the eigen values of A.

Step 6: Find the linear transformation $X = NY$.

Step 7: Find the canonical form $Y^T DT$.

Rank: The *rank* of the quadratic form Q is the total number of terms in the canonical form. It is denoted by r.

Index: The *index* of the quadratic form Q is the number of positive terms in the canonical form. It is denoted by i.

Signature: The *signature* of the quadratic form Q is the difference between the number of positive and negative terms in the canonical form. It is denoted by s.

Remark: If 'r' is the rank and 'i' is the index, then *signature* $s = i - (r - i) = 2i - r$.

1.7. Nature of Quadratic forms

Nature of a quadratic form Q can be defined by using the eigen values or the leading principal minors of the matrix A.

Find nature using Eigen values:

(i) The quadratic form is said to be *positive definite* if all the eigen values of the matrix A are positive.

(ii) The quadratic form is said to be *negative definite* if all the eigen values of the matrix A are negative.

(iii) The quadratic form is said to be *positive semidefinite* if all the eigen values of the matrix A are ≥ 0 with atleast one eigen value is 0.

(iv) The quadratic form is said to be *negative semidefinite* if all the eigen values of the matrix A are ≤ 0 with atleast one eigen value is 0.

(v) The quadratic form is said to be *indefinite* if all the eigen values are both positive and negative.

Find nature using Principal minors:

Let $A = \begin{bmatrix} a_{11} & a_{12} & a_{13} & \cdots & a_{1n} \\ a_{21} & a_{22} & a_{23} & \cdots & a_{2n} \\ \cdots & \cdots & \cdots & \cdots & \cdots \\ a_{n1} & a_{n2} & a_{n3} & \cdots & a_{nn} \end{bmatrix}$ be the matrix of the quadratic form Q. The leading principle minors of the matrix A are as follows.

Let $D_1 = |a_{11}|$, $D_2 = \begin{vmatrix} a_{11} & a_{12} \\ a_{21} & a_{22} \end{vmatrix}$, $D_3 = \begin{vmatrix} a_{11} & a_{12} & a_{13} \\ a_{21} & a_{22} & a_{23} \\ a_{31} & a_{32} & a_{33} \end{vmatrix}$, … , $D_n = |A|$.

(i) The quadratic form is said to be *positive definite* if all $D_i's$ are positive.

(ii) The quadratic form is said to be *negative definite* if all $D_i's$ are negative.

(iii) The quadratic form is said to be *positive semidefinite* if all $D_i's$ are ≥ 0 with atleast one $D_i = 0$.

(iv) The quadratic form is said to be *negative semidefinite* if all $D_i's$ are ≤ 0 with atleast one $D_i = 0$.

(v) The quadratic form is said to be *indefinite* if all $D_i's$ are both positive and negative.

Example 1: Reduce the matrix $\begin{bmatrix} 10 & -2 & -5 \\ -2 & 2 & 3 \\ -5 & 3 & 5 \end{bmatrix}$ to diagonal form.

[AU, April/May 2017]

Solution : Let $A = \begin{bmatrix} 10 & -2 & -5 \\ -2 & 2 & 3 \\ -5 & 3 & 5 \end{bmatrix}$.

To find eigen values:

The characteristic equation of A is $|A - \lambda I| = 0$.

$$\Rightarrow \qquad \lambda^3 - S_1\lambda^2 + S_2\lambda - S_3 = 0$$

Where S_1 = sum of the diagonal elements of A.

$= 10 + 2 + 5 = 17$

S_2 = sum of minors of the diagonal elements of A.

$$= \begin{vmatrix} 2 & 3 \\ 3 & 5 \end{vmatrix} + \begin{vmatrix} 10 & -5 \\ -5 & 5 \end{vmatrix} + \begin{vmatrix} 10 & -2 \\ -2 & 2 \end{vmatrix}$$

$$= (10 - 9) + (50 - 25) + (20 - 4) = 1 + 25 + 16 = 42$$

$S_3 = |A|$

$$= \begin{vmatrix} 10 & -2 & -5 \\ -2 & 2 & 3 \\ -5 & 3 & 5 \end{vmatrix}$$

$$= 10(10 - 9) + 2(-10 + 15) - 5(-6 + 10) = 10 + 10 - 20 = 0$$

∴ The characteristic equation is $\lambda^3 - 17\lambda^2 + 42\lambda = 0$

$$\Rightarrow \qquad \lambda(\lambda^2 - 17\lambda + 42) = 0$$

$$\lambda(\lambda - 3)(\lambda - 14) = 0$$

∴ The eigenvalues of A are $\lambda = 0, 3, 14$.

To find eigenvectors:

Let $X = \begin{bmatrix} x_1 \\ x_2 \\ x_3 \end{bmatrix}$ be the eigenvector of A corresponding to the eigen value λ.

Now, $(A - \lambda I)X = 0 \Rightarrow \begin{bmatrix} 10-\lambda & -2 & -5 \\ -2 & 2-\lambda & 3 \\ -5 & 3 & 5-\lambda \end{bmatrix} \begin{bmatrix} x_1 \\ x_2 \\ x_3 \end{bmatrix} = \begin{bmatrix} 0 \\ 0 \\ 0 \end{bmatrix}$

$$\Rightarrow \qquad (10 - \lambda)x_1 - 2x_2 - 5x_3 = 0$$

$$-2x_1 + (2 - \lambda)x_2 + 3x_3 = 0 \qquad \ldots \quad (1)$$

$$-5x_1 + 3x_2 + (5 - \lambda)x_3 = 0$$

Case (i) : When $\lambda = 0$, equation (1) becomes,

$$10x_1 - 2x_2 - 5x_3 = 0$$

$$-2x_1 + 2x_2 + 3x_3 = 0$$

$$-5x_1 + 3x_2 + 5x_3 = 0$$

Solving the first two equations by the rule of cross multiplication we get,

$$\frac{x_1}{-6+10} = \frac{x_2}{10-30} = \frac{x_3}{20-4}$$

$$\Rightarrow \quad \frac{x_1}{4} = \frac{x_2}{-20} = \frac{x_3}{16}$$

$$\Rightarrow \quad \frac{x_1}{1} = \frac{x_2}{-5} = \frac{x_3}{4}$$

$\therefore$ The eigenvector corresponding to $\lambda = 0$ is $X_1 = \begin{bmatrix} 1 \\ -5 \\ 4 \end{bmatrix}$.

Case (ii) : When $\lambda = 3$, equation (1) becomes,

$$7x_1 - 2x_2 - 5x_3 = 0$$

$$-2x_1 - x_2 + 3x_3 = 0$$

$$-5\, x_1 + 3x_2 + 2\, x_3 = 0$$

Solving the first two equations by the rule of cross multiplication we get,

$$\frac{x_1}{-6-5} = \frac{x_2}{10-21} = \frac{x_3}{-7-4}$$

$$\Rightarrow \quad \frac{x_1}{-11} = \frac{x_2}{-11} = \frac{x_3}{-11}$$

$$\Rightarrow \quad \frac{x_1}{1} = \frac{x_2}{1} = \frac{x_3}{1}$$

$\therefore$ The eigenvector corresponding to $\lambda = 3$ is $X_2 = \begin{bmatrix} 1 \\ 1 \\ 1 \end{bmatrix}$.

Case (iii) : When $\lambda = 14$, equation (1) becomes,

$$-4x_1 - 2x_2 - 5x_3 = 0$$

$$-2x_1 - 12x_2 + 3x_3 = 0$$

$$-5\, x_1 + 3x_2 - 9\, x_3 = 0$$

Solving the first two equations by the rule of cross multiplication we get,

$$\frac{x_1}{6+60} = \frac{x_2}{-10-12} = \frac{x_3}{-48+4}$$

$$\Rightarrow \quad \frac{x_1}{66} = \frac{x_2}{-22} = \frac{x_3}{-44}$$

$$\Rightarrow \quad \frac{x_1}{-3} = \frac{x_2}{1} = \frac{x_3}{2}$$

$\therefore$ The eigenvector corresponding to $\lambda = 14$ is $X_3 = \begin{bmatrix} -3 \\ 1 \\ 2 \end{bmatrix}$.

Hence the modal matrix is $M = \begin{bmatrix} 1 & 1 & -3 \\ -5 & 1 & 1 \\ 4 & 1 & 2 \end{bmatrix}$

To find $M^{-1} = \frac{1}{|M|}\,(cofactor\ A)^T$

Now, $|M| = \begin{vmatrix} 1 & 1 & -3 \\ -5 & 1 & 1 \\ 4 & 1 & 2 \end{vmatrix}$

$$= 1(2-1) - 1(-10-4) - 3(-5-4) = 1 + 14 + 27 = 42$$

$$cofactor\ A = \begin{bmatrix} 1 & 14 & -9 \\ -5 & 14 & 3 \\ 4 & 14 & 6 \end{bmatrix}$$

$$\therefore \quad M^{-1} = \frac{1}{42}\begin{bmatrix} 1 & -5 & 4 \\ 14 & 14 & 14 \\ -9 & 3 & 6 \end{bmatrix}$$

The required similarity transformation is $D = M^{-1}AM$

$$\text{Now, } AM = \begin{bmatrix} 10 & -2 & -5 \\ -2 & 2 & 3 \\ -5 & 3 & 5 \end{bmatrix}\begin{bmatrix} 1 & 1 & -3 \\ -5 & 1 & 1 \\ 4 & 1 & 2 \end{bmatrix}$$

$$= \begin{bmatrix} 0 & 3 & -42 \\ 0 & 3 & 14 \\ 0 & 3 & 28 \end{bmatrix}$$

$$M^{-1}AM = \frac{1}{42}\begin{bmatrix} 1 & -5 & 4 \\ 14 & 14 & 14 \\ -9 & 3 & 6 \end{bmatrix}\begin{bmatrix} 0 & 3 & -42 \\ 0 & 3 & 14 \\ 0 & 3 & 28 \end{bmatrix}$$

$$= \frac{1}{42}\begin{bmatrix} 0 & 0 & 0 \\ 0 & 126 & 0 \\ 0 & 0 & 588 \end{bmatrix} = \begin{bmatrix} 0 & 0 & 0 \\ 0 & 3 & 0 \\ 0 & 0 & 14 \end{bmatrix}$$, which is a diagonal matrix.

Example 2: The eigenvectors of a 3×3 real symmetric matrix A corresponding to eigenvalues 1,3 and 3 are $(1 \quad 0 \quad -1)^T, (1 \quad 0 \quad 1)^T$ and $(0 \quad 1 \quad 0)^T$ respectively. Find the matrix A by an orthogonal transformation. ***[AU, Nov/Dec 2017]***

Solution :

We know that the orthogonal transformation of a real symmetric matrix A can be done by using $N^TAN = D$, where N is the matrix of normalized eigenvectors and D is a diagonal matrix with eigenvalues of A as its diagonal elements.

$\therefore$ The normalized matrix is $N = \begin{bmatrix} \frac{1}{\sqrt{2}} & \frac{1}{\sqrt{2}} & 0 \\ 0 & 0 & 1 \\ \frac{-1}{\sqrt{2}} & \frac{1}{\sqrt{2}} & 0 \end{bmatrix}$ and $D = \begin{bmatrix} 1 & 0 & 0 \\ 0 & 3 & 0 \\ 0 & 0 & 3 \end{bmatrix}$

Now, $N^TAN = D \Rightarrow A = NDN^T$

$$ND = \begin{bmatrix} \frac{1}{\sqrt{2}} & \frac{1}{\sqrt{2}} & 0 \\ 0 & 0 & 1 \\ \frac{-1}{\sqrt{2}} & \frac{1}{\sqrt{2}} & 0 \end{bmatrix} \begin{bmatrix} 1 & 0 & 0 \\ 0 & 3 & 0 \\ 0 & 0 & 3 \end{bmatrix} = \begin{bmatrix} \frac{1}{\sqrt{2}} & \frac{3}{\sqrt{2}} & 0 \\ 0 & 0 & 3 \\ \frac{-1}{\sqrt{2}} & \frac{3}{\sqrt{2}} & 0 \end{bmatrix}$$

Hence, $$A = \begin{bmatrix} \frac{1}{\sqrt{2}} & \frac{3}{\sqrt{2}} & 0 \\ 0 & 0 & 3 \\ \frac{-1}{\sqrt{2}} & \frac{3}{\sqrt{2}} & 0 \end{bmatrix} \begin{bmatrix} \frac{1}{\sqrt{2}} & 0 & \frac{-1}{\sqrt{2}} \\ \frac{1}{\sqrt{2}} & 0 & \frac{1}{\sqrt{2}} \\ 0 & 1 & 0 \end{bmatrix}.$$

$\therefore$ $$A = \begin{bmatrix} 2 & 0 & 1 \\ 0 & 3 & 0 \\ 1 & 0 & 2 \end{bmatrix}.$$

Example 3: Reduce the quadratic form $3x^2 + 5y^2 + 3z^2 - 2yz + 2zx - 2xy$ to its canonical form through an orthogonal transformation.

[AU, Nov/Dec 2014, Nov/Dec 2015]

Solution : The matrix of the given quadratic form is $A = \begin{bmatrix} 3 & -1 & 1 \\ -1 & 5 & -1 \\ 1 & -1 & 3 \end{bmatrix}$.

To find eigen values:

The characteristic equation of A is $|A - \lambda I| = 0$.

$\Rightarrow$ $$\lambda^3 - S_1\lambda^2 + S_2\lambda - S_3 = 0$$

where $S_1 =$ sum of the diagonal elements of A.

$= 3 + 5 + 3 = 11$

$S_2 =$ sum of minors of the diagonal elements of A.

$$= \begin{vmatrix} 5 & -1 \\ -1 & 3 \end{vmatrix} + \begin{vmatrix} 3 & 1 \\ 1 & 3 \end{vmatrix} + \begin{vmatrix} 3 & -1 \\ -1 & 5 \end{vmatrix}$$

$= 15 - 1 + 9 - 1 + 15 - 1 = 36$

$S_3 = |A|$

$$= \begin{vmatrix} 3 & -1 & 1 \\ -1 & 5 & -1 \\ 1 & -1 & 3 \end{vmatrix}$$

$= 3(15 - 1) + 1(-3 + 1) + 1(1 - 5) = 42 - 2 - 4 = 36.$

$\therefore$ The characteristic equation is $\lambda^3 - 11\lambda^2 + 36\lambda - 36 = 0$

By synthetic division method,

$$(\lambda-2)(\lambda^2-9\lambda+18)=0$$

$$(\lambda-2)(\lambda-3)(\lambda-6)=0$$

2	1	−11	36	−36
	0	2	−18	36
	1	−9	18	0

∴ The eigenvalues of A are $\lambda = 2, 3, 6$.

To find eigenvectors:

Let $X = \begin{bmatrix} x_1 \\ x_2 \\ x_3 \end{bmatrix}$ be the eigenvector of A corresponding to the eigen value λ.

Now, $(A-\lambda I)X = 0$

$$\Rightarrow \begin{bmatrix} 3-\lambda & -1 & 1 \\ -1 & 5-\lambda & -1 \\ 1 & -1 & 3-\lambda \end{bmatrix}\begin{bmatrix} x_1 \\ x_2 \\ x_3 \end{bmatrix} = \begin{bmatrix} 0 \\ 0 \\ 0 \end{bmatrix}$$

$$\Rightarrow \left.\begin{aligned} (3-\lambda)x_1 - x_2 + x_3 &= 0 \\ -x_1 + (5-\lambda)x_2 - x_3 &= 0 \\ x_1 - x_2 + (3-\lambda)x_3 &= 0 \end{aligned}\right\} \qquad \ldots (1)$$

Case (i) : When $\lambda = 2$, equation (1) becomes,

$$x_1 - x_2 + x_3 = 0$$

$$-x_1 + 3x_2 - x_3 = 0$$

$$x_1 - x_2 + x_3 = 0$$

Solving the first two equations by the rule of cross multiplication we get,

$$\frac{x_1}{1-3} = \frac{x_2}{-1+1} = \frac{x_3}{3-1}$$

$$\Rightarrow \frac{x_1}{-2} = \frac{x_2}{0} = \frac{x_3}{2}$$

$$\Rightarrow \frac{x_1}{1} = \frac{x_2}{0} = \frac{x_3}{-1}$$

The eigenvector is $X_1 = \begin{bmatrix} 1 \\ 0 \\ -1 \end{bmatrix}$.

Case (ii) : When $\lambda = 3$, equation (1) becomes,

$$0x_1 - x_2 + x_3 = 0$$

$$-x_1 + 2x_2 - x_3 = 0$$

$$x_1 - x_2 + 0x_3 = 0$$

Solving the first two equations by the rule of cross multiplication we get,

$$\frac{x_1}{1-2} = \frac{x_2}{-1-0} = \frac{x_3}{0-1}$$

$$\Rightarrow \quad \frac{x_1}{-1} = \frac{x_2}{-1} = \frac{x_3}{-1}$$

$$\Rightarrow \quad \frac{x_1}{1} = \frac{x_2}{1} = \frac{x_3}{1}$$

The eigenvector is $X_2 = \begin{bmatrix} 1 \\ 1 \\ 1 \end{bmatrix}$.

Case (iii) : When $\lambda = 6$, equation (1) becomes,

$$-3x_1 - x_2 + x_3 = 0$$

$$-x_1 - x_2 - x_3 = 0$$

$$x_1 - x_2 - 3\,x_3 = 0$$

Solving the first two equations by the rule of cross multiplication we get,

$$\frac{x_1}{1+1} = \frac{x_2}{-1-3} = \frac{x_3}{3-1}$$

$$\Rightarrow \quad \frac{x_1}{2} = \frac{x_2}{-4} = \frac{x_3}{2}$$

$$\Rightarrow \quad \frac{x_1}{1} = \frac{x_2}{-2} = \frac{x_3}{1}$$

$\therefore$ The eigenvector is $X_3 = \begin{bmatrix} 1 \\ -2 \\ 1 \end{bmatrix}$.

The normalized form of X_1, X_2 and X_3 are, $\begin{bmatrix} \frac{1}{\sqrt{2}} \\ \frac{0}{\sqrt{2}} \\ \frac{-1}{\sqrt{2}} \end{bmatrix}$, $\begin{bmatrix} \frac{1}{\sqrt{3}} \\ \frac{1}{\sqrt{3}} \\ \frac{1}{\sqrt{3}} \end{bmatrix}$ and $\begin{bmatrix} \frac{1}{\sqrt{6}} \\ \frac{-2}{\sqrt{6}} \\ \frac{1}{\sqrt{6}} \end{bmatrix}$.

$\therefore$ The normalized matrix is $N = \begin{bmatrix} \frac{1}{\sqrt{2}} & \frac{1}{\sqrt{3}} & \frac{1}{\sqrt{6}} \\ 0 & \frac{1}{\sqrt{3}} & \frac{-2}{\sqrt{6}} \\ \frac{-1}{\sqrt{2}} & \frac{1}{\sqrt{3}} & \frac{1}{\sqrt{6}} \end{bmatrix}$.

$$\Rightarrow N^T = \begin{bmatrix} \frac{1}{\sqrt{2}} & 0 & \frac{-1}{\sqrt{2}} \\ \frac{1}{\sqrt{3}} & \frac{1}{\sqrt{3}} & \frac{1}{\sqrt{3}} \\ \frac{1}{\sqrt{6}} & \frac{-2}{\sqrt{6}} & \frac{1}{\sqrt{6}} \end{bmatrix}$$

$$\therefore \quad AN = \begin{bmatrix} 3 & -1 & 1 \\ -1 & 5 & -1 \\ 1 & -1 & 3 \end{bmatrix} \begin{bmatrix} \frac{1}{\sqrt{2}} & \frac{1}{\sqrt{3}} & \frac{1}{\sqrt{6}} \\ 0 & \frac{1}{\sqrt{3}} & \frac{-2}{\sqrt{6}} \\ \frac{-1}{\sqrt{2}} & \frac{1}{\sqrt{3}} & \frac{1}{\sqrt{6}} \end{bmatrix}$$

$$= \begin{bmatrix} \frac{2}{\sqrt{2}} & \frac{3}{\sqrt{3}} & \frac{6}{\sqrt{6}} \\ 0 & \frac{3}{\sqrt{3}} & \frac{-12}{\sqrt{6}} \\ \frac{-2}{\sqrt{2}} & \frac{3}{\sqrt{3}} & \frac{6}{\sqrt{6}} \end{bmatrix}$$

$$\therefore \quad N^T AN = \begin{bmatrix} \frac{1}{\sqrt{2}} & 0 & \frac{-1}{\sqrt{2}} \\ \frac{1}{\sqrt{3}} & \frac{1}{\sqrt{3}} & \frac{1}{\sqrt{3}} \\ \frac{1}{\sqrt{6}} & \frac{-2}{\sqrt{6}} & \frac{1}{\sqrt{6}} \end{bmatrix} \begin{bmatrix} \frac{2}{\sqrt{2}} & \frac{3}{\sqrt{3}} & \frac{6}{\sqrt{6}} \\ 0 & \frac{3}{\sqrt{3}} & \frac{-12}{\sqrt{6}} \\ \frac{-2}{\sqrt{2}} & \frac{3}{\sqrt{3}} & \frac{6}{\sqrt{6}} \end{bmatrix}$$

$$= \begin{bmatrix} 2 & 0 & 0 \\ 0 & 3 & 0 \\ 0 & 0 & 6 \end{bmatrix} = D.$$

Consider the orthogonal transformation $X = NY$ where $Y = \begin{bmatrix} y_1 \\ y_2 \\ y_3 \end{bmatrix}$.

$\therefore$ The canonical form is $= Y^T DT$

$$= [y_1 \quad y_2 \quad y_3] \begin{bmatrix} 2 & 0 & 0 \\ 0 & 3 & 0 \\ 0 & 0 & 6 \end{bmatrix} \begin{bmatrix} y_1 \\ y_2 \\ y_3 \end{bmatrix}$$

$$= [2y_1 \quad 3y_2 \quad 6y_3] \begin{bmatrix} y_1 \\ y_2 \\ y_3 \end{bmatrix}$$

$$= 2y_1^2 + 3y_2^2 + 6y_3^2.$$

Example 4: Reduce the quadratic form $2x^2 + 6y^2 + 2z^2 + 8xz$ to canonical form by orthogonal transformation.

Solution : The matrix of the given quadratic form is $A = \begin{bmatrix} 2 & 0 & 4 \\ 0 & 6 & 0 \\ 4 & 0 & 2 \end{bmatrix}$.

To find eigen values:

The characteristic equation of A is $|A - \lambda I| = 0$.

$\Rightarrow$ $$\lambda^3 - S_1\lambda^2 + S_2\lambda - S_3 = 0$$

where S_1 = sum of the diagonal elements of A.

$= 2 + 6 + 2 = 10$

S_2 = sum of minors of the diagonal elements of A.

$$= \begin{vmatrix} 6 & 0 \\ 0 & 2 \end{vmatrix} + \begin{vmatrix} 2 & 4 \\ 4 & 2 \end{vmatrix} + \begin{vmatrix} 2 & 0 \\ 0 & 6 \end{vmatrix}$$

$$= 12 - 0 + 4 - 16 + 12 - 0 = 12$$

$$S_3 = |A|$$

$$= \begin{vmatrix} 2 & 0 & 4 \\ 0 & 6 & 0 \\ 4 & 0 & 2 \end{vmatrix}$$

$$= 2(12 - 0) - 0(0 - 0) + 4(0 - 24) = 24 - 96 = -72.$$

$\therefore$ The characteristic equation is $\lambda^3 - 10\lambda^2 + 12\lambda + 72 = 0$

By synthetic division method,

$$(\lambda + 2)(\lambda^2 - 12\lambda + 36) = 0$$

$$(\lambda + 2)(\lambda - 6)(\lambda - 6) = 0$$

−2	1	−10	12	72
	0	− 2	24	−72
	1	−12	36	0

$\therefore$ The eigenvalues of A are $\lambda = -2, 6, 6$.

To find eigenvectors:

Let $X = \begin{bmatrix} x_1 \\ x_2 \\ x_3 \end{bmatrix}$ be the eigenvector of A corresponding to the eigen value λ.

Now, $(A - \lambda I)X = 0$

$$\Rightarrow \quad \begin{bmatrix} 2-\lambda & 0 & 4 \\ 0 & 6-\lambda & 0 \\ 4 & 0 & 2-\lambda \end{bmatrix} \begin{bmatrix} x_1 \\ x_2 \\ x_3 \end{bmatrix} = \begin{bmatrix} 0 \\ 0 \\ 0 \end{bmatrix}$$

$$\Rightarrow \quad \left.\begin{aligned} (2-\lambda)x_1 + 0x_2 + 4x_3 &= 0 \\ 0x_1 + (6-\lambda)x_2 + 0x_3 &= 0 \\ 4x_1 + 0x_2 + (2-\lambda)x_3 &= 0 \end{aligned}\right\} \qquad \ldots \quad (1)$$

Case (i) : When $\lambda = -2$, equation (1) becomes,

$$4x_1 + 0x_2 + 4x_3 = 0$$

$$0x_1 + 8x_2 + 0x_3 = 0$$

$$4x_1 + 0x_2 + 4x_3 = 0$$

Solving the first two equations by the rule of cross multiplication we get,

$$\frac{x_1}{0-32} = \frac{x_2}{0-0} = \frac{x_3}{32-0}$$

$$\Rightarrow \quad \frac{x_1}{-32} = \frac{x_2}{0} = \frac{x_3}{32}$$

$$\Rightarrow \quad \frac{x_1}{1} = \frac{x_2}{0} = \frac{x_3}{-1}$$

The eigenvector is $X_1 = \begin{bmatrix} 1 \\ 0 \\ -1 \end{bmatrix}$.

Case (ii) : When $\lambda = 6$, equation (1) becomes,

$$-4x_1 + 0x_2 + 4x_3 = 0$$

$$0x_1 + 0x_2 + 0x_3 = 0$$

$$4x_1 + 0x_2 - 4x_3 = 0$$

$$\Rightarrow \quad 4x_1 = 4x_3$$

$$\Rightarrow \quad \frac{x_1}{1} = \frac{x_3}{1}$$

The eigenvector is $X_2 = \begin{bmatrix} 1 \\ 0 \\ 1 \end{bmatrix}$.

Let $X_3 = \begin{bmatrix} l \\ m \\ n \end{bmatrix}$ be another eigen vector which is orthogonal to X_1 and X_2.

$\therefore \quad X_1^T X_3 = 0$ and $X_2^T X_3 = 0$

$$\Rightarrow \qquad l + 0m - n = 0$$

$$l + 0m + n = 0$$

Solving the above equations by the rule of cross multiplication we get,

$$\frac{l}{0} = \frac{m}{-2} = \frac{n}{0}$$

$$\Rightarrow \qquad \frac{x_1}{0} = \frac{x_2}{1} = \frac{x_3}{0}$$

$\therefore$ The eigenvector is $X_3 = \begin{bmatrix} 0 \\ 1 \\ 0 \end{bmatrix}$.

The normalized form of X_1, X_2 and X_3 are, $\begin{bmatrix} \frac{1}{\sqrt{2}} \\ \frac{0}{\sqrt{2}} \\ \frac{-1}{\sqrt{2}} \end{bmatrix}$, $\begin{bmatrix} \frac{1}{\sqrt{2}} \\ \frac{0}{\sqrt{3}} \\ \frac{1}{\sqrt{2}} \end{bmatrix}$ and $\begin{bmatrix} 0 \\ 1 \\ 0 \end{bmatrix}$.

$\therefore$ The normalized matrix is $N = \begin{bmatrix} \frac{1}{\sqrt{2}} & \frac{1}{\sqrt{2}} & 0 \\ 0 & 0 & 1 \\ \frac{-1}{\sqrt{2}} & \frac{1}{\sqrt{2}} & 0 \end{bmatrix}$.

$$\Rightarrow \qquad N^T = \begin{bmatrix} \frac{1}{\sqrt{2}} & 0 & \frac{-1}{\sqrt{2}} \\ \frac{1}{\sqrt{2}} & 0 & \frac{1}{\sqrt{2}} \\ 0 & 1 & 0 \end{bmatrix}$$

$$\therefore \qquad AN = \begin{bmatrix} 2 & 0 & 4 \\ 0 & 6 & 0 \\ 4 & 0 & 2 \end{bmatrix} \begin{bmatrix} \frac{1}{\sqrt{2}} & \frac{1}{\sqrt{2}} & 0 \\ 0 & 0 & 1 \\ \frac{-1}{\sqrt{2}} & \frac{1}{\sqrt{2}} & 0 \end{bmatrix}$$

$$= \begin{bmatrix} \frac{-2}{\sqrt{2}} & \frac{6}{\sqrt{2}} & 0 \\ 0 & 0 & 6 \\ \frac{2}{\sqrt{2}} & \frac{6}{\sqrt{6}} & 0 \end{bmatrix}.$$

$$\therefore \qquad N^T AN = \begin{bmatrix} \frac{1}{\sqrt{2}} & 0 & \frac{-1}{\sqrt{2}} \\ \frac{1}{\sqrt{2}} & 0 & \frac{1}{\sqrt{2}} \\ 0 & 1 & 0 \end{bmatrix} \begin{bmatrix} \frac{-2}{\sqrt{2}} & \frac{6}{\sqrt{2}} & 0 \\ 0 & 0 & 6 \\ \frac{2}{\sqrt{2}} & \frac{6}{\sqrt{6}} & 0 \end{bmatrix}$$

$$= \begin{bmatrix} -2 & 0 & 0 \\ 0 & 6 & 0 \\ 0 & 0 & 6 \end{bmatrix} = D.$$

Consider the orthogonal transformation $X = NY$ where $Y = \begin{bmatrix} y_1 \\ y_2 \\ y_3 \end{bmatrix}$.

$\therefore$ The canonical form is $= Y^T DT$

$$= [y_1 \quad y_2 \quad y_3] \begin{bmatrix} -2 & 0 & 0 \\ 0 & 6 & 0 \\ 0 & 0 & 6 \end{bmatrix} \begin{bmatrix} y_1 \\ y_2 \\ y_3 \end{bmatrix}$$

$$= [-2y_1 \quad 6y_2 \quad 6y_3] \begin{bmatrix} y_1 \\ y_2 \\ y_3 \end{bmatrix}$$

$$= -2y_1^2 + 6y_2^2 + 6y_3^2.$$

Example 5: Reduce the quadratic form $6x^2 + 3y^2 + 3z^2 - 4xy - 2yz + 4xz$ into a canonical form by an orthogonal reduction. Hence find its index, signature, rank and nature of the quadratic form. **[*AU, January 2014, Nov/Dec 2014, April/May 2015, Nov/Dec 2016, May/June 2016, Nov/Dec 2017*]**

Solution : The matrix of the given quadratic form is $A = \begin{bmatrix} 6 & -2 & 2 \\ -2 & 3 & -1 \\ 2 & -1 & 3 \end{bmatrix}$.

To find eigen values:

The characteristic equation of A is $|A - \lambda I| = 0$

$$\Rightarrow \qquad \lambda^3 - S_1\lambda^2 + S_2\lambda - S_3 = 0$$

where $S_1 =$ sum of the diagonal elements of A.

$= 6 + 3 + 3 = 12$

$S_2 =$ sum of minors of the diagonal elements of A.

$$= \begin{vmatrix} 3 & -1 \\ -1 & 3 \end{vmatrix} + \begin{vmatrix} 6 & 2 \\ 2 & 3 \end{vmatrix} + \begin{vmatrix} 6 & -2 \\ -2 & 3 \end{vmatrix}$$

$= 9 - 1 + 18 - 4 + 18 - 4 = 36$

$S_3 = |A|$

$$= \begin{vmatrix} 6 & -2 & 2 \\ -2 & 3 & -1 \\ 2 & -1 & 3 \end{vmatrix}$$

$= 6(9 - 1) + 2(-6 + 2) + 2(2 - 6) = 48 - 8 - 8 = 32$

$\therefore$ The characteristic equation is $\lambda^3 - 12\lambda^2 + 36\lambda - 32 = 0$

By synthetic division method,

$$\begin{array}{r|rrrr} 2 & 1 & -12 & 36 & -32 \\ & 0 & 2 & -20 & 32 \\ \hline & 1 & -10 & 16 & 0 \end{array}$$

$$(\lambda - 2)(\lambda^2 - 10\lambda + 16) = 0$$

$$(\lambda - 2)(\lambda - 2)(\lambda - 8) = 0$$

$\therefore$ The eigenvalues of A are $\lambda = 2, 2, 8$.

To find eigenvectors:

Let $X = \begin{bmatrix} x_1 \\ x_2 \\ x_3 \end{bmatrix}$ be the eigenvector of A corresponding to the eigen value λ.

Now, $(A - \lambda I)X = 0$

$$\Rightarrow \begin{bmatrix} 6-\lambda & -2 & 2 \\ -2 & 3-\lambda & -1 \\ 2 & -1 & 3-\lambda \end{bmatrix} \begin{bmatrix} x_1 \\ x_2 \\ x_3 \end{bmatrix} = \begin{bmatrix} 0 \\ 0 \\ 0 \end{bmatrix}$$

$$\Rightarrow \left.\begin{aligned} (6-\lambda)x_1 - 2x_2 + 2x_3 &= 0 \\ -2x_1 + (3-\lambda)x_2 - x_3 &= 0 \\ 2x_1 - x_2 + (3-\lambda)\,x_3 &= 0 \end{aligned}\right\} \quad \cdots \quad (1)$$

Case (i) : When $\lambda = 8$, equation (1) becomes,

$$-2x_1 - 2x_2 + 2x_3 = 0$$

$$-2x_1 - 5x_2 - x_3 = 0$$

$$2x_1 - x_2 - 5\,x_3 = 0$$

Solving the first two equations by the rule of cross multiplication we get,

$$\frac{x_1}{2+10} = \frac{x_2}{-4-2} = \frac{x_3}{10-4}$$

$$\Rightarrow \frac{x_1}{12} = \frac{x_2}{-6} = \frac{x_3}{6}$$

$$\Rightarrow \frac{x_1}{2} = \frac{x_2}{-1} = \frac{x_3}{1}$$

$\therefore$ The eigenvector corresponding to $\lambda = 8$ is $X_1 = \begin{bmatrix} 2 \\ -1 \\ 1 \end{bmatrix}$.

Case (ii) : When $\lambda = 2$, equation (1) becomes,

$$4x_1 - 2x_2 + 2x_3 = 0$$

$$-2x_1 + x_2 - x_3 = 0$$

$$2x_1 - x_2 + x_3 = 0$$

Here we get only one equation $2x_1 - x_2 + x_3 = 0$.

$\therefore$ By assigning arbitrary values for any two variables, we can find an independent eigenvector.

Take $x_1 = 0$, then we get $-x_2 + x_3 = 0$.

$$\Rightarrow \qquad x_2 = x_3$$

$$\Rightarrow \qquad \frac{x_2}{1} = \frac{x_3}{1}$$

The eigenvector is $X_2 = \begin{bmatrix} 0 \\ 1 \\ 1 \end{bmatrix}$.

Let $X_3 = \begin{bmatrix} l \\ m \\ n \end{bmatrix}$ be another eigen vector which is orthogonal to X_1 and X_2.

$$\therefore \qquad X_1^T X_3 = 0 \text{ and } X_2^T X_3 = 0$$

$$\Rightarrow \quad (2 \quad -1 \quad 1)\begin{pmatrix} l \\ m \\ n \end{pmatrix} = 0 \text{ and } (0 \quad 1 \quad 1)\begin{pmatrix} l \\ m \\ n \end{pmatrix} = 0$$

$$\Rightarrow \quad 2l - m + n = 0$$

$$0l + m + n = 0$$

Solving the above equations by the rule of cross multiplication we get,

$$\frac{l}{-2} = \frac{m}{-2} = \frac{n}{2}$$

$$\Rightarrow \qquad \frac{l}{1} = \frac{m}{1} = \frac{n}{-1}$$

$\therefore$ The eigenvector is $X_3 = \begin{bmatrix} 1 \\ 1 \\ -1 \end{bmatrix}$.

The normalized form of X_1, X_2 and X_3 are, $\begin{bmatrix} \frac{2}{\sqrt{6}} \\ \frac{-1}{\sqrt{6}} \\ \frac{1}{\sqrt{6}} \end{bmatrix}$, $\begin{bmatrix} 0 \\ \frac{1}{\sqrt{2}} \\ \frac{1}{\sqrt{2}} \end{bmatrix}$ and $\begin{bmatrix} \frac{1}{\sqrt{3}} \\ \frac{1}{\sqrt{3}} \\ \frac{-1}{\sqrt{3}} \end{bmatrix}$.

$\therefore$ The normalized matrix is $N = \begin{bmatrix} \frac{2}{\sqrt{6}} & 0 & \frac{1}{\sqrt{3}} \\ \frac{-1}{\sqrt{6}} & \frac{1}{\sqrt{2}} & \frac{1}{\sqrt{3}} \\ \frac{1}{\sqrt{6}} & \frac{1}{\sqrt{2}} & \frac{-1}{\sqrt{3}} \end{bmatrix}$.

$$\Rightarrow \quad N^T = \begin{bmatrix} \frac{2}{\sqrt{6}} & \frac{-1}{\sqrt{6}} & \frac{1}{\sqrt{6}} \\ 0 & \frac{1}{\sqrt{2}} & \frac{1}{\sqrt{2}} \\ \frac{1}{\sqrt{3}} & \frac{1}{\sqrt{3}} & \frac{-1}{\sqrt{3}} \end{bmatrix}$$

$$\therefore \quad AN = \begin{bmatrix} 6 & -2 & 2 \\ -2 & 3 & -1 \\ 2 & -1 & 3 \end{bmatrix} \begin{bmatrix} \frac{2}{\sqrt{6}} & 0 & \frac{1}{\sqrt{3}} \\ \frac{-1}{\sqrt{6}} & \frac{1}{\sqrt{2}} & \frac{1}{\sqrt{3}} \\ \frac{1}{\sqrt{6}} & \frac{1}{\sqrt{2}} & \frac{-1}{\sqrt{3}} \end{bmatrix}$$

$$= \begin{bmatrix} \frac{16}{\sqrt{6}} & 0 & \frac{2}{\sqrt{3}} \\ \frac{-8}{\sqrt{6}} & \frac{2}{\sqrt{2}} & \frac{2}{\sqrt{3}} \\ \frac{8}{\sqrt{6}} & \frac{2}{\sqrt{2}} & \frac{-2}{\sqrt{3}} \end{bmatrix}.$$

$$\therefore \quad N^T AN = \begin{bmatrix} \frac{2}{\sqrt{6}} & \frac{-1}{\sqrt{6}} & \frac{1}{\sqrt{6}} \\ 0 & \frac{1}{\sqrt{2}} & \frac{1}{\sqrt{2}} \\ \frac{1}{\sqrt{3}} & \frac{1}{\sqrt{3}} & \frac{-1}{\sqrt{3}} \end{bmatrix} \begin{bmatrix} \frac{16}{\sqrt{6}} & 0 & \frac{2}{\sqrt{3}} \\ \frac{-8}{\sqrt{6}} & \frac{2}{\sqrt{2}} & \frac{2}{\sqrt{3}} \\ \frac{8}{\sqrt{6}} & \frac{2}{\sqrt{2}} & \frac{-2}{\sqrt{3}} \end{bmatrix}$$

$$= \begin{bmatrix} 8 & 0 & 0 \\ 0 & 2 & 0 \\ 0 & 0 & 2 \end{bmatrix} = D.$$

Consider the orthogonal transformation $X = NY$ where $Y = \begin{bmatrix} y_1 \\ y_2 \\ y_3 \end{bmatrix}$.

$\therefore$ The canonical form is $= Y^T DT$

$$= [y_1 \quad y_2 \quad y_3] \begin{bmatrix} 8 & 0 & 0 \\ 0 & 2 & 0 \\ 0 & 0 & 2 \end{bmatrix} \begin{bmatrix} y_1 \\ y_2 \\ y_3 \end{bmatrix}$$

$$= [8y_1 \quad 2y_2 \quad 2y_3] \begin{bmatrix} y_1 \\ y_2 \\ y_3 \end{bmatrix}$$

$$= 8y_1^2 + 2y_2^2 + 2y_3^2.$$

We know, index i = number of positive square terms = 3

Rank r = number of the non-zero terms in canonical form = 3

Signature $s = 2i - r = (2)(3) - 3 = 3$.

Nature = positive definite (since all eigen values are positive).

Example 6: Reduce the quadratic form $x_1^2 + 5x_2^2 + x_3^2 + 2x_1x_2 + 2x_2x_3 + 6x_3x_1$ to the canonical form through orthogonal transformation and find its nature.

[AU, May/June 2014, April/May 2015, May/Jun 2016]

Solution : The matrix of the given quadratic form is $A = \begin{bmatrix} 1 & 1 & 3 \\ 1 & 5 & 1 \\ 3 & 1 & 1 \end{bmatrix}$.

To find eigen values:

The characteristic equation of A is $|A - \lambda I| = 0$.

$$\Rightarrow \qquad \lambda^3 - S_1\lambda^2 + S_2\lambda - S_3 = 0$$

where $S_1 =$ sum of the diagonal elements of A.

$= 1 + 5 + 1 = 7$

$S_2 =$ sum of minors of the diagonal elements of $|A|$.

$$= \begin{vmatrix} 5 & 1 \\ 1 & 1 \end{vmatrix} + \begin{vmatrix} 1 & 3 \\ 3 & 1 \end{vmatrix} + \begin{vmatrix} 1 & 1 \\ 1 & 5 \end{vmatrix}$$

$= 5 - 1 + 1 - 9 + 5 - 1 = 0$

$S_3 = |A|$

$$= \begin{vmatrix} 1 & 1 & 3 \\ 1 & 5 & 1 \\ 3 & 1 & 1 \end{vmatrix}$$

$= 1(5-1) - 1(1-3) + 3(1-15) = 4 + 2 - 42 = -36.$

$\therefore$ The characteristic equation is $\lambda^3 - 7\lambda^2 + 36 = 0$

By synthetic division method,

$$(\lambda + 2)(\lambda^2 - 9\lambda + 18) = 0$$

$$(\lambda + 2)(\lambda - 3)(\lambda - 6) = 0$$

$$\begin{array}{c|cccc} -2 & 1 & -7 & 0 & 36 \\ & 0 & -2 & 18 & -36 \\ \hline & 1 & -9 & 18 & 0 \end{array}$$

$\therefore$ The eigenvalues of A are $\lambda = -2, 3, 6$.

To find eigenvectors:

Let $X = \begin{bmatrix} x_1 \\ x_2 \\ x_3 \end{bmatrix}$ be the eigenvector of A corresponding to the eigen value λ.

Now, $(A - \lambda I)X = 0$

$$\Rightarrow \quad \begin{bmatrix} 1-\lambda & 1 & 3 \\ 1 & 5-\lambda & 1 \\ 3 & 1 & 1-\lambda \end{bmatrix} \begin{bmatrix} x_1 \\ x_2 \\ x_3 \end{bmatrix} = \begin{bmatrix} 0 \\ 0 \\ 0 \end{bmatrix}$$

$$\Rightarrow \quad \left.\begin{aligned} (1-\lambda)x_1 + x_2 + 3x_3 &= 0 \\ x_1 + (5-\lambda)x_2 + x_3 &= 0 \\ 3x_1 + x_2 + (1-\lambda)x_3 &= 0 \end{aligned}\right\} \qquad \ldots \qquad (1)$$

Case (i) : When $\lambda = -2$, equation (1) becomes,

$$3x_1 + x_2 + 3x_3 = 0$$

$$x_1 + 7x_2 + x_3 = 0$$
$$3x_1 + x_2 + 3x_3 = 0$$

Solving the first two equations by the rule of cross multiplication we get,

$$\frac{x_1}{1-21} = \frac{x_2}{3-3} = \frac{x_3}{21-1}$$

$$\Rightarrow \quad \frac{x_1}{-20} = \frac{x_2}{0} = \frac{x_3}{20}$$

$$\Rightarrow \quad \frac{x_1}{1} = \frac{x_2}{0} = \frac{x_3}{-1}$$

The eigenvector is $X_1 = \begin{bmatrix} 1 \\ 0 \\ -1 \end{bmatrix}$.

Case (ii) : When $\lambda = 3$, equation (1) becomes,

$$-2x_1 + x_2 + 3x_3 = 0$$
$$x_1 + 2x_2 + x_3 = 0$$
$$3x_1 + x_2 - 2x_3 = 0$$

Solving the first two equations by the rule of cross multiplication we get,

$$\frac{x_1}{1-6} = \frac{x_2}{3+2} = \frac{x_3}{-4-1}$$

$$\Rightarrow \quad \frac{x_1}{-5} = \frac{x_2}{5} = \frac{x_3}{-5}$$

$$\Rightarrow \quad \frac{x_1}{1} = \frac{x_2}{-1} = \frac{x_3}{1}$$

The eigenvector is $X_2 = \begin{bmatrix} 1 \\ -1 \\ 1 \end{bmatrix}$.

Case (iii) : When $\lambda = 6$, equation (1) becomes,

$$-5x_1 + x_2 + 3x_3 = 0$$
$$x_1 - x_2 + x_3 = 0$$
$$3x_1 + x_2 - 5x_3 = 0$$

Solving the first two equations by the rule of cross multiplication we get,

$$\frac{x_1}{1+3} = \frac{x_2}{3+5} = \frac{x_3}{5-1}$$

$$\Rightarrow \quad \frac{x_1}{4} = \frac{x_2}{8} = \frac{x_3}{4}$$

$$\Rightarrow \quad \frac{x_1}{1} = \frac{x_2}{2} = \frac{x_3}{1}$$

$\therefore$ The eigenvector corresponding to $\lambda = 6$ is $X_3 = \begin{bmatrix}1\\2\\1\end{bmatrix}$.

The normalized form of X_1, X_2 and X_3 are, $\begin{bmatrix}\frac{1}{\sqrt{2}}\\ \frac{0}{\sqrt{2}}\\ \frac{-1}{\sqrt{2}}\end{bmatrix}$, $\begin{bmatrix}\frac{1}{\sqrt{3}}\\ \frac{-1}{\sqrt{3}}\\ \frac{1}{\sqrt{3}}\end{bmatrix}$ and $\begin{bmatrix}\frac{1}{\sqrt{6}}\\ \frac{2}{\sqrt{6}}\\ \frac{1}{\sqrt{6}}\end{bmatrix}$.

$$\therefore \text{The normalized matrix is } N = \begin{bmatrix}\frac{1}{\sqrt{2}} & \frac{1}{\sqrt{3}} & \frac{1}{\sqrt{6}}\\ 0 & \frac{-1}{\sqrt{3}} & \frac{2}{\sqrt{6}}\\ \frac{-1}{\sqrt{2}} & \frac{1}{\sqrt{3}} & \frac{1}{\sqrt{6}}\end{bmatrix}.$$

$$\Rightarrow \quad N^T = \begin{bmatrix}\frac{1}{\sqrt{2}} & 0 & \frac{-1}{\sqrt{2}}\\ \frac{1}{\sqrt{3}} & \frac{-1}{\sqrt{3}} & \frac{1}{\sqrt{3}}\\ \frac{1}{\sqrt{6}} & \frac{2}{\sqrt{6}} & \frac{1}{\sqrt{6}}\end{bmatrix}$$

$$\therefore \quad AN = \begin{bmatrix}1 & 1 & 3\\ 1 & 5 & 1\\ 3 & 1 & 1\end{bmatrix}\begin{bmatrix}\frac{1}{\sqrt{2}} & \frac{1}{\sqrt{3}} & \frac{1}{\sqrt{6}}\\ 0 & \frac{-1}{\sqrt{3}} & \frac{2}{\sqrt{6}}\\ \frac{-1}{\sqrt{2}} & \frac{1}{\sqrt{3}} & \frac{1}{\sqrt{6}}\end{bmatrix} = \begin{bmatrix}\frac{-2}{\sqrt{2}} & \frac{3}{\sqrt{3}} & \frac{6}{\sqrt{6}}\\ 0 & \frac{-3}{\sqrt{3}} & \frac{12}{\sqrt{6}}\\ \frac{2}{\sqrt{2}} & \frac{3}{\sqrt{3}} & \frac{6}{\sqrt{6}}\end{bmatrix}.$$

$$\therefore \quad N^TAN = \begin{bmatrix}\frac{1}{\sqrt{2}} & 0 & \frac{-1}{\sqrt{2}}\\ \frac{1}{\sqrt{3}} & \frac{-1}{\sqrt{3}} & \frac{1}{\sqrt{3}}\\ \frac{1}{\sqrt{6}} & \frac{2}{\sqrt{6}} & \frac{1}{\sqrt{6}}\end{bmatrix}\begin{bmatrix}\frac{-2}{\sqrt{2}} & \frac{3}{\sqrt{3}} & \frac{6}{\sqrt{6}}\\ 0 & \frac{-3}{\sqrt{3}} & \frac{12}{\sqrt{6}}\\ \frac{2}{\sqrt{2}} & \frac{3}{\sqrt{3}} & \frac{6}{\sqrt{6}}\end{bmatrix} = \begin{bmatrix}-2 & 0 & 0\\ 0 & 3 & 0\\ 0 & 0 & 6\end{bmatrix}.$$

Consider the orthogonal transformation $X = NY$ where $Y = \begin{bmatrix}y_1\\y_2\\y_3\end{bmatrix}$.

$$\therefore \text{ The canonical form is } = Y^TDT$$

$$= \begin{bmatrix}y_1 & y_2 & y_3\end{bmatrix}\begin{bmatrix}-2 & 0 & 0\\ 0 & 3 & 0\\ 0 & 0 & 6\end{bmatrix}\begin{bmatrix}y_1\\y_2\\y_3\end{bmatrix}$$

$$= \begin{bmatrix}-2y_1 & 3y_2 & 6y_3\end{bmatrix}\begin{bmatrix}y_1\\y_2\\y_3\end{bmatrix}$$

$$= -2y_1^2 + 3y_2^2 + 6y_3^2.$$

Nature = indefinite (since the eigenvalues are both positive and negative).

EXERCISES

1. Reduce the matrix $\begin{bmatrix} -1 & 2 & -2 \\ 1 & 2 & 1 \\ -1 & -1 & 0 \end{bmatrix}$ to diagonal form.
2. Diagonalise the matrix $\begin{bmatrix} 0 & -2 & -2 \\ -1 & 1 & 2 \\ -1 & -1 & 2 \end{bmatrix}$.
3. The eigenvectors of a 3×3 real symmetric matrix A corresponding to eigenvalues 2,3 and 6 are $(1 \;\; 0 \;\; -1)^T, (1 \;\; 1 \;\; 1)^T$ and $(-1 \;\; 2 \;\; -1)^T$ respectively. Find the matrix A.
4. Obtain an orthogonal transformation which will transform the quadratic form $Q = 2x_1x_2 + 2x_2x_3 + 2x_3x_1$ into sum of squares.
5. Reduce the quadratic form below to its normal form by an orthogonal reduction $Q = 3x_1^2 + 2x_2^2 + 3x_3^2 - 2x_1x_2 - 2x_2x_3$.
6. Find the canonical form of the quadratic form $= 2x_1^2 + x_2^2 + x_3^2 + 2x_1x_2 - 4x_2x_3 - 2x_1x_3$.
7. Reduce the quadratic form to a canonical form by an orthogonal reduction $2x_1x_2 - 2x_2x_3 + 2x_3x_1$. Also discuss its nature.
8. Reduce the quadratic form $8x_1^2 + 7x_2^2 + 3x_3^2 - 12x_1x_2 - 8x_2x_3 + 4x_1x_3$ to the canonical form through an orthogonal transformation and hence, show that it is positive semi-definite.
9. Reduce the quadratic form $x_1^2 + 2x_2^2 + x_3^2 - 2x_1x_2 + 2x_2x_3$ to the canonical form through an orthogonal transformation and hence, show that it is positive semi-definite. Also give a non-zero set of values (x_1, x_2, x_3) which makes this quadratic form zero.

ANSWERS

1. $\begin{bmatrix} 0 & 0 & 0 \\ 0 & 3 & 0 \\ 0 & 0 & 14 \end{bmatrix}$ 2. Diagonalisation not possible 3. $\begin{bmatrix} 3 & -1 & 1 \\ -1 & 5 & -1 \\ 1 & -1 & 3 \end{bmatrix}$

4. $-2y_1^2 - y_2^2 - y_3^2$ 5. $y_1^2 + 3y_2^2 + 4y_3^2$ 6. $-y_1^2 + y_2^2 + 4y_3^2$

7. C.F.$= -2y_1^2 + y_2^2 + y_3^2$; Nature = indefinite. 8. $3y_2^2 + 15y_3^2$

9. C.F.$= y_2^2 + 3y_3^2$; Taking $y_1 = \sqrt{3}, y_2 = 0, y_3 = 0$ we get, $x_1 = 1, x_2 = 1, x_3 = -1$

1.8. Short Answer Questions

1. Find the eigenvalue of $\begin{bmatrix}1 & 4\\3 & 2\end{bmatrix}$ corresponding to the eigenvector $\begin{bmatrix}-4\\3\end{bmatrix}$.

Solution: Let $A = \begin{bmatrix}1 & 4\\3 & 2\end{bmatrix}$ and $X = \begin{bmatrix}-4\\3\end{bmatrix}$.

Now, $AX = \begin{bmatrix}1 & 4\\3 & 2\end{bmatrix}\begin{bmatrix}-4\\3\end{bmatrix} = \begin{bmatrix}8\\-6\end{bmatrix} = -2\begin{bmatrix}-4\\3\end{bmatrix} = \lambda X$.

Hence, the eigenvalue is $\lambda = -2$.

2. Find the sum and product of all the eigenvalues of $\begin{bmatrix}4 & -3 & 2\\-1 & 1 & 3\\0 & 5 & -2\end{bmatrix}$.

Solution:

We know that, sum of eigenvalues = sum of main diagonal elements of A

$$= 4 + 1 - 2 = 3$$

Also, product of the eigenvalues $= |A|$

$$= \begin{vmatrix}4 & -3 & 2\\-1 & 1 & 3\\0 & 5 & -2\end{vmatrix}$$

$$= 4(-2-15) + 3(2-0) + 2(-5-0)$$

$$= -68 + 6 - 10 = -72.$$

3. Two Eigenvalues of the matrix $A = \begin{bmatrix}8 & -6 & 2\\-6 & 7 & -4\\2 & -4 & 3\end{bmatrix}$ are 3 and 0. What is the third eigenvalue? What is the product of the eigenvalues of A? **[AU, *April/May 2017*]**

Solution: Given $\lambda_1 = 3, \lambda_2 = 0$ and $A = \begin{bmatrix}8 & -6 & 2\\-6 & 7 & -4\\2 & -4 & 3\end{bmatrix}$.

Let λ_3 be the third eigenvalue.

We know that, sum of eigenvalues = sum of main diagonal elements of A

$\Rightarrow \lambda_1 + \lambda_2 + \lambda_3 = 8 + 7 + 3$

$\Rightarrow 3 + 0 + \lambda_3 = 18$

$\Rightarrow \lambda_3 = 15$

Now, product of the eigenvalues $= \lambda_1 \times \lambda_2 \times \lambda_3 = 0$.

4. If 3 and 15 are the two eigenvalues of $A = \begin{bmatrix} 8 & -6 & 2 \\ -6 & 7 & -4 \\ 2 & -4 & 3 \end{bmatrix}$, find $|A|$ without expanding the determinant.

Solution: Given $\lambda_1 = 3, \lambda_2 = 15$ and $A = \begin{bmatrix} 8 & -6 & 2 \\ -6 & 7 & -4 \\ 2 & -4 & 3 \end{bmatrix}$.

Let λ_3 be the third eigenvalue.

We know that, sum of eigenvalues = sum of main diagonal elements of A

$\Rightarrow \quad \lambda_1 + \lambda_2 + \lambda_3 = 8 + 7 + 3$

$\Rightarrow \quad 3 + 15 + \lambda_3 = 18$

$\Rightarrow \quad \lambda_3 = 0$

Now, $|A|$ = product of the eigenvalues $= \lambda_1 \times \lambda_2 \times \lambda_3 = 0$.

Hence $|A| = 0$.

5. Given that 2 is an eigenvalue of the matrix $A = \begin{bmatrix} 6 & -2 & 2 \\ -2 & 3 & -1 \\ 2 & -1 & 3 \end{bmatrix}$, find the other two eigenvalues.

Solution: Given, eigenvalue $\lambda_1 = 2$ and $A = \begin{bmatrix} 6 & -2 & 2 \\ -2 & 3 & -1 \\ 2 & -1 & 3 \end{bmatrix}$.

Let λ_2 and λ_3 be the other two eigenvalues.

We know that, sum of eigenvalues = sum of main diagonal elements of A

$\Rightarrow \quad \lambda_1 + \lambda_2 + \lambda_3 = 6 + 3 + 3$

$\Rightarrow \quad 2 + \lambda_2 + \lambda_3 = 12$

$\Rightarrow \quad \lambda_2 + \lambda_3 = 10 \qquad \text{... (1)}$

Also, product of the eigenvalues = $|A|$

$$\Rightarrow \quad \lambda_1 \times \lambda_2 \times \lambda_3 = \begin{vmatrix} 6 & -2 & 2 \\ -2 & 3 & -1 \\ 2 & -1 & 3 \end{vmatrix}$$

$\Rightarrow \quad 2 \times \lambda_2 \times \lambda_3 = 6(9-1) + 2(-6+2) + 2(2-6) = 48 - 8 - 8 = 32$

$\Rightarrow \quad \lambda_2 \lambda_3 = 16$

Now, $(\lambda_2 - \lambda_3)^2 = (\lambda_2 + \lambda_3)^2 - 4\lambda_2\lambda_3 = 100 - 64 = 36$

$\therefore\ \lambda_2 - \lambda_3 = 6$ ... (2)

Solving (1) and (2) we get, $\lambda_2 = 8$ and $\lambda_3 = 2$.

Hence, the other two eigenvalues are 2 and 8.

6. Find the eigenvalues of A^3 if $A = \begin{bmatrix} 1 & 3 & -2 \\ 0 & 2 & -5 \\ 0 & 0 & 3 \end{bmatrix}$.

Solution: Given $A = \begin{bmatrix} 1 & 3 & -2 \\ 0 & 2 & -5 \\ 0 & 0 & 3 \end{bmatrix}$.

Since it is an upper triangular matrix, the eigenvalues A are 1,2,3.

$\therefore$ Eigenvalues of $A^3 = 1^3,\ 2^3,\ 3^3 = 1, 8, 27$.

7. The eigenvalues of the matrix $A = \begin{bmatrix} 2 & 2 & 1 \\ 1 & 3 & 1 \\ 1 & 2 & 2 \end{bmatrix}$ are equal to 1 each. Find the eigen values of A^{-1}.

Solution: Given that $A = \begin{bmatrix} 2 & 2 & 1 \\ 1 & 3 & 1 \\ 1 & 2 & 2 \end{bmatrix}$.

Also given that eigenvalues of $A = 1, 1, 1$.

$\therefore$ Eigenvalues of $A^{-1} = \frac{1}{1}, \frac{1}{1}, \frac{1}{1} = 1, 1, 1$.

8. Find the eigenvalues of $A - 2I$ and A^{-1} given that the two eigenvalues of $A = \begin{bmatrix} 6 & -2 & 2 \\ -2 & 3 & -1 \\ 2 & -1 & 3 \end{bmatrix}$ are 2 and 8.

Solution: Given $\lambda_1 = 2, \lambda_2 = 8$ and $A = \begin{bmatrix} 6 & -2 & 2 \\ -2 & 3 & -1 \\ 2 & -1 & 3 \end{bmatrix}$.

Let λ_3 be the third eigenvalue.

We know that, sum of eigenvalues = sum of main diagonal elements of A

$\Rightarrow\ \lambda_1 + \lambda_2 + \lambda_3 = 6 + 3 + 3$

$\Rightarrow\ 2 + 8 + \lambda_3 = 12$

$\Rightarrow\ \lambda_3 = 2$

$\therefore$ Eigenvalues of $A = 2, 2, 8$.

Now, the eigenvalues of $A-2I = 2-2, 2-2, 8-2 = 0, 0, 6$.

And the eigenvalues of $A^{-1} = \frac{1}{2}, \frac{1}{2}, \frac{1}{8}$.

9. **If the sum of two eigenvalues and trace of a matrix A are equal, find the value of $|A|$.**

Solution: **[AU, *Nov/Dec 2016*]**

Given that, sum of eigenvalues = trace of a matrix A.

$\therefore$ $|A|$ = product of the eigenvalues of A.

10. Two eigenvalues of the matrix $A = \begin{bmatrix} 2 & 2 & 1 \\ 1 & 3 & 1 \\ 1 & 2 & 2 \end{bmatrix}$ are equal and are $\frac{1}{5}$ times to the third.

Find the eigenvalues.

Solution: Given $A = \begin{bmatrix} 2 & 2 & 1 \\ 1 & 3 & 1 \\ 1 & 2 & 2 \end{bmatrix}$.

Let λ_2, λ_2 and λ_3 be the eigenvalues of the matrix A.

Also given $\lambda_1 = \lambda_2$, $\lambda_1 = \frac{1}{5}\lambda_3$ and $\lambda_2 = \frac{1}{5}\lambda_3$.

We know that, sum of eigenvalues = sum of main diagonal elements of A

$$\Rightarrow \quad \lambda_1 + \lambda_2 + \lambda_3 = 2+3+2$$

$$\Rightarrow \frac{1}{5}\lambda_3 + \frac{1}{5}\lambda_3 + \lambda_3 = 7$$

$$\Rightarrow \frac{7}{5}\lambda_3 = 7$$

$$\Rightarrow \quad \lambda_3 = 5 \quad \Rightarrow \quad \lambda_1 = 1 \text{ and } \lambda_2 = 1$$

$\therefore$ Eigenvalues of $A = 1, 1, 5$.

11. Find the constants a and b such that the matrix $\begin{bmatrix} a & 4 \\ 1 & b \end{bmatrix}$ has 3 and -2 as its eigenvalues. **[AU, *April/May 2017*]**

Solution: Given $\lambda_1 = 3$, $\lambda_2 = -2$ and $A = \begin{bmatrix} a & 4 \\ 1 & b \end{bmatrix}$.

We know that, sum of eigenvalues = sum of main diagonal elements of A

$$\Rightarrow \quad 3-2 = a+b$$

$$\Rightarrow \quad a+b = 1 \qquad \text{... (1)}$$

Also, product of the eigenvalues = $|A|$

$\Rightarrow \quad 3 \times (-2) \;= \begin{vmatrix} a & 4 \\ 1 & b \end{vmatrix}$

$\Rightarrow \quad -6 \;= ab - 4$

$\Rightarrow \quad ab = -2$

Now, $(a-b)^2 = (a+b)^2 - 4ab = 1 + 8 = 9$

$\therefore \quad a - b = 3 \qquad \ldots (2)$

Solving (1) and (2) we get, $a = 2$ and $b = -1$.

Hence, the matrix is $A = \begin{bmatrix} 2 & 4 \\ 1 & -1 \end{bmatrix}$.

12. Form the matrix whose eigenvalues are $\alpha - 5, \beta - 5, \gamma - 5$ where α, β, γ are the eigenvalues of $A = \begin{bmatrix} -1 & -2 & -3 \\ 4 & 5 & -6 \\ 7 & -8 & 9 \end{bmatrix}$.

Solution: Given $A = \begin{bmatrix} -1 & -2 & -3 \\ 4 & 5 & -6 \\ 7 & -8 & 9 \end{bmatrix}$.

Also given, α, β, γ are the eigenvalues of A.

$\Rightarrow \alpha - 5, \beta - 5, \gamma - 5$ are the eigenvalues of $A - 5I$.

$$\text{Now, } A - 5I = \begin{bmatrix} -1 & -2 & -3 \\ 4 & 5 & -6 \\ 7 & -8 & 9 \end{bmatrix} - 5\begin{bmatrix} 1 & 0 & 0 \\ 0 & 1 & 0 \\ 0 & 0 & 1 \end{bmatrix}.$$

$$= \begin{bmatrix} -1 & -2 & -3 \\ 4 & 5 & -6 \\ 7 & -8 & 9 \end{bmatrix} + \begin{bmatrix} -5 & 0 & 0 \\ 0 & -5 & 0 \\ 0 & 0 & -5 \end{bmatrix}$$

$$= \begin{bmatrix} -6 & -2 & -3 \\ 4 & 0 & -6 \\ 7 & -8 & 4 \end{bmatrix}.$$

13. State Caylely-Hamilton theorem. **[AU, *Nov/Dec 2014*]**

Solution: Every square matrix satisfies its own characteristic equation.

14. Verify Caylely-Hamilton theorem for the matrix $\begin{bmatrix} 3 & -1 \\ -1 & 5 \end{bmatrix}$.

Solution: The characteristic equation of A is $|A - \lambda I| = 0$.

$\Rightarrow \qquad \lambda^2 - S_1\lambda + S_2 = 0$

where S_1 = sum of the diagonal elements of A.

$= 3 + 5 = 8$

$S_2 = |A| = \begin{vmatrix} 3 & -1 \\ -1 & 5 \end{vmatrix} = 15 - 1 = 14.$

$\therefore$ The characteristic equation is $\lambda^2 - 8\lambda + 14 = 0$

Now we replace λ by A, $A^2 - 8A + 14I = 0$

Now, $A^2 = A.A = \begin{bmatrix} 3 & -1 \\ -1 & 5 \end{bmatrix}\begin{bmatrix} 3 & -1 \\ -1 & 5 \end{bmatrix} = \begin{bmatrix} 10 & -8 \\ -8 & 26 \end{bmatrix}$

$$\therefore \; A^2 - 8A + 14I = \begin{bmatrix} 10 & -8 \\ -8 & 26 \end{bmatrix} - 8\begin{bmatrix} 3 & -1 \\ -1 & 5 \end{bmatrix} + 14\begin{bmatrix} 1 & 0 \\ 0 & 1 \end{bmatrix}$$

$$= \begin{bmatrix} 0 & 0 \\ 0 & 0 \end{bmatrix}.$$

Hence Cayley-Hamilton theorem is verified.

15. If two eigenvectors of a real symmetric matrix of order 3 are $\begin{bmatrix} 2 \\ -1 \\ 1 \end{bmatrix}$ and $\begin{bmatrix} 0 \\ 1 \\ 1 \end{bmatrix}$, find the third eigenvector.

Solution: Given matrix is a real symmetric matrix, hence the third eigenvector is orthogonal to other two eigenvectors.

Let $X_1 = \begin{bmatrix} 2 \\ -1 \\ 1 \end{bmatrix}$ and $X_2 = \begin{bmatrix} 0 \\ 1 \\ 1 \end{bmatrix}$.

Let $X_3 = \begin{bmatrix} l \\ m \\ n \end{bmatrix}$ be the third eigenvector which is orthogonal to X_1 and X_2.

$$\therefore \quad X_1^T X_3 = 0 \text{ and } X_2^T X_3 = 0$$

$$\Rightarrow \quad (2 \;\; -1 \;\; 1)\begin{pmatrix} l \\ m \\ n \end{pmatrix} = 0 \text{ and } (0 \;\; 1 \;\; 1)\begin{pmatrix} l \\ m \\ n \end{pmatrix} = 0$$

$$\Rightarrow \quad 2l - m + n = 0$$

$$0l + m + n = 0$$

Solving the above equations by the rule of cross multiplication we get,

$$\frac{l}{-2} = \frac{m}{-2} = \frac{n}{2}$$

$$\Rightarrow \quad \frac{l}{1} = \frac{m}{1} = \frac{n}{-1}$$

$\therefore$ The third eigenvector is $X_3 = \begin{bmatrix} 1 \\ 1 \\ -1 \end{bmatrix}$.

16. Write the matrix of the quadratic form $2x^2 + 8z^2 + 4xy + 10xz - 2yz$.

Solution: The matrix of the quadratic form is

$$A = \begin{bmatrix} \text{coefficient of } x^2 & \frac{1}{2}\text{coefficient of } xy & \frac{1}{2}\text{coefficient of } xz \\ \frac{1}{2}\text{coefficient of } xy & \text{coefficient of } y^2 & \frac{1}{2}\text{coefficient of } yz \\ \frac{1}{2}\text{coefficient of } xz & \frac{1}{2}\text{coefficient of } yz & \text{coefficient of } z^2 \end{bmatrix} = \begin{bmatrix} 2 & 2 & 5 \\ 2 & 0 & -1 \\ 5 & -1 & 8 \end{bmatrix}.$$

17. Write down the matrix corresponding to the quadratic form $2x_1^2 + 5x_2^2 + 4x_1x_2 + 2x_3x_1$. **[*AU, Nov/Dec 2016*]**

Solution: The matrix of the quadratic form is

$$A = \begin{bmatrix} coefficient\ of\ x^2 & \frac{1}{2}coefficient\ of\ xy & \frac{1}{2}coefficient\ of\ xz \\ \frac{1}{2}coefficient\ of\ xy & coefficient\ of\ y^2 & \frac{1}{2}coefficient\ of\ yz \\ \frac{1}{2}coefficient\ of\ xz & \frac{1}{2}coefficient\ of\ yz & coefficient\ of\ z^2 \end{bmatrix}$$

$$= \begin{bmatrix} 2 & 2 & 1 \\ 2 & 5 & 0 \\ 1 & 0 & 0 \end{bmatrix}.$$

18. Find the nature of the quadratic form $2x^2 + 2xy + 3y^2$.

Solution: The matrix of the quadratic form is

$$A = \begin{bmatrix} coefficient\ of\ x^2 & \frac{1}{2}coefficient\ of\ xy \\ \frac{1}{2}coefficient\ of\ yx & coefficient\ of\ y^2 \end{bmatrix} = \begin{bmatrix} 2 & 1 \\ 1 & 3 \end{bmatrix}.$$

Now, $D_1 = |a_{11}| = |2| = 2$ (positive)

$$D_2 = |A| = \begin{vmatrix} 2 & 1 \\ 1 & 3 \end{vmatrix} = 6 - 1 = 5 \text{ (positive)}$$

∴ The quadratic form is positive definite.

19. Prove that $x^2 - y^2 + 4z^2 + 4xy + 6xz + 2yz$ is indefinite.

Solution: The matrix of the quadratic form is

$$A = \begin{bmatrix} coefficient\ of\ x^2 & \frac{1}{2}coefficient\ of\ xy & \frac{1}{2}coefficient\ of\ xz \\ \frac{1}{2}coefficient\ of\ xy & coefficient\ of\ y^2 & \frac{1}{2}coefficient\ of\ yz \\ \frac{1}{2}coefficient\ of\ xz & \frac{1}{2}coefficient\ of\ yz & coefficient\ of\ z^2 \end{bmatrix}$$

$$= \begin{bmatrix} 1 & 2 & 3 \\ 2 & -1 & 1 \\ 3 & 1 & 4 \end{bmatrix}.$$

Now, $D_1 = |a_{11}| = |1| = 1$ (positive)

$$D_2 = \begin{vmatrix} a_{11} & a_{12} \\ a_{21} & a_{22} \end{vmatrix} = \begin{vmatrix} 1 & 2 \\ 2 & -1 \end{vmatrix} = -1 - 4 = -5 \text{ (negative)}$$

$$D_3 = |A| = \begin{vmatrix} 1 & 2 & 3 \\ 2 & -1 & 1 \\ 3 & 1 & 4 \end{vmatrix} = 1(-4-1) - 2(8-3) + 3(2+3) = 0 \text{ (zero)}$$

∴ The quadratic form is indefinite.

20. What is the nature of the quadratic form $x^2 + y^2 + z^2$ in four variables?

Solution: **[AU, Dec2015/Jan 2016]**

Given quadratic form is $x^2 + y^2 + z^2 + 0t^2$.

The matrix of the quadratic form is $A = \begin{bmatrix} 1 & 0 & 0 & 0 \\ 0 & 1 & 0 & 0 \\ 0 & 0 & 1 & 0 \\ 0 & 0 & 0 & 0 \end{bmatrix}$.

$\therefore$ The eigenvalues A are 1,1,1,0.

$\therefore$ The eigenvalues are positive and one eigenvalue is zero.

$\therefore$ The nature of the quadratic form is positive semidefinite.

21. Find the index, rank and signature of the quadratic form $x_1^2 + 2x_2^2 - 3x_3^2$.

Solution: We know, index i = number of positive square terms = 2

Rank r = number of the non-zero terms in canonical form = 3

Signature $s = 2i - r = (2)(2) - 3 = 1$.

22. Write down the quadratic form corresponding to the matrix $\begin{bmatrix} 2 & 0 & -2 \\ 0 & 2 & 1 \\ -2 & 1 & -2 \end{bmatrix}$.

Solution: Let $A = \begin{bmatrix} a_{11} & a_{12} & a_{13} \\ a_{21} & a_{22} & a_{23} \\ a_{31} & a_{32} & a_{33} \end{bmatrix} = \begin{bmatrix} 2 & 0 & -2 \\ 0 & 2 & 1 \\ -2 & 1 & -2 \end{bmatrix}$.

The quadratic form $= a_{11}x^2 + a_{22}y^2 + a_{33}z^2 + 2a_{12}xy + 2a_{23}yz + 2a_{13}xz$

$= 2x^2 + 2y^2 - 2z^2 + 2yz - 4xz$.

23. Give the nature of a quadratic form whose matrix is $\begin{bmatrix} -1 & 0 & 0 \\ 0 & -1 & 0 \\ 0 & 0 & -2 \end{bmatrix}$.

[AU, April/May 2015]

Solution: Given matrix is a triangular matrix. Therefore its eigen values are its diagonal elements.

$\therefore$ The eigen values are $-1, -1, -2$.

All the eigen values are negative.

$\therefore$ Nature = negative definite.

CHAPTER 2

VECTOR CALCULUS

2.1. Introduction

Vector calculus or vector analysis, is a branch of Mathematics concerned with differentiation and integration of vector fields, primarily in 3-dimensional Euclidean space. Vectors are quantities which have both magnitude (size) and direction. In everyday language speed and velocity mean the same thing, but in Mathematics speed is a scalar and velocity is a vector. One of the most common uses of vectors is in the description of velocity. By using vectors, physicists describe the movement of a car in motion using a simple line on a geometric plane. This same principle is also applied by navigators to chart the movements of airplanes and ships. Vector calculus plays an important role in differential geometry and in the study of partial differential equations. It is used extensively in Physics and engineering, especially in the description of Fluid mechanics, Thermodynamics, Control/ Signals.

Representation of a Vector

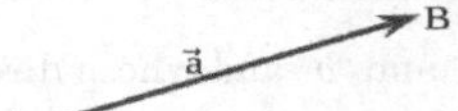

A vector can be represented graphically by a directed straight line segment AB of finite length, drawn in a fixed direction AB. The starting arbitrary point A in space is called the *origin* or the *initial point* of the vector, and the ending point B is called the *terminal point* or *terminus*. The length AB is called the magnitude of the vector denoted by $\overrightarrow{AB}$. Thus a directed line segment like $\overrightarrow{AB}$ is a vector. Normally, we represent $\overrightarrow{AB}$ by $\vec{a}$ and its magnitude by $|\vec{a}|$ or simply a.

Modulus of a Vector

The *modulus* of a vector is the non-negative number, which measures its length. The modulus of a vector $\overrightarrow{AB}$ or $\vec{a}$ is denoted by $|\overrightarrow{AB}|$ or $|\vec{a}|$ or a.

Position Vector

If the vector $\overrightarrow{OP}$ represents the position of the point P in space relative to the point O, then $\overrightarrow{OP}$ is called the *position vector* of P referred to O as the origin.

A vector whose modulus is unity is called a *unit vector.*

Note : If $P(x, y, z)$ is any point in space, then its position vector is given by

$\overrightarrow{OP} = \vec{r} = x\vec{i} + y\vec{j} + z\vec{k}$, where $\vec{i}, \vec{j}, \vec{k}$ are the unit vectors along the coordinate axes OX, OY, OZ respectively. The modulus of $\vec{r}$ is given by $|\vec{r}| = r = \sqrt{x^2 + y^2 + z^2}$.

Scalar Product: Let $\vec{a}$ and $\vec{b}$ be two vectors. The *scalar product* or *dot product* of $\vec{a}$ and $\vec{b}$ is defined to be $\vec{a} \cdot \vec{b} = ab\cos\theta$, where θ is the angle between the two vectors when drawn from a common origin.

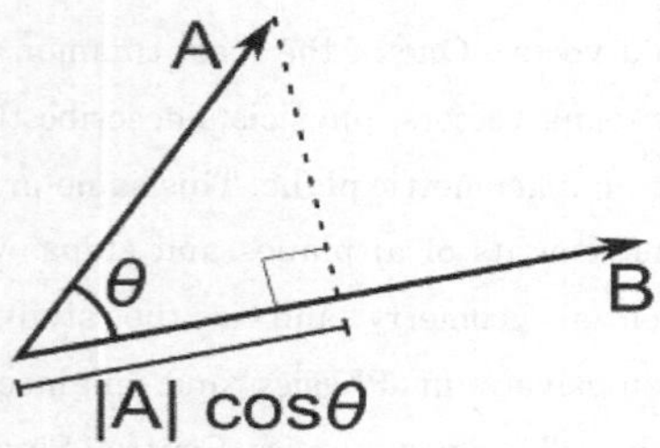

Vector Product: Let $\vec{a}$ and $\vec{b}$ be two vectors. The *vector product* or *cross product* of $\vec{a}$ and $\vec{b}$ is a vector perpendicular to both $\vec{a}$ and $\vec{b}$, whose magnitude is $ab\sin\theta$, where $0 \leq \theta \leq \pi$ is the angle between $\vec{a}$ and $\vec{b}$ and whose direction is along a unit vector $\hat{n}$ such that $\vec{a}$, $\vec{b}$, $\hat{n}$ form a right handed system. Thus $\vec{a} \times \vec{b} = ab\sin\theta\,\hat{n}$ $(0 \leq \theta \leq \pi)$.

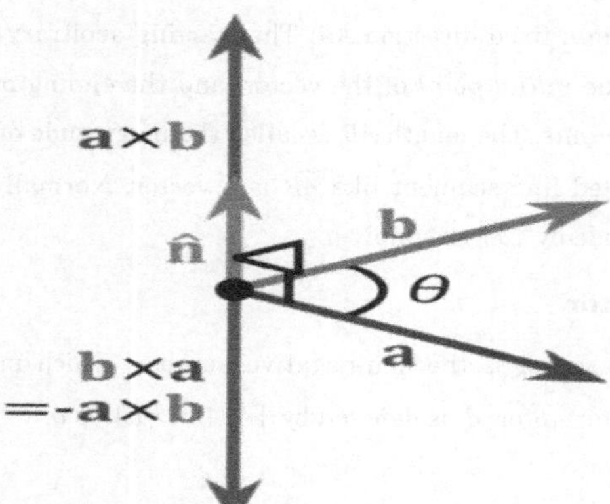

2.2. Vector Identities

1. $\vec{a} \cdot \vec{b} = \vec{b} \cdot \vec{a}$
2. $\vec{a} \times \vec{b} = -\vec{b} \times \vec{a}$.
3. $\vec{a} \cdot \vec{a} = |\vec{a}|^2 = a^2$
4. If $\vec{a} \cdot \vec{b} = 0$, then $\vec{a} = \vec{0}$ or $\vec{b} = \vec{0}$ or $\vec{a}$ and $\vec{b}$ are perpendicular vectors.
5. If $\vec{a}$ and $\vec{b}$ are parallel, then $\vec{a} \times \vec{b} = \vec{0}$.
6. $\vec{i} \cdot \vec{i} = \vec{j} \cdot \vec{j} = \vec{k} \cdot \vec{k} = 1$.
7. $\vec{i} \cdot \vec{j} = \vec{j} \cdot \vec{k} = \vec{k} \cdot \vec{i} = 0$.
8. $\vec{i} \times \vec{i} = \vec{j} \times \vec{j} = \vec{k} \times \vec{k} = \vec{0}$.
9. $\vec{i} \times \vec{j} = \vec{k},\ \vec{j} \times \vec{k} = \vec{i},\ \vec{k} \times \vec{i} = \vec{j}$.
10. $\vec{j} \times \vec{i} = -\vec{k},\ \vec{k} \times \vec{j} = -\vec{i},\ \vec{i} \times \vec{k} = -\vec{j}$.
11. If $\vec{a} = a_1\vec{i} + a_2\vec{j} + a_3\vec{k}$ and $\vec{b} = b_1\vec{i} + b_2\vec{j} + b_3\vec{k}$ then $\vec{a} \cdot \vec{b} = a_1 b_1 + a_2 b_2 + a_3 b_3$.
12. If $\vec{a} = a_1\vec{i} + a_2\vec{j} + a_3\vec{k}$ and $\vec{b} = b_1\vec{i} + b_2\vec{j} + b_3\vec{k}$, then

$$\vec{a} \times \vec{b} = \begin{vmatrix} \vec{i} & \vec{j} & \vec{k} \\ a_1 & a_2 & a_3 \\ b_1 & b_2 & b_3 \end{vmatrix}.$$

Scalar and Vector Point Functions

A variable quantity whose value at any point in a region of space depends upon the position of the point, it is called *point function.* There are two types of point functions.

Scalar Point Function: Let R be a region of space at each point of which a scalar $\varphi = \varphi(x,y,z)$ is given, then φ is called a *scalar point function* and R is called a *scalar field.*

The temperature distributions in a medium, the distribution of atmospheric pressure in space are examples of scalar point function.

Vector Point Function: Let R be a region of space at each point of which a vector $\vec{v} = \vec{v}(x,y,z)$ is given, then $\vec{v}$ is called a *vector point function* and R is called a *vector field.*

The velocity of a moving fluid at any instant, the gravitational force are examples of vector point function.

2.3. Vector Differential Operator ∇

We now consider an operator ∇ (to be read as 'del') which is useful in defining three quantities known as the gradient, the divergence and the curl, that are useful in engineering applications. The operator $\nabla$is defined as $\nabla = \vec{i}\frac{\partial}{\partial x} + \vec{j}\frac{\partial}{\partial y} + \vec{k}\frac{\partial}{\partial z}$

∇ is called the *vector differential operator,* as it behaves like a vector (though not a vector) with $\frac{\partial}{\partial x}, \frac{\partial}{\partial y}, \frac{\partial}{\partial z}$ as coefficients of $\vec{i}, \vec{j}, \vec{k}$ respectively.

When writing $\nabla = \sum \vec{i}\frac{\partial}{\partial x}$, it should be noted that $\vec{i}$ and $\frac{\partial}{\partial x}$ are written as the first and second factors respectively.

2.4. Gradient of a Scalar Point Function

Gradient: Let $\varphi(x,y,z)$ be a scalar point function defined in a certain region of space. Then the vector point function given by

$\nabla\varphi = \left[\vec{i}\frac{\partial}{\partial x} + \vec{j}\frac{\partial}{\partial y} + \vec{k}\frac{\partial}{\partial z}\right]\varphi = \vec{i}\frac{\partial\varphi}{\partial x} + \vec{j}\frac{\partial\varphi}{\partial y} + \vec{k}\frac{\partial\varphi}{\partial z}$ is defined as the *gradient* of φ and shortly denoted as *grad* φ. Thus $grad\varphi = \vec{i}\frac{\partial\varphi}{\partial x} + \vec{j}\frac{\partial\varphi}{\partial y} + \vec{k}\frac{\partial\varphi}{\partial z} = \nabla\varphi$.

Thus we observe that the gradient of a scalar point function is a vector point function.

Note: $\nabla^2 = \nabla . \nabla = \frac{\partial^2}{\partial x^2} + \frac{\partial^2}{\partial y^2} + \frac{\partial^2}{\partial z^2}$ is the *Laplacian operator.*

Properties of Gradient

(i) If φ is constant scalar point function, then $\nabla\varphi = \vec{0}$.

(ii) $\nabla(\varphi_1 \pm \varphi_2) = \nabla\varphi_1 \pm \nabla\varphi_2$.

(iii) $\nabla(c_1\varphi_1 + c_2\varphi_2) = c_1\nabla\varphi_1 + c_2\nabla\varphi_2$, where c_1 and c_2 are constants.

(iv) $\nabla(\varphi_1\varphi_2) = \varphi_1\nabla\varphi_2 + \varphi_2\nabla\varphi_1$.

(v) $\nabla\left(\frac{\varphi_1}{\varphi_2}\right) = \frac{\varphi_2\nabla\varphi_1 - \varphi_1\nabla\varphi_2}{{\varphi_2}^2}$.

Geometrical Interpretation of Gradient

If a surface $\varphi(x,y,z) = 0$ is drawn through any point P such that at each point on the surface, the function has the same value as at P, then such surface is called *level surface* through P. For example, if $\varphi(x,y,z)$ represents potential at the point (x,y,z), the equipotential surface $\varphi(x,y,z) = c$ is a level surface.Through any point, passes one and only level surface. Moreover, no two level surfaces can intersect.

Let $\vec{r}$ and $\vec{r} + d\vec{r}$ be the position vectors of P and Q respectively, then $\overrightarrow{PQ} = d\vec{r}$.

$$\therefore \ \nabla\varphi . d\vec{r} = \left(\frac{\partial\varphi}{\partial x}\vec{\imath} + \frac{\partial\varphi}{\partial y}\vec{\jmath} + \frac{\partial\varphi}{\partial z}\vec{k}\right) . (dx\vec{\imath} + dy\vec{\jmath} + dz\vec{k})$$

$$= \frac{\partial\varphi}{\partial x}dx + \frac{\partial\varphi}{\partial y}dy + \frac{\partial\varphi}{\partial z}dz$$

$$= d\varphi = 0 \qquad [\because \ \varphi(x,y,z) = c]$$

$\therefore \ \nabla\varphi . dr = 0$. Thus $\nabla\varphi$ is perpendicular to every $d\vec{r}$ lying in the surface. Hence $\nabla\varphi$ is normal to the surface $\varphi(x,y,z) = c$.

Directional Derivative

The *directional derivative* of φ in the direction of $\vec{a}$ is the rate of change of φ in the direction of $\vec{a}$.

Thus the directional derivative of φ in the direction of $\vec{a}$ given by $\dfrac{\nabla\varphi . \vec{a}}{|\vec{a}|}$.

Angle between Surfaces

If θ is the angle between the surfaces $\varphi_1 = c_1$ and $\varphi_2 = c_2$, then

$$\cos\theta = \frac{\nabla\varphi_1 . \nabla\varphi_2}{|\nabla\varphi_1||\nabla\varphi_2|}.$$

Result: The surfaces $\varphi_1 = c_1$ and $\varphi_2 = c_2$ cut *orthogonally* if $\nabla\varphi_1 . \nabla\varphi_2 = 0$.

Example 1 : Find $grad\varphi$ at $(1,-2,-1)$ for $\varphi(x,y,z) = 3x^2y - y^3z^2$.

Solution: $\varphi(x,y,z) = 3x^2y - y^3z^2$.

$$\therefore \frac{\partial\varphi}{\partial x} = 6xy, \ \frac{\partial\varphi}{\partial y} = 3x^2 - 3y^2z^2, \ \frac{\partial\varphi}{\partial z} = -2y^3z.$$

Hence $\nabla\varphi = \dfrac{\partial\varphi}{\partial x}\vec{\imath} + \dfrac{\partial\varphi}{\partial y}\vec{\jmath} + \dfrac{\partial\varphi}{\partial z}\vec{k}$

$$= 6xy\vec{\imath} + (3x^2 - 3y^2z^2)\vec{\jmath} - 2y^3z\vec{k}.$$

$\therefore \nabla\varphi$ at $(1,-2,-1) = -12\vec{\imath} - 9\vec{\jmath} - 16\vec{k}$.

Example 2 : Find $\nabla\varphi$ if $\varphi = \log(x^2 + y^2 + z^2)$.

Solution: $\varphi = \log(x^2 + y^2 + z^2)$

$$\therefore \nabla\varphi = \frac{\partial\varphi}{\partial x}\vec{i} + \frac{\partial\varphi}{\partial y}\vec{j} + \frac{\partial\varphi}{\partial z}\vec{k}$$

$$= \frac{1}{x^2+y^2+z^2}\,2x\vec{i} + \frac{1}{x^2+y^2+z^2}\,2y\vec{j} + \frac{1}{x^2+y^2+z^2}\,2z\vec{k}$$

$$= \frac{2}{x^2+y^2+z^2}(x\vec{i} + y\vec{j} + z\vec{k}) = \frac{2\vec{r}}{x^2+y^2+z^2}.$$

Example 3 : If $\vec{r} = x\vec{i} + y\vec{j} + z\vec{k}$, then find $\nabla r, \nabla r^2$ and $\nabla\left(\frac{1}{r}\right)$.

Solution: $|\vec{r}| = \sqrt{x^2 + y^2 + z^2}$ i.e., $r^2 = x^2 + y^2 + z^2$

$\therefore\ 2r\frac{\partial r}{\partial x} = 2x$. Hence $\frac{\partial r}{\partial x} = \frac{x}{r}$.

Similarly, $\frac{\partial r}{\partial y} = \frac{y}{r}$ and $\frac{\partial r}{\partial z} = \frac{z}{r}$.

(i) $$\nabla r = \frac{\partial r}{\partial x}\vec{i} + \frac{\partial r}{\partial y}\vec{j} + \frac{\partial r}{\partial z}\vec{k}$$

$$= \frac{x}{r}\vec{i} + \frac{y}{r}\vec{j} + \frac{z}{r}\vec{k}$$

$$= \frac{1}{r}(x\vec{i} + y\vec{j} + z\vec{k}) = \frac{\vec{r}}{r}.$$

(ii) $$\nabla r^2 = \frac{\partial}{\partial x}(r^2)\vec{i} + \frac{\partial}{\partial y}(r^2)\vec{j} + \frac{\partial}{\partial z}(r^2)\vec{k}$$

$$= 2r\frac{\partial r}{\partial x}\vec{i} + 2r\frac{\partial r}{\partial y}\vec{j} + 2r\frac{\partial r}{\partial z}\vec{k}$$

$$= 2r\frac{x}{r}\vec{i} + 2r\frac{y}{r}\vec{j} + 2r\frac{z}{r}\vec{k}$$

$$= 2(x\vec{i} + y\vec{j} + z\vec{k}) = 2\vec{r}.$$

(iii) $$\nabla\left(\frac{1}{r}\right) = \frac{\partial}{\partial x}\left(\frac{1}{r}\right)\vec{i} + \frac{\partial}{\partial y}\left(\frac{1}{r}\right)\vec{j} + \frac{\partial}{\partial z}\left(\frac{1}{r}\right)\vec{k}$$

$$= \left(\frac{-1}{r^2}\right)\frac{\partial r}{\partial x}\vec{i} + \left(\frac{-1}{r^2}\right)\frac{\partial r}{\partial y}\vec{j} + \left(\frac{-1}{r^2}\right)\frac{\partial r}{\partial z}\vec{k}$$

$$= \frac{-1}{r^2}\left[\frac{x}{r}\vec{i} + \frac{y}{r}\vec{j} + \frac{z}{r}\vec{k}\right]$$

$$= \frac{-1}{r^3}(x\vec{i} + y\vec{j} + z\vec{k})$$

$$= -\frac{\vec{r}}{r^3}.$$

Example 4: Prove that $\nabla r^n = n\, r^{n-2}\, \vec{r}$.

Solution: $\vec{r} = x\vec{i} + y\vec{j} + z\vec{k}.$

$\therefore\ r^2 = x^2 + y^2 + z^2$.

Hence $2r\,\frac{\partial r}{\partial x} = 2x.$

i.e., $\frac{\partial r}{\partial x} = \frac{x}{r}$.

Similarly, $\frac{\partial r}{\partial y} = \frac{y}{r}$ and $\frac{\partial r}{\partial z} = \frac{z}{r}$.

Now, $grad\, r^n = \frac{\partial}{\partial x}(r^n)\vec{i} + \frac{\partial}{\partial y}(r^n)\vec{j} + \frac{\partial}{\partial z}(r^n)\vec{k}$

$= n\,r^{n-1}\,\frac{\partial r}{\partial x}\vec{i} + n\,r^{n-1}\,\frac{\partial r}{\partial y}\vec{j} + n\,r^{n-1}\,\frac{\partial r}{\partial z}\vec{k}$

$= n\,r^{n-1}\left[\frac{x}{r}\,\vec{i} + \frac{y}{r}\vec{j} + \frac{z}{r}\vec{k}\right]$

$= n\,r^{n-2}\left(x\vec{i} + y\vec{j} + z\vec{k}\right)$

$= n\,r^{n-2}\vec{r}.$

Example 5: Prove that $\nabla f(r) = \frac{f'(r)}{r}\vec{r}$.

Solution: $\vec{r} = x\vec{i} + y\vec{j} + z\vec{k}.$

$\therefore\ \frac{\partial r}{\partial x} = \frac{x}{r},\ \frac{\partial r}{\partial y} = \frac{y}{r}$ and $\frac{\partial r}{\partial z} = \frac{z}{r}$.

$\therefore\ \nabla f(r) = \frac{\partial}{\partial x}f(r)\vec{i} + \frac{\partial}{\partial y}f(r)\vec{j} + \frac{\partial}{\partial z}f(r)\vec{k}$

$= f'(r)\,\frac{\partial r}{\partial x}\vec{i} + f'(r)\,\frac{\partial r}{\partial y}\vec{j} + f'(r)\,\frac{\partial r}{\partial z}\vec{k}$

$= f'(r)\left[\frac{x}{r}\vec{i} + \frac{y}{r}\vec{j} + \frac{z}{r}\vec{k}\right]$

$= \frac{f'(r)}{r}\left(x\vec{i} + y\vec{j} + z\vec{k}\right)$

$= \frac{f'(r)}{r}\,\vec{r}$.

Example 6: Show that $\nabla(\vec{a}\,.\,\vec{r}) = \vec{a}$ if $\vec{a} = 2\vec{i} - 3\vec{j} + \vec{k}$.

Solution: $\vec{a}\,.\,\vec{r} = (2\vec{i} - 3\vec{j} + \vec{k})\,.\,(x\vec{i} + y\vec{j} + z\vec{k})$

$$= 2x - 3y + z.$$

$$\therefore \nabla(\vec{a}\,.\,\vec{r}) = \nabla(2x - 3y + z)$$

$$= \frac{\partial}{\partial x}(2x - 3y + z)\vec{i} + \frac{\partial}{\partial y}(2x - 3y + z)\vec{j} + \frac{\partial}{\partial z}(2x - 3y + z)\vec{k}$$

$$= 2\vec{i} - 3\vec{j} + \vec{k} = \vec{a}.$$

Example 7: Find the directional derivative of $4x^2z + xy^2z$ at $(1, -1, 2)$ in the direction of $2\vec{i} - \vec{j} + 3\vec{k}$. ***[AU, Nov/Dec 2015]***

Solution: Let $\varphi = 4x^2z + xy^2z$

and $\vec{a} = 2\vec{i} - \vec{j} + 3\vec{k}$.

$$\nabla\varphi = \frac{\partial\varphi}{\partial x}\vec{i} + \frac{\partial\varphi}{\partial y}\vec{j} + \frac{\partial\varphi}{\partial z}\vec{k}$$

$$= (8xz + y^2z)\vec{i} + (2xyz)\vec{j} + (4x^2 + xy^2)\vec{k}$$

$$\therefore \nabla\varphi \text{ at } (1, -1, 2) = 18\vec{i} - 4\vec{j} + 5\vec{k}\ .$$

$$\text{Directional derivative} = \frac{\nabla\varphi\,.\,\vec{a}}{|\vec{a}|} = \frac{(18\vec{i} - 4\vec{j} + 5\vec{k})\,.\,(2\vec{i} - \vec{j} + 3\vec{k})}{\sqrt{4+1+9}} = \frac{36+4+15}{\sqrt{14}} = \frac{55}{\sqrt{14}}.$$

Example 8: Find the directional derivative of $x^2yz + 4xz^2 + xyz$ at $(1, 2, 3)$ in the direction of $2\vec{i} + \vec{j} - \vec{k}$.

Solution: Let $\varphi = x^2yz + 4xz^2 + xyz$ and $\vec{a} = 2\vec{i} + \vec{j} - \vec{k}$.

$$\nabla\varphi = \frac{\partial\varphi}{\partial x}\vec{i} + \frac{\partial\varphi}{\partial y}\vec{j} + \frac{\partial\varphi}{\partial z}\vec{k}$$

$$= (2xyz + 4z^2 + yz)\vec{i} + (x^2z + xz)\vec{j} + (x^2y + 8xz + xy)\vec{k}$$

$$\therefore \nabla\varphi \text{ at } (1, 2, 3) = 54\vec{i} + 6\vec{j} + 28\vec{k}\ .$$

$$\text{Directional derivative} = \frac{\nabla\varphi\,.\,\vec{a}}{|\vec{a}|} = \frac{(54\vec{i} + 6\vec{j} + 28\vec{k})\,.\,(2\vec{i} + \vec{j} - \vec{k})}{\sqrt{4+1+1}}$$

$$= \frac{108+6-28}{\sqrt{6}} = \frac{86}{\sqrt{6}}.$$

Example 9: Find the directional derivative of $\varphi = x^2yz + 4xz^2$ at the point $P(1,-2,-1)$ in the direction of PQ, where Q is $(3,-3,-2)$.

Solution: $\varphi = x^2yz + 4xz^2$.

Then $\nabla\varphi = \frac{\partial\varphi}{\partial x}\vec{\imath} + \frac{\partial\varphi}{\partial y}\vec{\jmath} + \frac{\partial\varphi}{\partial z}\vec{k}$

$= (2xyz + 4z^2)\vec{\imath} + x^2z\vec{\jmath} + (x^2y + 8xz)\vec{k}$

$\therefore$ $\nabla\varphi$ at $(1,-2,-1) = 8\vec{\imath} - \vec{\jmath} - 10\vec{k}$.

Let $\vec{a} = \overrightarrow{PQ} = \overrightarrow{OQ} - \overrightarrow{OP}$

$= (3\vec{\imath} - 3\vec{\jmath} - 2\vec{k}) - (\vec{\imath} - 2\vec{\jmath} - \vec{k}) = 2\vec{\imath} - \vec{\jmath} - \vec{k}$.

$\therefore$ The directional derivative of φ in the direction of $\vec{a}$ is

$$\frac{\nabla\varphi \cdot \vec{a}}{|\vec{a}|} = \frac{(8\vec{\imath} - \vec{\jmath} - 10\vec{k}) \cdot (2\vec{\imath} - \vec{\jmath} - \vec{k})}{\sqrt{4+1+1}} = \frac{27}{\sqrt{6}}.$$

Example 10: In what direction from $(-1,1,2)$ is the directional derivative of $\varphi = xy^2z^3$ a maximum? and what is its magnitude?

Solution: $\varphi = xy^2z^3$

$\nabla\varphi = \frac{\partial\varphi}{\partial x}\vec{\imath} + \frac{\partial\varphi}{\partial y}\vec{\jmath} + \frac{\partial\varphi}{\partial z}\vec{k}$

$= y^2z^3\vec{\imath} + 2xyz^3\vec{\jmath} + 3xy^2z^2\vec{k}$

$\therefore$ $\nabla\varphi$ at $(-1,1,2) = 8\vec{\imath} - 16\vec{\jmath} - 12\vec{k}$.

$|\nabla\varphi| = \sqrt{64 + 256 + 144} = \sqrt{464}$.

The directional derivative is maximum along the direction of $\nabla\varphi = 8\vec{\imath} - 16\vec{\jmath} - 12\vec{k}$ and magnitude of the maximum directional derivative is $|\nabla\varphi| = \sqrt{464}$.

Example 11: Prove that the directional derivative of $\varphi = x^3y^2z$ at $(1,2,3)$ is maximum along the direction $9\vec{\imath} + 3\vec{\jmath} + \vec{k}$. Also find the maximum directional derivative.

Solution: $\varphi = x^3y^2z$.

$\nabla\varphi = \frac{\partial\varphi}{\partial x}\vec{\imath} + \frac{\partial\varphi}{\partial y}\vec{\jmath} + \frac{\partial\varphi}{\partial z}\vec{k}$

$= 3x^2y^2z\vec{\imath} + 2x^3yz\vec{\jmath} + x^3y^2\vec{k}$.

$\therefore$ $\nabla\varphi$ at $(1,2,3) = 36\vec{\imath} + 12\vec{\jmath} + 4\vec{k} = 4(9\vec{\imath} + 3\vec{\jmath} + \vec{k})$.

$\therefore$ The directional derivative of $\varphi = x^3y^2z$ at $(1,2,3)$ is maximum along the direction $9\vec{\imath} + 3\vec{\jmath} + \vec{k}$.

The magnitude of the maximum directional derivative = $|\nabla\varphi| = 4\sqrt{81+9+1} = 4\sqrt{91}$.

Example 12: Find a unit vector normal to the surface $x^2y + 2xz = 4$ at the point $(2,-2,3)$.

Solution: Let $\varphi = x^2y + 2xz - 4$.

Then $\nabla\varphi = \frac{\partial\varphi}{\partial x}\vec{\imath} + \frac{\partial\varphi}{\partial y}\vec{\jmath} + \frac{\partial\varphi}{\partial z}\vec{k}$

$= (2xy + 2z)\vec{\imath} + x^2\vec{\jmath} + 2x\vec{k}$.

$\therefore$ $\nabla\varphi$ at $(2,-2,3) = -2\vec{\imath} + 4\vec{\jmath} + 4\vec{k}$

$\therefore$ $|\nabla\varphi| = \sqrt{4+16+16} = 6$.

Hence unit normal vector is $\hat{n} = \frac{\nabla\varphi}{|\nabla\varphi|} = \frac{1}{6}\left(-2\vec{\imath} + 4\vec{\jmath} + 4\vec{k}\right)$.

Example 13: Find a unit normal vector to the surface $x = t^3 - t^2,\ y = t,\ z = t^2$ at $t = 1$.

Solution: The given surface can be written in Cartesian form as $x = y^3 - z$.

i.e., $x - y^3 + z = 0$.

Let $\varphi = x - y^3 + z$ and the particular point is $(0,1,1)$.

Now $\nabla\varphi = \frac{\partial\varphi}{\partial x}\vec{\imath} + \frac{\partial\varphi}{\partial y}\vec{\jmath} + \frac{\partial\varphi}{\partial z}\vec{k}$

$\therefore$ $\nabla\varphi = \vec{\imath} - 3y^2\vec{\jmath} + \vec{k}$.

$\therefore$ $\nabla\varphi$ at $(0,1,1) = \vec{\imath} - 3\vec{\jmath} + \vec{k}$

$\therefore$ $|\nabla\varphi| = \sqrt{1+9+1} = \sqrt{11}$

$\therefore$ Unit normal vector is $\hat{n} = \frac{\nabla\varphi}{|\nabla\varphi|} = \frac{1}{\sqrt{11}}\left(\vec{\imath} - 3\vec{\jmath} + \vec{k}\right)$.

Example 14: Find the angle between the surfaces $x^2 + y^2 + z^2 = 9$ and $z = x^2 + y^2 - 3$ at the point $(2, -1, 2)$. **[AU, *April/May 2017*]**

Solution: Let $\varphi_1 = x^2 + y^2 + z^2 - 9$ and $\varphi_2 = x^2 + y^2 - z - 3$.

$$\text{Then } \nabla\varphi_1 = \frac{\partial\varphi_1}{\partial x}\vec{i} + \frac{\partial\varphi_1}{\partial y}\vec{j} + \frac{\partial\varphi_1}{\partial z}\vec{k}$$

$$= 2x\vec{i} + 2y\vec{j} + 2z\vec{k}$$

$$\text{and } \nabla\varphi_2 = \frac{\partial\varphi_2}{\partial x}\vec{i} + \frac{\partial\varphi_2}{\partial y}\vec{j} + \frac{\partial\varphi_2}{\partial z}\vec{k}$$

$$= 2x\vec{i} + 2y\vec{j} - \vec{k}.$$

$\therefore$ $\nabla\varphi_1$ at $(2, -1, 2) = 4\vec{i} - 2\vec{j} + 4\vec{k}$ and $\nabla\varphi_2$ at $(2, -1, 2) = 4\vec{i} - 2\vec{j} - \vec{k}$.

$$\text{Then } \cos\theta = \frac{\nabla\varphi_1 \cdot \nabla\varphi_2}{|\nabla\varphi_1|\,|\nabla\varphi_2|} = \frac{(4\vec{i}-2\vec{j}+4\vec{k}) \cdot (4\vec{i}-2\vec{j}-\vec{k})}{\sqrt{16+4+16}\,\sqrt{16+4+1}}$$

$$= \frac{16+4-4}{\sqrt{36}\,\sqrt{21}} = \frac{16}{6\sqrt{21}}.$$

$$\therefore \theta = \cos^{-1}\left(\frac{8}{3\sqrt{21}}\right).$$

Example 15: Find the angle between the surfaces $x \log z = y^2 - 1$ and $x^2 y = 2 - z$ at the point $(1, 1, 1)$.

Solution: Let $\varphi_1 = y^2 - x\log z - 1$ and $\varphi_2 = x^2 y + z - 2$.

$$\text{Then } \nabla\varphi_1 = \frac{\partial\varphi_1}{\partial x}\vec{i} + \frac{\partial\varphi_1}{\partial y}\vec{j} + \frac{\partial\varphi_1}{\partial z}\vec{k}$$

$$= -\log z\,\vec{i} + 2y\vec{j} - \frac{x}{z}\vec{k}$$

$$\text{and } \nabla\varphi_2 = \frac{\partial\varphi_2}{\partial x}\vec{i} + \frac{\partial\varphi_2}{\partial y}\vec{j} + \frac{\partial\varphi_2}{\partial z}\vec{k}$$

$$= 2xy\vec{i} + x^2\vec{j} + \vec{k}.$$

$\therefore$ $\nabla\varphi_1$ at $(1, 1, 1) = 2\vec{j} - \vec{k}$ and $\nabla\varphi_2$ at $(1, 1, 1) = 2\vec{i} + \vec{j} + \vec{k}$.

$$\text{Then } \cos\theta = \frac{\nabla\varphi_1 \cdot \nabla\varphi_2}{|\nabla\varphi_1|\,|\nabla\varphi_2|} = \frac{(2\vec{j}-\vec{k}) \cdot (2\vec{i}+\vec{j}+\vec{k})}{\sqrt{0+4+1}\,\sqrt{4+1+1}}$$

$$= \frac{1}{\sqrt{5}\,\sqrt{6}} = \frac{1}{\sqrt{30}}.$$

$$\therefore \theta = \cos^{-1}\left(\frac{1}{\sqrt{30}}\right).$$

Example 16: Find the angle between the normal's to the surface $x^2 = yz$ at the points $(1,1,1)$ and $(2,4,1)$. **[AU, Nov/Dec 2014]**

Solution: Let $\varphi = x^2 - yz$.

$$\text{Then } \nabla\varphi = \frac{\partial\varphi}{\partial x}\vec{i} + \frac{\partial\varphi}{\partial y}\vec{j} + \frac{\partial\varphi}{\partial z}\vec{k} = 2x\vec{i} - z\vec{j} - y\vec{k}.$$

Let $\nabla\varphi_1 = \nabla\varphi$ at $(1,1,1) = 2\vec{i} - \vec{j} - \vec{k}$

and $\nabla\varphi_2 = \nabla\varphi$ at $(2,4,1) = 4\vec{i} - \vec{j} - 4\vec{k}$.

$$\therefore \cos\theta = \frac{\nabla\varphi_1 \cdot \nabla\varphi_2}{|\nabla\varphi_1|\,|\nabla\varphi_2|}$$

$$= \frac{(2\vec{i}-\vec{j}-\vec{k})\cdot(4\vec{i}-\vec{j}-4\vec{k})}{\sqrt{4+1+1}\sqrt{16+1+16}} = \frac{13}{\sqrt{6}\sqrt{33}} = \frac{13}{3\sqrt{12}}.$$

$$\therefore \theta = \cos^{-1}\left(\frac{13}{3\sqrt{12}}\right).$$

Example 17: Find the values of a and b so that the surfaces $ax^2 - byz = (a+2)x$ and $4x^2y + z^3 = 4$ cut orthogonally at $(1,-1,2)$.

Solution: The given surfaces are

$$\varphi_1 = ax^2 - byz - (a+2)x = 0 \qquad (1)$$

$$\text{and} \quad \varphi_2 = 4x^2y + z^3 = 4 \qquad (2)$$

$$\nabla\varphi_1 = \frac{\partial\varphi_1}{\partial x}\vec{i} + \frac{\partial\varphi_1}{\partial y}\vec{j} + \frac{\partial\varphi_1}{\partial z}\vec{k}$$

$$= \left(2ax - (a+2)\right)\vec{i} - bz\vec{j} - by\vec{k}.$$

$\therefore \nabla\varphi_1$ at $(1,-1,2) = (a-2)\vec{i} - 2b\vec{j} + b\vec{k}$

$$\text{and} \quad \nabla\varphi_2 = \frac{\partial\varphi_2}{\partial x}\vec{i} + \frac{\partial\varphi_2}{\partial y}\vec{j} + \frac{\partial\varphi_2}{\partial z}\vec{k}$$

$$= 8xy\vec{i} + 4x^2\vec{j} + 3z^2\vec{k}$$

$\therefore \nabla\varphi_2$ at $(1,-1,2) = -8\vec{i} + 4\vec{j} + 12\vec{k}$.

The two surfaces are orthogonal and so $\nabla\varphi_1 \cdot \nabla\varphi_2 = 0$

$$\text{i.e.,} \quad \left[(a-2)\vec{i} - 2b\vec{j} + b\vec{k}\right]\cdot\left[-8\vec{i} + 4\vec{j} + 12\vec{k}\right] = 0$$

$$-8(a-2) - 8b + 12b = 0$$

$$-8a + 4b = -16$$

$$2a - b = 4 \qquad (3)$$

Since the point $(1, -1, 2)$ lies on (1), we have

$a + 2b - (a + 2) = 0$

$\therefore\ b = 1.$

$\therefore$ (3) gives $a = 5/2$.

Example 18: Find the values of *'a'* and *'b'*, if the surfaces $ax^3 - by^2z = (a+3)x^2$ and $4x^2y - z^3 = 11$ may cut orthogonally at the point $(2, -1, -3)$.

Solution: ***[AU May/June 2016]***

The given surfaces are

$$\varphi_1 = \lambda x^3 - \mu y^2 z - (\lambda + 3)x^2 = 0 \qquad (1)$$

and $$\varphi_2 = 4x^2y - z^3 = 11 \qquad (2)$$

Now, $\nabla\varphi_1 = \frac{\partial\varphi_1}{\partial x}\vec{i} + \frac{\partial\varphi_1}{\partial y}\vec{j} + \frac{\partial\varphi_1}{\partial z}\vec{k} = (3\lambda x^2 - (\lambda+3)2x)\vec{i} + (-2\mu yz)\vec{j} + (-\mu y^2)\vec{k}$.

$\therefore\ \nabla\varphi_1$ at $(2, -1, -3) = \left(12\lambda - 4(\lambda+3)\right)\vec{i} - 6\mu\vec{j} - \mu\vec{k}$

$$= (8\lambda - 12)\vec{i} - 6\mu\vec{j} - \mu\vec{k}\ .$$

Also, $\nabla\varphi_2 = \frac{\partial\varphi_2}{\partial x}\vec{i} + \frac{\partial\varphi_2}{\partial y}\vec{j} + \frac{\partial\varphi_2}{\partial z}\vec{k} = 8xy\vec{i} + 4x^2\vec{j} - 3z^2\vec{k}$.

$\therefore\ \nabla\varphi_2$ at $(2, -1, -3) = -16\vec{i} + 16\vec{j} - 27\vec{k}$.

The two surfaces are orthogonal and so $\nabla\varphi_1 . \nabla\varphi_2 = 0$.

i.e., $\left[(8\lambda - 12)\vec{i} - 6\mu\vec{j} - \mu\vec{k}\right] . \left[-16\vec{i} + 16\vec{j} - 27\vec{k}\right] = 0$

$$-16(8\lambda - 12) - 96\mu + 27\mu = 0$$

$$-128\lambda + 192 - 69\mu = 0$$

$$\text{i.e.,} 128\lambda + 69\mu = 192. \qquad (3)$$

Since the point $(2, -1, -3)$ lies on (1), we have

$$8\lambda + 3\mu - 4(\lambda + 3) = 0.$$

$$\text{i.e., } 4\lambda + 3\mu = 12. \qquad (4)$$

Solving (3) and (4), we have $\mu = \frac{64}{9}$ and $\lambda = \frac{-7}{3}$.

Example 19: Find φ given $\varphi = 2x\vec{i} + 4y\vec{j} + 8z\vec{k}$.

Solution: Given that $\nabla\varphi = 2x\vec{i} + 4y\vec{j} + 8z\vec{k}$

i.e., $\frac{\partial\varphi}{\partial x}\vec{i} + \frac{\partial\varphi}{\partial y}\vec{j} + \frac{\partial\varphi}{\partial z}\vec{k} = 2x\vec{i} + 4y\vec{j} + 8z\vec{k}$

Hence $\frac{\partial\varphi}{\partial x} = 2x$, $\frac{\partial\varphi}{\partial y} = 4y$ and $\frac{\partial\varphi}{\partial z} = 8z$.

Integrating the above equations with respect to the corresponding variables, we get

$\varphi = \int 2x\,dx$ + a function not containing x,

$\varphi = \int 4y\,dy$ + a function not containing y,

and $\varphi = \int 8z\,dz$ + a function not containing z.

i.e., $\varphi = x^2$ + a function not containing x,

$\varphi = 2y^2$ + a function not containing y,

and $\varphi = 4z^2$ + a function not containing z.

∴ The general form of φ is

$\varphi = x^2 + 2y^2 + 4z^2 +$ a function not containing x, y, z.

i.e., $\varphi = x^2 + 2y^2 + 4z^2 + c$.

Note: The general term is obtained by combining all possible terms of φ in which repeated term is taken only once.

Example 20: If $\nabla\varphi = 2xyz^3\vec{i} + x^2z^3\vec{j} + 3x^2yz^2\vec{k}$ find $\varphi(x,y,z)$ given that $\varphi(1,-2,2) = 4$. **[*AU, May/June 2016*]**

Solution: Given $\nabla\varphi = 2xyz^3\vec{i} + x^2z^3\vec{j} + 3x^2yz^2\vec{k}$

i.e., $\frac{\partial\varphi}{\partial x}\vec{i} + \frac{\partial\varphi}{\partial y}\vec{j} + \frac{\partial\varphi}{\partial z}\vec{k} = 2xyz^3\vec{i} + x^2z^3\vec{j} + 3x^2yz^2\vec{k}$

Hence $\frac{\partial\varphi}{\partial x} = 2xyz^3$, $\frac{\partial\varphi}{\partial y} = x^2z^3$ and $\frac{\partial\varphi}{\partial z} = 3x^2yz^2$.

Now, $\frac{\partial\varphi}{\partial x} = 2xyz^3$ gives $\varphi = x^2yz^3 +$ a function not containing x.

$\frac{\partial\varphi}{\partial y} = x^2z^3$ gives $\varphi = x^2yz^3$ + a function not containing y.

$\frac{\partial \varphi}{\partial z} = 3x^2yz^2$ gives $\varphi = x^2yz^3 +$ a function not containing z.

Combining the above solutions, we get

$$\varphi = x^2yz^3 + c \qquad (1)$$

Put $(1,-2,2)$ in (1), $\varphi(1,-2,2) = -16 + c$

i.e., $4 = -16 + c$

$\therefore \quad c = 20.$

$\therefore$ (1) gives $\varphi = x^2yz^3 + 20$.

Example 21: Find the equation of the tangent plane at $(1,0,2)$ to the surface $x^2y + 2xz^2 = 8$.

Solution: Let $\varphi = x^2y + 2xz^2 - 8$.

The vector corresponding to the point $(1,0,2)$ is $\vec{a} = \vec{\imath} + 0\vec{\jmath} + 2\vec{k} = \vec{\imath} + 2\vec{k}$.

Now, $\nabla\varphi = \frac{\partial \varphi}{\partial x}\vec{\imath} + \frac{\partial \varphi}{\partial y}\vec{\jmath} + \frac{\partial \varphi}{\partial z}\vec{k} = (2xy + 2z^2)\vec{\imath} + x^2\vec{\jmath} + 4xz\vec{k}$

$\nabla\varphi$ at $(1,0,2) = 8\vec{\imath} + \vec{\jmath} + 8\vec{k}$.

Let $\vec{r} = x\,\vec{\imath} + y\vec{\jmath} + z\vec{k}$.

Then the equation of the tangent plane is

$$(\vec{r} - \vec{a}) . \nabla\varphi = 0$$

i.e., $[(x\,\vec{\imath} + y\vec{\jmath} + z\vec{k}) - (\vec{\imath} + 2\vec{k})] . [8\vec{\imath} + \vec{\jmath} + 8\vec{k}] = 0$

i.e., $[(x-1)\vec{\imath} + y\vec{\jmath} + (z-2)\vec{k}] . [8\vec{\imath} + \vec{\jmath} + 8\vec{k}] = 0$

i.e., $8(x-1) + y + 8(z-2) = 0$

Hence $8x + y + 8z = 24$.

Example 22: Find the equation of the normal line to the surface $xyz = 4$ at the point (1, 2, 2).

Solution: Let $\varphi = xyz = 4$.

The vector corresponding to the point $(1,2,2)$ is $\vec{a} = \vec{\imath} + 2\vec{\jmath} + 2\vec{k}$.

Now, $\nabla\varphi = \frac{\partial\varphi}{\partial x}\vec{i} + \frac{\partial\varphi}{\partial y}\vec{j} + \frac{\partial\varphi}{\partial z}\vec{k} = yz\vec{i} + xz\vec{j} + xy\vec{k}$.

$\therefore \nabla\varphi$ at $(1,2,2) = 4\vec{i} + 2\vec{j} + 2\vec{k}$.

Let $\vec{r} = x\vec{i} + y\vec{j} + z\vec{k}$.

Then the equation of the normal is

$$(\vec{r} - \vec{a}) \times \nabla\varphi = \vec{0}.$$

i.e., $[(x\vec{i} + y\vec{j} + z\vec{k}) - (\vec{i} + 2\vec{j} + 2\vec{k})] \times [4\vec{i} + 2\vec{j} + 2\vec{k}] = \vec{0}$

i.e. ,$[(x-1)\vec{i} + (y-2)\vec{j} + (z-2)\vec{k}] \times [4\vec{i} + 2\vec{j} + 2\vec{k}] = \vec{0}$

i.e., $\begin{vmatrix} \vec{i} & \vec{j} & \vec{k} \\ x-1 & y-2 & z-2 \\ 4 & 2 & 2 \end{vmatrix} = \vec{0}$

$[2(y-2) - 2(z-2)]\vec{i} - [2(x-1) - 4(z-2)]\vec{j} + [2(x-1) - 4(y-2)]\vec{k} = \vec{0}$

i.e.,$2(y-2) - 2(z-2) = 0, 4(z-2) - 2(x-1) = 0$

and $2(x-1) - 4(y-2) = 0$.

i.e., $y - 2 = z - 2,\ z - 2 = \frac{x-1}{2}$ and $y - 2 = \frac{x-1}{2}$

i.e. ,$\frac{x-1}{2} = y - 2 = z - 2$

i.e., $\frac{x-1}{2} = \frac{y-2}{1} = \frac{z-2}{1}$.

EXERCISES

1. Find *grad* φ at $(1, -1, 2)$ if $\varphi = x^2 - y^2 - z^2 - 2$.
2. Find $\nabla\varphi$ if $\varphi = xyz$.
3. Prove that $\nabla r^3 = 3r\,\vec{r}$.
4. Prove that $\nabla(\vec{a} \cdot \vec{r}) = \vec{a}$ for any constant vector $\vec{a}$.
5. Prove that $\nabla(\log r) = \frac{\vec{r}}{r^2}$.
6. Find $\nabla(e^r)$ where $\vec{r} = x\vec{i} + y\vec{j} + z\vec{k}$.
7. Find the directional derivative of $\varphi = xy + yz + zx$ at (1, 2, 0) in the direction $\vec{i} + 2\vec{j} + 2\vec{k}$.

8. Find the directional derivative of $x^2yz + 4xz^2$ at $(1,-2,-1)$ in the direction of $2\vec{\imath} - \vec{\jmath} - 2\vec{k}$.

9. In what direction from the point $(2,1,-1)$ is the directional derivative of $\varphi = x^3yz^3$ a maximum? What is the magnitude of this maximum?

10. In what direction from the point $(1,3,2)$ is the directional derivative of $\varphi = 2xz - y^2$ a maximum? What is the magnitude of this maximum?

11. Find the unit vector normal to the surface $x^2 + y^2 = z + 10$ at $(1,1,1)$.

12. Find a unit normal vector to the surface $x = t, y = t^2, z = t^3$ at $t = 1$.

13. Find the angle between the normals to the surface $xy = z^2$ at the point $(1,1,1)$ and $(4,1,2)$.

14. Find the constants a and b so that the surfaces $5x^2 - 2yz - 9x = 0$ and $ax^2y + bz^3 = 4$ may cut orthogonally at the point $(1,-1,2)$.

15. Show that the level surfaces of scalars $y^2 + z^2 - x$ and $\log(y^2 + z^2) + 4x$ form orthogonal families.

16. Find φ given $\nabla\varphi = (2xy - z^2)\vec{\imath} + (x^2 + 2yz)\vec{\jmath} + (y^2 - 2zx)\vec{k}$.

17. If $grad\ f = 2xyz\vec{\imath} + x^2z\vec{\jmath} + x^2y\vec{k}$, find f.

18. If $\nabla\varphi = 2xyz^3\vec{\imath} + x^2z^3\vec{\jmath} + 3x^2yz^2\vec{k}$, find φ if $\varphi(-1,2,2) = 4$.

19. Find the equation of the tangent plane to the surface $2xz^2 - 3xy - 4x = 7$ at the point $(1,-1,2)$.

20. Find the equation of the normal to the surface $xz^2 + x^2z = z - 1$ at the point $(1,-3,2)$.

ANSWERS

1. $2\vec{\imath} + 2\vec{\jmath} - 4\vec{k}$ 2. $yz\vec{\imath} + xz\vec{\jmath} - xy\vec{k}$ 6. $\frac{e^r}{r}\vec{r}$ 7. $\frac{10}{3}$ 8. $\frac{37}{3}$

9. $-12\vec{\imath} - 8\vec{\jmath} + 24\vec{k}$; 28 10. $4\vec{\imath} - 6\vec{\jmath} + 2\vec{k}$; $2\sqrt{14}$ 11. $\frac{2\vec{\imath}+2\vec{\jmath}-\vec{k}}{3}$

12. $\frac{1}{\sqrt{3}}(-\vec{\imath} - \vec{\jmath} + \vec{k})$ 13. $\cos^{-1}\left(\frac{13}{\sqrt{6}\sqrt{33}}\right)$

14. $a = 4, b = 1$ 16. $\varphi = x^2y - xz^2 + y^2z + c$

17. $f = x^2yz + c$ 18. $\varphi = x^2yz^3 - 12$

19. $2x + y + z = 6$ 20. $4x + 3z = 25$; $\frac{x-4}{4} = \frac{y}{0} = \frac{z-3}{3}$.

2.5. The Divergence and Curl of a Vector

Divergence

If $\vec{F}$ is a differentiable vector point function defined in some region of space, then the *divergence* of $\vec{F}$ is defined as

$$div\,\vec{F} = \nabla.\vec{F} = \vec{\imath}\,.\frac{\partial\vec{F}}{\partial x} + \vec{\jmath}.\frac{\partial\vec{F}}{\partial y} + \vec{k}\,.\frac{\partial\vec{F}}{\partial z} = \sum\vec{\imath}\,.\frac{\partial\vec{F}}{\partial x}$$

Note: 1. If $\vec{F} = F_1\vec{\imath} + F_2\vec{\jmath} + F_3\vec{k}$, then $\vec{F} = \nabla\,.\,\vec{F} = \frac{\partial F_1}{\partial x} + \frac{\partial F_2}{\partial y} + \frac{\partial F_3}{\partial z}$.

2. The divergence of a vector point function is a scalar point function.

3. $\nabla\,.\,\vec{F}$ is a scalar point function where as $\vec{F}\,.\nabla$ is an operator.

Physical Interpretation of Divergence

Consider a fluid having density $\rho = \rho(x,y,z,t)$ and velocity $\vec{v} = \vec{v}(x,y,z,t)$ at a point (x,y,z) at time t. Let $\vec{F} = \rho\vec{v}$, then $\vec{F}$ is a vector having the same direction as $\vec{v}$ and magnitude $\rho|\vec{v}|$. It is known as *flux*. Its direction gives the direction of the fluid flow, and its magnitude gives the mass of the fluid crossing per unit time a unit area placed perpendicular to the direction of flow.

Consider the motion of the fluid having velocity $\vec{F} = F_x\vec{\imath} + F_y\vec{\jmath} + F_z\vec{k}$ at a point $P(x,y,z)$. Consider a small parallelepiped with edges $\delta_x, \delta_y, \delta_z$ parallel to the axis with one of its corners at P. The mass of the fluid entering through the face G_1 per unit time is $F_y\delta_x\delta_z$ and that flowing out through the opposite face G_2 is

$$F_{y+\delta y}\,\delta_x\delta_z = \left(F_y + \frac{\partial F_y}{\partial y}\delta_y\right)\delta_x\delta_z \quad \text{by using Taylor's series.}$$

Therefore the net decrease in the mass of fluid flowing across these two faces is $\left(F_y + \frac{\partial F_y}{\partial y}\delta_y\right)\delta_x\delta_z - F_y\delta_x\delta_z = \frac{\partial F_y}{\partial y}\,\delta_x\delta_y\delta_z$. Similarly, considering the other two pairs of faces, we get the total decrease in the mass of fluid inside the parallelepiped per unit time $= \left(\frac{\partial F_x}{\partial x} + \frac{\partial F_y}{\partial y} + \frac{\partial F_z}{\partial z}\right)\delta_x\,\delta_y\,\delta_z$. Dividing this by the volume $\delta_x\,\delta_y\,\delta_z$ of the parallelepiped, we have the rate of loss of fluid per unit time $= \frac{\partial F_x}{\partial x} + \frac{\partial F_y}{\partial y} + \frac{\partial F_z}{\partial z} = \text{div}\,\vec{F}$.

Hence div $\vec{F}$ gives the rate of outflow per unit volume at a point of the fluid.

Solenoidal Vector

If $\vec{F}$ is a vector such that $\nabla . \vec{F} = 0$ at all points in a given region, then it is said to be a *solenoidal vector* in that region.

Curl of a Vector

If $\vec{F}$ is a differentiable vector point function defined in some region of space, then the *curl* of $\vec{F}$ is defined as $\vec{F} = \nabla \times \vec{F} = \sum \vec{i} \times \frac{\partial \vec{F}}{\partial x}$.

Note: 1.If $\vec{F} = F_1\vec{i} + F_2\vec{j} + F_3\vec{k}$, then $curl\, \vec{F} = \nabla \times \vec{F} = \begin{vmatrix} \vec{i} & \vec{j} & \vec{k} \\ \frac{\partial}{\partial x} & \frac{\partial}{\partial y} & \frac{\partial}{\partial z} \\ F_1 & F_2 & F_3 \end{vmatrix}$.

2. The curl of a vector point function is a vector point function.

Physical Interpretation of Curl

Consider a rigid body rotating about a fixed axis through O with uniform angular velocity

Irrotational Vector

If $\vec{F}$ is a vector such that $\nabla \times \vec{F} = \vec{0}$ at all points in a given region, then it is said to be an *irrotational vector* in that region.

Scalar Potential of an Irrotational Vector

If $\vec{F}$ is irrotational, then it can be expressed as the gradient of a scalar point function. Let $\vec{F} = F_1\vec{i} + F_2\vec{j} + F_3\vec{k}$. Since $\vec{F}$ is irrotational, $\text{curl}\,\vec{F} = \vec{0}$.

i.e. $\begin{vmatrix} \vec{i} & \vec{j} & \vec{k} \\ \frac{\partial}{\partial x} & \frac{\partial}{\partial y} & \frac{\partial}{\partial z} \\ F_1 & F_2 & F_3 \end{vmatrix} = \vec{0}.$

i.e. $\left(\frac{\partial F_3}{\partial y} - \frac{\partial F_2}{\partial z}\right)\vec{i} - \left(\frac{\partial F_3}{\partial x} - \frac{\partial F_1}{\partial z}\right)\vec{j} + \left(\frac{\partial F_2}{\partial x} - \frac{\partial F_1}{\partial y}\right)\vec{k} = \vec{0}$

$$\therefore \frac{\partial F_3}{\partial y} = \frac{\partial F_2}{\partial z}\ ;\ \frac{\partial F_3}{\partial x} = \frac{\partial F_1}{\partial z}\ ;\ \frac{\partial F_2}{\partial x} = \frac{\partial F_1}{\partial y} \qquad (1)$$

Equations in (1) are satisfied when

$$F_1 = \frac{\partial \varphi}{\partial x},\ F_2 = \frac{\partial \varphi}{\partial y},\ F_3 = \frac{\partial \varphi}{\partial z}$$

$\therefore \qquad \vec{F} = \frac{\partial \varphi}{\partial x}\vec{i} + \frac{\partial \varphi}{\partial y}\vec{j} + \frac{\partial \varphi}{\partial z}\vec{k} = \nabla\varphi.$

If $\vec{F}$ is irrotational and $\vec{F} = \nabla\varphi$, then φ is called the *scalar potential* of $\vec{F}$ and $\vec{F}$ is called a *conservative force.*

Properties of Divergence and Curl

Property 1: $(\vec{f} \pm \vec{g}) = div\,\vec{f} \pm div\,\vec{g}$.

Proof: $div\,(\vec{f} \pm \vec{g}) = \nabla.(\vec{f} \pm \vec{g})$

$$= \vec{i}\,.\,\frac{\partial}{\partial x}(\vec{f} \pm \vec{g}) + \vec{j}\,.\,\frac{\partial}{\partial y}(\vec{f} \pm \vec{g}) + \vec{k}\,.\,\frac{\partial}{\partial z}(\vec{f} \pm \vec{g})$$

$$= \vec{i}\,.\left(\frac{\partial \vec{f}}{\partial x} \pm \frac{\partial \vec{g}}{\partial x}\right) + \vec{j}\,.\left(\frac{\partial \vec{f}}{\partial y} \pm \frac{\partial \vec{g}}{\partial y}\right) + \vec{k}\,.\left(\frac{\partial \vec{f}}{\partial z} \pm \frac{\partial \vec{g}}{\partial z}\right)$$

$$= \left(\vec{i}\,.\,\frac{\partial \vec{f}}{\partial x} + \vec{j}\,.\,\frac{\partial \vec{f}}{\partial y} + \vec{k}\,.\,\frac{\partial \vec{f}}{\partial z}\right) \pm \left(\vec{i}\,.\,\frac{\partial \vec{g}}{\partial x} + \vec{j}\,.\,\frac{\partial \vec{g}}{\partial y} + \vec{k}\,.\,\frac{\partial \vec{g}}{\partial z}\right)$$

$$= \nabla.\vec{f} \pm \nabla.\vec{g} = div\,\vec{f} \pm div\,\vec{g}\,.$$

Property 2: Let $\vec{f}$ be a vector point function and φ a scalar point function.

Then $div\,(\varphi\,\vec{f}) = (grad\,\varphi)\,.\vec{f} + \varphi(div\,\vec{f})$.

Proof: Let $\vec{f} = f_1\vec{i} + f_2\vec{j} + f_3\vec{k}$.

$$div\,(\varphi\,\vec{f}) = \nabla.(\varphi\,\vec{f}) = \Sigma\vec{i}\,.\,\frac{\partial}{\partial x}(\varphi\vec{f})$$

$$= \Sigma\vec{i}\,.\left[\varphi\,\frac{\partial \vec{f}}{\partial x} + \frac{\partial \varphi}{\partial x}\vec{f}\right]$$

$$= \Sigma\vec{i}\,.\left(\varphi\,\frac{\partial \vec{f}}{\partial x}\right) + \Sigma\vec{i}\,.\left(\frac{\partial \varphi}{\partial x}\vec{f}\right)$$

$$= \varphi\,\Sigma\vec{i}\,.\,\frac{\partial \vec{f}}{\partial x} + \left(\Sigma\frac{\partial \varphi}{\partial x}\vec{i}\right).\,\vec{f}$$

$$= \varphi\,(div\,\vec{f}) + (grad\,\varphi)\,.\vec{f}$$

Property 3: $div\,(\vec{f} \times \vec{g}) = \vec{g}\,.\,curl\,\vec{f} - \vec{f}\,.curl\,\vec{g}$.

Proof: $div(\vec{f} \times \vec{g}) = \nabla.(\vec{f} \times \vec{g})$

$$= \Sigma\vec{i}\,.\,\frac{\partial}{\partial x}(\vec{f} \times \vec{g})$$

$$= \Sigma\vec{i}\,.\left(\frac{\partial \vec{f}}{\partial x} \times \vec{g} + \vec{f} \times \frac{\partial \vec{g}}{\partial x}\right)$$

$$= \sum \vec{i} \,.\left(\frac{\partial \vec{f}}{\partial x} \times \vec{g}\right) + \sum \vec{i} \,.\left(\vec{f} \times \frac{\partial \vec{g}}{\partial x}\right)$$

$$= \sum \vec{i} \,.\left(\frac{\partial \vec{f}}{\partial x} \times \vec{g}\right) - \sum \vec{i} \,.\left(\frac{\partial \vec{g}}{\partial x} \times \vec{f}\right) \qquad [\because\ \vec{a} \times \vec{b} = -\vec{b} \times \vec{a}]$$

$$= \sum \left(\vec{i} \times \frac{\partial \vec{f}}{\partial x}\right) . \vec{g} - \sum \left(\vec{i} \times \frac{\partial \vec{g}}{\partial x}\right) . \vec{f} \quad [\because\ \vec{a} \,.(\vec{b} \times \vec{c}) = (\vec{a} \times \vec{b}) \,. \vec{c}]$$

$$= (\nabla \times \vec{f}) \,. \vec{g} - (\nabla \times \vec{g}) \,. \vec{f}$$

$$= curl\, \vec{f} \,. \vec{g} - curl\, \vec{g} \,. \vec{f}$$

$$= \vec{g} \,.\, curl\, \vec{f} - \vec{f} \,. curl\, \vec{g} \,.$$

Property 4: $div\, grad\, \varphi = \nabla^2 \varphi$.

Proof: $\nabla \varphi = \frac{\partial \varphi}{\partial x} \vec{i} + \frac{\partial \varphi}{\partial y} \vec{j} + \frac{\partial \varphi}{\partial z} \vec{k}$.

$$div\, grad\, \varphi = \nabla \,.\, \nabla \varphi$$

$$= \nabla . \left(\frac{\partial \varphi}{\partial x} \vec{i} + \frac{\partial \varphi}{\partial y} \vec{j} + \frac{\partial \varphi}{\partial z} \vec{k}\right)$$

$$= \frac{\partial}{\partial x}\left(\frac{\partial \varphi}{\partial x}\right) + \frac{\partial}{\partial y}\left(\frac{\partial \varphi}{\partial y}\right) + \frac{\partial}{\partial z}\left(\frac{\partial \varphi}{\partial z}\right) = \frac{\partial^2 \varphi}{\partial x^2} + \frac{\partial^2 \varphi}{\partial y^2} + \frac{\partial^2 \varphi}{\partial z^2}$$

$$= \nabla^2 \varphi. \qquad \left[\because\ \nabla^2 = \frac{\partial^2}{\partial x^2} + \frac{\partial^2}{\partial y^2} + \frac{\partial^2}{\partial z^2}\right]$$

Property 5: $(\vec{f} \pm \vec{g}) = curl\, \vec{f} \pm curl\, \vec{g}$.

Proof: $curl\, (\vec{f} \pm \vec{g}) = \nabla \times (\vec{f} \pm \vec{g})$

$$= \sum \vec{i} \times \frac{\partial}{\partial x}(\vec{f} \pm \vec{g})$$

$$= \sum \vec{i} \times \frac{\partial \vec{f}}{\partial x} \pm \sum \vec{i} \times \frac{\partial \vec{g}}{\partial x}$$

$$= \nabla \times \vec{f} \pm \nabla \times \vec{g}$$

$$= curl\, \vec{f} \pm curl\, \vec{g} \,.$$

Property 6: $curl\, (\vec{f} \times \vec{g}) = (\vec{g} \,.\, \nabla)\vec{f} + \vec{f}\, div\, \vec{g} - (\vec{f} \,.\, \nabla)\vec{g} - \vec{g}\, div\, \vec{f}$.

Proof: $curl\, (\vec{f} \times \vec{g}) = \nabla \times (\vec{f} \times \vec{g})$

$$= \sum \vec{i} \times \frac{\partial}{\partial x}(\vec{f} \times \vec{g})$$

$$= \sum \vec{i} \times \left[\vec{f} \times \frac{\partial \vec{g}}{\partial x} + \frac{\partial \vec{f}}{\partial x} \times \vec{g}\right]$$

$$= \sum \vec{i} \times \left(\vec{f} \times \frac{\partial \vec{g}}{\partial x}\right) + \sum \vec{i} \times \left(\frac{\partial \vec{f}}{\partial x} \times \vec{g}\right)$$

$$= \sum \left[\left(\vec{i} \cdot \frac{\partial \vec{g}}{\partial x}\right)\vec{f} - (\vec{i} \cdot \vec{f})\frac{\partial \vec{g}}{\partial x}\right] + \sum \left[(\vec{i} \cdot \vec{g})\frac{\partial \vec{f}}{\partial x} - \left(\vec{i} \cdot \frac{\partial \vec{f}}{\partial x}\right)\vec{g}\right]$$

$$= \vec{f}\left(\sum \vec{i} \cdot \frac{\partial \vec{g}}{\partial x}\right) - \vec{g} \sum \left(\vec{i} \cdot \frac{\partial \vec{f}}{\partial x}\right) + \sum (\vec{i} \cdot \vec{g})\frac{\partial \vec{f}}{\partial x} - \sum (\vec{i} \cdot \vec{f})\frac{\partial \vec{g}}{\partial x}$$

$$= \vec{f}\, div\, \vec{g} - \vec{g}\, div\, \vec{f} + (\vec{g} \cdot \nabla)\vec{f} - (\vec{f} \cdot \nabla)\vec{g}.$$

Property 7: $div\,(curl\,\vec{f}) = \nabla \cdot (\nabla \times \vec{f}) = 0.$

Proof : Let $\vec{f} = f_1\vec{i} + f_2\vec{j} + f_3\vec{k}$.

Then $curl\vec{f} = \nabla \times \vec{f} = \begin{vmatrix} \vec{i} & \vec{j} & \vec{k} \\ \frac{\partial}{\partial x} & \frac{\partial}{\partial y} & \frac{\partial}{\partial z} \\ f_1 & f_2 & f_3 \end{vmatrix}$

$$= \left(\frac{\partial f_3}{\partial y} - \frac{\partial f_2}{\partial z}\right)\vec{i} - \left(\frac{\partial f_3}{\partial x} - \frac{\partial f_1}{\partial z}\right)\vec{j} + \left(\frac{\partial f_2}{\partial x} - \frac{\partial f_1}{\partial y}\right)\vec{k}$$

$$\therefore div\; curl\,\vec{f} = \nabla \cdot (\nabla \times \vec{f}) = \frac{\partial}{\partial x}\left(\frac{\partial f_3}{\partial y} - \frac{\partial f_2}{\partial z}\right) - \frac{\partial}{\partial y}\left(\frac{\partial f_3}{\partial x} - \frac{\partial f_1}{\partial z}\right) + \frac{\partial}{\partial z}\left(\frac{\partial f_2}{\partial x} - \frac{\partial f_1}{\partial y}\right)$$

$$= \frac{\partial^2 f_3}{\partial x \partial y} - \frac{\partial^2 f_2}{\partial x \partial z} - \frac{\partial^2 f_3}{\partial y \partial x} + \frac{\partial^2 f_1}{\partial y \partial z} + \frac{\partial^2 f_2}{\partial z \partial x} - \frac{\partial^2 f_1}{\partial z \partial y}$$

$$= \mathbf{0}.$$

Property 8: $rad\, \varphi = \nabla \times (\nabla\varphi) = \vec{0}$.

Proof: $grad\, \varphi = \nabla\varphi = \frac{\partial \varphi}{\partial x}\vec{i} + \frac{\partial \varphi}{\partial y}\vec{j} + \frac{\partial \varphi}{\partial z}\vec{k}$

$$\therefore curl\; grad\, \varphi = \nabla \times (\nabla\varphi) = \begin{vmatrix} \vec{i} & \vec{j} & \vec{k} \\ \frac{\partial}{\partial x} & \frac{\partial}{\partial y} & \frac{\partial}{\partial z} \\ \frac{\partial \varphi}{\partial x} & \frac{\partial \varphi}{\partial y} & \frac{\partial \varphi}{\partial z} \end{vmatrix}$$

$$= \vec{i}\left(\frac{\partial^2 \varphi}{\partial y \partial z} - \frac{\partial^2 \varphi}{\partial z \partial y}\right) - \vec{j}\left(\frac{\partial^2 \varphi}{\partial x \partial z} - \frac{\partial^2 \varphi}{\partial z \partial x}\right) + \vec{k}\left(\frac{\partial^2 \varphi}{\partial x \partial y} - \frac{\partial^2 \varphi}{\partial y \partial x}\right)$$

$$= 0\vec{i} - 0\vec{j} + 0\vec{k}$$

$$= \vec{0}\,.$$

Property 9: $grad\,(\vec{f}\,.\,\vec{g}) = \vec{f} \times curl\,\vec{g} + \vec{g} \times curl\,\vec{f} + (\vec{f}\,.\,\nabla)\vec{g} + (\vec{g}\,.\,\nabla)\vec{f}$

Proof: $grad\,(\vec{f}\,.\,\vec{g}) = \sum \vec{\imath} \frac{\partial}{\partial x}(\vec{f}\,.\,\vec{g})$

$$= \sum \vec{\imath}\left(\vec{f}\,.\frac{\partial \vec{g}}{\partial x} + \frac{\partial \vec{f}}{\partial x}\,.\,\vec{g}\right)$$

$$= \sum\left(\vec{f}\,.\frac{\partial \vec{g}}{\partial x}\right)\vec{\imath} + \sum\left(\frac{\partial \vec{f}}{\partial x}\,.\,\vec{g}\right)\vec{\imath} \qquad (1)$$

Now, $\vec{f} \times \left(\vec{\imath} \times \frac{\partial \vec{g}}{\partial x}\right) = \left(\vec{f}\,.\frac{\partial \vec{g}}{\partial x}\right)\vec{\imath} - (\vec{f}\,.\vec{\imath})\,\frac{\partial \vec{g}}{\partial x}$

$\therefore \left(\vec{f}\,.\frac{\partial \vec{g}}{\partial x}\right)\vec{\imath} = \vec{f} \times \left(\vec{\imath} \times \frac{\partial \vec{g}}{\partial x}\right) + (\vec{f}\,.\vec{\imath})\,\frac{\partial \vec{g}}{\partial x}$

$\therefore \sum\left(\vec{f}\,.\frac{\partial \vec{g}}{\partial x}\right)\vec{\imath} = \sum \vec{f} \times \left(\vec{\imath} \times \frac{\partial \vec{g}}{\partial x}\right) + \sum(\vec{f}\,.\vec{\imath})\,\frac{\partial \vec{g}}{\partial x}$

$$= \vec{f} \times \left(\sum \vec{\imath} \times \frac{\partial \vec{g}}{\partial x}\right) + \sum \vec{f}\,.\,\vec{\imath}\,\frac{\partial \vec{g}}{\partial x}$$

$$= \vec{f} \times \left(\sum \vec{\imath} \times \frac{\partial \vec{g}}{\partial x}\right) + \left(\vec{f}\,.\,\sum \vec{\imath}\frac{\partial}{\partial x}\right)\vec{g}$$

$$= \vec{f} \times curl\,\vec{g} + (\vec{f}\,.\,\nabla)\vec{g} \qquad (2)$$

Similarly, $\sum\left(\frac{\partial \vec{f}}{\partial x}\,.\,\vec{g}\right)\vec{\imath} = \vec{g} \times curl\,\vec{f} + (\vec{g}\,.\,\nabla)\,\vec{f} \qquad (3)$

Substituting (2) and (3) in (1), we have

$grad\,(\vec{f}\,.\,\vec{g}) = \vec{f} \times curl\,\vec{g} + \vec{g} \times curl\,\vec{f} + (\vec{f}\,.\,\nabla)\vec{g} + (\vec{g}\,.\,\nabla)\vec{f}$.

Property 10: $curl(\varphi\vec{f}) = grad\,\varphi \times \vec{f} + \varphi\, curl\,\vec{f}$.

Proof: $curl(\varphi\vec{f}) = \nabla \times (\varphi\vec{f})$

$$= \sum \vec{\imath} \times \frac{\partial}{\partial x}\,(\varphi\vec{f})$$

$$= \sum\left[\vec{\imath} \times \left(\frac{\partial \varphi}{\partial x}\vec{f} + \varphi\,\frac{\partial \vec{f}}{\partial x}\right)\right]$$

$$= \left(\sum \frac{\partial \varphi}{\partial x}\vec{\imath}\right) \times \vec{f} + \varphi\left(\sum \vec{\imath} \times \frac{\partial \vec{f}}{\partial x}\right)$$

$$= \nabla\varphi \times \vec{f} + \varphi\,(\nabla \times \vec{f})$$

$$= grad\,\varphi \times \vec{f} + \varphi\, curl\,\vec{f}\,.$$

Property 11: $curl\left(curl\,\vec{f}\right) = grad\,div\,\vec{f} - \nabla^2\vec{f}$. **[AU, May/June 2016]**

Proof : Let $\vec{f} = f_1\vec{i} + f_2\vec{j} + f_3\vec{k}$.

$$\text{Then } curl\,\vec{f} = \nabla\times\vec{f} = \begin{vmatrix} \vec{i} & \vec{j} & \vec{k} \\ \frac{\partial}{\partial x} & \frac{\partial}{\partial y} & \frac{\partial}{\partial z} \\ f_1 & f_2 & f_3 \end{vmatrix}$$

$$= \left(\frac{\partial f_3}{\partial y} - \frac{\partial f_2}{\partial z}\right)\vec{i} - \left(\frac{\partial f_3}{\partial x} - \frac{\partial f_1}{\partial z}\right)\vec{j} + \left(\frac{\partial f_2}{\partial x} - \frac{\partial f_1}{\partial y}\right)\vec{k}$$

$$\therefore\ curl\,curl\,\vec{f} = \nabla\times\left(\nabla\times\vec{f}\right)$$

$$= \begin{vmatrix} \vec{i} & \vec{j} & \vec{k} \\ \frac{\partial}{\partial x} & \frac{\partial}{\partial y} & \frac{\partial}{\partial z} \\ \left(\frac{\partial f_3}{\partial y} - \frac{\partial f_2}{\partial z}\right) & \left(\frac{\partial f_1}{\partial z} - \frac{\partial f_3}{\partial x}\right) & \left(\frac{\partial f_2}{\partial x} - \frac{\partial f_1}{\partial y}\right) \end{vmatrix}$$

$$= \Sigma\left[\frac{\partial}{\partial y}\left(\frac{\partial f_2}{\partial x} - \frac{\partial f_1}{\partial y}\right) - \frac{\partial}{\partial z}\left(\frac{\partial f_1}{\partial z} - \frac{\partial f_3}{\partial x}\right)\right]\vec{i}$$

$$= \Sigma\left[\frac{\partial^2 f_2}{\partial y\partial x} - \frac{\partial^2 f_1}{\partial y^2} - \frac{\partial^2 f_1}{\partial z^2} + \frac{\partial^2 f_3}{\partial z\partial x}\right]\vec{i}$$

$$= \Sigma\left[\left(\frac{\partial^2 f_2}{\partial y\partial x} + \frac{\partial^2 f_3}{\partial z\partial x}\right) - \left(\frac{\partial^2 f_1}{\partial y^2} + \frac{\partial^2 f_1}{\partial z^2}\right)\right]\vec{i}$$

$$= \Sigma\left[\frac{\partial}{\partial x}\left(\frac{\partial f_2}{\partial y} + \frac{\partial f_3}{\partial z}\right) - \left(\frac{\partial^2 f_1}{\partial y^2} + \frac{\partial^2 f_1}{\partial z^2}\right)\right]\vec{i}$$

$$= \Sigma\left[\frac{\partial}{\partial x}\left(\frac{\partial f_1}{\partial x} + \frac{\partial f_2}{\partial y} + \frac{\partial f_3}{\partial z}\right) - \left(\frac{\partial^2 f_1}{\partial x^2} + \frac{\partial^2 f_1}{\partial y^2} + \frac{\partial^2 f_1}{\partial z^2}\right)\right]\vec{i}$$

$$= \Sigma\left[\frac{\partial}{\partial x}\left(\nabla\,.\,\vec{f}\right) - \nabla^2 f_1\right]\vec{i} = \Sigma\frac{\partial}{\partial x}\left(\nabla\,.\,\vec{f}\right)\vec{i} - \Sigma\left(\nabla^2 f_1\right)\vec{i}$$

$$= \nabla\left(\nabla\,.\,\vec{f}\right) - \nabla^2\vec{f} = grad\left(div\,\vec{f}\right) - \nabla^2\vec{f}.$$

Example 1: Find curl$\vec{F}$ if $\vec{F} = yz\vec{i} + zx\vec{j} + xy\vec{k}$. **[AU Nov/Dec 2014]**

Solution : $curl\,\vec{F} = \nabla\times\vec{F}$

$$= \begin{vmatrix} \vec{i} & \vec{j} & \vec{k} \\ \frac{\partial}{\partial x} & \frac{\partial}{\partial y} & \frac{\partial}{\partial z} \\ yz & zx & xy \end{vmatrix} = \vec{i}(x-x) - \vec{j}(y-y) + \vec{k}(z-z) = 0\vec{i} - 0\vec{j} + 0\vec{k} = \vec{0}.$$

Example 2: Find divergence and curl of $3\,x^2\,\vec{\imath} + 5\,x\,y^2\,\vec{\jmath} + x\,y\,z^3\,\vec{k}$ at the point $(1,2,3)$.

Solution: Let $\vec{f} = 3x^2\vec{\imath} + 5xy^2\vec{\jmath} + xyz^3\vec{k}$.

(i) $$div\,\vec{f} = \nabla\,.\,\vec{f} = \frac{\partial f_1}{\partial x} + \frac{\partial f_2}{\partial y} + \frac{\partial f_3}{\partial z}$$

$$= 6x + 10xy + 3xyz^2.$$

$\therefore$ $div\,\vec{f}$ at $(1,2,3) = 6 + 20 + 54 = 80.$

(ii) $$curl\,\vec{f} = \nabla \times \vec{f} = \begin{vmatrix} \vec{\imath} & \vec{\jmath} & \vec{k} \\ \frac{\partial}{\partial x} & \frac{\partial}{\partial y} & \frac{\partial}{\partial z} \\ 3x^2 & 5xy^2 & xyz^3 \end{vmatrix}$$

$$= \vec{\imath}(xz^3 - 0) - \vec{\jmath}(yz^3 - 0) + \vec{k}(5y^2 - 0)$$

$$= x\,z^3\vec{\imath} - y\,z^3\vec{\jmath} + 5\,y^2\vec{k}$$

$\therefore$ $curl\,\vec{f}$ at $(1,2,3) = 27\vec{\imath} - 54\vec{\jmath} + 20\vec{k}.$

Example 3: Find $div\ grad\varphi$ if $\varphi = xy + yz + zx$.

Solution : $$grad\,\varphi = \frac{\partial\varphi}{\partial x}\vec{\imath} + \frac{\partial\varphi}{\partial y}\vec{\jmath} + \frac{\partial\varphi}{\partial z}\vec{k}$$

$$= (y+z)\vec{\imath} + (x+z)\vec{\jmath} + (y+x)\vec{k}$$

$$\therefore\ div\ grad\,\varphi = \frac{\partial}{\partial x}(y+z) + \frac{\partial}{\partial y}(x+z) + \frac{\partial}{\partial z}(y+x) = 0.$$

Example 4: When $\varphi = x^3 + y^3 + z^3 - 3xyz$, find $\nabla\,\varphi$, $\nabla\,.\,\nabla\,\varphi$ and $\nabla \times \nabla\,\varphi$ at $(1,2,3)$.

Solution : $$\nabla\varphi = \frac{\partial\varphi}{\partial x}\vec{\imath} + \frac{\partial\varphi}{\partial y}\vec{\jmath} + \frac{\partial\varphi}{\partial z}\vec{k}$$

$$= (3x^2 - 3yz)\vec{\imath} + (3y^2 - 3xz)\vec{\jmath} + (3z^2 - 3xy)\vec{k}$$

$\therefore$ $\nabla\varphi$ at $(1,2,3) = -15\vec{\imath} + 3\vec{\jmath} + 21\vec{k}$.

Also, $\nabla\,.\,\nabla\,\varphi = \frac{\partial}{\partial x}(3x^2 - 3yz) + \frac{\partial}{\partial y}(3y^2 - 3xz) + \frac{\partial}{\partial z}(3z^2 - 3xy) = 6x + 6y + 6z$.

$\therefore$ $(\nabla\,.\,\nabla\,\varphi)$ at $(1,2,3) = 6 + 12 + 18 = 36.$

We have $$\nabla \times \nabla\,\varphi = \begin{vmatrix} \vec{\imath} & \vec{\jmath} & \vec{k} \\ \frac{\partial}{\partial x} & \frac{\partial}{\partial y} & \frac{\partial}{\partial z} \\ 3x^2 - 3yz & 3y^2 - 3xz & 3z^2 - 3xy \end{vmatrix}$$

$$= \vec{\imath}(-3x + 3x) - \vec{\jmath}(-3y + 3y) + \vec{k}(-3z + 3z) = \vec{0}\,.$$

$\therefore$ $(\nabla \times \nabla\,\varphi)$ at $(1,2,3) = \vec{0}.$

Example 5: Find $div\,\vec{r}$ and $curl\,\vec{r}$.

Solution : Let $\vec{r} = x\vec{i} + y\vec{j} + z\vec{k}$.

Hence $div\,\vec{r} = \frac{\partial}{\partial x}(x) + \frac{\partial}{\partial y}(y) + \frac{\partial}{\partial z}(z) = 1 + 1 + 1 = 3.$

$$\text{Also, } curl\,\vec{r} = \begin{vmatrix} \vec{i} & \vec{j} & \vec{k} \\ \frac{\partial}{\partial x} & \frac{\partial}{\partial y} & \frac{\partial}{\partial z} \\ x & y & z \end{vmatrix}$$

$$= \vec{i}(0-0) - \vec{j}(0-0) + \vec{k}(0-0) = \vec{0}.$$

Example 6: If $\vec{F} = x^3\vec{i} + y^3\vec{j} + z^3\vec{k}$, find $div\,curl\,\vec{F}$

$$\textbf{Solution :}\quad curl\,\vec{F} = \begin{vmatrix} \vec{i} & \vec{j} & \vec{k} \\ \frac{\partial}{\partial x} & \frac{\partial}{\partial y} & \frac{\partial}{\partial z} \\ x^3 & y^3 & z^3 \end{vmatrix}$$

$$= \vec{i}(0-0) - \vec{j}(0-0) + \vec{k}(0-0) = \vec{0}.$$

$$\therefore div\,curl\,\vec{F} = \nabla \,.\, \vec{0} = 0.$$

Example 7: Find $div\,\vec{F}$ and $curl\,\vec{F}$, where $\vec{F} = grad\,(x^3 + y^3 + z^3 - 3xyz)$.

Solution:

Give that $\vec{F} = grad\,(x^3 + y^3 + z^3 - 3xyz)$

$$= (3x^2 - 3yz)\vec{i} + (3y^2 - 3xz)\vec{j} + (3z^2 - 3xy)\vec{k}$$

$$\therefore div\,\vec{F} = \nabla \,.\, \vec{F} = \frac{\partial}{\partial x}(3x^2 - 3yz) + \frac{\partial}{\partial y}(3y^2 - 3xz) + \frac{\partial}{\partial z}(3z^2 - 3xy)$$

$$= 6x + 6y + 6z = 6\,(x + y + z),$$

$$curl\,\vec{F} = \nabla \times \vec{F} = \begin{vmatrix} \vec{i} & \vec{j} & \vec{k} \\ \frac{\partial}{\partial x} & \frac{\partial}{\partial y} & \frac{\partial}{\partial z} \\ 3x^2 - 3yz & 3y^2 - 3xz & 3z^2 - 3xy \end{vmatrix}$$

$$= \vec{i}(-3x + 3x) - \vec{j}(-3y + 3y) + \vec{k}(-3z + 3z)$$

$$= 0\vec{i} + 0\vec{j} + 0\vec{k} = \vec{0}.$$

Example 8: Find *curl curl* $\vec{f}$ at the point (1, 1, 1) if $\vec{f} = x^2y\vec{\imath} + xz\vec{\jmath} + 2yz\vec{k}$.

Solution: $$curl\,\vec{f} = \nabla \times \vec{f} = \begin{vmatrix} \vec{\imath} & \vec{\jmath} & \vec{k} \\ \frac{\partial}{\partial x} & \frac{\partial}{\partial y} & \frac{\partial}{\partial z} \\ x^2y & xz & 2yz \end{vmatrix}$$

$$= \vec{\imath}\,(2z - x) - \vec{\jmath}(0 - 0) + \vec{k}\,(z - x^2)$$

$$= (2z - x)\vec{\imath} - 0\vec{\jmath} + (z - x^2)\vec{k}.$$

$$\therefore\ curl\,curl\,\vec{f} = \nabla \times (\nabla \times \vec{f}) = \begin{vmatrix} \vec{\imath} & \vec{\jmath} & \vec{k} \\ \frac{\partial}{\partial x} & \frac{\partial}{\partial y} & \frac{\partial}{\partial z} \\ 2z - x & 0 & z - x^2 \end{vmatrix}$$

$$= \vec{\imath}\,(0 - 0) - \vec{\jmath}\,(-2x - 2) + \vec{k}\,(0 - 0)$$

$$= 2(x + 1)\vec{\jmath}.$$

$\therefore$ *curl curl* $\vec{f}$ at $(1, 1, 1) = 4\vec{\jmath}$.

Example 9: If $\vec{a}$ is a constant vector, then prove that

(i) $div\,(\vec{r} \times \vec{a}) = 0$

(ii) $curl\,(\vec{r} \times \vec{a}) = -2\vec{a}$.

Solution: Let $\vec{a} = a_1\vec{\imath} + a_2\vec{\jmath} + a_3\vec{k}$ and $\vec{r} = x\vec{\imath} + y\vec{\jmath} + z\vec{k}$.

Then $$\vec{r} \times \vec{a} = \begin{vmatrix} \vec{\imath} & \vec{\jmath} & \vec{k} \\ x & y & z \\ a_1 & a_2 & a_3 \end{vmatrix}$$

$$= \vec{\imath}(ya_3 - za_2) - \vec{\jmath}(xa_3 - a_1z) + \vec{k}(xa_2 - ya_1).$$

(i) $$div\,(\vec{r} \times \vec{a}) = \frac{\partial}{\partial x}(ya_3 - za_2) - \frac{\partial}{\partial y}(xa_3 - a_1z) + \frac{\partial}{\partial z}(xa_2 - ya_1) = 0.$$

(ii) $$curl\,(\vec{r} \times \vec{a}) = \begin{vmatrix} \vec{\imath} & \vec{\jmath} & \vec{k} \\ \frac{\partial}{\partial x} & \frac{\partial}{\partial y} & \frac{\partial}{\partial z} \\ ya_3 - za_2 & -(xa_3 - a_1z) & xa_2 - ya_1 \end{vmatrix}$$

$$= \vec{\imath}(-a_1 - a_1) - \vec{\jmath}(a_2 + a_2) + \vec{k}(-a_3 - a_3)$$

$$= -2a_1\vec{\imath} - 2a_2\vec{\jmath} - 2a_3\vec{k}$$

$$= -2(a_1\vec{\imath} + a_2\vec{\jmath} + a_3\vec{k}) = -2\vec{a}.$$

Example 10 : If $\vec{v} = \vec{w} \times \vec{r}$, prove that $\vec{w} = \frac{1}{2}\, curl\, \vec{v}$, where $\vec{w}$ is a constant vector.

Proof : Let $\vec{w} = w_1\vec{i} + w_2\vec{j} + w_3\vec{k}$ and $\vec{r} = x\vec{i} + y\vec{j} + z\vec{k}$.

$$\text{Now, } \vec{v} = \vec{w} \times \vec{r} = \begin{vmatrix} \vec{i} & \vec{j} & \vec{k} \\ w_1 & w_2 & w_3 \\ x & y & z \end{vmatrix} = \vec{i}(w_2z - w_3y) - \vec{j}(w_1z - w_3x) + \vec{k}(w_1y - w_2x)$$

$$\text{Hence } curl\, \vec{v} = \nabla \times \vec{v} = \begin{vmatrix} \vec{i} & \vec{j} & \vec{k} \\ \frac{\partial}{\partial x} & \frac{\partial}{\partial y} & \frac{\partial}{\partial z} \\ w_2z - w_3y & -(w_1z - w_3x) & w_1y - w_2x \end{vmatrix}$$

$$= \vec{i}\,(w_1 + w_1) - \vec{j}\,(-w_2 - w_2) + \vec{k}\,(w_3 + w_3)$$

$$= 2w_1\vec{i} + 2w_2\vec{j} + 2w_3\vec{k}$$

$$= 2(w_1\vec{i} + w_2\vec{j} + w_3\vec{k}) = 2\,\vec{w}$$

$$\therefore \vec{w} = \frac{1}{2}\, curl\, \vec{v}\, .$$

Example 11: If $\varphi = x^2 - y^2$, find $\nabla^2\varphi$.

Solution: $\nabla^2\varphi = \dfrac{\partial^2\varphi}{\partial x^2} + \dfrac{\partial^2\varphi}{\partial y^2} + \dfrac{\partial^2\varphi}{\partial z^2}$.

Here $\dfrac{\partial\varphi}{\partial x} = 2x$ and $\dfrac{\partial^2\varphi}{\partial x^2} = 2$

$\dfrac{\partial\varphi}{\partial y} = -2y$ and $\dfrac{\partial^2\varphi}{\partial y^2} = -2$

$\dfrac{\partial\varphi}{\partial z} = 0$ and $\dfrac{\partial^2\varphi}{\partial z^2} = 0.$

$\therefore\ \nabla^2\varphi = 2 - 2 + 0 = 0.$

Example 12: Find $\nabla^2 r^2$ and hence prove that $\nabla^2 r^n = n(n+1)r^{n-2}$.

Solution: Let $\vec{r} = x\vec{i} + y\vec{j} + z\vec{k}$ ***[AU, Nov/Dec 2015]***

$r = |\vec{r}| = \sqrt{x^2 + y^2 + z^2}$

$\therefore\ r^2 = x^2 + y^2 + z^2$.

$\therefore\ 2r\dfrac{\partial r}{\partial x} = 2x \Rightarrow \dfrac{\partial r}{\partial x} = \dfrac{x}{r}$.

Similarly, $\dfrac{\partial r}{\partial y} = \dfrac{y}{r}$ and $\dfrac{\partial r}{\partial z} = \dfrac{z}{r}$.

Now, $\frac{\partial}{\partial x}(r^n) = nr^{n-1}\,\frac{\partial r}{\partial x} = n\,r^{n-1}\,\frac{x}{r} = n\,r^{n-2}x$

$$\frac{\partial^2}{\partial x^2}(r^n) = \frac{\partial}{\partial x}(nr^{n-2}x)$$

$$= n\left[r^{n-2} + (n-2)r^{n-3}\frac{\partial r}{\partial x}\,x\right]$$

$$= n\left[r^{n-2} + (n-2)r^{n-3}\frac{x}{r}\,x\right]$$

$$= n[r^{n-2} + (n-2)r^{n-4}x^2]\,.$$

Thus $\frac{\partial^2}{\partial x^2}(r^n)= nr^{n-2} + n(n-2)r^{n-4}x^2$.

Similarly, $\frac{\partial^2}{\partial y^2}(r^n)= nr^{n-2} + n(n-2)r^{n-4}y^2$,

and $\frac{\partial^2}{\partial z^2}(r^n) = nr^{n-2} + n(n-2)r^{n-4}z^2$.

Now, $\nabla^2 r^n = \left(\frac{\partial^2}{\partial x^2}+\frac{\partial^2}{\partial y^2}+\frac{\partial^2}{\partial z^2}\right)(r^n)$

$$= \frac{\partial^2}{\partial x^2}(r^n) + \frac{\partial^2}{\partial y^2}(r^n) + \frac{\partial^2}{\partial z^2}(r^n)$$

$$= 3nr^{n-2} + n(n-2)r^{n-4}[x^2+y^2+z^2]$$

$$= 3nr^{n-2} + n(n-2)r^{n-4}r^2$$

$$= nr^{n-2}[3+n-2]$$

$$= n(n+1)r^{n-2}\,.$$

Put $n = 2$, we get $\nabla^2 r^2 = 2(2+1)r^{\,2-2} = 6$.

Example 13: Show that $\nabla^2 r^n = n(n+1)r^{n-2}$ where $r^2 = x^2+y^2+z^2$. Hence find the value of $\nabla^2\left(\frac{1}{r}\right)$. ***[AU, April/May 2015]***

Solution : We know that $\nabla^2 r^n = n(n+1)r^{n-2}$.

Put $n = -1$ $\quad\therefore\ \nabla^2\left(\frac{1}{r}\right) = 0$.

Example 14: Prove that $\nabla^2(r^n\vec{r}) = n(n+3)r^{n-2}\vec{r}$.

Solution : $\vec{r} = \vec{i} + y\vec{j} + z\vec{k}$.

Then $\frac{\partial \vec{r}}{\partial x} = \vec{i}$, $\frac{\partial \vec{r}}{\partial y} = \vec{j}$ and $\frac{\partial \vec{r}}{\partial z} = \vec{k}$.

Also, $r = |\vec{r}| = \sqrt{x^2 + y^2 + z^2}$.

$\therefore\ r^2 = x^2 + y^2 + z^2$.

Hence $\frac{\partial r}{\partial x} = \frac{x}{r}$, $\frac{\partial r}{\partial y} = \frac{y}{r}$ and $\frac{\partial r}{\partial z} = \frac{z}{r}$.

Now, $\frac{\partial}{\partial x}(r^n \vec{r}) = nr^{n-1} \frac{\partial r}{\partial x} \vec{r} + r^n \frac{\partial \vec{r}}{\partial x}$

$$= nr^{n-1} \frac{x}{r} \vec{r} + r^n \vec{i} = nr^{n-2} x\vec{r} + r^n \vec{i}$$

$$\frac{\partial^2}{\partial x^2}(r^n \vec{r}) = n\left[(n-2)\, r^{n-3} \frac{\partial r}{\partial x}\, x\vec{r} + r^{n-2}\vec{r} + r^{n-2}\, x\, \frac{\partial \vec{r}}{\partial x}\right] + nr^{n-1} \frac{\partial r}{\partial x}\, \vec{i}$$

$$= n(n-2) r^{n-3} \frac{x}{r}\, x\,\vec{r} + n\, r^{n-2}\, \vec{r} + nr^{n-2}\, x\, \vec{i} + n\, r^{n-1} \frac{x}{r} \vec{i}$$

$$= n(n-2)\, r^{n-4}\, x^2\, \vec{r} + n\, r^{n-2}\, \vec{r} + 2\, n\, r^{n-2}\, x\vec{i}$$

Similarly, $\frac{\partial^2}{\partial y^2}(r^n \vec{r}) = n(n-2)r^{n-4}y^2\vec{r} + nr^{n-2}\vec{r} + 2nr^{n-2}y\,\vec{j}$

and $\frac{\partial^2}{\partial z^2}(r^n \vec{r}) = n(n-2)r^{n-4}z^2\vec{r} + nr^{n-2}\vec{r} + 2nr^{n-2}z\,\vec{k}$.

Now, $\nabla^2(r^n\vec{r}) = \frac{\partial^2}{\partial x^2}(r^n\vec{r}) + \frac{\partial^2}{\partial y^2}(r^n\vec{r}) + \frac{\partial^2}{\partial z^2}(r^n\vec{r})$

$$= n\,(n-2)r^{n-4}\,\vec{r}(x^2 + y^2 + z^2) + 3nr^{n-2}\,\vec{r} + 2nr^{n-2}(x\vec{i} + y\vec{j} + z\vec{k})$$

$$= n\,(n-2)r^{n-4}\,\vec{r}r^2 + 3nr^{n-2}\,\vec{r} + 2nr^{n-2}\vec{r}$$

$$= nr^{n-2}\vec{r}[n-2+3+2] = n(n+3)r^{n-2}\vec{r}.$$

Example 15: Prove that $\nabla^2 f(r) = f''(r) + \frac{2}{r} f'(r)$.

Solution: Let $\vec{r} = x\vec{i} + y\vec{j} + z\vec{k}$.

$\therefore\ r = |\vec{r}| = \sqrt{x^2 + y^2 + z^2}$.

i.e., $r^2 = x^2 + y^2 + z^2$.

Hence $\frac{\partial r}{\partial x} = \frac{x}{r}$, $\frac{\partial r}{\partial y} = \frac{y}{r}$ and $\frac{\partial r}{\partial z} = \frac{z}{r}$.

Now, $\frac{\partial}{\partial x} f(r) = f'(r) \frac{\partial r}{\partial x} = f'(r) \frac{x}{r}$

$$\therefore \quad \frac{\partial^2}{\partial x^2} f(r) = f''(r)\frac{\partial r}{\partial x}\left(\frac{x}{r}\right) + f'(r)\left[\frac{r - x\frac{\partial r}{\partial x}}{r^2}\right]$$

$$= f''(r)\frac{x^2}{r^2} + \frac{f'(r)}{r^2}\left[r - \frac{x^2}{r}\right]$$

Similarly, $\frac{\partial^2}{\partial y^2} f(r) = f''(r)\,\frac{y^2}{r^2} + \frac{f'(r)}{r^2}\left[r - \frac{y^2}{r}\right]$

and $\frac{\partial^2}{\partial z^2} f(r) = f''(r)\,\frac{z^2}{r^2} + \frac{f'(r)}{r^2}\left[r - \frac{z^2}{r}\right].$

$$\therefore \quad \nabla^2 f(r) = \frac{\partial^2}{\partial x^2} f(r) + \frac{\partial^2}{\partial y^2} f(r) + \frac{\partial^2}{\partial z^2} f(r)$$

$$= \frac{f''(r)}{r^2}(x^2 + y^2 + z^2) + \frac{f'(r)}{r^2}\left[3r - \frac{1}{r}(x^2 + y^2 + z^2)\right]$$

$$= \frac{f''(r)}{r^2}r^2 + \frac{f'(r)}{r^2}\left[3r - \frac{1}{r}r^2\right]$$

$$= f''(r) + \frac{f'(r)}{r^2}\,2r$$

$$= f''(r) + \frac{2}{r}f'(r).$$

Example 16: Show that $3y^4z^2\vec{i} + 4x^3z^2\vec{j} - 3x^2y^2\vec{k}$ is solenoidal.

Solution: Let $\vec{f} = 3y^4z^2\vec{i} + 4x^3z^2\vec{j} - 3x^2y^2\vec{k}$.

$$\therefore div\,\vec{f} = \nabla \cdot \vec{f} = \frac{\partial}{\partial x}(3y^4z^2) + \frac{\partial}{\partial y}(4x^3z^2) + \frac{\partial}{\partial z}(-3x^2y^2)$$

$$= 0 + 0 + 0 = 0.$$

$\therefore \vec{f}$ is solenoidal.

Example 17: Prove that $\vec{f} = (y^2\cos x + z^3)\vec{i} + (2y\sin x - 4)\vec{j} + 3xz^2\vec{k}$ is irrotational. Hence find its scalar potential.

Solution: Given that $\vec{f} = (y^2\cos x + z^3)\vec{i} + (2y\sin x - 4)\vec{j} + 3xz^2\vec{k}$

$$\therefore curl\,\vec{f} = \nabla \times \vec{f} = \begin{vmatrix} \vec{i} & \vec{j} & \vec{k} \\ \frac{\partial}{\partial x} & \frac{\partial}{\partial y} & \frac{\partial}{\partial z} \\ y^2\cos x + z^3 & 2y\sin x - 4 & 3xz^2 \end{vmatrix}$$

$$= \vec{i}[0 - 0] - \vec{j}[3z^2 - 3z^2] + \vec{k}[2y\cos x - 2y\cos x]$$

$$= 0\vec{\imath} + 0\vec{\jmath} + 0\vec{k} = \vec{0}.$$

Hence $\vec{f}$ is irrotational.

Let $\vec{f} = \nabla\varphi$.

$$\therefore \frac{\partial\varphi}{\partial x}\vec{\imath} + \frac{\partial\varphi}{\partial y}\vec{\jmath} + \frac{\partial\varphi}{\partial z}\vec{k} = (y^2\cos x + z^3)\vec{\imath} + (2y\sin x - 4)\vec{\jmath} + 3xz^2\vec{k}.$$

Hence $\frac{\partial\varphi}{\partial x} = y^2\cos x + z^3$, $\frac{\partial\varphi}{\partial y} = 2y\sin x - 4$ and $\frac{\partial\varphi}{\partial z} = 3xz^2$.

$\therefore$ $\frac{\partial\varphi}{\partial x} = y^2\cos x + z^3$ gives $\varphi = y^2\sin x + z^3 x$ + a function not containing x,

$\frac{\partial\varphi}{\partial y} = 2y\sin x - 4$ gives $\varphi = y^2\sin x - 4y$ + a function not containing y,

and $\frac{\partial\varphi}{\partial z} = 3xz^2$ gives $\varphi = xz^3$ + a function not containing z.

Combining the above functions, we have, $\varphi = y^2\sin x + z^3 x - 4y + c$.

Example 18: Show that $\vec{F} = (y^2 - z^2 + 3yz - 2x)\vec{\imath} + (3xz + 2xy)\vec{\jmath} + (3xy - 2xz + 2z)\vec{k}$ is both solenoidal and irrotational.

Solution: Let $div\,\vec{F} = \nabla\cdot\vec{F}$

$$= \frac{\partial}{\partial x}(y^2 - z^2 + 3yz - 2x) + \frac{\partial}{\partial y}(3xz + 2xy) + \frac{\partial}{\partial z}(3xy - 2xz + 2z)$$

$$= -2 + 2x - 2x + 2$$

$$= 0.$$

$\therefore \vec{F}$ is solenoidal.

Also,

$$curl\,\vec{F} = \nabla\times\vec{F} = \begin{vmatrix} \vec{\imath} & \vec{\jmath} & \vec{k} \\ \frac{\partial}{\partial x} & \frac{\partial}{\partial y} & \frac{\partial}{\partial z} \\ y^2 - z^2 + 3yz - 2x & 3xz + 2xy & 3xy - 2xz + 2z \end{vmatrix}$$

$$= \vec{\imath}(3x - 3x) - \vec{\jmath}(3y - 2z + 2z - 3y) + \vec{k}(3z + 2y - 2y - 3z)$$

$$= 0\vec{\imath} + 0\vec{\jmath} + 0\vec{k} = \vec{0}.$$

$\therefore \vec{F}$ is irrotational.

Example 19: Show that $\vec{F} = (x^2 - y^2 + x)\vec{\imath} - (2xy + y)\vec{\jmath}$ is irrotational(conservative field) and hence find its scalar potential. ***[AU, Nov/Dec 2017, Nov/Dec 2014]***

Solution:

$$curl\,\vec{F} = \nabla \times \vec{F} = \begin{vmatrix} \vec{\imath} & \vec{\jmath} & \vec{k} \\ \frac{\partial}{\partial x} & \frac{\partial}{\partial y} & \frac{\partial}{\partial z} \\ x^2 - y^2 + x & -2xy - y & 0 \end{vmatrix}$$

$$= \vec{\imath}(0 - 0) - \vec{\jmath}(0 - 0) + \vec{k}(-2y + 2y)$$

$$= 0\vec{\imath} - 0\vec{\jmath} + 0\vec{k} = \vec{0}\,.$$

$\therefore\ \vec{F}$ is irrotational.

If φ is the scalar potential, then $\nabla\varphi = \vec{F}$.

$$\therefore\ \frac{\partial\varphi}{\partial x}\vec{\imath} + \frac{\partial\varphi}{\partial y}\vec{\jmath} + \frac{\partial\varphi}{\partial z}\vec{k} = (x^2 - y^2 + x)\vec{\imath} - (2xy + y)\vec{\jmath}$$

Hence $\frac{\partial\varphi}{\partial x} = x^2 - y^2 + x, \frac{\partial\varphi}{\partial y} = -(2xy + y)$ and $\frac{\partial\varphi}{\partial z} = 0$.

Now, $\frac{\partial\varphi}{\partial x} = x^2 - y^2 + x$ gives $\varphi = \frac{x^3}{3} - xy^2 + \frac{x^2}{2} +$ a function not containing x,

$\frac{\partial\varphi}{\partial y} = -(2xy + y)$ gives $\varphi = -xy^2 - \frac{y^2}{2} +$ a function not containing y,

and $\frac{\partial\varphi}{\partial z} = 0$ gives $\varphi = c +$ a function not containing z.

Combining the above functions, $\varphi = \frac{x^3}{3} - xy^2 + \frac{x^2}{2} - \frac{y^2}{2} + c.$

Example 20: Find the value of m if $\vec{F} = (x + 2y)\vec{\imath} + (my + 4z)\vec{\jmath} + (5z + 6x)\vec{k}$ is solenoidal.

Solution:

Given that $\vec{F}$ is solenoidal.

$\therefore\ div\,\vec{F} = 0.$

i.e., $\frac{\partial}{\partial x}(x + 2y) + \frac{\partial}{\partial y}(my + 4z) + \frac{\partial}{\partial z}(5z + 6x) = 0.$

i.e., $1 + m + 5 = 0 \quad \therefore\ m = -6.$

Example 21: If $\vec{v} = (x+3y)\vec{i} + (y-2z)\vec{j} + (x+\lambda z)\vec{k}$ is solenoidal, find the value of λ.

Solution: Since $\vec{v}$ is solenoidal,

we have $div\,\vec{v} = 0$.

$$\therefore \quad \frac{\partial}{\partial x}(x+3y) + \frac{\partial}{\partial y}(y-2z) + \frac{\partial}{\partial z}(x+\lambda z) = 0$$

i.e., $\quad 1+1+\lambda = 0$

$\therefore \quad \lambda = -2.$

Example 22: Find the constants a, b, c so that $\vec{f} = (x+2y+az)\vec{i} + (bx-3y-z)\vec{j} + (4x+cy+2z)\vec{k}$ is irrotational. For those values of a, b, c find its scalar potential.

Solution: ***[AU, April/May 2017]***

Since $\vec{f}$ is irrotational,

we have $curl\,\vec{f} = 0$.

$$\therefore \begin{vmatrix} \vec{i} & \vec{j} & \vec{k} \\ \frac{\partial}{\partial x} & \frac{\partial}{\partial y} & \frac{\partial}{\partial z} \\ x+2y+az & bx-3y-z & 4x+cy+2z \end{vmatrix} = \vec{0}$$

i.e., $\quad \vec{i}(c+1) - \vec{j}(4-a) + \vec{k}(b-2) = \vec{0}$

$\quad (c+1)\vec{i} + (a-4)\vec{j} + (b-2)\vec{k} = \vec{0}$

Hence $c+1 = 0,\ a-4 = 0$ and $b-2 = 0.$

i.e., $c = -1,\ a = 4$ and $b = 2.$

$\therefore \vec{f} = (x+2y+4z)\vec{i} + (2x-3y-z)\vec{j} + (4x-y+2z)\vec{k}.$

Let $\nabla\varphi = \vec{f}.$

$$\therefore \frac{\partial\varphi}{\partial x}\vec{i} + \frac{\partial\varphi}{\partial y}\vec{j} + \frac{\partial\varphi}{\partial z}\vec{k} = (x+2y+4z)\vec{i} + (2x-3y-z)\vec{j} + (4x-y+2z)\vec{k}$$

Hence $\frac{\partial\varphi}{\partial x} = x+2y+4z,\ \frac{\partial\varphi}{\partial y} = 2x-3y-z$ and $\frac{\partial\varphi}{\partial z} = 4x-y+2z.$

Now, $\frac{\partial\varphi}{\partial x} = x+2y+4z$ gives $\varphi = \frac{x^2}{2} + 2xy + 4xz +$ a function not containing x,

$\frac{\partial\varphi}{\partial y} = 2x-3y-z$ gives $\varphi = 2xy - \frac{3}{2}y^2 - yz +$ a function not containing y, and $\frac{\partial \varphi}{\partial z} = 4x - y + 2z$ gives $\varphi = 4xz - yz + z^2 +$ a function not containing z

Combining the above functions,

$$\varphi = \frac{x^2}{2} + 2xy + 4xz - \frac{3}{2}y^2 - yz + z^2 + c.$$

Example 23: Determine $f(r)$ so that the vector $f(r)\vec{r}$ is both solenoidal and irrotational.

Solution: Let $\vec{r} = x\vec{i} + y\vec{j} + z\vec{k}$.

$$\therefore\ f(r)\vec{r} = f(r)\,x\vec{i} + f(r)\,y\vec{j} + f(r)\,z\vec{k}$$

Since $f(r)\,\vec{r}$ is solenoidal, we have $\nabla \cdot (f(r)\,\vec{r}) = 0$.

$\therefore \nabla f(r) \cdot \vec{r} + (\nabla \cdot \vec{r})\, f(r) = 0$, by Property 2 (1)

Now, $\nabla \cdot \vec{r} = \frac{\partial}{\partial x}(x) + \frac{\partial}{\partial y}(y) + \frac{\partial}{\partial z}(z) = 1 + 1 + 1 = 3$

and $\nabla f(r) = \frac{\partial}{\partial x} f(r)\vec{i} + \frac{\partial}{\partial y} f(r)\vec{j} + \frac{\partial}{\partial z} f(r)\vec{k}$

$$= f'(r)\frac{\partial r}{\partial x}\vec{i} + f'(r)\frac{\partial r}{\partial y}\vec{j} + f'(r)\frac{\partial r}{\partial z}\vec{k}$$

$$= f'(r)\left[\frac{x}{r}\vec{i} + \frac{y}{r}\vec{j} + \frac{z}{r}\vec{k}\right] = f'(r)\frac{\vec{r}}{r} \qquad (2)$$

$\therefore$ (1) gives $f'(r)\frac{\vec{r}}{r} \cdot \vec{r} + 3\,f(r) = 0.$

i.e., $f'(r)\frac{r^2}{r} + 3\,f(r) = 0.$

$\therefore\ \frac{f'(r)}{f(r)} = \frac{-3}{r}.$

Integrating, $\log f(r) = -3 \log r + \log c.$

i.e., $\log f(r) + \log r^3 = \log c$

i.e., $\log (r^3 f(r)) = \log c$

$\therefore\ r^3 f(r) = c$

$\therefore\ f(r) = \frac{c}{r^3}. \qquad (3)$

Also, it is given that $f(r)\vec{r}$ is irrotational. $\therefore$ $curl\, f(r)\,\vec{r} = \vec{0}$

$\therefore$ $grad\, f(r) \times \vec{r} + f(r)\, curl\, \vec{r} = \vec{0}$, by Property 10.

i.e., $\nabla f(r) \times \vec{r} + f(r)\, \nabla \times \vec{r} = \vec{0}$

i.e., $f'(r)\frac{\vec{r}}{r} \times \vec{r} + f(r)\vec{0} = \vec{0}$ $[\because \nabla \times \vec{r} = \vec{0}]$

i.e., $\vec{0} + \vec{0} = \vec{0}$ $[\because \vec{r} \times \vec{r} = \vec{0}]$

This is true for all values of $f(r)$. (4)

From (3) and (4), we have $f(r)\vec{r}$ is both solenoidal and irrotational if $f(r) = \frac{c}{r^3}$.

Example 24: If $\vec{F}$ is solenoidal, prove that $curl\, curl\, curl\, curl\, \vec{F} = \nabla^4 \vec{F}$.

Solution: Since $\vec{F}$ is solenoidal, $\nabla . \vec{F} = 0$. (1)

By Property 11, $curl\, curl\, \vec{F} = grad\, div\, \vec{F} - \nabla^2 \vec{F}$

$= \nabla(\nabla . \vec{F}) - \nabla^2 \vec{F}$ (2)

$= 0 - \nabla^2 \vec{F}$ [by (1)]

$\therefore$ $curl\, curl\, curl\, curl\, \vec{F} = curl\, curl\, (-\nabla^2 \vec{F})$

$= -\, curl\, curl\, (\nabla^2 \vec{F})$

$= -[\nabla(\nabla . \nabla^2 \vec{F}) - \nabla^2 (\nabla^2 \vec{F})]$ [by (2)]

$= -\left[\nabla\left(\nabla^2(\nabla . \vec{F})\right) - \nabla^4 \vec{F}\right]$

$= -[0 - \nabla^4 \vec{F}]$

$= \nabla^4 \vec{F}$. [by (1)]

Example 25: If $\vec{f}$ and $\vec{g}$ are irrotational, show that $\vec{f} \times \vec{g}$ is solenoidal.

Solution: Given that $\vec{f}$ and $\vec{g}$ are irrotational.

$\therefore$ $curl\, \vec{f} = \vec{0}$ and $curl\, \vec{g} = \vec{0}$.

Now, $div\, (\vec{f} \times \vec{g}) = \vec{g} . curl\, \vec{f} - \vec{f} . curl\, \vec{g}$

$= \vec{g} . \vec{0} - \vec{f} . \vec{0} = 0 - 0 = 0$.

$\therefore$ $\vec{f} \times \vec{g}$ is solenoidal.

Example 26: Find the value of n such that the vector $r^n\vec{r}$ is both solenoidal and irrotational. ***[AU, Nov/Dec 2017, May/June 2014]***

Solution: Let $\vec{r} = x\vec{i} + y\vec{j} + z\vec{k}$.

Then $r^2 = |\vec{r}|^2 = x^2 + y^2 + z^2$.

$\therefore \frac{\partial r}{\partial x} = \frac{x}{r}, \frac{\partial r}{\partial y} = \frac{y}{r}, \frac{\partial r}{\partial z} = \frac{z}{r}$.

Now, $r^n\vec{r} = r^n(x\vec{i} + y\vec{j} + z\vec{k})$

$= r^n x\vec{i} + r^n y\vec{j} + r^n z\vec{k}$.

$\therefore div\,(r^n\vec{r}) = \nabla . r^n\vec{r}$

$$= \frac{\partial}{\partial x}(r^n x) + \frac{\partial}{\partial y}(r^n y) + \frac{\partial}{\partial z}(r^n z)$$

$$= r^n + nr^{n-1}\frac{\partial r}{\partial x}x + r^n + nr^{n-1}\frac{\partial r}{\partial y}y + r^n + nr^{n-1}\frac{\partial r}{\partial z}z$$

$$= 3r^n + nr^{n-1}\left[\frac{\partial r}{\partial x}x + \frac{\partial r}{\partial y}y + \frac{\partial r}{\partial z}z\right]$$

$$= 3r^n + nr^{n-1}\left[\frac{x}{r}x + \frac{y}{r}y + \frac{z}{r}z\right]$$

$$= 3r^n + nr^{n-1}\frac{1}{r}[x^2 + y^2 + z^2]$$

$$= 3r^n + nr^{n-1}\frac{1}{r}r^2 = 3r^n + nr^n = (3+n)r^n.$$

Hence, $r^n\vec{r}$ is solenoidal iff $div\,(r^n\vec{r}) = 0$

i.e., iff $(n+3)\,r^n = 0$

i.e., iff $n + 3 = 0$

i.e., iff $n = -3$.

Now, $curl\,(r^n\vec{r}) = \nabla \times r^n\vec{r}$

$$= \begin{vmatrix} \vec{i} & \vec{j} & \vec{k} \\ \frac{\partial}{\partial x} & \frac{\partial}{\partial y} & \frac{\partial}{\partial z} \\ r^n x & r^n y & r^n z \end{vmatrix}$$

$$= \vec{i}\left(\frac{\partial r^n z}{\partial y} - \frac{\partial r^n y}{\partial z}\right) - \vec{j}\left(\frac{\partial r^n z}{\partial x} - \frac{\partial r^n x}{\partial z}\right) + \vec{k}\left(\frac{\partial r^n y}{\partial x} - \frac{\partial r^n x}{\partial y}\right)$$

$$= \vec{i}\left(znr^{n-1}\frac{\partial r}{\partial y} - ynr^{n-1}\frac{\partial r}{\partial z}\right) - \vec{j}\left(znr^{n-1}\frac{\partial r}{\partial x} - xnr^{n-1}\frac{\partial r}{\partial z}\right) + \vec{k}\left(ynr^{n-1}\frac{\partial r}{\partial x} - xnr^{n-1}\frac{\partial r}{\partial y}\right)$$

$$= \vec{i}\left(znr^{n-1}\frac{y}{r} - ynr^{n-1}\frac{z}{r}\right) - \vec{j}\left(znr^{n-1}\frac{x}{r} - xnr^{n-1}\frac{z}{r}\right) + \vec{k}\left(ynr^{n-1}\frac{x}{r} - xnr^{n-1}\frac{y}{r}\right)$$

$= 0\vec{i} - 0\vec{j} + 0\vec{k} = \vec{0}$.

$\therefore$ $r^n\vec{r}$ is irrotational for all n.

EXERCISES

1. If $\vec{f} = xz^3\vec{i} - 2xyz\vec{j} + xz\vec{k}$, find $div\,\vec{f}$ and $curl\,\vec{f}$ at $(1, 2, 0)$.
2. If $\vec{A} = x^2\vec{i} + y^2\vec{j} + z^2\vec{k}$, then find $\nabla.\vec{A}$ and $\nabla \times \vec{A}$.
3. If $\vec{F} = xz^3\vec{i} - 2x^2yz\vec{j} + 2yz^4\vec{k}$, find $div\,\vec{F}$ and $url\,\vec{F}$.
4. If $\vec{f} = (x + y + 1)\vec{i} + \vec{j} - (x + y)\vec{k}$, prove that $\vec{f}.\,curl\,\vec{f} = 0$.
5. Calculate the *curl* of the vector $\vec{f} = xyz\,\vec{i} + 3x^2y\,\vec{j} + (xz^2 - y^2z)\vec{k}$ at $(1, -1, 1)$.
6. Find the *curl curl* $\vec{f}$ at the point $(1, 0, 2)$ if $\vec{f} = x^2y\vec{i} - 2xz\vec{j} + 2yz\vec{k}$.
7. Find $div\left(\frac{\vec{r}}{r}\right)$.
8. Prove that $\nabla.(r^3\vec{r}) = 6\,r^3$.
9. Find $\nabla(\nabla.\,\vec{f})$ where $\vec{f} = \frac{\vec{r}}{r}$.
10. Prove that $div\,(u\,grad\,v) = u\,\nabla^2u + grad\,u\,.grad\,v$.
11. Prove that $\nabla \times [f(r)\vec{r}] = \vec{0}$.
12. Prove that $\vec{f} = z\vec{i} + x\vec{j} + y\vec{k}$ is solenoidal.
13. Show that $\vec{F} = (6xy + z^3)\vec{i} + (3x^2 - z)\vec{j} + (3xz^2 - y)\vec{k}$ is irrotational. Also find φ such that $\vec{F} = \nabla\varphi$.
14. Find a such that $(3x - 2y + x)\vec{i} + (4x + ay - z)\vec{j} + (x - y + 2z)\vec{k}$ is solenoidal.
15. Determine the constants a and b such that $(2xy + 3yz)\vec{i} + (x^2 + axz - 4z^2)\vec{j} + (3xy + 2byz)\vec{k}$ is irrotational vector.
16. Find the constants a, b and c so that $\vec{f} = (axy + bz^3)\vec{i} + (3x^2 - z)\vec{j} + (3xz^2 - y)\vec{k}$ is irrotational.

ANSWERS

1. 1; $4\vec{i}$

2. $(x + y + z)$; $\vec{0}$

3. $z^3 - 2x^2z + 8yz^3$; $(2z^4 + 2x^2y)\vec{i} + 3xz^2\vec{j} - 4xyz\vec{k}$

5. $2\vec{i} - 2\vec{j} - 7\vec{k}$

6. $4\vec{j}$

7. $\frac{2}{r}$

9. $-2\,r^{-3}\vec{r}$

13. $\varphi = xz^3 - yz + 3x^2y + c$

14. $a = -5$

15. $a = 3,\ b = 4$

16. $a = 6,\ b = 1,\ c = 1$

Vector Integration

Let $\vec{f}(t)$ and $\vec{F}(t)$ be two vector functions of a scalar variable t such that $\frac{d}{dt}\vec{F}(t) = \vec{f}(t)$, then $\vec{F}(t)$ is called an integral of $\vec{f}(t)$ with respect to t and we write $\int \vec{f}(t)\,dt = \vec{F}(t)$. The definite integral of $\vec{f}(t)$ between the limits $t = a$ and $t = b$ is written as $\int_C \vec{f}(t)d\vec{t} = \left[\vec{F}(t)\right]_a^b = \vec{F}(b) - \vec{F}(a)$.

2.6 Line Integrals

Any integral which to be evaluated along a curve is called a line integral.

Let $\vec{F}(P)$ be a continuous vector point function defined at every point of a curve C in space. Divide the curve C into n parts by the points $A = P_0, P_1, P_2, \ldots, P_n = B$ and let $\vec{r_o}, \vec{r_1}, \vec{r_2}, \ldots, \vec{r_n}$ be the position vectors of these points. Let Q_i be any point on the arc $P_{i-1}P_i$. Then the limit of the sum $\sum_{i=1}^{n} \vec{F}(Q_i).\delta\vec{r_i}$ where $\delta\vec{r_i} = \vec{r_i} - \vec{r_{i-1}}$ as $n \to \infty$ and every $|\delta\vec{r_i}| \to 0$, if it exists, is called a line integral of $\vec{F}$ along C and is denoted by $\int_C \vec{F}.d\vec{r}$. Clearly it is a scalar. It is called the *tangential line integral* of $\vec{F}$ along C.

Note

(i) If A and B are two points in the curve C, then the line integral $\int_A^B \vec{F}.d\vec{r}$ depends not only on the curve C but also on the terminal points A and B.

(ii) If $\vec{F} = F_1\vec{i} + F_2\vec{j} + F_3\vec{k}$, then the line integral is

$$\int_C \vec{F}.d\vec{r} = \int_C (F_1 dx + F_2 dy + F_3 dz).$$

(iii) $\int_C \vec{F} \times d\vec{r}$ and $\int_C \varphi\, d\vec{r}$ are also line integrals, where φ is a scalar point function.

(iv) If $\vec{F}$ represents velocity of a fluid particle, then the line integral $\int_C \vec{F}.d\vec{r}$ is called the *circulation* of $\vec{F}$ around the curve C.

Work Done by a Force

Let $\vec{F}$ represent the force acting on a particle moving along an arc AB. The *work done* during a small displacement $d\vec{r}$ is $\vec{F}.d\vec{r}$.

∴ The total work done by the force $\vec{F}$ during displacement from A to B is given by $\int_A^B \vec{F}.d\vec{r}$.

Conservative Vector

If the value of the line integral $\int_A^B \vec{F}\,.d\vec{r}$ does not depend on the curve C, but only on the terminal points A and B, then $\vec{F}$ is called a *conservative vector*.

Condition for $\vec{F}$ to be Conservative

If $\vec{F}$ is an irrotational vector, it is conservative.

Proof:

Since $\vec{F}$ is an irrotational, it can be expressed as $\nabla\varphi$. i.e.,$\vec{F} = \nabla\varphi$.

$$\int_A^B \vec{F}\,.d\vec{r} = \int_A^B \nabla\varphi\,.d\vec{r}$$

$$= \int_A^B \left(\frac{\partial\varphi}{\partial x}\vec{\imath} + \frac{\partial\varphi}{\partial y}\vec{\jmath} + \frac{\partial\varphi}{\partial z}\vec{k}\right).(dx\,\vec{\imath} + dy\vec{\jmath} + dz\vec{k})$$

$$= \int_A^B \left(\frac{\partial\varphi}{\partial x}dx + \frac{\partial\varphi}{\partial y}dy + \frac{\partial\varphi}{\partial z}dz\right) = \int_A^B d\varphi$$

$= [\,\varphi\,]_A^B$, whatever is the path of integration.

$= \varphi(B) - \varphi(A)$.

$\therefore$ $\vec{F}$ is conservative.

Note: If $\vec{F}$ is irrotational (and hence conservative) and C is a closed curve, then $\oint_C \vec{F}\,.d\vec{r} = 0$. [Since $\varphi(A) = \varphi(B)$, as A and B coincide.]

Example 1: If $\vec{f} = (3x^2 + 6y)\vec{\imath} - 14yz\vec{\jmath} + 20xz^2\vec{k}$, evaluate $\int \vec{f}\,.d\vec{r}$ from $(0,0,0)$ to $(1,1,1)$ along the curve $x = t,\ y = t^2,\ z = t^3$.

Solution: The given curve is $x = t,\ y = t^2,\ z = t^3$.

From the end points $(0,0,0)$ and$(1,1,1)$, we have t varies from $t = 0$ to $t = 1$. Also $dx = dt,\ dy = 2t\,dt,\ dz = 3\,t^2 dt$.

$$\therefore\ \int \vec{f}\,.d\vec{r} = \int_{(0,0,0)}^{(1,1,1)} (f_1 dx + f_2 dy + f_3 dz)$$

$$= \int_{(0,0,0)}^{(1,1,1)} [(3x^2 + 6y)dx - 14yz\,dy + 20xz^2\,dz]$$

$$= \int_0^1 [(3t^2 + 6t^2)dt - 14t^5(2t\,dt) + 20t^7(3\,t^2\,dt)]$$

$$= \int_0^1 [9t^2 - 28t^6 + 60t^9]\,dt$$

$$= \left[9\frac{t^3}{3} - 28\frac{t^7}{7} + 60\frac{t^{10}}{10}\right]_0^1$$

$$= [3t^3 - 4t^7 + 6t^{10}]_0^1$$

$$= [(3 - 4 + 6) - 0] = 5.$$

Example 2: If $\vec{f} = x^2\vec{\imath} + xy^2\vec{\jmath}$, evaluate the line integral $\int \vec{f} . d\vec{r}$ from $(0,0)$ to $(1,1)$ along the path $y = x$. ***[AU, May/June 2005]***

Solution: The given curve is $x = y$ from $(0,0)$ to $(1,1)$.

$\therefore dx = dy$ and x varies from 0 to 1.

Now, $\vec{f} . d\vec{r} = (x^2\vec{\imath} + xy^2\vec{\jmath}) . (dx\,\vec{\imath} + dy\vec{\jmath} + dz\,\vec{k})$

$= x^2\,dx + x\,y^2\,dy$

$= x^2\,dx + x\,x^2\,dx = (x^2 + x^3)\,dx$

$\therefore \int \vec{f} . d\vec{r} = \int_0^1 (x^2 + x^3)\,dx$

$$= \left[\frac{x^3}{3} + \frac{x^4}{4}\right]_0^1$$

$$= \left(\frac{1}{3} + \frac{1}{4}\right) - 0 = \frac{7}{12}.$$

Example 3: If $\vec{F} = 3xy\vec{\imath} - y^2\vec{\jmath}$, evaluate $\int \vec{f} . d\vec{r}$ where C is the arc of the parabola $y = 2x^2$ from $(0,0)$ to $(1,1)$.

Solution: The given curve is $y = 2x^2$.

$\therefore dy = 4x\,dx$ and x varies from 0 to 1.

Now, $\vec{F} . d\vec{r} = (3xy\vec{\imath} - y^2\vec{\jmath}) . (dx\vec{\imath} + dy\vec{\jmath} + dz\vec{k})$

$= 3xy\,dx - y^2\,dy$

$= 6x^3\,dx - 4\,x^4(4x\,dx)$

$= (6x^3 - 16\,x^5)\,dx$.

$\therefore \int \vec{f} . d\vec{r} = \int_0^1 (6x^3 - 16\,x^5)\,dx$

$$= \left[\frac{3}{2}x^4 - \frac{8}{3}x^6\right]_0^1 = \left(\frac{3}{2} - \frac{8}{3}\right) - 0 = -\frac{7}{6}.$$

Example 4: Evaluate $\int \vec{f}\,.\,d\vec{r}$ where $\vec{f} = (2xy + z^3)\vec{\imath} + x^2\vec{\jmath} + 3xz^2\vec{k}$ along the straight line joining the points $(1,-2,1)$ and $(3,1,4)$.

Solution: The equation of the line joining $(1,-2,1)$ and $(3,1,4)$ is

$$\frac{x-1}{3-1} = \frac{y+2}{1+2} = \frac{z-1}{4-1}$$

$$\text{i.e., } \frac{x-1}{2} = \frac{y+2}{3} = \frac{z-1}{3} = t \text{ (say)} \qquad (1)$$

Hence the parametric equations of the straight line are $x = 2t+1$, $y = 3t-2$, $z = 3t+1$ and t varies from 0 to 1.

$\therefore\ dx = 2dt,\ dy = 3dt,\ dz = 3\,dt.$

$$\vec{f}\,.\,d\vec{r} = \left[(2xy + z^3)\vec{\imath} + x^2\vec{\jmath} + 3xz^2\vec{k}\right].\left(dx\vec{\imath} + dy\vec{\jmath} + dz\vec{k}\right)$$

$$= (2xy + z^3)\,dx + x^2\,dy + 3xz^2dz$$

$$= [2\,(2t+1)(3t-2) + (3t+1)^3]\,2\,dt + (2t+1)^2\,3\,dt$$
$$+3(2t+1)(3t+1)^2 3\,dt$$

$$= (216\,t^3 + 279t^2 + 98\,t + 6)\,dt.$$

$$\therefore\ \int \vec{f}\,.\,d\vec{r} = \int_0^1 (216\,t^3 + 279t^2 + 98\,t + 6)\,dt$$

$$= [54t^4 + 93t^3 + 49t^2 + 6t]_0^1$$

$$= (54 + 93 + 49 + 6) - 0 = 202.$$

Example 5: Evaluate $\int \vec{f}\,.\,d\vec{r}$ where $\vec{f} = (x^2 + y^2)\vec{\imath} - 2xy\vec{\jmath}$ along the curve C is the rectangle in the XY –plane bounded by $x = 0, x = a, y = 0, y = b$.

Solution:

The given curve C is a rectangle and let it be given by the points

$$O = (0,0),\ A = (a,0),\ B = (a,b) \text{ and } D = (0,b).$$

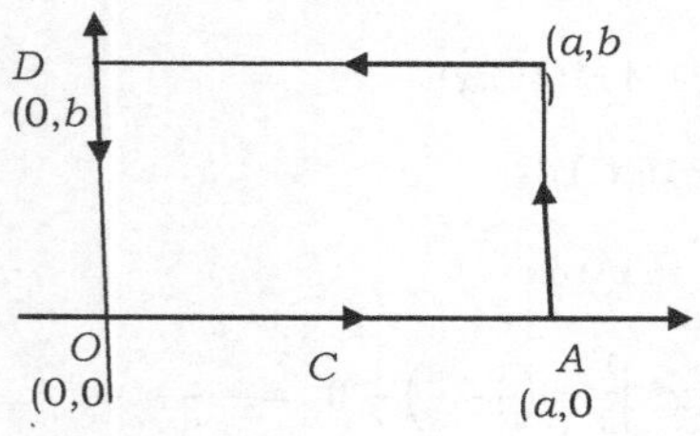

$$\therefore \int \vec{f} \cdot d\vec{r} = \int_{OA} \vec{f} \cdot d\vec{r} + \int_{AB} \vec{f} \cdot d\vec{r} + \int_{BD} \vec{f} \cdot d\vec{r} + \int_{DO} \vec{f} \cdot d\vec{r} \quad ...(1)$$

Now, $\vec{f} \cdot d\vec{r} = [(x^2 + y^2)\vec{i} - 2xy\vec{j}] \cdot (dx\vec{i} + dy\vec{j} + dz\vec{k})$

$$= (x^2 + y^2)dx - 2xy\,dy$$

(i) Along OA

$y = 0, dy = 0$ and x varies from 0 to a.

$$\therefore \int_{OA} \vec{f} \cdot d\vec{r} = \int_0^a x^2\,dx$$

$$= \left[\frac{x^3}{3}\right]_0^a$$

$$= \frac{a^3}{3}.$$

(ii) Along AB

$x = a,\ dx = 0$ and y varies from 0 to b.

$$\therefore \int_{AB} \vec{f} \cdot d\vec{r} = \int_0^b (-2ay)\,dy$$

$$= -2a\left[\frac{y^2}{2}\right]_0^b$$

$$= -a\,b^2.$$

(iii) Along BD

$y = b, dy = 0$ and x varies from a to 0.

$$\therefore \int_{BD} \vec{f} \cdot d\vec{r} = \int_a^0 (x^2 + b^2)\,dx$$

$$= \left[\frac{x^3}{3} + b^2 x\right]_a^0$$

$$= 0 - \left(\frac{a^3}{3} + b^2 a\right)$$

$$= -\left(\frac{a^3}{3} + b^2 a\right).$$

(iii) Along DO

$x = 0,\ dx = 0$ and y varies from b to 0.

$$\therefore \int_{DO} \vec{f} \cdot d\vec{r} = \int_b^0 0\, dy = 0.$$

$\therefore$ (1) gives

$$\int \vec{f} \cdot d\vec{r} = \frac{a^3}{3} - ab^2 - \frac{a^3}{3} - ab^2$$

$$= -2ab^2.$$

Example 6: If $\vec{f} = (2x + y^2)\vec{\imath} + (3y - 4x)\vec{\jmath}$, evaluate $\int \vec{f} \cdot d\vec{r}$ around a triangle ABC in the xy −plane with $A(0,0), B(2,0), C(2,1)$.

Solution:

Let C denote the boundary of the given triangle ABC.

$$\therefore \int \vec{f} \cdot d\vec{r} = \int_{AB} \vec{f} \cdot d\vec{r} + \int_{BC} \vec{f} \cdot d\vec{r} + \int_{CA} \vec{f} \cdot d\vec{r} \qquad \text{... (1)}$$

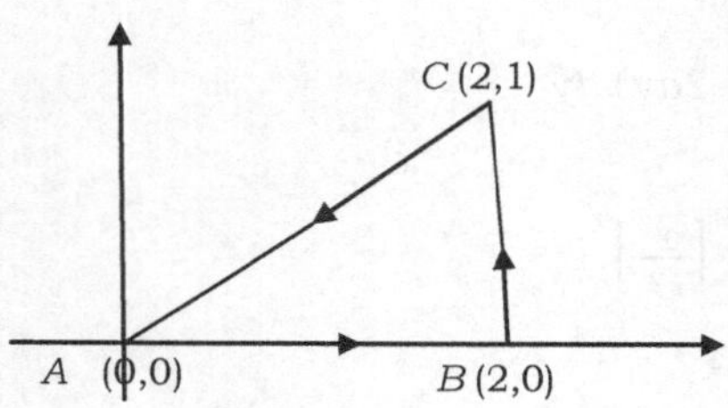

Now, $\vec{f} \cdot d\vec{r} = [(2x + y^2)\vec{\imath} + (3y - 4x)\vec{\jmath}] \cdot (dx\vec{\imath} + dy\vec{\jmath} + dz\vec{k})$

$$= (2x + y^2)dx + (3y - 4x)\, dy.$$

(i) Along AB

$y = 0,\ dy = 0$ and x varies from 0 to 2.

$$\therefore \int_{AB} \vec{f} \cdot d\vec{r} = \int_0^2 2x\, dx$$

$$= [x^2]_0^2$$

$$= 4.$$

(ii) Along BC

$x = 2,\ dx = 0$ and y varies from 0 to 1.

$$\therefore \int_{BC} \vec{f} \cdot d\vec{r} = \int_0^1 (3y - 8)\, dy$$

$$= \left[3\frac{y^2}{2} - 8y\right]_0^1$$

$$= \left(\frac{3}{2} - 8\right) - 0 = \frac{-13}{2}.$$

(iii) Along CA

The line CA is $\frac{x-2}{0-2} = \frac{y-1}{0-1}$

i.e., $\frac{x-2}{-2} = \frac{y-1}{-1} = t$ (say)

i.e.,$x = -2t + 2,\ y = -t + 1$

$\therefore\ dx = -2\,dt,\ dy = -dt$ and t varies from 0 to 1.

$$\therefore\ \vec{f} \cdot d\vec{r} = [2(-2t + 2) + (-t + 1)^2](-2dt) + [3(-t + 1) - 4(-2t + 2)](-dt)$$

$$= (-2t^2 + 7t - 5)\, dt.$$

$$\therefore\ \int \vec{f} \cdot d\vec{r} = \int_0^1 (-2t^2 + 7t - 5)\, dt$$

$$= \left[-\frac{2t^3}{3} + \frac{7t^2}{2} - 5t\right]_0^1$$

$$= \left(-\frac{2}{3} + \frac{7}{2} - 5\right) - 0$$

$$= -\frac{13}{6}.$$

$\therefore$ (1) gives

$$\int \vec{f} \cdot d\vec{r} = 4 - \frac{13}{2} - \frac{13}{6}$$

$$= -\frac{15}{3}.$$

Example 7: If $\vec{A} = 2xy\,\vec{i} + (x^2 + 2yz)\vec{j} + (y^2 + 1)\vec{k}$, find the value $\int \vec{A}\,.\,d\vec{r}$ of around the unit circle with centre at the origin in the xy −plane.

Solution: The given curve is $x^2 + y^2 = 1,\ z = 0$.

i.e., $x = \cos\theta$, $y = \sin\theta$, $z = 0$ and θ varies from 0 to 2π.

$\therefore dx = -\sin\theta\ d\theta$, $dy = \cos\theta\ d\theta$, $dz = 0$.

Now, $\vec{A}\,.\,d\vec{r} = \left[2xy\,\vec{i} + (x^2 + 2yz)\vec{j} + (y^2 + 1)\vec{k}\right].\left(dx\vec{i} + dy\vec{j} + dz\vec{k}\right)$

$$= 2xy\,dx + (x^2 + 2yz)\,dy + (y^2 + 1)dz$$

$$= 2\cos\theta\sin\theta\,(-\sin\theta\ d\theta) + (\cos^2\theta + 0)(\cos\theta\ d\theta) + 0$$

$$= (-2\cos\theta\sin^2\theta + \cos^3\theta)\ d\theta.$$

$$\therefore\ \int \vec{A}\,.\,d\vec{r} = \int_0^{2\pi} (\cos^3\theta - 2\cos\theta\sin^2\theta)\ d\theta$$

$$= \int_0^{2\pi} [\cos^3\theta - 2\cos\theta\,(1 - \cos^2\theta)]\ d\theta$$

$$= \int_0^{2\pi} (3\cos^3\theta - 2\cos\theta)\ d\theta$$

$$= \int_0^{2\pi} \left[\frac{3}{4}(\cos 3\theta + 3\cos\theta) - 2\cos\theta\right] d\theta$$

$$= \int_0^{2\pi} \left[\frac{3}{4}\cos 3\theta + \frac{1}{4}\cos\theta\right] d\theta$$

$$= \left[\frac{3}{4}\frac{\sin 3\theta}{3} + \frac{1}{4}\sin\theta\right]_0^{2\pi} = 0.$$

Example 8: Find the work done, when a force $\vec{f} = (x^2 - y^2 + x)\vec{i} - (2xy + y)\vec{j}$ moves a particle from the origin to the point (1, 1) along $y^2 = x$.

Solution: The given curve is $y^2 = x$ from (0, 0) to (1, 1).

$\therefore\ 2y\,dy = dx$ and y varies from 0 to 1.

$$\therefore\ \vec{f}\,.\,d\vec{r} = [(x^2 - y^2 + x)\vec{i} - (2xy + y)\vec{j}].\left(dx\vec{i} + dy\vec{j} + dz\vec{k}\right)$$

$$= (x^2 - y^2 + x)\,dx - (2xy + y)\,dy$$

$$= (y^4 - y^2 + y^2)2y\,dy - (2y^3 + y)\,dy$$

$$= (2\,y^5 - 2y^3 - y)\,dy.$$

$\therefore$ Work done $= \int \vec{f} \, . \, d\vec{r}$

$$= \int_0^1 (2\,y^5 - 2y^3 - y)\,dy$$

$$= \left[\frac{y^6}{3} - \frac{y^4}{2} - \frac{y^2}{2}\right]_0^1$$

$$= \left(\frac{1}{3} - \frac{1}{2} - \frac{1}{2}\right) - 0$$

$$= -\frac{2}{3}.$$

Example 9: Find the work done in moving a particle in the force field $\vec{f} = 3x^2\vec{\imath} + (2xz - y)\vec{\jmath} + z\vec{k}$ along the curve $x^2 = 4y$, $3x^3 = 8z$ from $x = 0$ to $x = 2$.

Solution: The given curve is $y = \frac{x^2}{4}$, $z = \frac{3}{8}x^3$ and x varies from 0 to 2.

$\therefore$ $dy = \frac{x}{2}\,dx$, $dz = \frac{9}{8}x^2 dx$.

$$\vec{f} \,.\, d\vec{r} = \left[3x^2\vec{\imath} + (2xz - y)\vec{\jmath} + z\vec{k}\right] . \left(dx\vec{\imath} + dy\vec{\jmath} + dz\vec{k}\right)$$

$$= 3x^2\,dx + (2xz - y)\,dy + z\,dz$$

$$= 3x^2\,dx + \left(2x\frac{3}{8}x^3 - \frac{x^2}{4}\right)\left(\frac{x}{2}dx\right) + \frac{3}{8}x^3\left(\frac{9}{8}x^2 dx\right)$$

$$= \left(3\,x^2 + \frac{3}{8}x^5 - \frac{x^3}{8} + \frac{27}{64}x^5\right) dx$$

$$= \left(3\,x^2 - \frac{x^3}{8} + \frac{51}{64}x^5\right) dx.$$

$\therefore$ Work done $= \int \vec{f} \,.\, d\vec{r}$

$$= \int_0^2 \left(3\,x^2 - \frac{x^3}{8} + \frac{51}{64}x^5\right) dx$$

$$= \left[x^3 - \frac{x^4}{32} + \frac{51}{384}x^6\right]_0^2$$

$$= \left(8 - \frac{1}{2} + \frac{51}{6}\right) - 0$$

$$= 16\,.$$

Example 10: If $\vec{F} = (4xy - 3x^2z^2)\vec{\imath} + 2x^2\vec{\jmath} - 2x^3z\vec{k}$, check whether the integral is $\int \vec{f}\,.d\vec{r}$ independent of the path C.

Solution: $curl\,\vec{F} = \nabla \times \vec{F} = \begin{vmatrix} \vec{\imath} & \vec{\jmath} & \vec{k} \\ \frac{\partial}{\partial x} & \frac{\partial}{\partial y} & \frac{\partial}{\partial z} \\ 4xy - 3x^2z^2 & 2x^2 & -2x^3z \end{vmatrix}$

$$= \vec{\imath}(0-0) - \vec{\jmath}(-6x^2z + 6x^2z) + \vec{k}(4x - 4x)$$

$$= \vec{0}\,.$$

$\therefore \vec{F}$ is conservative and so $\int \vec{f}\,.d\vec{r}$ is independent of the path C.

Example 11: Evaluate $\int \vec{\varphi}\,.d\vec{r}$ where $\varphi = 2xy^2z + x^2y$ and C is the curve $x = t$, $y = t^2$, $z = t^3$ from $t = 0$ to $t = 1$.

Solution : $\varphi\, d\vec{r} = (2xy^2z + x^2y)(dx\vec{\imath} + dy\vec{\jmath} + dz\vec{k})$

$$= (2t^8 + t^4)(dt\,\vec{\imath} + 2t\,dt\,\vec{\jmath} + 3t^2dt\vec{k})$$

$$= (2t^8 + t^4)dt\,\vec{\imath} + (4t^9 + 2t^5)dt\,\vec{\jmath} + (6t^{10} + 3t^6)dt\,\vec{k}$$

$$\therefore \int \vec{\varphi}\,.d\vec{r} = \int_0^1 [(2t^8 + t^4)dt\,\vec{\imath} + (4t^9 + 2t^5)dt\,\vec{\jmath} + (6t^{10} + 3t^6)dt\,\vec{k}]$$

$$= \vec{\imath}\int_0^1 (2t^8 + t^4)dt + \vec{\jmath}\int_0^1 (4t^9 + 2t^5)dt + \vec{k}\int_0^1 (6t^{10} + 3t^6)dt$$

$$= \vec{\imath}\left[\frac{2t^9}{9} + \frac{t^5}{5}\right]_0^1 + \vec{\jmath}\left[\frac{4t^{10}}{10} + \frac{2t^6}{6}\right]_0^1 + \vec{k}\left[\frac{6t^{11}}{11} + \frac{3t^7}{7}\right]_0^1$$

$$= \frac{19}{45}\vec{\imath} + \frac{11}{15}\vec{\jmath} + \frac{75}{77}\vec{k}\,.$$

EXERCISES

1. Evaluate $\int \vec{f}\,.d\vec{r}$ where $\vec{f} = 3x^2\vec{\imath} + (2xz - y)\vec{\jmath} - z\vec{k}$ from $t = 0$ to $t = 1$ along the curve $x = 2t^2,\ y = t,\ z = 4t^3$.
2. Evaluate $\int \vec{f}\,.d\vec{r}$ along the curve $x = \cos t,\ y = \sin t\,,\ z = t$from $t = 0$ to $t = 2\pi$, given $\vec{F} = z\vec{\imath} + x\,\vec{\jmath} + y\,\vec{k}$.
3. Evaluate $\int \vec{f}\,.d\vec{r}$ where $\vec{f} = x^2\vec{\imath} + xy\,\vec{\jmath}$ from the point $(0,0)$ to $(1,1)$ along the parabola $y^2 = x$.
4. If $\vec{f} = x^2\vec{\imath} + y^3\vec{\jmath}$, evaluate $\int \vec{f}\,.d\vec{r}$ where C is the curve $y = x^3$ in the xy –plane from the point $(1,1)$ to $(2,8)$.

5. If $\vec{F} = (x-y)\vec{i} + (x+y)\vec{j}$, evaluate $\int \vec{f}\,.d\vec{r}$ around the closed curve C which is given by the equations $x = y^2$ and $y = x^2$.
6. Find the work done in moving a particle by the force $\vec{f} = 3x^2\vec{i} + (2xz - y)\vec{j} + z\vec{k}$ along the straight line from $(0,0,0)$to $(2,1,3)$.
7. Evaluate $\int \vec{f}\,.d\vec{r}$ where $\vec{F} = y\vec{i} + z\vec{j} + x\vec{k}$ and C is the triangle in the xy –plane with vertices at $(0,0)$, $(2,0)$ and $(2,1)$.
8. Find the work done in moving a particle in the force field $\vec{f} = 3x^2\vec{i} + (2xz - y)\vec{j} - z\vec{k}$ from $t = 0$ to $t = 1$ along the curve: $x = 2t^2$, $y = t$, $z = 4t^3$.
9. Find the work done in moving a particle in the force field $\vec{f} = 3x^2\vec{i} + (2xy - y)\vec{j} + z\vec{k}$ along the straight line from $(0,0,0)$ to $(2,1,3)$.
10. Prove that $\vec{f} = (y^2\cos x + z^3)\vec{i} + (2y\sin x - 4)\vec{j} + (3xz^2 + 2)\vec{k}$ is (a) conservative field (b) find scalar potential of $\vec{f}$ (c) find work done in moving an object in this field from $P_1(0,1,-1)$ to $P_2\left(\frac{\pi}{2},-1,2\right)$.
11. Show that $\vec{F} = (e^x z - 2xy)\vec{i} - (x^2 - 1)\vec{j} + (e^x + z)\vec{k}$ is a conservative field. Hence evaluate $\int \vec{f}\,.d\vec{r}$ where the end points of C are $(0,1,-1)$ to $(2,3,0)$.
12. If $\vec{f} = xy\,\vec{i} - z\vec{j} + x^2\vec{k}$, evaluate, $\int \vec{f}\,.d\vec{r}$ where C is the curve $x = t^2$, $y = 2t$, $z = t^3$ from $(0,0,0)$ to $(1,2,1)$.

ANSWERS

1. $\frac{13}{6}$ 2. 3π 3. $\frac{7}{12}$ 4. $\frac{12313}{12}$

5. $\frac{2}{3}$ 6. 16 7. -1 8. $\frac{13}{6}$

9. 16 10. $\varphi = y^2\sin x + xz^3 - 4y + z^2 + c$, Work done $= 12 + 5\pi$

11. $-\frac{19}{2}$ 12. $\frac{-9}{10}\vec{i} - \frac{2}{3}\vec{j} + \frac{7}{5}\vec{k}$

2.7. Surface Integrals

An integral which is evaluated over a surface is called a *surface integral.*

Definition: Consider a surface S. Let $\hat{n}$ denote the unit outward normal to the surface S. Let $\vec{f}$ be a vector point function defined in some region containing the surface S. Then the *surface integral* of $\vec{f}$ is defined to be

$$\iint_S \vec{f}.\hat{n}\, dS$$

The surface integral $\iint_S \vec{f}.\hat{n}\, dS$ can also be written as $\iint_S \vec{f}.d\vec{S}$.

Note: $\iint_S \varphi\, dS$, $\iint_S \vec{f} \times d\vec{S}$, $\iint_S \vec{u}.\vec{v}\, dS$, $\iint_S \vec{u} \times \vec{v}\, dS$ are also surface integrals.

Evaluation of a Surface Integral

To evaluate a surface integral, we convert it into a double integral by projecting the given surface S onto one of the coordinate planes.

If R is the projection of S on the xy –plane, then

$$\iint_S \vec{f}.\hat{n}\, dS = \iint_R \frac{\vec{f}.\hat{n}}{|\hat{n}.\vec{k}|}\, dxdy.$$

If R_1 is the projection of S on the yz –plane, then

$$\iint_S \vec{f}.\hat{n}\, dS = \iint_{R_1} \frac{\vec{f}.\hat{n}}{|\hat{n}.\vec{i}|}\, dydz.$$

If R_2 is the projection of S on the xz –plane, then

$$\iint_S \vec{f}.\hat{n}\, dS = \iint_{R_2} \frac{\vec{f}.\hat{n}}{|\hat{n}.\vec{j}|}\, dxdz.$$

Physical Meaning: Let $\vec{f}$ denote the velocity vector of a moving fluid in which a fixed surface is S. Then the surface integral $\int_S \vec{f}.\hat{n}\, dS$ will represent the total flux of the vector $\vec{f}$ through the whole surface S.

Example 1: Evaluate $\iint_S \vec{f}.\hat{n}\, dS$ if $\vec{f} = (x + y^2)\vec{i} - 2x\vec{j} + 2yz\vec{k}$ and S is the surface of the plane $2x + y = 6$ in the first octant.

Solution:

Given that S is the surface of the plane $2x + y + 2z = 6$ in the first octant. Let R be the projection of S on the xy –plane. Then R is the plane $2x + y = 6$ in the first quadrant. (see figure)

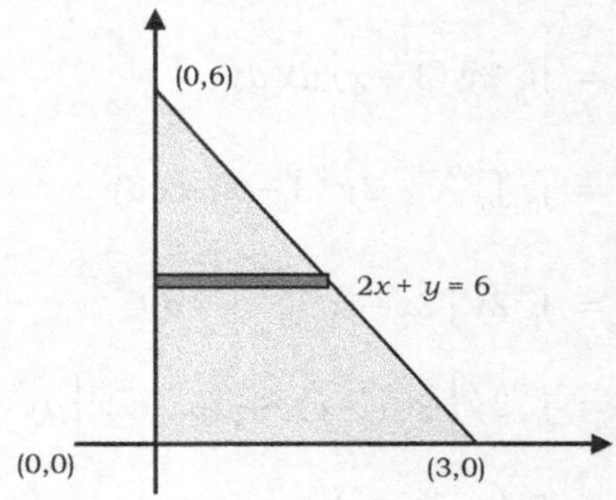

Then $\iint_S \vec{f}.\hat{n}\, dS = \iint_R \frac{\vec{f}.\hat{n}}{|\hat{n}.\vec{k}|}\; dxdy$ (1)

Let $\varphi = 2x + y + 2z - 6.$

Then $\nabla\varphi = \frac{\partial\varphi}{\partial x}\vec{i} + \frac{\partial\varphi}{\partial y}\vec{j} + \frac{\partial\varphi}{\partial z}\vec{k}$

$= 2\vec{i} + \vec{j} + 2\vec{k}$

and $|\nabla\varphi| = \sqrt{4+1+4}$

$= 3.$

$\therefore \hat{n} = \frac{\nabla\varphi}{|\nabla\varphi|} = \frac{2\vec{i}+\vec{j}+2\vec{k}}{3}.$

Now $, \vec{f}.\hat{n} = [(x+y^2)\vec{i} - 2x\vec{j} + 2yz\vec{k}].\left(\frac{2\vec{i}+\vec{j}+2\vec{k}}{3}\right)$

$= \frac{2}{3}(x+y^2) - \frac{2}{3}x + \frac{4}{3}\,yz$

$= \frac{2}{3}(y^2 + 2yz)$

$\hat{n}.\vec{k} = \left(\frac{2\vec{i}+\vec{j}+2\vec{k}}{3}\right).\vec{k} = \frac{2}{3}.$

$\therefore \frac{\vec{f}.\hat{n}}{|\hat{n}.\vec{k}|} = y^2 + 2yz = y^2 + 2y\left(\frac{6-2x-y}{2}\right)$

$= y^2 + 6y - 2xy - y^2$

$= 6y - 2xy = 2y(3-x).$

In R, x varies from 0 to $\frac{1}{2}(6-y)$ and y varies from 0 to 6.

$\therefore$ (1) gives $\iint_S \vec{f}.\hat{n}\, dS = \iint_R 2y\,(3-x)\, dx\, dy$

$$= \int_0^6 \int_0^{\frac{1}{2}(6-y)} 2y\,(3-x)\, dx\, dy$$

$$= \int_0^6 2y\,[\,3x - x^2\,]_0^{\frac{1}{2}(6-y)}\, dy$$

$$= \int_0^6 2y\left[\frac{3}{2}(6-y) - \frac{1}{4}(6-y)^2\right] dy$$

$$= \int_0^6 \left(3y^2 - \frac{y^3}{2}\right) dy$$

$$= \left[y^3 - \frac{y^4}{8}\right]_0^6$$

$$= (216 - 162) - 0$$

$$= 54.$$

Example 2: Evaluate $\iint_S curl\,\vec{f}.\hat{n}\, dS$ where $\vec{f} = y^2\vec{i} + y\vec{j} - xz\vec{k}$ and S is the upper half of the sphere $x^2+y^2+z^2=a^2$ and $z \geq 0$.

Solution: Given that S is the upper half of the sphere $x^2+y^2+z^2=a^2$ and $z \geq 0$. (1)

Let R be the projection of S on the xy –plane.Then R is the circle $x^2+y^2=a^2$.

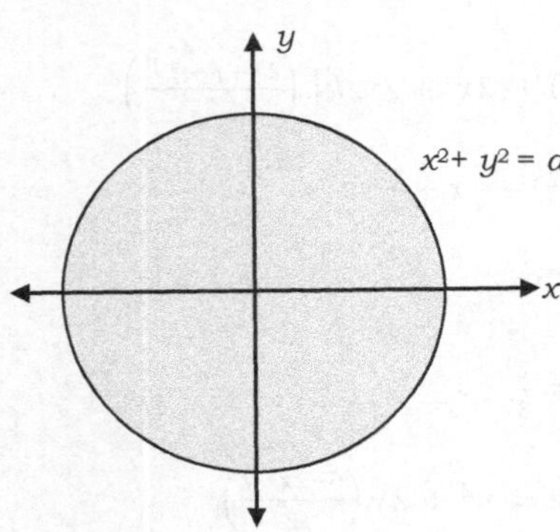

Then $\iint_S curl\,\vec{f}.\hat{n}\, dS = \iint_R \frac{\vec{f}.\vec{n}}{|\vec{n}.\vec{k}|}\, dx\, dy$ (2)

Let $\varphi = x^2+y^2+z^2-a^2$.

Then $\nabla\varphi = 2x\vec{i} + 2y\vec{j} + 2z\vec{k}$

$$= 2(x\vec{i} + y\vec{j} + z\vec{k})$$

$$|\nabla\varphi| = \sqrt{4x^2 + 4y^2 + 4z^2}$$

$$= 2\sqrt{x^2 + y^2 + z^2}$$

$$= 2a \quad \text{by (1)}$$

$$\therefore \hat{n} = \frac{\nabla\varphi}{|\nabla\varphi|} = \frac{2(x\vec{i} + y\vec{j} + z\vec{k})}{2a}$$

$$= \frac{1}{a}(x\vec{i} + y\vec{j} + z\vec{k})$$

$$curl\ \vec{f} = \begin{vmatrix} \vec{i} & \vec{j} & \vec{k} \\ \frac{\partial}{\partial x} & \frac{\partial}{\partial y} & \frac{\partial}{\partial z} \\ y^2 & y & -xz \end{vmatrix}$$

$$= \vec{i}(0-0) - \vec{j}(-z-0) + \vec{k}(0-2y)$$

$$= z\vec{j} - 2y\vec{k}$$

$$curl\ \vec{f}.\hat{n} = (z\vec{j} - 2y\vec{k}).\frac{1}{a}(x\vec{i} + y\vec{j} + z\vec{k})$$

$$= \frac{1}{a}(zy - 2yz) = -\frac{yz}{a}$$

$$\hat{n}.\vec{k} = \frac{1}{a}(x\vec{i} + y\vec{j} + z\vec{k}).\vec{k} = \frac{z}{a}$$

$$\therefore \frac{curl\ \vec{f}.\hat{n}}{|\hat{n}.\vec{k}|} = \frac{\frac{-yz}{a}}{\frac{z}{a}} = -y.$$

In R, y varies from $-\sqrt{a^2 - x^2}$ to $\sqrt{a^2 - x^2}$ and x varies from $-a$ to a.

$\therefore$ (2) gives $\iint_S \vec{f}.\hat{n}\, dS = \int_{-a}^{a} \int_{-\sqrt{a^2-x^2}}^{\sqrt{a^2-x^2}} (-y)dy\, dx$

$$= -\int_{-a}^{a} \left[\frac{y^2}{2}\right]_{-\sqrt{a^2-x^2}}^{\sqrt{a^2-x^2}} dx = -\frac{1}{2}\int_{-a}^{a} [(a^2 - x^2) - (a^2 - x^2)]\, dx$$

$$= -\frac{1}{2}\int_{-a}^{a} 0\, dx = 0.$$

Example 3: Evaluate $\iint_S \vec{f}.d\vec{S}$ if $\vec{f} = yz\vec{i} + 2y^2\vec{j} + xz^2\vec{k}$ and S is the surface of the cylinder $x^2 + y^2 = 9$ contained in the first octant between the planes $z = 0$ and $z = 2$.

Solution: Given surface S is the cylinder

$$x^2 + y^2 = 9, \quad z = 0 \text{ and } z = 2. \qquad (1)$$

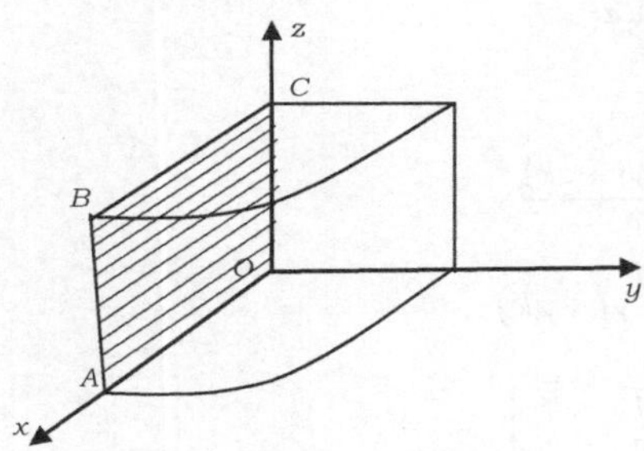

Let R be the projection of S on the xz –plane. Then R is the rectangular region $OABC$, as shown in the figure.

$$\therefore \iint_S \vec{f}.d\vec{S} = \iint_R \frac{\vec{f}.\hat{n}}{|\hat{n}.\vec{j}|}\,dxdz \qquad (2)$$

Let $\varphi = x^2 + y^2 - 9$. Then $\nabla\varphi = 2x\vec{i} + 2y\vec{j}$.

$|\nabla\varphi| = \sqrt{4x^2 + 4y^2} = 2\sqrt{x^2 + y^2} = 6$, by (1)

$$\therefore \hat{n} = \frac{\nabla\varphi}{|\nabla\varphi|} = \frac{2x\vec{i} + 2y\vec{j}}{6} = \frac{1}{3}(x\vec{i} + y\vec{j})$$

$$\vec{f}.\hat{n} = (yz\vec{i} + 2y^2\vec{j} + xz^2\vec{k}).\frac{1}{3}(x\vec{i} + y\vec{j}) = \frac{1}{3}(xyz + 2y^3).$$

$$\therefore \hat{n}.\vec{j} = \frac{1}{3}(x\vec{i} + y\vec{j}).\vec{j} = \frac{1}{3}y.$$

$$\therefore \frac{\vec{f}.\hat{n}}{|\hat{n}.\vec{j}|} = \frac{\frac{1}{3}(xyz + 2y^3)}{\frac{1}{3}y} = xz + 2y^2 = xz + 2(9 - x^2), \qquad \text{by(1).}$$

In R, x varies from 0 to 3 and z varies from 0 to 2.

$$\therefore \text{(2) gives } \iint_S \vec{f}.d\vec{S} = \int_0^2\int_0^3 [xz + 2(9 - x^2)]\,dxdz$$

$$= \int_0^2 \left[\frac{x^2z}{2} + 18x - \frac{2x^3}{3}\right]_0^3 dz = \int_0^2 \left(\frac{9z}{2} + 54 - 18\right) dz$$

$$= \int_0^2 \left(\frac{9z}{2} + 36\right) dz = \left[\frac{9}{2}\frac{z^2}{2} + 36z\right]_0^2 = (9 + 72) - 0 = 81.$$

Example 4: Evaluate $\iint_S \vec{F}.\hat{n}\, d$ where $\vec{F} = z\,\vec{i} + x\,\vec{j} - 3y^2 z\,\vec{k}$ and S is the surface of the cylinder $x^2 + y^2 = 16$ included in the first octant between $z = 0$ and $z = 5$.

Solution: **[AU, *Nov/Dec 2017*]**

Given surface S is the cylinder

$$x^2 + y^2 = 16, \quad z = 0 \text{ and } z = 5. \qquad (1)$$

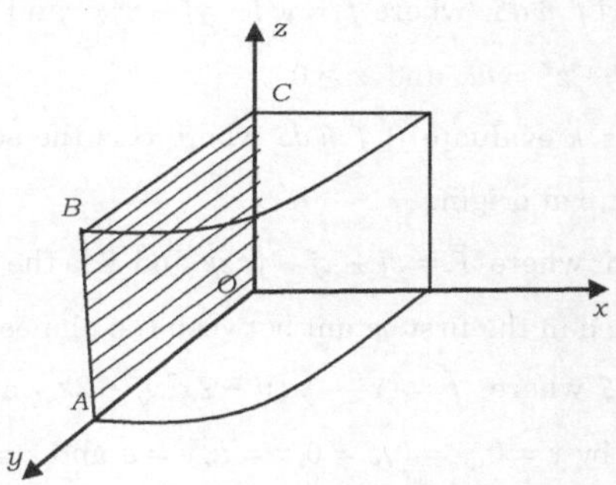

Let R be the projection of S on the yz –plane. Then R is the rectangular region $OABC$, as shown in the figure.

$$\therefore \iint_S \vec{f}.d\vec{S} = \iint_R \frac{\vec{f}.\hat{n}}{|\hat{n}.\vec{\imath}|}\, dydz \qquad (2)$$

Let $\varphi = x^2 + y^2 - 16$. Then $\nabla\varphi = 2x\vec{\imath} + 2y\vec{\jmath}$.

$$|\nabla\varphi| = \sqrt{4x^2 + 4y^2} = 2\sqrt{x^2 + y^2} = 8, \text{ by (1)}$$

$$\therefore \hat{n} = \frac{\nabla\varphi}{|\nabla\varphi|} = \frac{2x\vec{\imath} + 2y\vec{\jmath}}{8} = \frac{1}{4}(x\vec{\imath} + y\vec{\jmath})$$

$$\vec{f}.\hat{n} = (z\,\vec{i} + x\,\vec{j} - 3y^2 z\,\vec{k}).\frac{1}{4}(x\vec{\imath} + y\vec{\jmath}) = \frac{1}{4}(zx + xy) = \frac{x}{4}(z + y).$$

$$\therefore \hat{n}.\vec{\imath} = \frac{1}{4}(x\vec{\imath} + y\vec{\jmath}).\vec{\jmath} = \frac{1}{4}x.$$

$$\therefore \frac{\vec{f}.\hat{n}}{|\hat{n}.\vec{\imath}|} = \frac{\frac{x}{4}(y+z)}{\frac{1}{4}x} = y + z$$

In R, x varies from 0 to 4 and z varies from 0 to 5.

$$\therefore \text{(2) gives } \iint_S \vec{f}.d\vec{S} = \int_0^5 \int_0^4 [y + z]\, dydz$$

$$= \int_0^5 \left[\frac{y^2}{2} + yz\right]_0^4 dz = \int_0^5 (8 + 4z)dz$$

$$= \left[8z + \frac{4z^2}{2}\right]_0^5 = (40 + 50) - 0 = 90.$$

EXERCISES

1. Evaluate $\iint_S \vec{f}.\hat{n}\, dS$where $\vec{F} = 12x^2y\vec{\imath} - 3yz\vec{\jmath} + 2z\vec{k}$ and S is the portion of the plane $x + y + z = 1$ included in the first octant. **[*AU, May/June 2007*]**
2. Evaluate $\iint_S \vec{f}.\hat{n}\, dS$ where $\vec{f} = 18z\vec{\imath} - 12\vec{\jmath} + 3y\vec{k}$ and S is the part of the plane $2x + 3y + 6z = 12$ which is in the first octant.
3. Evaluate $\iint_S curl\, \vec{f}.\hat{n}\, dS$ where $\vec{f} = y^2\vec{\imath} + y\vec{\jmath} - x^2\vec{k}$ and S is the upper half of the sphere $x^2 + y^2 + z^2 = a^2$ and $z \geq 0$.
4. If $\vec{F} = x^3\vec{\imath} + y^3\vec{\jmath} + z^3\vec{k}$ evaluate $\iint_S \vec{f}.\hat{n}\, dS$ where S is the surface of the sphere of radius a with centre at origin.
5. Evaluate $\iint_S \vec{f}.\hat{n}\, dS$ where $\vec{F} = z\vec{\imath} + x\vec{\jmath} - y^2z\vec{k}$ and S is the surface of the cylinder $x^2 + y^2 = 1$ included in the first octant between the planes $z = 0$ and $z = 2$.
6. Evaluate $\iint_S \vec{f}.\hat{n}\, dS$ where $\vec{f} = (x^3 - yz)\vec{\imath} - 2x^2y\vec{\jmath} + 2\vec{k}$ and S is the surface of the cube bounded by $x = 0, y = 0, z = 0, x = a, y = a$ and $z = a$.
7. If $\vec{F} = 2xy\vec{\imath} + yz^2\vec{\jmath} + xz\vec{k}$ and S is the surface of the parallelopiped bounded by the planes $x = 0, x = 2, y = 0, y = 1, z = 0, z = 3$, evaluate $\iint_S \vec{f}.d\vec{S}$.

ANSWERS

1. $\frac{49}{120}$ 2. 24 3. 0 4. $\frac{12}{5}\pi a^5$ 5. 3

6. $\frac{1}{3}a^5$ 7. 30.

2.8. Volume Integrals

An integral which is evaluated over a volume bounded by a surface is called a *volume integral.*

Definition: Let V be a region in space enclosed by a closed surface. Let $\vec{F}$ be a vector point function. Then the triple integral $\iiint_V \vec{F}\, dV$ is known as *volume integral* or *space integral.*

Note: **(i)** If φ is a scalar point function, then $\iiint_V \varphi\, dV$ is also a volume integral.

(ii) If $\vec{F} = F_1\vec{\imath} + F_2\vec{\jmath} + F_3\vec{k}$, then

$$\iiint_V \vec{F}\, dV = \vec{\imath} \iiint_V F_1\, dxdydz + \vec{\jmath} \iiint_V F_2\, dxdydz + \vec{k} \iiint_V F_3\, dxdydz.$$

Example 1: Evaluate $\iiint_V (4 - 4y + 2z)\, dV$, where V is the region bounded by the cylinder $x^2 + y^2 = 4$, $z = 0$ and $z = 3$.

Solution:

$$\iiint_V (4 - 4y + 2z)\, dV = \int_{-2}^{2} \int_{-\sqrt{4-x^2}}^{\sqrt{4-x^2}} \int_0^3 (4 - 4y + 2z) dzdydx$$

$$= \int_{-2}^{2} \int_{-\sqrt{4-x^2}}^{\sqrt{4-x^2}} [4z - 4yz + z^2]_0^3\, dydx$$

$$= \int_{-2}^{2} \int_{-\sqrt{4-x^2}}^{\sqrt{4-x^2}} (21 - 12y)\, dydx$$

$$= \int_{-2}^{2} [21y - 6y^2]_{-\sqrt{4-x^2}}^{\sqrt{4-x^2}}\, dx$$

$$= \int_{-2}^{2} \Big[\left(21\sqrt{4 - x^2} - 6(4 - x^2)\right) - \left(-21\sqrt{4 - x^2} - 6(4 - x^2)\right)\Big]\, dx$$

$$= 42 \int_{-2}^{2} \sqrt{4 - x^2}\; dx$$

$$= 84 \int_0^2 \sqrt{4 - x^2}\, dx \qquad [\because \sqrt{4 - x^2} \text{ is an even function}]$$

$$= 84 \left[\frac{x}{2}\sqrt{4 - x^2} + \frac{4}{2}\sin^{-1}\left(\frac{x}{2}\right)\right]_0^2 = 84\, [2 \sin^{-1}(1) - 0]$$

$$= 84 \left[2\left(\frac{\pi}{2}\right)\right] = 84\pi.$$

Example 2: If $\vec{f} = x^2\vec{\imath} + z\vec{\jmath} + yz\vec{k}$, $\iiint_V \nabla.\vec{f}\, dV$ where V is the volume bounded by the cube $x = \pm 1$, $y = \pm 1$, $z = \pm 1$.

Solution: $\nabla.\vec{f} = \frac{\partial}{\partial x}(x^2) + \frac{\partial}{\partial y}(z) + \frac{\partial}{\partial z}(yz) = 2x + 0 + y = 2x + y.$

$$\therefore \iiint_V \nabla.\vec{f}\, dV = \int_{-1}^{1} \int_{-1}^{1} \int_{-1}^{1} (2x + y)\, dxdydz$$

$$= \int_{-1}^{1} \int_{-1}^{1} [x^2 + xy]_{-1}^{1} dydz$$

$$= \int_{-1}^{1} \int_{-1}^{1} [(1 + y) - (1 - y)]\, dydz$$

$$= \int_{-1}^{1} \int_{-1}^{1} 2y\, dy\, dz$$

$$= \int_{-1}^{1} [y^2]_{-1}^{1}\, dz = \int_{-1}^{1} 0\, dz = 0.$$

Example 3: Evaluate $\iiint_V div\,\vec{F}\,dV$, where $\vec{F} = x\vec{\imath} - y\vec{\jmath} + (z^2 - 1)\vec{k}$ and V is the region bounded by the planes $z = 0, z = 1, x^2 + y^2 = 4$.

Solution: $div\,\vec{F} = \frac{\partial}{\partial x}(x) + \frac{\partial}{\partial y}(-y) + \frac{\partial}{\partial z}(z^2 - 1) = 1 - 1 + 2z = 2z.$

$$\therefore \iiint_V div\,\vec{F}\,dV = \int_{-2}^{2}\int_{-\sqrt{4-x^2}}^{\sqrt{4-x^2}}\int_0^1 2z\,dzdydx$$

$$= \int_{-2}^{2}\int_{-\sqrt{4-x^2}}^{\sqrt{4-x^2}} [z^2]_0^1\,dy\,dx$$

$$= \int_{-2}^{2}\int_{-\sqrt{4-x^2}}^{\sqrt{4-x^2}} dydx = \int_{-2}^{2}[y]_{-\sqrt{4-x^2}}^{\sqrt{4-x^2}}dx$$

$$= \int_{-2}^{2} 2\sqrt{4-x^2}\,dx = 4\int_0^2 \sqrt{4-x^2}\,dx$$

$$= 4\left[\frac{x}{2}\sqrt{4-x^2} + \frac{4}{2}\sin^{-1}\left(\frac{x}{2}\right)\right]_0^2 = 8\left(\frac{\pi}{2}\right) = 4\pi.$$

Example 4: If $\vec{F} = (2x^2 - 3z)\vec{\imath} - 2xy\,\vec{\jmath} - 4x\,\vec{k}$, evaluaet $\iiint_V \nabla\times\vec{F}\,dV$ where V is the region bounded by $x = 0, y = 0, z = 0$ and $2x + 2y + z = 4$.

Solution: $\nabla\times\vec{F} = \begin{vmatrix} \vec{\imath} & \vec{\jmath} & \vec{k} \\ \frac{\partial}{\partial x} & \frac{\partial}{\partial y} & \frac{\partial}{\partial z} \\ 2x^2 - 3z & -2xy & -4x \end{vmatrix} = \vec{\imath}\,(0) - \vec{\jmath}\,(-4+3) + \vec{k}(-2y - 0) = \vec{\jmath} - 2y\vec{k}.$

$$\iiint_V \nabla\times\vec{F}\,dV = \int_0^2\int_0^{2-x}\int_0^{4-2x-2y}(\vec{\jmath} - 2y\,\vec{k})\,dz\,dy\,dx$$

$$= \int_0^2\int_0^{2-x}[z\,\vec{\jmath} - 2yz\,\vec{k}]_0^{4-2x-2y}\,dy\,dx$$

$$= \int_0^2\int_0^{2-x}[(4-2x-2y)\,\vec{\jmath} - 2y(4-2x-2y)\,\vec{k}]\,dy\,dx$$

$$= \int_0^2\left[(4y - 2xy - y^2)\vec{\jmath} - \left(4y^2 - 2xy^2 - \frac{4}{3}y^3\right)\vec{k}\right]_0^{2-x}$$

$$= \int_0^2\{[4(2-x) - 2x(2-x) - (2-x)^2]\vec{\jmath} - [4(2-x)^2 - 2x\,(2-x)^2 -$$

432−x3kdx

$$= \int_0^2\left[(4 - 4x + x^2)\vec{\jmath} - \frac{\vec{k}}{3}(80 - 120x + 60x^2 - 10x^3)\right]dx$$

$$= \left[\left(4x - 2x^2 + \frac{x^3}{3}\right)\vec{\jmath} - \frac{10}{3}\vec{k}\left(8x - 12\frac{x^2}{2} + 6\frac{x^3}{3} - \frac{x^4}{4}\right)\right]_0^2$$

$$= \left(8 - 8 + \frac{8}{3}\right)\vec{\jmath} - \frac{10}{3}\vec{k}(16 - 24 + 16 - 4) = \frac{8}{3}(\vec{\jmath} - 5\vec{k}).$$

EXERCISES

1. Evaluate $\iiint_V \nabla \times \vec{F}\, dV$, where $\vec{F} = x^2\vec{i} + y^2\vec{j} + z^2\vec{k}$ and V is the volume enclosed by the cube $x = 0, x = 1, y = 0, y = 1, z = 0$ and $z = 1$.

2. Evaluate $\iiint_V div\, \vec{F}\, dV$ where $\vec{F} = 2x^2\vec{i} - y^2\vec{j} + 4xz^2\vec{k}$ and V is the region in the first octant bounded by the cylinder $y^2 + z^2 = 9$ and $x = 0$.

3. If $\vec{F} = 2xz\,\vec{i} - x\,\vec{j} + y^2\,\vec{k}$, evaluate $\iiint_V \vec{F}\, dV$ where V is the region bounded by the planes $x = y = z = 0, x = y = z = 1$.

4. If $\vec{F} = (2x^2 - 3z)\,\vec{i} - 2xy\,\vec{j} - 4x\,\vec{k}$, evaluate $\iiint_V \nabla.\vec{F}\, dV$ where V is the region bounded by $x = 0, y = 0, z = 0$ and $2x + 2y + z = 4$.

5. If $\varphi = 4y + 2xz$, evaluate $\iiint_V \varphi\, dV$ over the region in the first octant bounded by $x^2 + y^2 = 9, z = 0,\ z = 2$.

6. Evaluate $\iiint_V \vec{F}\, dV$, for $\vec{F} = x\vec{i} + y\vec{j} + z\vec{k}$ where V is the region bounded by the surface $x = 0, y = 0, y = 6, z = 4$ and $z = x^2$.

ANSWERS

1. 3   2. 180   3. $\frac{1}{2}\vec{i} - \frac{1}{2}\vec{j} + \frac{1}{3}\vec{k}$ 4. $\frac{8}{3}$ 5. 108   6. $24\vec{i} + 96\vec{j} + \frac{384}{5}\vec{k}$.

2.9. Green's Theorem in a Plane

Theorem: (Green's Theorem) **[AU, Nov/Dec 2015]**

If C is a simple closed curve in the xy – plane and R is the region bounded by C, then

$$\int_C M\, dx + N\, dy = \iint_R \left(\frac{\partial N}{\partial x} - \frac{\partial M}{\partial y}\right) dxdy,$$

Where M and N are continuously differentiable functions of x and y inside and on C.

Corollary`1: The area A of the plane region R bounded by the simple closed curve C is $A = \frac{1}{2}\int_C (xdy - ydx)$.

Corollary 2: If $\frac{\partial N}{\partial x} = \frac{\partial M}{\partial y}$, then by Green's theorem $\int_C Mdx + Ndy = 0$.

Example 1: Verify Green's theorem in the plane for $\int_C (3x^2 - 8y^2)dx + (4y - 6xy)dy$ where C is the boundary of the region defined by $x = 0, y = 0,\ x + y = 1$.

Solution: ***[AU, Nov/Dec 2014]***

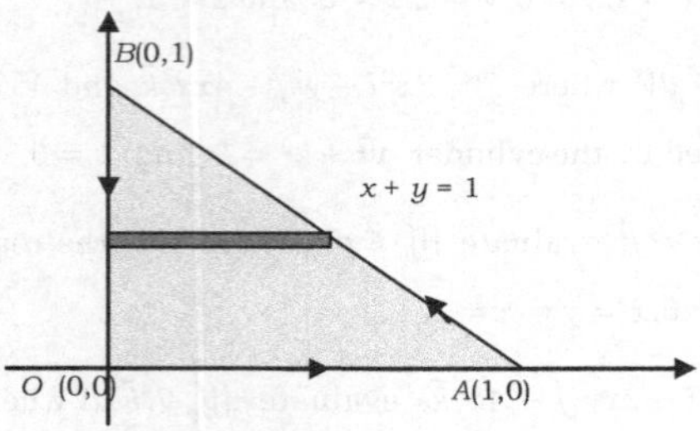

The Green's theorem is

$$\int_C Mdx + Ndy = \iint_R \left(\frac{\partial N}{\partial x} - \frac{\partial M}{\partial y}\right) dxdy.$$

Consider $\int_C (3x^2 - 8y^2)dx + (4y - 6xy)dy$.

Let $M = 3x^2 - 8y^2$ and $N = 4y - 6xy$.

$\therefore \quad \frac{\partial M}{\partial y} = -16y$ and $\frac{\partial N}{\partial x} = -6y$.

$\therefore \quad \frac{\partial N}{\partial x} - \frac{\partial M}{\partial y} = -6y + 16y = 10y.$

In R, x varies from 0 to $1 - y$ and y varies from 0 to 1.

$$\therefore \quad \iint_R \left(\frac{\partial N}{\partial x} - \frac{\partial M}{\partial y}\right) dxdy = \int_0^1 \int_0^{1-y} 10ydxdy$$

$$= \int_0^1 10y[x]_0^{1-y} dy = \int_0^1 10y[1 - y]dy$$

$$= 10 \int_0^1 (y - y^2)dy = 10\left[\frac{y^2}{2} - \frac{y^3}{3}\right]_0^1$$

$$= 10\left[\frac{1}{2} - \frac{1}{3}\right] = \frac{5}{3}. \qquad (1)$$

To evaluate $\int_C Mdx + Ndy$

The curve is $C = OA + AB + BO$.

$$\therefore \quad \int_C Mdx + Ndy = \int_{OA} + \int_{AB} + \int_{BO} [Mdx + Ndy] \qquad (2)$$

(i) Along the straight line OA

$y = 0$ and x varies from 0 to 1. $\therefore\ dy = 0.$

$$Mdx + Ndy = 3x^2\,dx$$

$$\therefore\ \int_{OA} Mdx + Ndy = \int_0^1 3x^2\,dx$$

$$= \left[\frac{3\,x^3}{3}\right]_0^1 = 1.$$

(ii) Along the straight line AB

$y = 1 - x$ and x varies from 1 to 0. $\therefore\ dy = -dx$

$$Mdx + Ndy = [3x^2 - 8(1-x)^2]dx + [4(1-x) - 6x(1-x)](-dx)$$

$$= [3x^2 - 8 - 8x^2 + 16x - 4 + 4x + 6x - 6x^2]dx$$

$$= [-12 + 26x - 11x^2]dx$$

$$\therefore\ \int_{AB} Mdx + Ndy = \int_1^0 [-12 + 26x - 11x^2]\,dx$$

$$= \left[-12x + 26\frac{x^2}{2} - 11\frac{x^3}{3}\right]_1^0$$

$$= 0 - \left(-12 + 13 - \frac{11}{3}\right) = \frac{8}{3}.$$

(iii) Along the straight line BO

$x = 0$ and y varies from 1 to 0. $\therefore\ dx = 0.$

$$Mdx + Ndy = 4ydy$$

$$\therefore\ \int_{BO} Mdx + Ndy = \int_1^0 4ydy$$

$$= \left[4\frac{y^2}{2}\right]_1^0 = 0 - 2 = -2.$$

$$\therefore\ (2) \Rightarrow \int_C Mdx + Ndy = 1 + \frac{8}{3} - 2 = \frac{5}{3}. \qquad (3)$$

From (1) and (3), $\int_C Mdx + Ndy = \iint_R \left(\frac{\partial N}{\partial x} - \frac{\partial M}{\partial y}\right) dxdy.$

Hence Green's theorem is verified.

Example 2: Verify Green's theorem in the plane with respect to $\int_C (x^2 - y^2)dx + 2xydy$, where C is the boundary of the rectangle in the $xy-$ plane bounded by the lines $x = 0, x = a, y = 0$ and $y = b$.

Solution:

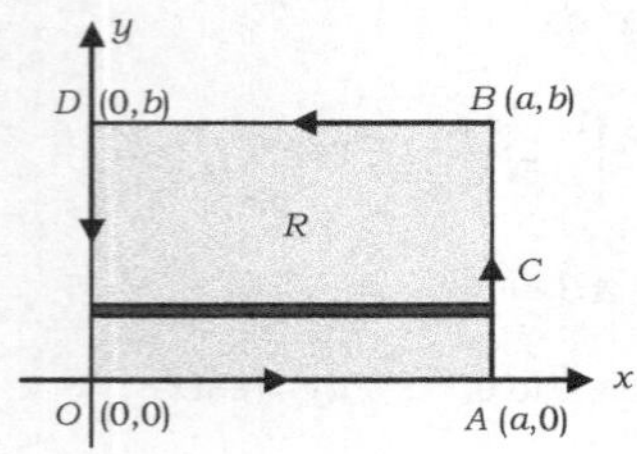

The Green's theorem is

$$\int_C Mdx + Ndy = \iint_R \left(\frac{\partial N}{\partial x} - \frac{\partial M}{\partial y}\right) dxdy.$$

Consider $\int_C (x^2 - y^2)dx + 2xydy.$

Let $M = x^2 - y^2$ and $N = 2xy.$

$$\frac{\partial M}{\partial y} = -2y \quad \text{and} \quad \frac{\partial N}{\partial x} = 2y.$$

$$\therefore \frac{\partial N}{\partial x} - \frac{\partial M}{\partial y} = 2y - (-2y) = 4y.$$

In the region R, x varies from 0 to a and y varies from 0 to b.

$$\therefore \iint_R \left(\frac{\partial N}{\partial x} - \frac{\partial M}{\partial y}\right) dxdy = \int_0^b \int_0^a 4y\, dx\, dy$$

$$= \int_0^b [4yx]_0^a dy = \int_0^b 4y[a - 0]dy = 4a \int_0^b y\, dy = 4a \left[\frac{y^2}{2}\right]_0^b$$

$$= 2ab^2. \qquad (1)$$

To evaluate $\int_C Mdx + Ndy.$

The curve is $C : OA + AB + BD + DO.$

$$\therefore \int_C Mdx + Ndy = \int_{OA} + \int_{AB} + \int_{BD} + \int_{DO} [Mdx + Ndy]. \qquad (2)$$

(i) Along the straight line OA

$y = 0$ and x varies from 0 to a. $\therefore dy = 0$.

$Mdx + Ndy = (x^2 - 0)dx = x^2 dx$.

$$\therefore \int_{OA} Mdx + Ndy = \int_0^a x^2\,dx$$

$$= \left[\frac{x^3}{3}\right]_0^a$$

$$= \frac{a^3}{3}.$$

(ii) Along the straight line AB

$x = a$ and y varies from 0 to b. $\therefore dx = 0$.

$Mdx + Ndy = 2aydy$.

$$\therefore \int_{AB} Mdx + Ndy = \int_0^b 2aydy = 2a\left[\frac{y^2}{2}\right]_0^b = ab^2.$$

(iii) Along the straight line BD

$y = b$ and x varies from a to 0. $\therefore dy = 0$.

$Mdx + Ndy = (x^2 - b^2)dx$.

$$\therefore \int_{BD} Mdx + Ndy = \int_a^0 (x^2 - b^2)dx = \left[\frac{x^3}{3} - b^2x\right]_a^0 = ab^2 - \frac{a^3}{3}.$$

(iv) Along the straight line DO

$x = 0$ and y varies from b to 0. $\therefore dx = 0$.

$\therefore M\,dx + N\,dy = 0$

$$\therefore \int_{BD} Mdx + Ndy = \int_b^0 0\,dy = 0$$

$\therefore$ (2) gives $\int_C Mdx + Ndy = \frac{a^3}{3} + ab^2 + ab^2 - \frac{a^3}{3} + 0 = 2ab^2$. (3)

From (1) and (3), $\int_C Mdx + Ndy = \iint_R \left(\frac{\partial N}{\partial x} - \frac{\partial M}{\partial y}\right) dxdy$.

Hence Green's theorem is verified.

Example 3: Using Green's theorem in a plane evaluate $\int_C [x^2(1+y)dx + (x^3+y^3)dy]$, where C is the square bounded by $x = \pm 1,\ y = \pm 1$. **[*AU, May/June 2016*]**

Solution:

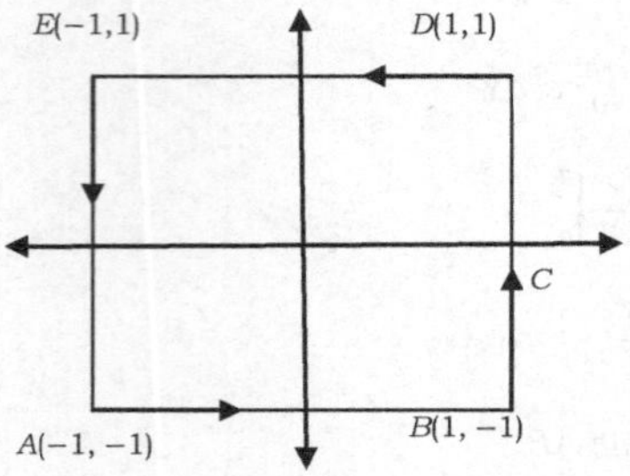

The Green's theorem is

$$\int_C Mdx + Ndy = \iint_R \left(\frac{\partial N}{\partial x} - \frac{\partial M}{\partial y}\right) dxdy.$$

Consider $\int_C x^2(1+y)dx + (y^3+x^3)dy$.

Let $M = x^2(1+y)$ and $N = y^3 + x^3$.

$$\therefore \quad \frac{\partial M}{\partial y} = x^2 \text{ and } \frac{\partial N}{\partial x} = 3x^2.$$

$$\therefore \quad \frac{\partial N}{\partial x} - \frac{\partial M}{\partial y} = 3x^2 - x^2 = 2x^2.$$

In R, x varies from -1 to 1 and y varies from -1 to 1.

$$\therefore \quad \iint_R \left(\frac{\partial N}{\partial x} - \frac{\partial M}{\partial y}\right) dxdy = \int_{-1}^{1}\int_{-1}^{1} 2x^2\, dxdy$$

$$= \int_{-1}^{1} \left[\frac{2x^3}{3}\right]_{-1}^{1} dy$$

$$= \int_{-1}^{1} \left[\frac{2}{3} - \left(\frac{-2}{3}\right)\right] dy$$

$$= \frac{4}{3}\int_{-1}^{1} dy = \frac{4}{3}\,[y]_{-1}^{1}$$

$$= \frac{4}{3}\,[1-(-1)] = \frac{8}{3}$$

Example 4: Evaluate $\int_C (x^2 + xy)dx + (x^2 + y^2)dy$ where C is the square formed by the lines $x = \pm 1, y = \pm 1$.

Solution: ***[AU, Nov/Dec 2017]***

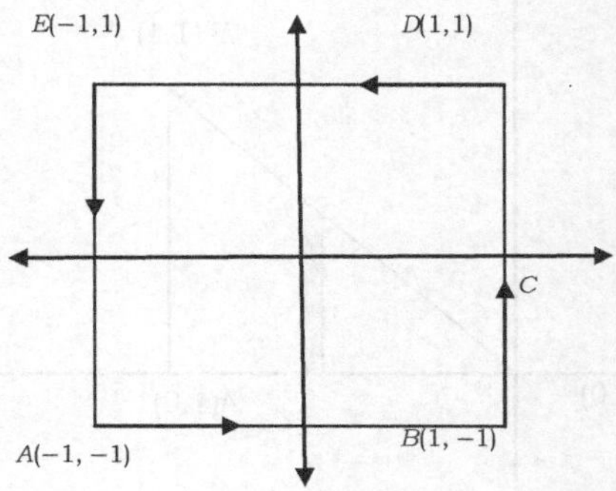

The Green's theorem is

$$\int_C Mdx + Ndy = \iint_R \left(\frac{\partial N}{\partial x} - \frac{\partial M}{\partial y}\right) dxdy.$$

Consider $\int_C (x^2 + xy)dx + (x^2 + y^2)dy.$

Let $M = x^2 + xy$ and $N = x^2 + y^2$.

$$\therefore \quad \frac{\partial M}{\partial y} = x \text{ and } \frac{\partial N}{\partial x} = 2x.$$

$$\therefore \quad \frac{\partial N}{\partial x} - \frac{\partial M}{\partial y} = 2x - x = x.$$

In R, x varies from -1 to 1 and y varies from -1 to 1.

$$\therefore \quad \iint_R \left(\frac{\partial N}{\partial x} - \frac{\partial M}{\partial y}\right) dxdy = \int_{-1}^{1}\int_{-1}^{1} x\, dxdy$$

$$= \int_{-1}^{1} \left[\frac{x^2}{2}\right]_{-1}^{1} dy$$

$$= \int_{-1}^{1} \left[\frac{1}{2} - \left(\frac{1}{2}\right)\right] dy$$

$$= \int_{-1}^{1} 0\, dy = 0.$$

Example 5: Apply Green's theorem to evaluate in the plane for $\int_C (xy - x^2)dx + x^2 y dy$ along the closed curve C formed by $y = 0, x = 1$ and $y = x$.

Solution: **[AU, *Nov/Dec 2017*]**

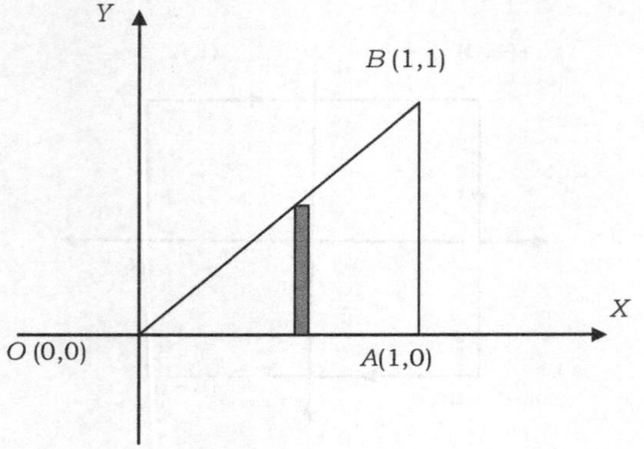

The Green's theorem is

$$\int_C Mdx + Ndy = \iint_R \left(\frac{\partial N}{\partial x} - \frac{\partial M}{\partial y}\right) dxdy.$$

Consider $\int_C (xy - x^2)dx + x^2 y dy$.

Let $M = xy - x^2$ and $N = x^2 y$.

$$\therefore \quad \frac{\partial M}{\partial y} = x \text{ and } \frac{\partial N}{\partial x} = 2xy.$$

$$\therefore \quad \frac{\partial N}{\partial x} - \frac{\partial M}{\partial y} = 2xy - x.$$

In R, x varies from 0 to 1 and y varies from 0 to x.

$$\therefore \quad \iint_R \left(\frac{\partial N}{\partial x} - \frac{\partial M}{\partial y}\right) dxdy = \int_0^1 \int_0^x (2xy - x)\, dy\, dx$$

$$= \int_0^1 \left[2x\frac{y^2}{2} - xy\right]_0^x dx$$

$$= \int_0^1 (x^3 - x^2)\, dx$$

$$= \left[\frac{x^4}{4} - \frac{x^3}{3}\right]_0^1 = \frac{1}{4} - \frac{1}{3} = \frac{-1}{12}.$$

Example 6: Evaluate using Green's theorem in the plane for $\int_C (xy + y^2)dx + x^2 dy$ where C is the closed curve of the region bounded by $y = x$ and $y = x^2$.

Solution:

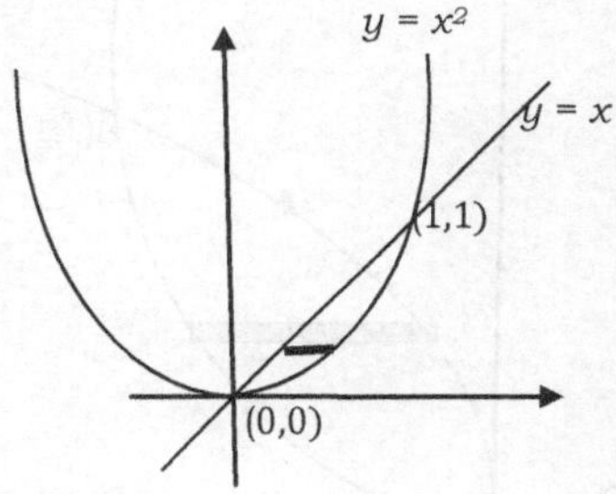

The Green's theorem is

$$\int_C Mdx + Ndy = \iint_R \left(\frac{\partial N}{\partial x} - \frac{\partial M}{\partial y}\right) dxdy.$$

Where R is the region enclosed by the curve C.

In the region R (refer figure) x varies from y to $\sqrt{y}$ and y varies from 0 to 1.

Let $M = xy + y^2$ and $N = x^2$.

$\therefore \quad \frac{\partial M}{\partial y} = x + 2y$ and $\frac{\partial N}{\partial x} = 2x.$

$\therefore \quad \frac{\partial N}{\partial x} - \frac{\partial M}{\partial y} = 2x - (x + 2y) = x - 2y.$

$\therefore$ By Green's Theorem,

$$\int_C (xy + y^2)dx + x^2 dy = \int_0^1 \int_y^{\sqrt{y}} (x - 2y)dxdy$$

$$= \int_0^1 \left[\frac{x^2}{2} - 2yx\right]_y^{\sqrt{y}} dy$$

$$= \int_0^1 \left[\left(\frac{y}{2} - 2y\sqrt{y}\right) - \left(\frac{y^2}{2} - 2y^2\right)\right] dy$$

$$= \int_0^1 \left[\frac{y}{2} - 2y^{\frac{3}{2}} + \frac{3}{2}y^2\right] dy$$

$$= \left[\frac{y^2}{4} - 2\frac{y^{\frac{5}{2}}}{\frac{5}{2}} + \frac{3}{2}\frac{y^3}{3}\right]_0^1 = \left(\frac{1}{4} - \frac{4}{5} + \frac{1}{2}\right) - 0 = \frac{-1}{20}.$$

Example 7: Verify Green's theorem in the plane for $\int_C (3x^2 - 8y^2)dx + (4y - 6xy)dy$ here C is the boundary of the region defined by $x = y^2,\ y = x^2$.

Solution:

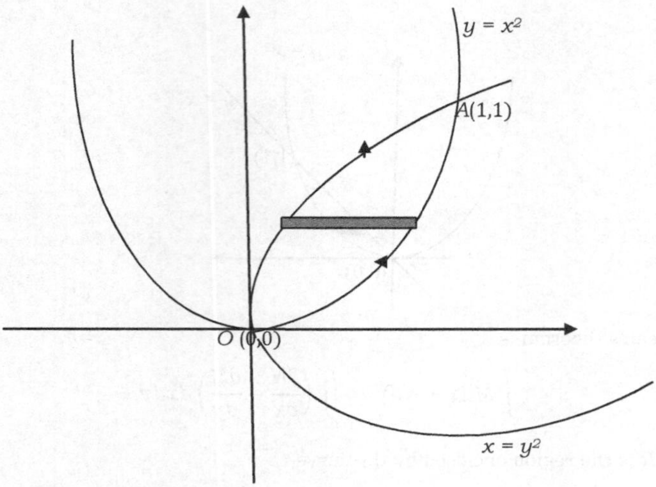

The Green's theorem is

$$\int_C M dx + N dy = \iint_R \left(\frac{\partial N}{\partial x} - \frac{\partial M}{\partial y}\right) dxdy.$$

Consider $\int_C (3x^2 - 8y^2)dx + (4y - 6xy)dy$.

Let $M = 3x^2 - 8y^2$ and $N = 4y - 6xy$

$$\therefore \quad \frac{\partial M}{\partial y} = -16y \text{ and } \frac{\partial N}{\partial x} = -6y$$

$$\therefore \quad \frac{\partial N}{\partial x} - \frac{\partial M}{\partial y} = -6y - (-16y) = 10y.$$

Now, $\iint_R \left(\frac{\partial N}{\partial x} - \frac{\partial M}{\partial y}\right) dxdy = \iint_R 10y\, dxdy$

$$= \int_0^1 \int_{y^2}^{\sqrt{y}} 10y\, dx\, dy = \int_0^1 [10yx]_{y^2}^{\sqrt{y}}\ dy$$

$$= \int_0^1 10y\left(\sqrt{y} - y^2\right) dy = 10 \int_0^1 \left(y^{3/2} - y^3\right)dy$$

$$= 10\left[\frac{y^{5/2}}{5/2} - \frac{y^4}{4}\right]_0^1 = 10\left[\frac{2}{5} - \frac{1}{4}\right] = \frac{3}{2}. \qquad (1)$$

To evaluate $\int_C Mdx + Ndy$ Here $C : OA + AO$

$$\therefore \int_C Mdx + Ndy = \int_{OA} Mdx + Ndy + \int_{AO} Mdx + Ndy. \qquad (2)$$

(i) Along the curve OA

$y = x^2,\ dy = 2xdx$ and x varies from 0 to 1.

$$\therefore \quad Mdx + Ndy = (3x^2 - 8x^4)dx + (4x^2 - 6x^3)2xdx$$

$$= (3x^2 - 8x^4 + 8x^3 - 12x^4)dx$$

$$= (3x^2 + 8x^3 - 20x^4)dx.$$

$$\therefore \int_{OA} Mdx + Ndy = \int_0^1 (3x^2 + 8x^3 - 20x^4)dx$$

$$= \left[3\frac{x^3}{3} + 8\frac{x^4}{4} - 20\frac{x^5}{5}\right]_0^1$$

$$= [(1 + 2 - 4) - 0]$$

$$= -1.$$

(ii) Along the curve AO.

$x = y^2, dx = 2ydy$ and y varies from 0 to 1.

$$\therefore Mdx + Ndy = (3y^4 - 8y^2)(2ydy) + (4y - 6y^3)dy$$

$$= (6y^5 - 16y^3 + 4y - 6y^3)dy = (4y - 22y^3 + 6y^5)dy.$$

$$\therefore \int_{AO} Mdx + Ndy = \int_1^0 (4y - 22y^3 + 6y^5)dy$$

$$= \left[4\frac{y^2}{2} - 22\frac{y^4}{4} + 6\frac{y^6}{6}\right]_1^0 = \left[0 - \left(2 - \frac{11}{2} + 1\right)\right] = \frac{5}{2}.$$

$$\therefore \text{(2) gives } \int_C Mdx + Ndy = -1 + \frac{5}{2} = \frac{3}{2}. \qquad (3)$$

From (1) and (3),

$$\int_C Mdx + Ndy = \iint_R \left(\frac{\partial N}{\partial x} - \frac{\partial M}{\partial y}\right) dxdy.$$

Hence Green's theorem is verified.

Example 8: Evaluate $\int_C (\sin x - y)dx - \cos x\, dy$, where C is the triangle with vertices $(0,0), \left(\frac{\pi}{2}, 0\right)$ *and* $\left(\frac{\pi}{2}, 1\right)$.

Solution:

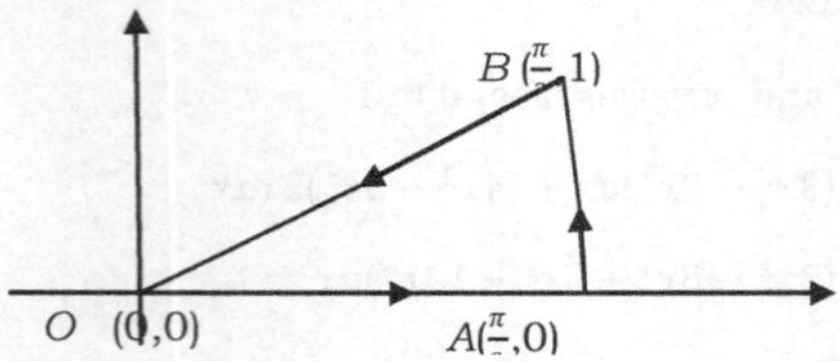

The Green's theorem is

$$\int_C M dx + N dy = \iint_R \left(\frac{\partial N}{\partial x} - \frac{\partial M}{\partial y}\right) dx\, dy. \qquad (1)$$

The equations of the line OB is $\frac{y-0}{1-0} = \frac{x-0}{\frac{\pi}{2}-0}$

i.e., $y = \frac{2}{\pi}x. \quad \therefore x = \frac{\pi}{2}y.$

Let $M = \sin x - y$ and $N = -\cos x$.

$\therefore \quad \frac{\partial M}{\partial y} = -1$ and $\frac{\partial N}{\partial x} = \sin x.$

$\therefore \quad \frac{\partial N}{\partial x} - \frac{\partial M}{\partial y} = \sin x + 1.$

In the region R, x varies from $\frac{\pi}{2}y$ to $\frac{\pi}{2}$ and y varies from 0 to 1.

$$\therefore (1) \text{ gives } \int_C (\sin x - y)dx - \cos x\, dy = \int_0^1 \int_{\frac{\pi}{2}y}^{\frac{\pi}{2}} (\sin x + 1)dxdy$$

$$= \int_0^1 [-\cos x + x]_{\frac{\pi}{2}y}^{\frac{\pi}{2}} dy$$

$$= \int_0^1 \left[\left(\frac{\pi}{2}\right) - \left(-\cos\frac{\pi y}{2} + \frac{\pi}{2}y\right)\right] dy$$

$$= \left[\frac{\pi}{2}y + \frac{\sin\frac{\pi y}{2}}{\pi/2} - \frac{\pi}{2}\frac{y^2}{2}\right]_0^1 = \left(\frac{\pi}{2} + \frac{2}{\pi} - \frac{\pi}{4}\right) - (0) = \frac{\pi}{4} + \frac{2}{\pi}.$$

Example 9: Using Green's theorem, evaluate $\int_C (y - \sin x)dx + \cos x\, dy$, where C is the triangle formed by $y = 0, x = \frac{\pi}{2}, y = \frac{2x}{\pi}$. **[*AU, April/May 2015*]**

Solution:

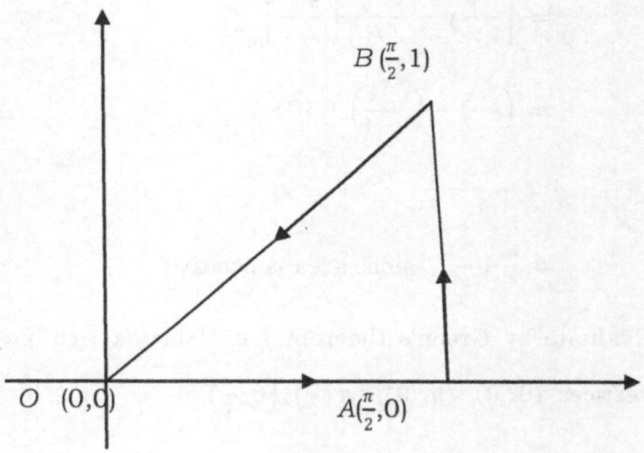

The Green's theorem is

$$\int_C Mdx + Ndy = \iint_R \left(\frac{\partial N}{\partial x} - \frac{\partial M}{\partial y}\right) dxdy. \quad (1)$$

The equations of the line OB is $\frac{y-0}{1-0} = \frac{x-0}{\frac{\pi}{2}-0}$

i.e., $y = \frac{2}{\pi}x. \quad \therefore x = \frac{\pi}{2}y.$

Let $M = y - \sin x$ and $N = \cos x$.

$\therefore \frac{\partial M}{\partial y} = 1$ and $\frac{\partial N}{\partial x} = -\sin x.$

$\therefore \frac{\partial N}{\partial x} - \frac{\partial M}{\partial y} = -\sin x - 1.$

In the region R, x varies from $\frac{\pi}{2}y$ to $\frac{\pi}{2}$ and y varies from 0 to 1.

$$\therefore (1) \text{ gives} \int_C (y - \sin x)dx + \cos x\, dy = \int_0^1 \int_{\frac{\pi}{2}y}^{\frac{\pi}{2}} (-\sin x - 1)dxdy$$

$$= \int_0^1 [\cos x - x]_{\frac{\pi}{2}y}^{\frac{\pi}{2}}\, dy$$

$$= \int_0^1 \left[\left(\cos\frac{\pi}{2} - \frac{\pi}{2}\right) - \left(\cos\frac{\pi y}{2} - \frac{\pi}{2}y\right)\right] dy$$

$$= \left[-\frac{\pi}{2}y - \frac{\sin\frac{\pi y}{2}}{\pi/2} + \frac{\pi}{2}\frac{y^2}{2}\right]_0^1$$

$$= \left(-\frac{\pi}{2} - \frac{2}{\pi} + \frac{\pi}{4}\right) - (0)$$

$$= -\frac{\pi}{4} - \frac{2}{\pi}$$

$$= \frac{\pi}{4} + \frac{2}{\pi} \quad \text{since area is positive.}$$

Example 10: Evaluate by Green's theorem $\int_C e^{-x}(\sin y\, dx + \cos y\, dy)$, where C is the rectangle with vertices $(0, 0)$, $(\pi, 0)$, $\left(\pi, \frac{\pi}{2}\right), \left(0, \frac{\pi}{2}\right)$.

Solution:

The Green's theorem is $\int_C M dx + N dy = \iint_R \left(\frac{\partial N}{\partial x} - \frac{\partial M}{\partial y}\right) dxdy$.(1)

Let $M = e^{-x}\sin y$ and $N = e^{-x}\cos y$.

$\therefore \frac{\partial M}{\partial y} = e^{-x}\cos y$ and $\frac{\partial N}{\partial x} = -e^{-x}\cos y$

$\therefore \frac{\partial N}{\partial x} - \frac{\partial M}{\partial y} = -e^{-x}\cos y - e^{-x}\cos y = -2e^{-x}\cos y.$

In the region R, x varies from 0 to π and y varies from 0 to $\frac{\pi}{2}$. $\therefore$ (1)gives

$$\int_C e^{-x}(\sin y\, dx + \cos y\, dy) = \int_0^{\pi/2}\int_0^{\pi}(-2e^{-x}\cos y)dxdy$$

$$= -2\int_0^{\pi/2}\cos y\,[-e^{-x}]_0^{\pi} dy$$

$$= -2\int_0^{\pi/2}\cos y\,[-e^{-\pi} + 1]dy$$

$$= -2\,(1 - e^{-\pi})\int_0^{\pi/2}\cos y\, dy$$

$$= 2\,(e^{-\pi} - 1)[\sin y]_0^{\pi/2}$$

$$= 2\,(e^{-\pi} - 1)(1 - 0) = 2\,(e^{-\pi} - 1).$$

Example 11: Using Green's theorem evaluate $\int_C (x^2 - y^2)dx + 2xydy$ where C is the region bounded by $y = x^2$ and $y^2 = x$.

Solution:

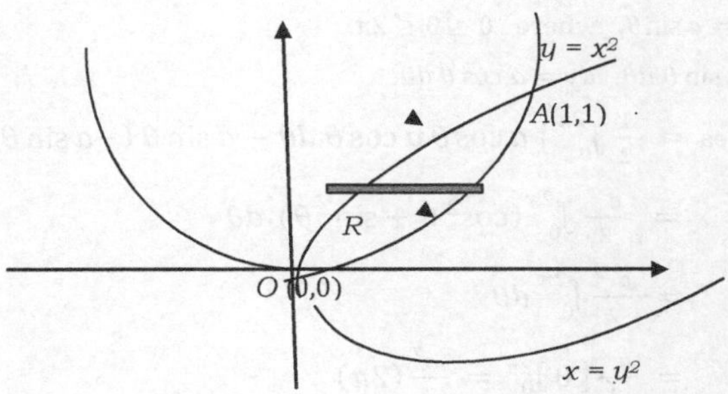

The Green's theorem is

$$\int_C Mdx + Ndy = \iint_R \left(\frac{\partial N}{\partial x} - \frac{\partial M}{\partial y}\right) dxdy. \qquad (1)$$

Let $M = x^2 - y^2$ and $N = 2xy$.

$$\therefore \frac{\partial M}{\partial y} = -2y \text{ and } \frac{\partial N}{\partial x} = 2y. \quad \therefore \frac{\partial N}{\partial x} - \frac{\partial M}{\partial y} = 2y - (-2y) = 4y.$$

In the region R, x varies from

y^2 to $\sqrt{y}$ and y varies from 0 to 1.

$\therefore$ (1) gives

$$\int_C (x^2 - y^2)dx + 2xydy = \int_0^1 \int_{y^2}^{\sqrt{y}} 4y \, dx \, dy$$

$$= \int_0^1 4y \, [\, x \,]_{y^2}^{\sqrt{y}} \, dy$$

$$= \int_0^1 4y \left(\sqrt{y} - y^2\right) dy$$

$$= 4 \int_0^1 \left[y^{\frac{3}{2}} - y^3 \right] dy$$

$$= 4 \left[\frac{y^{\frac{5}{2}}}{\frac{5}{2}} - \frac{y^4}{4} \right]_0^1 = 4\left[\left(\frac{2}{5} - \frac{1}{4}\right) - 0\right] = \frac{3}{5}.$$

Example 12: Using Green's theorem, find the area of a circle $x^2 + y^2 = a^2$.

Solution: By Green's theorem, ***[AU, April/May 2017]***

$$\text{Area} = \frac{1}{2}\int_C x\,dy - y\,dx \qquad (1)$$

The parametric form of a circle of radius a is

$x = a\cos\theta, y = a\sin\theta$, where $0 \le \theta \le 2\pi$.

Then $dx = -a\sin\theta\, d\theta$, $dy = a\cos\theta\, d\theta$.

$$\therefore \text{(1) gives area} = \frac{1}{2}\int_0^{2\pi} [a\cos\theta\, a\cos\theta\, d\theta - a\sin\theta\,(-a\sin\theta)]\, d\theta$$

$$= \frac{a^2}{2}\int_0^{2\pi} (\cos^2\theta + \sin^2\theta)\, d\theta$$

$$= \frac{a^2}{2}\int_0^{2\pi} d\theta$$

$$= \frac{a^2}{2}[\theta]_0^{2\pi} = \frac{a^2}{2}(2\pi)$$

$$= \pi a^2$$

Example 13: Using Green's theorem, find the area of the hypocycloid

$x^{2/3} + y^{2/3} = a^{2/3}, a > 0$.

Solution:

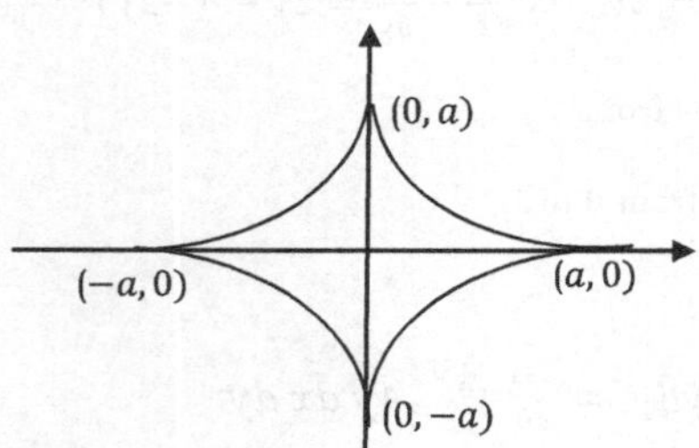

By Green's theorem,

$$\text{Area} = \frac{1}{2}\int_C x\,dy - y\,dx \qquad (1)$$

The parametric equations of the hypocycloid are

$x = a\cos^3\theta, y = a\sin^3\theta$, where $0 \le \theta \le 2\pi$.

Then $dx = -3a\cos^2 d\theta\sin\theta\, d\theta$, $dy = 3a\sin^2\theta\cos\theta\, d\theta$.

$$\therefore \text{(1) gives area} = \frac{1}{2}\int_0^{2\pi} [a\cos^3\theta\, 3a\sin^2\theta\,(\cos\theta) d\theta$$

$$-a\sin^3\theta\, 3a\cos^2\theta(-\sin\theta)d\theta]$$

$$= \frac{3a^2}{2}\int_0^{2\pi}[\cos^4\theta\sin^2\theta + \sin^4\theta\cos^2\theta]d\theta$$

$$= \frac{3a^2}{2}\int_0^{2\pi}\cos^2\theta\sin^2\theta\,(\cos^2\theta+\sin^2\theta)d\theta = \frac{3a^2}{2}$$

$$\int_0^{2\pi}\cos^2\theta\sin^2\theta\, d\theta$$

$$= \frac{3a^2}{2}\int_0^{2\pi}\frac{\sin^2 2\theta}{4}d\theta \qquad \left[\because \sin\theta\cos\theta = \frac{\sin 2\theta}{2}\right]$$

$$= \frac{3a^2}{2}\int_0^{2\pi}\left[\frac{1-\cos 4\theta}{8}\right]d\theta \qquad \left[\because \sin^2 x = \frac{1-\cos 2x}{2}\right]$$

$$= \frac{3a^2}{16}\left[\theta - \frac{\sin 4\theta}{4}\right]_0^{2\pi} = \frac{3a^2}{16}[2\pi] = \frac{3a^2\pi}{8}.$$

Example 14: Prove that the area bounded by a simple closed curve C is given by $\frac{1}{2}\int_C(xdy - ydx)$. Hence find area of the ellipse $x = a\cos\theta, y = b\,sin\theta$

Solution:

The Green's theorem is

$$\int_C M\,dx + N\,dy = \iint_R\left(\frac{\partial N}{\partial x} - \frac{\partial M}{\partial y}\right)dxdy \qquad (1)$$

Consider $\int_C(xdy - ydx)$,

Let $M = -y$ and $N = x$. Then $\frac{\partial M}{\partial y} = -1$ and $\frac{\partial N}{\partial x} = 1$.

$\therefore$ (1) gives $\int_C(-ydx + xdy) = \iint_R(1+1)dxdy$

i.e., $\int_C(xdy - ydx) \quad = 2\iint_R dxdy$

$= 2 \times$ Area of the region R.

$\therefore$ Required area $= \frac{1}{2}\int_C(xdy - ydx)$. (2)

For an ellipse, $x = a\cos\theta, y = b\sin\theta$ and θ varies from 0 to 2π.

Then $dx = -a\sin\theta\, d\theta,\ dy = b\cos\theta\, d\theta$.

$\therefore$ (2) gives area of the ellipse

$$= \frac{1}{2}\int_0^{2\pi}[a\cos\theta\; b\cos\theta\, d\theta - b\,\sin\theta\,(-a\sin\theta)d\theta]$$

$$= \frac{ab}{2}\int_0^{2\pi}(\cos^2\theta + \sin^2\theta)d\theta = \frac{ab}{2}\int_0^{2\pi}d\theta = \frac{ab}{2}[\theta]_0^{2\pi} = \pi ab.$$

EXERCISES

1. Verify Green's theorem for $\int_C (xy + y^2)dx + x^2 dy$ where C is the closed curve of the region bounded by $y = x$ and $y = x^2$.
2. Verify Green's theorem for the integral $\int_C (x - y)dx + (x + y)dy$ taken around the boundary area in the first quadrant between the curves $y = x^2$ and $y^2 = x$.
3. Verify Green's theorem in the plane for $\int_C (x^2 dx + xy\, dy)$ where C is the curve in the XY plane given by $x = 0, y = 0, x = a$ and $y = a$.
4. Verify Green's theorem in the xy – plane for $\int_C x^2(1 + y)dx + (y^3 + x^3)dy$ where C is the square formed by $x = -a, x = a, y = -a, y = a$.
5. Verify Green's theorem in the xy –plane for $\int_C [\sin(x - y)dx - \cos y\, dy]$ where C is the triangle with vertices $(0,0), \left(\frac{\pi}{2}, 0\right)$ and $\left(\frac{\pi}{2}, 1\right)$.
6. Verify Green's theorem in the plane for $\int_C (xy - x^2)dx + x^2 y\, dy$ along the closed curve C formed by $y = 0, x = 1$ and $x = y$.
7. Verify Green's theorem in the plane for $\int_C (x^2 - 2xy)\, dx + (x^2 y + 3)dy$ where C is the boundary of the region defined by $y^2 = 8x$ and $x = 2$.
8. Using Green's theorem evaluate $\int_C x^2 y dx + y^3 dy$ where C is the closed path formed by $y = x$ and $y^3 = x^2$ from (0, 0) to (1, 1).
9. Using Green's theorem, evaluate $\int_C (2x^2 - y^2)dx + (x^2 + y^2)dy$ where C is the boundary of the square enclosed by the lines $x = 0, y = 0, x = 2$ and $y = 3$.
10. Using Green's theorem, evaluate $\int_C (y - \sin x)dx + \cos x\, dy,$ where C is the triangle bounded by $y = 0, x = \frac{\pi}{2}, y = \frac{2x}{\pi}$.
11. Prove the area bounded by a simple closed curve C is given by $\frac{1}{2}\int_C (xdy - ydx)$. Hence find the area of the circle $x^2 + y^2 = a^2$.
12. Find the area bounded between the curves $y^2 = 4x$ and $x^2 = 4y$ using Green's theorem.
13. Find the area of the region bounded by $y = x^2$ and $y = x + 2$.

ANSWERS

7. $\frac{1}{44}$ 9. 30 10. $-\left(\frac{\pi}{4} + \frac{2}{\pi}\right)$

11. πa^2 12. $\frac{16}{3}$ 13. $\frac{9}{2}$

2.10. Stoke's Theorem

Theorem: (*Stoke's Theorem*) **[*AU, Nov/Dec 2014, May/June 2016*]**

If S is an open two sided surface bounded by a simple closed curve C and if $\vec{F}$ is a vector point function with continuous first order partial derivatives on S, then

$\int_C \vec{F}.d\vec{r} = \iint_S curl\,\vec{F}.\hat{n}\,dS$ where C is described in the anticlockwise direction.

Example 1: Verify Stoke's theorem for $\vec{F} = (x^2 - y^2)\vec{i} + 2xy\vec{j}$ in the rectangular region of $z = o$ plane bounded by the lines $x = 0, x = a, y = 0$and $y = b$.

Solution: **[*AU, April/May 2017, May/June 2014*]**

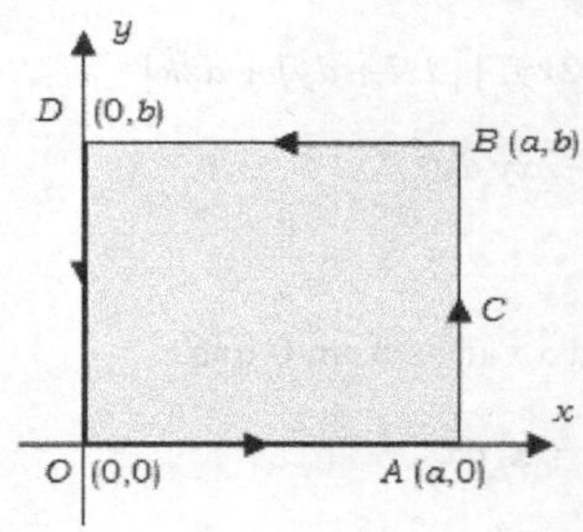

Stoke's theorem is $\int_C \vec{F}.d\vec{r} = \iint_S curl\,\vec{F}.\hat{n}\,dS$

Given $\vec{F} = (x^2 - y^2)\vec{i} + 2xy\vec{j}$

$$Curl\,\vec{F} = \begin{vmatrix} \vec{i} & \vec{j} & \vec{k} \\ \dfrac{\partial}{\partial x} & \dfrac{\partial}{\partial y} & \dfrac{\partial}{\partial z} \\ x^2 - y^2 & 2xy & 0 \end{vmatrix}$$

$$= \vec{i}(0-0) - \vec{j}(0-0) + \vec{k}(2y + 2y)$$

$$= 4\,y\,\vec{k}.$$

In S, $\hat{n} = \vec{k}$ and $dS = dxdy$.

$$\therefore\ curl\,\vec{F}.\hat{n} = 4\,y\vec{k}.\vec{k} = 4\,y$$

$$\therefore \iint_S curl\,\vec{F}.\hat{n}\,dS = \int_0^b \int_0^a 4\,y\,dx\,dy$$

$$= \int_0^b 4\,y\,[x]_0^a\,dy$$

$$= 4\,a\int_0^b y\,dy$$

$$= 4a\left[\frac{y^2}{2}\right]_0^b$$

$$= 2ab^2. \qquad (1)$$

To evaluate $\int_C \vec{F}.d\vec{r}$:

The curve C is $C: OA + AB + BD + DO$

$$\therefore \quad \int_C \vec{F}.d\vec{r} = \int_{OA} + \int_{AB} + \int_{BD} + \int_{DO} \vec{F}.d\vec{r} \qquad (2)$$

$$\vec{F}.d\vec{r} = [(x^2 - y^2)\vec{i} + 2xy\vec{j}]\,[dx\vec{i} + dy\vec{j} + dz\vec{k}]$$

$$= (x^2 - y^2)dx + 2xy\,dy.$$

(i) *Along OA*

$y = 0, dy = 0$ and x varies from 0 to a.

$$\therefore \vec{F}.d\vec{r} = (x^2 - y^2)dx = x^2 dx$$

$$\therefore \int_{OA} \vec{F}.d\vec{r} = \int_0^a x^2\,dx = \left[\frac{x^3}{3}\right]_0^a = \frac{a^3}{3}.$$

(ii) *Along AB*

$x = a, dx = 0$ and y varies from 0 to b.

$$\vec{F}.d\vec{r} = 2xy\,dy = 2ay\,dy$$

$$\therefore \int_{AB} \vec{F}.d\vec{r} = \int_0^b 2ay\,dy = [ay^2]_0^b = ab^2.$$

(ii) *Along BD*

$y = b, dy = 0$ and x varies from a to 0.

$$\vec{F}.d\vec{r} = (x^2 - y^2)dx = (x^2 - b^2)dx$$

$$\therefore \int_{BD} \vec{F}.d\vec{r} = \int_a^0 (x^2 - b^2)dx$$

$$= \left[\frac{x^3}{3} - b^2x\right]_a^0$$

$$= 0 - \left(\frac{a^3}{3} - b^2a\right) = b^2a - \frac{a^3}{3}.$$

(iv) Along DO

$x = 0, dx = 0$ and y varies from b to 0.

$\vec{F}.d\vec{r} = 2xy\,dy = 0$

$\int_{DO} \vec{F}.d\vec{r} = \int_b^0 0\,dy = 0$

$\therefore$ (2) gives $\int_C \vec{F}.d\vec{r} = \frac{a^3}{3} + ab^2 + b^2a - \frac{a^3}{3} + 0$

$= 2ab^2.$ (3)

From (1) and (3),

$\int_C \vec{F}.d\vec{r} = \iint_S curl\,\vec{F}.\hat{n}\,dS.$

Hence Stoke's theorem is verified.

Example 2: Verify Stoke's theorem for $\vec{F} = (x^2 + y^2)\vec{\imath} - 2xy\vec{\jmath}$ taken around the rectangle bounded by the lines $x = \pm a,\ y = 0,\ y = b.$

Solution:

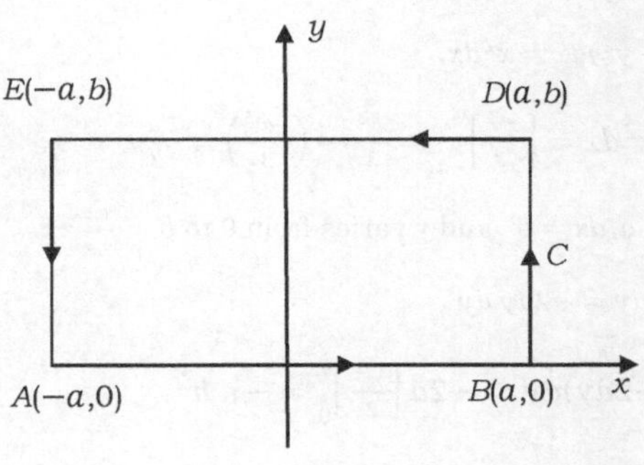

Stoke's theorem is $\int_C \vec{F}.d\vec{r} = \iint_S curl\,\vec{F}.\,\hat{n}\,dS.$

Given $\vec{F} = (x^2 + y^2)\vec{\imath} - 2xy\vec{\jmath}$

$$curl\,\vec{F} = \begin{vmatrix} \vec{\imath} & \vec{\jmath} & \vec{k} \\ \frac{\partial}{\partial x} & \frac{\partial}{\partial y} & \frac{\partial}{\partial z} \\ x^2 + y^2 & -2xy & 0 \end{vmatrix}$$

$$= \vec{i}(0-0) - \vec{j}(0-0) + \vec{k}(-2y-2y) = -4y\vec{k}.$$

In the given region, $\hat{n} = \vec{k}$ and $dS = dxdy$.

$\therefore\ curl\,\vec{F}.\hat{n} = -4y\vec{k}.\vec{k} = -4y.$

Now $\iint_S curl\,\vec{F}.\hat{n}\,dS = \int_0^b \int_{-a}^{a} (-4y)\,dxdy = -4\int_0^b y[x]_{-a}^{a}\,dy$

$$= -4\int_0^b y[a+a]\,dy = -8a\int_0^b y\,dy$$

$$= -8a\left[\frac{y^2}{2}\right]_0^b = -4ab^2. \qquad (1)$$

To evaluatete $\int_C \vec{F}.d\vec{r}$:

The given cure C is $C: AB + BD + DE + EA$

$$\therefore \int_C \vec{F}.d\vec{r} = \int_{AB} + \int_{BD} + \int_{DE} + \int_{EA} \vec{F}.d\vec{r} \qquad (2)$$

Now $\vec{F}.d\vec{r} = [(x^2+y^2)\vec{i} - 2xy\vec{j}].[dx\vec{i} + dy\vec{j} + dz\vec{k}]$

$$= (x^2+y^2)dx - 2xy\,dy.$$

(*i*) *Along AB* $\quad y = 0,\ dy = 0$ and x varies from $-a$ to a.

$$\therefore \vec{F}.d\vec{r} = (x^2+y^2)dx = x^2dx.$$

$$\therefore \int_{AB} \vec{F}.d\vec{r} = \int_{-a}^{a} x^2dx = \left[\frac{x^3}{3}\right]_{-a}^{a} = \frac{a^3}{3} - \left(\frac{-a^3}{3}\right) = \frac{2a^3}{3}.$$

(*ii*) *Along BD* $x = a, dx = 0$ and y varies from 0 to b.

$$\therefore \quad \vec{F}.d\vec{r} = -2xydy = -2ay\,dy$$

$$\therefore \int_{BD} \vec{F}.d\vec{r} = \int_0^b (-2ay)dy = -2a\left[\frac{y^2}{2}\right]_0^b = -ab^2.$$

(*iii*) *Along DE* $y = b,\ dy = 0$ and x varies from a *to* $-a$.

$$\vec{F}.d\vec{r} = (x^2+y^2)dx = (x^2+b^2)dx$$

$$\therefore \quad \int_{DE} \vec{F}.d\vec{r} = \int_a^{-a} (x^2+b^2)dx$$

$$= \left[\frac{x^3}{3} + b^2x\right]_a^{-a}$$

$$= \left(\frac{-a^3}{3} - b^2a\right) - \left(\frac{a^3}{3} + b^2a\right) = -2b^2a - \frac{2a^3}{3}.$$

(iv)Along EA $x = -a,\ dx = 0$ and y varies from b to 0.

$$\vec{F}.d\vec{r} = -2xy\,dy = 2ay\,dy$$

$$\therefore \int_{EA} \vec{F}.d\vec{r} = \int_b^0 2ay\,dy = [a\,y^2]_b^0 = 0 - a\,b^2 = -ab^2.$$

$$\therefore \text{(2) gives } \int_C \vec{F}.d\vec{r} = \frac{2\,a^3}{3} - a\,b^2 - 2\,b^2\,a - \frac{2\,a^3}{3} - a\,b^2$$

$$= -4\,ab^2. \qquad (3)$$

From (1) and (3), $\int_C \vec{F}.d\vec{r} = \iint_S curl\,\vec{F}.\hat{n}\,dS$.

Hence Stokes theorem is verified.

Example 3: Verify Stoke's theorem for $\vec{F} = (y - z + 2)\vec{\imath} + (yz + 4)\vec{\jmath} - xz\vec{k}$ over the open surface of the cube $x = 0, y = 0, z = 0, x = 1, y = 1, z = 1$ not included in the XOY plane.

Solution:

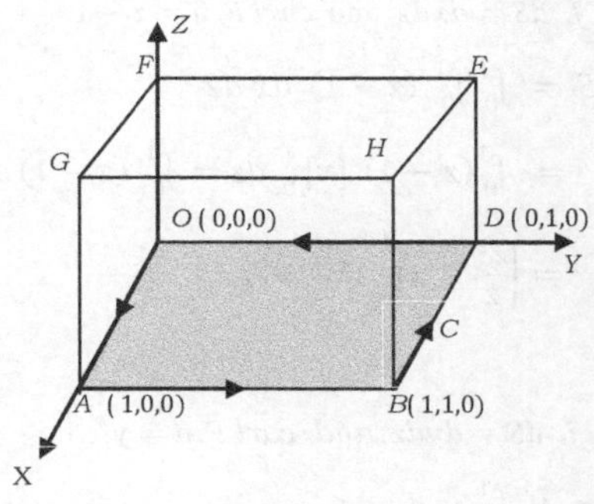

Stoke's theorem is $\int_C \vec{F}.d\vec{r} = \iint_S curl\,\vec{F}.\hat{n}\,dS$.

Given $\vec{F} = (y - z + 2)\vec{\imath} + (yz + 4)\vec{\jmath} - xz\vec{k}$

$$curl\,\vec{F} = \begin{vmatrix} \vec{\imath} & \vec{\jmath} & \vec{k} \\ \frac{\partial}{\partial x} & \frac{\partial}{\partial y} & \frac{\partial}{\partial z} \\ y - z + 2 & yz + 4 & -xz \end{vmatrix}$$

$$= \vec{\imath}(0 - y) - \vec{\jmath}(-z + 1) + \vec{k}(0 - 1)$$

$$= -y\vec{\imath} + (z - 1)\vec{\jmath} - \vec{k}.$$

The given surface S is

$S : OAGF + DEHB + ODEF + ABHG + EFGH$

$\therefore \iint_S curl\, \vec{F}.\hat{n}\, dS = \iint_{OAGF} + \iint_{DEHB} + \iint_{ODEF} \quad + \iint_{ABHG} + \iint_{EFGH} curl\, \vec{F}.\hat{n}\, dS$ (1)

(i) On the face OAGF

$y = 0, dy = 0,\ \hat{n} = -\vec{j},\ dS = dxdz$

$curl\, \vec{F}.\hat{n} = -(z-1) = 1 - z.$

$$\therefore \iint_{OAGF} curl\, \vec{F}.\hat{n} dS = \int_0^1\int_0^1 (1-z)\, dxdz$$

$$= \int_0^1 (1-z)\ [\, x\,]_0^1\ dz = \int_0^1 (1-z)\, dz$$

$$= \left[z - \frac{z^2}{2}\right]_0^1 = 1 - \frac{1}{2} = \frac{1}{2}.$$

(ii) On the face DEHB

$y = 0, dy = 0,\ \hat{n} = \vec{j},\ dS = dxdz$ and $curl\, \vec{F}.\hat{n} = z - 1.$

$$\therefore \quad \iint_{DEHB} curl\, \vec{F}.\hat{n}\, dS = \int_0^1\int_0^1 (z-1)\ dx\, dz$$

$$= \int_0^1 (z-1)\ [x]_0^1\ dz = \int_0^1 (z-1)\, dz$$

$$= \left[\frac{z^2}{2} - z\right]_0^1 = \frac{1}{2} - 1 = -\frac{1}{2}.$$

(iii) On the face ODEF

$x = 0,\ dx = 0,\ \hat{n} = -\vec{i},\ dS = dydz$ *and* $curl\, \vec{F}.\hat{n} = y.$

$$\therefore \iint_{ODEF} curl\, \vec{F}.\hat{n} dS = \int_0^1\int_0^1 y\ dy\, dz$$

$$= \int_0^1 \left[\frac{y^2}{2}\right]_0^1 dz = \int_0^1 \frac{1}{2}\ dz.$$

$$= \frac{1}{2}[\, z\,]_0^1 = \frac{1}{2}$$

(iv) On the face ABHG

$x = 1,\ dx = 0,\ \hat{n} = \vec{i},\ dS = dydz$ *and* $curl\, \vec{F}.\hat{n} = -y.$

$$\therefore \iint_{ABHG} curl\, \vec{F}.\hat{n} dS = \int_0^1\int_0^1 (-y)\, dy\, dz$$

$$= -\int_0^1 \left[\frac{y^2}{2}\right]_0^1 dz = -\int_0^1 \frac{1}{2}\ dz$$

$$= -\frac{1}{2}\,[\,z\,]_0^1 = -\frac{1}{2}.$$

(v) On the face EFGH

$$z = 1,\ dz = 0,\ \hat{n} = \vec{k},\ dS = dxdy \quad and \quad curl\,\vec{F}.\hat{n} = -1.$$

$$\therefore \quad \iint_{EFGH} curl\,\vec{F}.\hat{n}dS = \int_0^1\int_0^1(-1)\,dxdy$$

$$= -\int_0^1 [\,x\,]_0^1\,dy = -\int_0^1 dy$$

$$= -[\,y\,]_0^1 = -1.$$

$\therefore$ (1) gives $\iint_S curl\,\vec{F}.\hat{n}\,dS = \frac{1}{2} - \frac{1}{2} + \frac{1}{2} - \frac{1}{2} - 1 = -1$ (2)

To evaluate $\int_C \vec{F}.d\vec{r}$:

The given curve C is $C : OABDO = OA + AB + BD + DO$

$\therefore \int_C \vec{F}.d\vec{r} = \int_{OA} + \int_{AB} + \int_{BD} + \int_{DO} \vec{F}.d\vec{r}$ (3)

Now $\vec{F}.d\vec{r} = \left[(y - z + 2)\vec{\imath} + (yz + 4)\vec{\jmath} - xz\vec{k}\right].\left[dx\vec{\imath} + dy\vec{\jmath} + dz\vec{k}\right]$

$$= (y - z + 2)dx + (yz + 4)dy - xzdz.$$

(i) Along OA, $y = 0, z = 0$ and x varies from 0 to 1.

$$\vec{F}.d\vec{r} = 2dx$$

$$\therefore \int_{OA} \vec{F}.d\vec{r} = \int_0^1 2dx = 2[x]_0^1 = 2.$$

(ii) Along AB, $x = 1, z = 0$ and y varies from 0 to 1.

$$\vec{F}.d\vec{r} = (yz + 4)dy = 4dy$$

$$\therefore \int_{AB} \vec{F}.d\vec{r} = \int_0^1 4dy = 4[y]_0^1 = 4.$$

(iii) Along BD, $y = 1, z = 0$ and x varies from 1 to 0.

$$\vec{F}.d\vec{r} = (y - z + 2)dx = 3dx$$

$$\therefore \int_{BD} \vec{F}.d\vec{r} = \int_1^0 3dx = 3[x]_1^0 = -3.$$

(iv) Along DO, $x = 0, z = 0$ and y varies from 1 to 0.

$$\vec{F}.d\vec{r} = (yz + 4)dy = 4dy$$

$$\therefore \int_{DO} \vec{F}.d\vec{r} = \int_1^0 4dy = 4[y]_1^0 = -4.$$

$\therefore$(3) gives $\int_C \vec{F}.d\vec{r} = 2 + 4 - 3 - 4 = -1\,.$ (4)

From (2) and (4),

$\int_C \vec{F}.d\vec{r} = \iint_S curl\,\vec{F}.\hat{n}\,dS.$

Hence Stoke's theorem is verified.

Example 4: Verify Stoke's theorem for $\vec{F} = y^2z\vec{i} + z^2x\vec{j} + x^2y\vec{k}$ where S is the open surface of the cube formed by the planes $x = \pm a$, $y = \pm a$ and $z = \pm a$ in which the plane $z = -a$ is cut.

Solution:

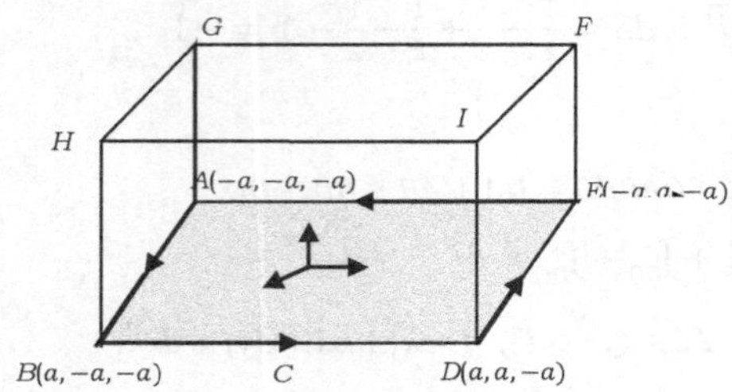

Stoke's theorem is $\int_C \vec{F}.d\vec{r} = \iint_S curl\,\vec{F}.\hat{n}\,dS.$

Given $\vec{F} = y^2z\vec{i} + z^2x\vec{j} + x^2y\vec{k}$

$$Curl\,\vec{F} = \begin{vmatrix} \vec{i} & \vec{j} & \vec{k} \\ \frac{\partial}{\partial x} & \frac{\partial}{\partial y} & \frac{\partial}{\partial z} \\ y^2z & z^2x & x^2y \end{vmatrix}$$

$$= \vec{i}(x^2 - 2zx) - \vec{j}(2xy - y^2) + \vec{k}(z^2 - 2yz).$$

The surface S is

$$S : ABHG + EDIF + AEFG + BDIH + HIFG$$

$$\therefore \iint_S curl\,\vec{F}.\hat{n}\,dS = \iint_{ABHG} + \iint_{EDIF} + \iint_{AEFG} + \iint_{BDIH} + \iint_{HIFG} curl\,\vec{F}.\hat{n}dS \quad (1)$$

(i) On the face ABHG

$y = -a$, $\hat{n} = -\vec{j}$, $dS = dxdz$ and $curl\,\vec{F}.\hat{n} = 2xy - y^2 = -2ax - a^2.$

$$\therefore \iint_{ABHG} curl\,\vec{F}.\hat{n}\,dS = \int_{-a}^{a}\int_{-a}^{a}(-2ax - a^2)\,dx\,dz$$

$$= \int_{-a}^{a}\left[-ax^2 - a^2x\right]_{-a}^{a} dz$$

$$= \int_{-a}^{a}[(-2a^3) - 0]\, dz$$

$$= -2a^3[z]_{-a}^{a} = -2a^3(a+a) = -4\,a^4.$$

(ii) On the face EDIF

$y = a,\ \hat{n} = \hat{j},\ dS = dxdz$ and $curl\,\vec{F}.\hat{n} = -(2xy - y^2) = a^2 - 2ax.$

$$\therefore \quad \iint_{EDIF} curl\,\vec{F}.\hat{n}\,dS = \int_{-a}^{a}\int_{-a}^{a}(a^2 - 2ax)\,dx\,dz$$

$$= \int_{-a}^{a}\left[a^2x - ax^2\right]_{-a}^{a} dz$$

$$= \int_{-a}^{a}[0 + 2a^3]\,dz = 2a^3[z]_{-a}^{a}$$

$$= 2a^3(a+a) = 4a^4.$$

(iii) *On the face AEFG*

$x = -a,\ \hat{n} = -\hat{i},\ dS = dydz$ and $curl\,\vec{F}.\hat{n} = -(x^2 - 2zx) = -a^2 - 2az.$

$$\therefore \quad \iint_{AEFG} curl\,\vec{F}.\hat{n}\,dS = \int_{-a}^{a}\int_{-a}^{a}(-a^2 - 2az)\,dy\,dz$$

$$= \int_{-a}^{a}\left[-a^2y - 2azy\right]_{-a}^{a} dz$$

$$= \int_{-a}^{a}[(-a^3 - 2a^2z) - (a^3 + 2a^2z)]\,dz$$

$$= \int_{-a}^{a}(-2a^3 - 4a^2z)\,dz$$

$$= [-2a^3z - 2a^2z^2]_{-a}^{a}$$

$$= (-4a^4) - (0)$$

$$= -4\,a^4.$$

(iv) On the face BDIH

$x = a,\ \hat{n} = \hat{i},\ dS = dydz$ and $curl\,\vec{F}.\hat{n} = (x^2 - 2zx) = a^2 - 2az.$

$$\therefore \iint_{BDIH} curl\,\vec{F}.\hat{n}\,dS = \int_{-a}^{a}\int_{-a}^{a}(a^2 - 2az)\,dy\,dz$$

$$= \int_{-a}^{a} [a^2y - 2azy]_{-a}^{a} \; dz$$

$$= \int_{-a}^{a} [(a^3 - 2a^2z) - (-a^3 + 2a^2z)]\, dz$$

$$= \int_{-a}^{a} (2a^3 - 4\,a^2z)\, dz$$

$$= [2a^3z - 2a^2z^2]_{-a}^{a} = 0 - (-4a^4) = 4\,a^4.$$

(v) On the face HIFG

$z = a,\ \hat{n} = \vec{k},\ dS = dxdy$ and $curl\,\vec{F}.\hat{n} = (z^2 - 2yz) = a^2 - 2ay.$

$$\therefore \quad \iint_{HIFG} curl\,\vec{F}.\hat{n}\, dS = \int_{-a}^{a} \int_{-a}^{a} (a^2 - 2ay)\, dxdy$$

$$= \int_{-a}^{a} [a^2x - 2ay\,x]_{-a}^{a} \; dy$$

$$= \int_{-a}^{a} [(a^3 - 2a^2y) - (-a^3 + 2a^2y)]\, dy$$

$$= \int_{-a}^{a} (2a^3 - 4\,a^2y)\, dy$$

$$= [2a^3y - 2a^2y^2]_{-a}^{a} = 0 - (-4a^4) = 4\,a^4.$$

$\therefore$ (1) gives $\iint_S curl\,\vec{F}.\hat{n}\, dS = -4\,a^4 + 4\,a^4 - 4\,a^4 + 4\,a^4 + 4\,a^4 = 4\,a^4$ (2)

To evaluate $\int_C \vec{F}.d\vec{r}$

$C : ABDEA = AB + BD + DE + EA$

$$\therefore \quad \int_C \vec{F}.d\vec{r} = \int_{AB} + \int_{BD} + \int_{DE} + \int_{EA} \vec{F}.d\vec{r} \qquad (3)$$

$\vec{F}.d\vec{r} = y^2zdx + z^2xdy + x^2ydz.$

(i) Along AB, $y = -a,\ z = -a,\ dy = 0,\ dz = 0$ and x varies from $-a$ to a.

$\vec{F}.d\vec{r} = y^2\, z\, dx = -\,a^3 dx.$

$$\therefore \int_{AB} \vec{F}.d\vec{r} = \int_{-a}^{a} (-a^3)\; dx = -a^3[x]_{-a}^{a} = -a^3(a + a) = -2a^4.$$

(ii) Along BD, $x = a,\ z = -a,\ dx = 0,\ dz = 0$ and y varies from $-a$ to a.

$\vec{F}.d\vec{r} = z^2x\; dy = a^3 dy$

$$\therefore \int_{BD} \vec{F}.d\vec{r} = \int_{-a}^{a} a^3 \; dy = a^3\,[\,y\,]_{-a}^{a} = 2a^4.$$

(iii) *Along DE*, $y = a,\ z = -a, dy = 0,\ dz = 0$ and x varies from a to $-a$.

$\vec{F}.d\vec{r} = y^2 z dx = \ -a^3 dx.$

$\therefore \int_{DE} \vec{F}.d\vec{r} = \int_a^{-a} (-a^3)\ dx = \ -a^3 [\,x\,]_a^{-a} = -a^3(-2a) = 2a^4.$

(iv) *Along EA*, $x = -a, z = -a, dx = 0, dz = 0$ and y varies from a to $-a$.

$$\vec{F}.d\vec{r} = z^2 x\, dy = -a^3 dy$$

$$\therefore \int_{EA} \vec{F}.d\vec{r} = \int_a^{-a} (-a^3)\ dy = -a^3\ [\,y\,]_a^{-a} = -a^3(-2a) = 2a^4.$$

$\therefore (3) \Rightarrow \int_C \vec{F}.d\vec{r} = \ -2a^4 + 2a^4 + 2a^4 + 2a^4 \quad = \ 4a^4.$ (4)

From (2) and (4), $\int_C \vec{F}.d\vec{r} = \iint_S curl\, \vec{F}.\hat{n}\, dS.$

Hence Stoke's theorem is verified.

Example 5: Using Stoke's theorem, calculate $\int_C \vec{F}.d\vec{r}$ where $\vec{F} = y^2\vec{i} + x^2\vec{j} - (x+z)\vec{k}$ and C is the boundary of the triangle with vertices at $(0,0,0), (1,0,0), (1,1,0)$.

Solution: ***[AU, May/June 2013]***

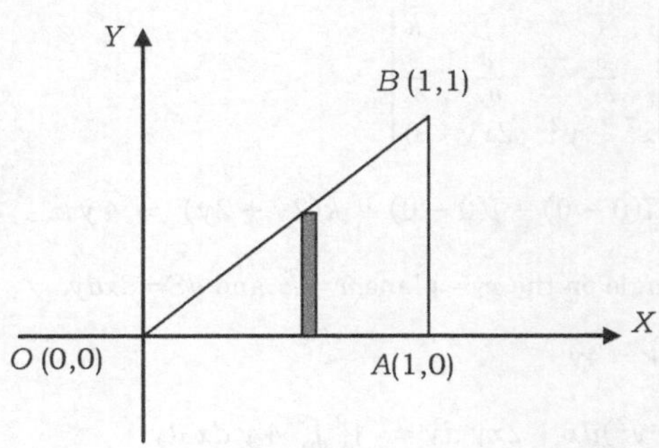

Stoke's theorem is $\int_C \vec{F}.d\vec{r} = \iint_S curl\, \vec{F}.\hat{n}\, dS$ (1)

Given $\vec{F} = y^2\vec{i} + x^2\vec{j} - (x+z)\vec{k}$

$$curl\, \vec{F} = \begin{vmatrix} \vec{i} & \vec{j} & \vec{k} \\ \frac{\partial}{\partial x} & \frac{\partial}{\partial y} & \frac{\partial}{\partial z} \\ y^2 & x^2 & -(x+z) \end{vmatrix}$$

$$= \vec{i}(0-0) - \vec{j}(-1-0) + \vec{k}(2x-2y)$$

$$= \vec{j} + (2x-2y)\vec{k}.$$

Since the triangle formed by the given points lie on the xy –plane, $\hat{n} = \vec{k}$ and $dS = dx\, dy$.

Then $curl\,\vec{F}.\hat{n} = 2x - 2y.$

$\therefore$ (1) gives $\int_C \vec{F}.d\vec{r} = \int_0^1 \int_0^x (2x-2y)\, dy\, dx$

$$= \int_0^1 [\,2xy - y^2]_0^x\, dx$$

$$= \int_0^1 (2x^2 - x^2)\, dx$$

$$= \int_0^1 x^2\, dx = \left[\frac{x^3}{3}\right]_0^1 = \frac{1}{3}.$$

Example 6: Evaluate $\int_C (x^2 - y^2)dx + 2xy\, dy$ by Stoke's theorem, where C is the rectangle $0 \le x \le a$ and $0 \le y \le b$ in the xy –plane.

Solution: By Stoke's theorem $\int_C \vec{F}.d\vec{r} = \iint_S curl\,\vec{F}.\hat{n}\, dS$ (1)

Let $\vec{F} = (x^2 - y^2)\vec{i} + 2xy\,\vec{j}$

Then $curl\,\vec{F} = \begin{vmatrix} \vec{i} & \vec{j} & \vec{k} \\ \frac{\partial}{\partial x} & \frac{\partial}{\partial y} & \frac{\partial}{\partial z} \\ x^2 - y^2 & 2xy & 0 \end{vmatrix}$

$$= \vec{i}(0-0) - \vec{j}(0-0) + \vec{k}(2y+2y) = 4\,y\,\vec{k}.$$

Since S is a rectangle on the xy – plane, $\hat{n} = \vec{k}$ and $dS = dxdy$.

$\therefore curl\,\vec{F}.\hat{n} = 4y\vec{k}.\vec{k} = 4y.$

(1) gives $\int_C (x^2 - y^2)dx + 2xy\, dy = \int_0^b \int_0^a 4\,y\, dx\, dy$

$$= \int_0^b 4\,y\,[\,x\,]_0^a\, dy = \int_0^b 4\,a\,y\, dy = 2a\,[y^2]_0^b = 2\,a\,b^2.$$

Example 7: By using Stoke's theorem prove that $\int_C \vec{r}\,.d\vec{r} = 0$ where

$\vec{r} = x\vec{i} + y\vec{j} + z\vec{k}$ and C is any curve.

Solution: By Stoke's theorem $\int_C \vec{r}.d\vec{r} = \iint_S curl\ \vec{r}.\,\hat{n}\, dS = \iint_S \vec{0}.\,\hat{n}\, dS = 0.$

EXERCISES

1. Verify Stoke's theorem for the function $\vec{F} = x^2\vec{\imath} + xy\vec{\jmath}$ integrated over the square in the plane $z = 0$ whose sides are along the lines $x = 0, y = 0, x = a, y = a$.
2. 2. Verify Stoke's theorem for $\vec{F} = x^2\vec{\imath} + xy\vec{\jmath}$ where S is the square $0 \le x \le a$, $0 \le y \le a$ in the $XY-$ plane.
3. Verify stoke's theorem for $\vec{F} = (y - z + 2)\vec{\imath} + (yz + 4)\vec{\jmath} - xz\vec{k}$ where S is the open surface of the cube formed by $x = 0, x = 2, y = 0, y = 2, z = 0$ and $z = 2$.
4. Verify Stoke's theorem for $\vec{F} = xy\vec{\imath} - 2yz\vec{\jmath} - zx\vec{k}$, where S is the open surface of the rectangular parallelepiped formed by the planes $x = 0,\ x = 1,\ y = 0,\ y = 1, z = 3$ above the XOY $-$plane.
5. Verify Stoke's theorem for the vector function $\vec{F} = y^2\vec{\imath} + y\vec{\jmath} - zx\vec{k}$ and S is the upper half of the sphere $x^2 + y^2 + z^2 = a^2$ and $z \ge 0$.
6. Evaluate by Stoke's theorem, $\int_C \sin z\, dx - \cos x\, dy + \sin y\, dz$ where C is the boundary of the rectangle $0 \le x \le \pi, 0 \le y \le 1,\ z = 3$.

ANSWERS

6. 2

2.11. Gauss Divergence Theorem

Theorem: (Gauss Divergence Theorem) ***[AU, Nov/Dec 2017]***

If S is a closed surface enclosing a region of space with volume V and if $\vec{F}$ is a vector point function with continuous first order partial derivatives in V, then

$$\iint_S \vec{F}.\hat{n}\, dS = \iiint_V div\, \vec{F}\, dV,$$

Where $\hat{n}$ is the outward drawn unit normal vector of the surface S.

Example 1: Verify Gauss divergence theorem for $\vec{F} = x^2\vec{\imath} + y^2\vec{\jmath} + z^2\vec{k}$ taken over the cube bounded by the planes$x = 0,\ y = 0, z = 0, x = 1, y = 1$ and $= 1$.

Solution: ***[AU, Nov/Dec 2017, May/June 2014]***

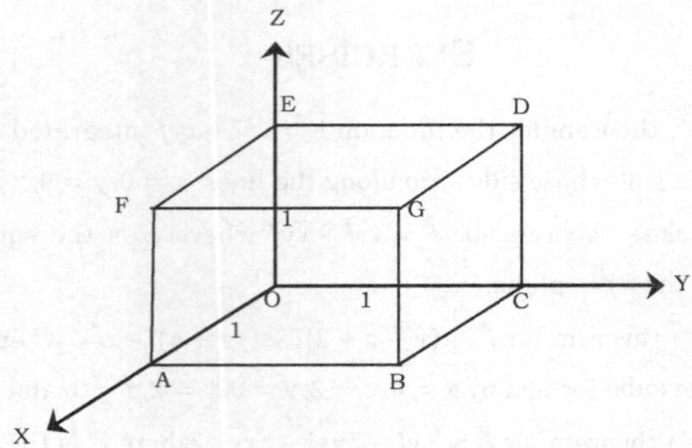

Gauss divergence theorem is $\iint_S \vec{F}.\hat{n}\, dS = \iiint_V div\, \vec{F}\, dV$.

$\vec{F} = x^2\vec{\imath} + y^2\vec{\jmath} + z^2\vec{k}$.

$$\therefore div\, \vec{F} = \frac{\partial}{\partial x}(x^2) + \frac{\partial}{\partial y}(y^2) + \frac{\partial}{\partial z}(z^2)$$

$$= 2x + 2y + 2z = 2(x + y + z).$$

$$\therefore \iiint_V \vec{F}\, dV = \int_0^1 \int_0^1 \int_0^1 2(x + y + z)dxdydz$$

$$= 2\int_0^1 \int_0^1 \left[\frac{x^2}{2} + xy + xz\right]_0^1 dydz$$

$$= 2\int_0^1 \int_0^1 \left[\frac{1}{2} + y + z\right] dydz$$

$$= 2\int_0^1 \left[\frac{1}{2}y + \frac{y^2}{2} + zy\right]_0^1 dz$$

$$= 2\int_0^1 [1 + z]\, dz = 2\left[z + \frac{z^2}{2}\right]_0^1 = 2\left[1 + \frac{1}{2}\right]$$

$$= 2\,\frac{3}{2} = 3. \qquad (1)$$

To evaluate $\iint_S \vec{F}.\hat{n}\, dS$.

The surface S consists of six faces, namely *OABC*, *EFGD*, *OCDE*, *ABGF*, *OAFE*, *CBGD*.

$$\therefore \iint_S \vec{F}.\hat{n}\, dS = \iint_{OABC} + \iint_{EFGD} + \iint_{OCDE} + \iint_{ABGF} + \iint_{OAFE} + \iint_{CBGD} \vec{F}.\hat{n}\, dS \qquad (2)$$

(i) On the face OABC

$z = 0, \hat{n} = -\vec{k}, dS = dxdy$ and x varies from 0 to 1 and y varies 0 to 1.

$\therefore \vec{F}.\hat{n} = [x^2\vec{i} + y^2\vec{j} + z^2\vec{k}].(-\vec{k})$

$= -z^2 = 0. \quad [\because z = 0]$

$\therefore \iint_{OABC} \vec{F}.\hat{n}\, dS = \int_0^1 \int_0^1 0\, dxdy = 0.$

(ii) On the face EFGD

$z = 1, \hat{n} = \vec{k}, dS = dxdy$ and x varies from 0 to 1 and y varies 0 to 1.

$\therefore \quad \vec{F}.\hat{n} = z^2 = 1.$

$\therefore \quad \iint_{EFGD} \vec{F}.\hat{n}\, dS = \int_0^1 \int_0^1 dxdy = \int_0^1 [x]_0^1\, dy = [y]_0^1 = 1.$

(iii) On the face OCDE

$x = 0, \hat{n} = -\vec{i}, dS = dydz$ and y varies from 0 to 1 and z varies from 0 to 1.

$\therefore \quad \vec{F}.\hat{n} = -x^2 = 0.$

$\therefore \quad \iint_{OCDE} \vec{F}.\hat{n}\, dS = \int_0^1 \int_0^1 0\, dydz = 0.$

(iv) On the face ABGF

$x = 1, \hat{n} = \vec{i}, dS = dydz$ and y varies from 0 to 1 and z varies from 0 to 1.

$\therefore \quad \vec{F}.\hat{n} = x^2 = 1.$

$\therefore \iint_{ABGF} \vec{F}.\hat{n}\, dS = \int_0^1 \int_0^1 1dydz = 1.$

(v) On the face OAFE

$y = 0, \hat{n} = -\vec{j}, dS = dxdz$ and x varies from 0 to 1 and z varies from 0 to 1.

$\therefore \vec{F}.\hat{n} = -y^2 = 0.$

$$\therefore \iint_{OAFE} \vec{F}.\hat{n}\, dS = \int_0^1 \int_0^1 0\, dx\, dz = 0.$$

(vi) On the face CBGD

$y = 1,\ \hat{n} = \vec{j},\ dS = dxdz$ and x varies from 0 to 1 and z varies from 0 to 1.

$\therefore\ \vec{F}.\hat{n} = y^2 = 1.$

$\therefore\ \iint_{CBGD} \vec{F}.\hat{n}\, dS = \int_0^1 \int_0^1 1\, dx\, dz = 1.$

$\therefore$ (2) gives $\iint_S \vec{F}.\hat{n}\, dS = 1 + 1 + 1 = 3$. (3)

From (1) and (3),

$\iint_S \vec{F}.\hat{n}\, dS = \iiint_V div\, \vec{F}\, dV.$

Hence divergence theorem is verified.

Example 2: Verify Gauss divergence theorem for $\vec{F} = (x^2 - yz)\vec{i} + (y^2 - zx)\vec{j} + (z^2 - xy)\vec{k}$ taken over the rectangular parallelepiped $0 \le x \le a,\ 0 \le y \le b,\ 0 \le z \le c.$

Solution:

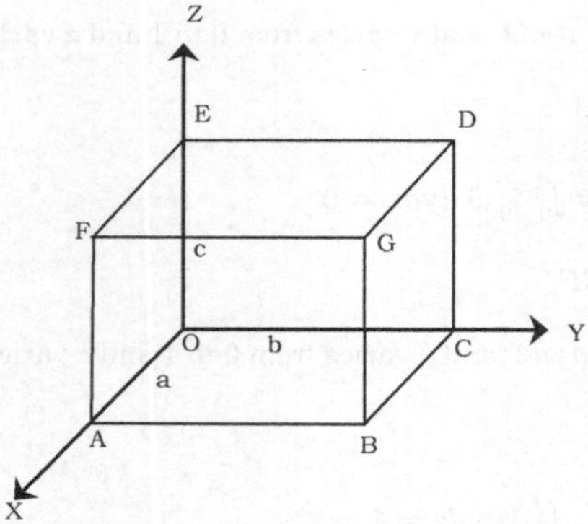

Gauss divergence theorem is $\iint_S \vec{F}.\hat{n}\, dS = \iiint_V div\, \vec{F}\, dV.$

$\vec{F} = (x^2 - yz)\vec{i} + (y^2 - zx)\vec{j} + (z^2 - xy)\vec{k}.$

$$\therefore\quad div\, \vec{F} = \frac{\partial}{\partial x}(x^2 - yz) + \frac{\partial}{\partial y}(y^2 - zx) + \frac{\partial}{\partial z}(z^2 - xy)$$

$$= 2x + 2y + 2 = 2(x + y + z).$$

$$\therefore\ \iiint_V \vec{F}\, dV = \int_0^c \int_0^b \int_0^a 2(x + y + z) dxdydz$$

$$= 2\int_0^c \int_0^b \left[\frac{x^2}{2} + xy + xz\right]_0^a dydz$$

$$= 2\int_0^c \int_0^b \left[\frac{a^2}{2} + ay + az\right] dydz$$

$$= 2\int_0^c \left[\frac{a^2}{2}y + \frac{ay^2}{2} + azy\right]_0^b dz$$

$$= 2\int_0^c \left[\frac{a^2 b}{2} + \frac{ab^2}{2} + abz\right] dz$$

$$= 2\left[\frac{a^2 bz}{2} + \frac{ab^2 z}{2} + \frac{abz^2}{2}\right]_0^c = 2\left[\frac{a^2 bc}{2} + \frac{ab^2 c}{2} + \frac{abc^2}{2}\right]$$

$$= 2\,\frac{abc}{2}(a+b+c) = abc\,(a+b+c). \quad (1)$$

To evaluate $\iint_S \vec{F}.\hat{n}\, dS$.

The surface S consists of six faces, namely *OABC, EFGD, OCDE, ABGF, OAFE, CBGD.*

$$\therefore \iint_S \vec{F}.\hat{n}\, dS = \iint_{OABC} + \iint_{EFGD} + \iint_{OCDE} + \iint_{ABGF} + \iint_{OAFE} + \iint_{CBGD} \vec{F}.\hat{n}\, dS \quad (2)$$

(i) On the face OABC

$z = 0, \hat{n} = -\vec{k}\ dS = dxdy$ and x varies from 0 to a and y varies 0 to b.

$$\therefore \vec{F}.\hat{n} = [(x^2 - yz)\vec{i} + (y^2 - zx)\vec{j} + (z^2 - xy)\vec{k}].(-\vec{k})$$

$$= -(z^2 - xy) = xy. \qquad [\because z = 0]$$

$$\therefore \iint_{OABC} \vec{F}.\hat{n}\, dS = \int_0^b \int_0^a xy\, dxdy$$

$$= \int_0^b \left[\frac{x^2 y}{2}\right]_0^a dy = \int_0^b \frac{a^2 y}{2}\, dy$$

$$= \frac{a^2}{2}\left[\frac{y^2}{2}\right]_0^b = \frac{a^2 b^2}{4}.$$

(ii) *On the face EFGD* $z = c,\ \hat{n} = \vec{k}, dS = dxdy$

and x varies from 0 to a and y varies 0 to b.

$$\therefore \quad \vec{F}.\hat{n} = z^2 - xy = c^2 - xy.$$

$$\therefore \iint_{EFGD} \vec{F}.\hat{n}\, dS = \int_0^b \int_0^a (c^2 - xy)\, dxdy$$

$$= \int_0^b \left[c^2 x - \frac{x^2 y}{2}\right]_0^a dy = \int_0^b \left[c^2 a - \frac{a^2}{2}y\right] dy$$

$$= \left[c^2 a\, y - \frac{a^2}{2}\frac{y^2}{2}\right]_0^b = abc^2 - \frac{a^2b^2}{4}.$$

(*iii*) *On the face OCDE* $x = 0,\ \hat{n} = -\vec{i},\ dS = dydz$

and y varies from 0 to b and z varies from 0 to c.

$\therefore \quad \vec{F}.\hat{n} = -(x^2 - yz) = yz\,.$

$\therefore \quad \iint_{OCDE} \vec{F}.\hat{n}\, dS = \int_0^c \int_0^b yz\, dydz$

$$= \int_0^c \left[\frac{y^2 z}{2}\right]_0^b dz = \int_0^c \left[\frac{b^2}{2} z\right] dz$$

$$= \frac{b^2}{2}\left[\frac{z^2}{2}\right]_0^c = \frac{b^2c^2}{4}.$$

(*iv*) *On the face ABGF* $x = a,\ \hat{n} = \vec{i},\ dS = dy\, dz$

and y varies from 0 to b and z varies from 0 to c.

$\therefore \quad \vec{F}.\hat{n} = x^2 - yz = a^2 - yz.$

$$\therefore \quad \iint_{ABGF} \vec{F}.\hat{n}\, dS = \int_0^c \int_0^b (a^2 - yz)\, dy\, dz$$

$$= \int_0^c \left[a^2 y - \frac{y^2 z}{2}\right]_0^b dz$$

$$= \int_0^c \left[a^2 b - \frac{b^2}{2} z\right] dz$$

$$= \left[a^2 b\, z - \frac{b^2}{2}\frac{z^2}{2}\right]_0^c = a^2 b\, c - \frac{b^2c^2}{4}.$$

(*v*) *On the face OAFE* $y = 0,\ \hat{n} = -\vec{j},\ dS = dx\, dz$

and x varies from 0 to a and z varies from 0 to c.

$\therefore \vec{F}.\hat{n} = -(y^2 - zx) = zx.$

$$\therefore \quad \iint_{OAFE} \vec{F}.\hat{n}\, dS = \int_0^c \int_0^a zx\, dx\, dz$$

$$= \int_0^c \left[z\, \frac{x^2}{2}\right]_0^a dz$$

$$= \int_0^c \left[z\, \frac{a^2}{2}\right] dz$$

$$= \left[\frac{z^2}{2}\frac{a^2}{2}\right]_0^c = \frac{a^2c^2}{4}.$$

(v) *On the face CBGD* $y = b$, $\hat{n} = \vec{j}$, $dS = dxdz$ and x varies from 0 to a and z varies from 0 to c.

$\therefore \vec{F}.\hat{n} = y^2 - zx = b^2 - zx.$

$$\therefore \iint_{CBGD} \vec{F}.\hat{n}\, dS = \int_0^c \int_0^a (b^2 - zx)\, dx\, dz$$

$$= \int_0^c \left[b^2 x - z\frac{x^2}{2} \right]_0^a dz$$

$$= \int_0^c \left[b^2 a - z\frac{a^2}{2} \right] dz$$

$$= \left[b^2 a z - \frac{z^2}{2}\frac{a^2}{2} \right]_0^c = a b^2 c - \frac{a^2 c^2}{4}.$$

$\therefore$ (2) gives $\iint_S \vec{F}.\hat{n}\, dS = \frac{a^2b^2}{4} + abc^2 - \frac{a^2b^2}{4} + \frac{b^2c^2}{4} + a^2bc$

$$- \frac{b^2c^2}{4} + \frac{a^2c^2}{4} + ab^2c - \frac{a^2c^2}{4}$$

$$= abc^2 + a^2bc + ab^2c = a\, b\, c\, (a + b + c). \quad (3)$$

From (1) and (3), $\iint_S \vec{F}.\hat{n}\, dS = \iiint_V div\, \vec{F}\, dV$.

Hence divergence theorem is verified.

Example 3: Verify Gauss divergence theorem for $\vec{F} = 4xz\,\vec{i} - y^2\vec{j} + yz\vec{k}$ taken over the cube bounded by the planes $x = 0,\ x = 1, y = 0, y = 1, z = 0$ and $z = 1$.

Solution: ***[AU, April/May 2017, Apr/May 2015, Nov/Dec 2015]***

Gauss divergence theorem is

$\iint_S \vec{F}.\hat{n}\, dS = \iiint_V div\, \vec{F}\, dV.$

$\vec{F} = 4xz\,\vec{i} - y^2\vec{j} + yz\vec{k}.$

$\therefore div\, \vec{F} = \frac{\partial}{\partial x}(4xz) + \frac{\partial}{\partial y}(-y^2) + \frac{\partial}{\partial z}(yz)$

$= 4z - 2y + y = 4z - y.$

$$\therefore \iiint_V \vec{F}\, dV = \int_0^1 \int_0^1 \int_0^1 (4z - y)dxdydz$$

$$= \int_0^1 \int_0^1 [4zx - yx]_0^1\, dydz$$

$$= \int_0^1 \int_0^1 [4z - y]\ dydz$$

$$= \int_0^1 \left[4zy - \frac{y^2}{2}\right]_0^1 dz$$

$$= \int_0^1 \left[4z - \frac{1}{2}\right] dz$$

$$= \left[4\frac{z^2}{2} - \frac{z}{2}\right]_0^1$$

$$= 2 - \frac{1}{2}$$

$$= \frac{3}{2}. \quad (1)$$

To evaluate $\iint_S \vec{F}.\hat{n}\, dS$.

The surface S consists of six faces, namely *OABC*, *EFGD*, *OCDE*, *ABGF*, *OAFE*, *CBGD*

$$\therefore \iint_S \vec{F}.\hat{n}\, dS = \iint_{OABC} + \iint_{EFGD} + \iint_{OCDE} + \iint_{ABGF} + \iint_{OAFE} + \iint_{CBGD} \vec{F}.\hat{n}\, dS \quad (2)$$

(i) On the face OABC

$z = 0, \hat{n} = -\vec{k}, dS = dxdy$ and x varies from 0 to 1 and y varies 0 to 1.

$$\therefore \vec{F}.\hat{n} = \left[4xz\,\vec{i} - y^2\vec{j} + yz\vec{k}\right].(-\vec{k})$$

$$= -yz = 0. \quad [\because z = 0]$$

$$\therefore \iint_{OABC} \vec{F}.\hat{n}\, dS = \int_0^1 \int_0^1 0\ dxdy = 0.$$

(ii) On the face EFGD

$z = 1,\ \hat{n} = \vec{k}\ , dS = dxdy$ and x varies from 0 to 1 and y varies 0 to 1.

$$\therefore \quad \vec{F}.\hat{n} = yz = y.$$

$$\therefore \quad \iint_{EFGD} \vec{F}.\hat{n}\, dS = \int_0^1 \int_0^1 y\ dxdy = \int_0^1 y[x]_0^1\ dy$$

$$= \left[\frac{y^2}{2}\right]_0^1 = \frac{1}{2}.$$

(iii) On the face OCDE

$x = 0, \hat{n} = -\vec{\imath}, dS = dydz$ and y varies from 0 to 1 and z varies from 0 to 1.

$\therefore \quad \vec{F}.\hat{n} = -x^2 = 0.$

$\therefore \quad \iint_{OCDE} \vec{F}.\hat{n}\, dS = \int_0^1 \int_0^1 0\, dydz = 0.$

(iv) On the face ABGF

$x = 1,\ \hat{n} = \vec{\imath}, dS = dydz$ and y varies from 0 to 1 and z varies from 0 to 1.

$\therefore \quad \vec{F}.\hat{n} = 4xz = 4z.$

$$\therefore \iint_{ABGF} \vec{F}.\hat{n}\, dS = \int_0^1 \int_0^1 4zdydz = \int_0^1 4z\,[y]_0^1\, dz = 4\left[\frac{z^2}{2}\right]_0^1 = 2.$$

(v) On the face OAFE

$y = 0, \hat{n} = -\vec{\jmath}, dS = dxdz$ and x varies from 0 to 1 and z varies from 0 to 1.

$\therefore \vec{F}.\hat{n} = y^2 = 0.$

$\therefore \iint_{OAFE} \vec{F}.\hat{n}\, dS = \int_0^1 \int_0^1 0\, dx\, dz = 0.$

(vi) On the face CBGD

$y = 1,\ \hat{n} = \vec{\jmath},\ dS = dxdz$ and x varies from 0 to 1 and z varies from 0 to 1.

$\therefore \vec{F}.\hat{n} = -y^2 = -1.$

$$\therefore \iint_{CBGD} \vec{F}.\hat{n}\, dS = \int_0^1 \int_0^1 (-1)\, dx\, dz = -\int_0^1 [x]_0^1\, dz$$

$$= -[z]_0^1 = -1.$$

$\therefore$ (2) gives $\iint_S \vec{F}.\hat{n}\, dS = 0 + \frac{1}{2} + 0 + 2 + 0 - 1 = \frac{3}{2}.$ (3)

From (1) and (3), $\iint_S \vec{F}.\hat{n}\, dS = \iiint_V div\,\vec{F}\, dV.$

Hence divergence theorem is verified.

Example 4: Verify divergence theorem for $\vec{F} = x^2\vec{i} + z\vec{j} + yz\vec{k}$ over the cube formed by $x = \pm 1,\ y = \pm 1, z = \pm 1$.

Solution:

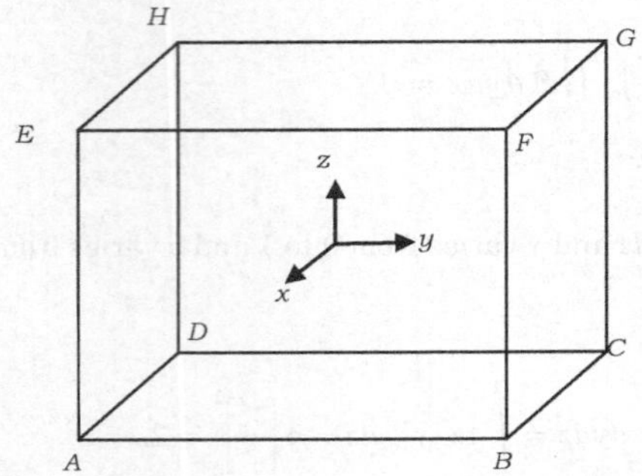

Gauss divergence theorem is

$\iint_S \vec{F}.\hat{n}\, dS = \iiint_V div\, \vec{F}\, dV.$

$\vec{F} = x^2\vec{i} + z\vec{j} + yz\vec{k}.$

$\therefore div\, \vec{F} = \frac{\partial}{\partial x}(x^2) + \frac{\partial}{\partial y}(z) + \frac{\partial}{\partial z}(yz)$

$= 2x + 0 + y = 2x + y.$

x varies from -1 to 1, y varies from

-1 to 1 and z varies from -1 to 1.

$$\therefore \iiint_V div\, \vec{F}\, dV = \int_{-1}^{1}\int_{-1}^{1}\int_{-1}^{1}(2x + y)\, dxdydz$$

$$= \int_{-1}^{1}\int_{-1}^{1}\left[\frac{2x^2}{2} + yx\right]_{-1}^{1} dydz$$

$$= \int_{-1}^{1}\int_{-1}^{1}[(1 + y) - (1 - y)]\, dydz$$

$$= \int_{-1}^{1}\int_{-1}^{1} 2y\ dydz$$

$$= \int_{-1}^{1}[y^2]_{-1}^{1}\ dz = \int_{-1}^{1} 0\, dz = 0. \qquad (1)$$

To evaluate $\iint_S \vec{F}.\hat{n}\, dS$:

$$\iint_S \vec{F}.\hat{n}\, dS = \iint_{ABCD} + \iint_{EFGH} + \iint_{DCGH} + \iint_{ABFE} + \iint_{ADHE} + \iint_{BCGF} \vec{F}.\hat{n}\, dS \quad (2)$$

(i) On the face ABCD

$z = -1,\ \hat{n} = -\vec{k},\ dS = dxdy$ and x varies from -1 to 1 and y varies from -1 to 1.

$\therefore\ \vec{F}.\hat{n} = (x^2\vec{\imath} + z\vec{\jmath} + yz\vec{k}).(-\vec{k}) = -yz = y$

$$\therefore \iint_{ABCD} \vec{F}.\hat{n}\, dS = \int_{-1}^{1}\int_{-1}^{1} y\, dx\, dy = \int_{-1}^{1} [yx]_{-1}^{1}\, dy$$

$$= \int_{-1}^{1} y(1+1)dy = 2\int_{-1}^{1} y\, dy = 2\left[\frac{y^2}{2}\right]_{-1}^{1} = 0.$$

(ii) On the face EFGH

$z = 1, \hat{n} = \vec{k}$, $dS = dxdy$ and x varies from -1 to 1 and y varies from -1 to 1.

$\therefore\quad \vec{F}.\hat{n} = yz = y$

$$\therefore \iint_{EFGH} \vec{F}.\hat{n}\, dS = \int_{-1}^{1}\int_{-1}^{1} y\, dx\, dy$$

$= 0 \qquad [\because$ the integrand is an odd function of y]

(iii) On the face DCGH

$x = -1,\ \hat{n} = -\vec{\imath},\ dS = dydz$ and y varies from -1 to 1 and z varies from -1 to 1.

$\therefore\quad \vec{F}.\hat{n} = -x^2 = -1.$

$$\therefore \iint_{DCGH} \vec{F}.\hat{n}\, dS = \int_{-1}^{1}\int_{-1}^{1}(-1)\, dy\, dz = -\int_{-1}^{1}[y]_{-1}^{1}\, dz$$

$$= -\int_{-1}^{1} 2\, dz = -2[z]_{-1}^{1} = -4.$$

(iv) On the face ABFE

$x = 1, \hat{n} = \vec{\imath},\ dS = dydz$ and y varies from -1 to 1 and z varies from -1 to 1.

$\therefore\quad \vec{F}.\hat{n} = x^2 = 1.$

$$\therefore \iint_{ABFE} \vec{F}.\hat{n}\, dS = \int_{-1}^{1}\int_{-1}^{1} dy\, dz$$

$= \int_{-1}^{1} [y]_{-1}^{1}\ dz = 2\int_{-1}^{1} dz$

$= 2[z]_{-1}^{1} = 4.$

(*v*) *On the face ADHE*

$y = -1, \hat{n} = -\vec{j},\ dS = dxdz$ and x varies from -1 to 1 and z varies from -1 to 1.

$\therefore \quad \vec{F}.\hat{n} = -z.$

$\therefore \iint_{ADHE} \vec{F}.\hat{n}\, dS = \int_{-1}^{1}\int_{-1}^{1}(-z)dx\, dz$

$= 0.$ [$\because$ the integrand is an odd function of z]

(*vi*) *On the face BCGF*

$y = 1, \hat{n} = \vec{j},\ dS = dxdz$ and x varies from -1 to 1 and z varies from -1 to 1.

$\therefore \vec{F}.\hat{n} = z.$

$\therefore \iint_{BCGF} \vec{F}.\hat{n}\, dS = \int_{-1}^{1}\int_{-1}^{1} z\, dx\, dz = 0.$ [$\because$ the integrand is an odd function of z]

$\therefore$ (2) gives $\iint_S \vec{F}.\hat{n}\, dS = 0 + 0 + 0 + 0 + 4 - 4 = 0.$ (3)

From (1) and (3), $\iint_S \vec{F}.\hat{n}\, dS = \iiint_V div\, \vec{F}\, dV.$

Hence Gauss divergence theorem is verified.

Example 5: Evaluate $\iint_S \vec{r}.\hat{n}\, dS$ where S is the surface of the cube bounded by the planes $x = -1,\ y = -1,\ z = -1,\ x = 1,\ y = 1,\ z = 1.$

Solution: By Gauss divergence theorem,

$\iint_S \vec{r}.\hat{n}\, dS = \iiint_V div\, \vec{r}\, dV$ (1)

Let $\vec{r} = x\vec{i} + y\vec{j} + z\vec{k}$.

Then $div\, \vec{r} = 1 + 1 + 1 = 3.$

$\therefore$ (1) gives $\iint_S \vec{r}.\hat{n}\, dS = \int_{-1}^{1}\int_{-1}^{1}\int_{-1}^{1} 3dx\, dy\, dz$

$= 3\int_{-1}^{1}\int_{-1}^{1} [x]_{-1}^{1}\ dy\, dz$

$= 3\int_{-1}^{1}\int_{-1}^{1} 2\ dy\, dz = 6\int_{-1}^{1} [y]_{-1}^{1}\ dz$

$= 6\int_{-1}^{1} 2\ dz = 12\,[z]_{-1}^{1} = 12 \times 2 = 24.$

Example 6: Prove that for a closed surface S, $\iint_S \vec{r}.\hat{n}\, dS = 3V$ where V is the volume enclosed by S.

Solution: By Gauss divergence theorem,

$\iint_S \vec{r}.\hat{n}\, dS = \iiint_V div\, \vec{r}\, dV$

$= \iiint_V 3\, dV [\because\ div\, \vec{r} = 3\,] = 3\iiint_V dV = 3V.$

Example 7: Show that $\iint_S curl\,\vec{F}.\hat{n}\,dS = 0$ where S is any closed surface.

Solution: By Gauss divergence theorem,

$$\iint_S curl\,\vec{F}.\hat{n}\,dS = \iiint_V div\left(curl\,\vec{F}\right) dV$$

$$= \iiint_V 0\,dV\left[\because\ div\left(curl\,\vec{F}\right) = 0\right] \quad = 0.$$

EXERCISES

1. Verify Gauss divergence theorem for $\vec{F} = x^2\vec{\imath} + y^2\vec{\jmath} + z^2\vec{k}$, where S is the surface of the cuboid formed by the planes $x = 0, x = a, y = 0, y = b, z = 0$ and $z = c$.

2. Verify Gauss divergence theorem for $\vec{F} = (x^2 - yz)\vec{\imath} + (y^2 - zx)\vec{\jmath} + (z^2 - xy)\vec{k}$ taken over the cuboid $0 \le x \le a, 0 \le y \le a$ and $0 \le y \le a$.

3. Verify divergence theorem for the vector function $\vec{F} = (x^3 - yz)\vec{\imath} - 2x^2y\vec{\jmath} + 2\vec{k}$ over the cube bounded by $x = 0, y = 0, z = 0, x = a,\ y = a$ and $z = a$.

4. Verify Gauss divergence theorem for $\vec{F} = (2x - z)\vec{\imath} + x^2y\vec{\jmath} - xz^2\vec{k}$taken over the region bounded by $x = 0, x = 1, y = 0, y = 1, z = 0$ and $z = 1$.

5. Given the vector $\vec{F} = (x^2 - y^2)\vec{\imath} + 2xy\vec{\jmath} + (y^2 - xy)\vec{k}$. Verify the divergence theorem over the cube with centre at the origin and of side length a.

6. Evaluate $\iint_S \vec{F}.\hat{n}\,dS$ using Gauss divergence theorem, where $\vec{F} = 2xy\vec{\imath} + yx^2\vec{\jmath} + xz\vec{k}$and S is the rectangular parallelepiped bounded by $x = 0, y = 0, z = 0, x = 2$, $y = 1,\ z = 3$.

7. Evaluate by using divergence theorem $\iint_S \vec{F}.d\vec{S}$where $\vec{F} = (x^2 - yz)\vec{\imath} + (y^2 - zx)\vec{\jmath} + (z^2 - xy)\vec{k}$ and S is the surface of the rectangular parallelopiped bounded by the planes $0 \le x \le a, 0 \le y \le b$ and $0 \le z \le c$.

8. If S is any closed surface enclosing a volume V and $\vec{F} = x\vec{\imath} + 2y\vec{\jmath} + 3z\vec{k}$, show that $\iint_S \vec{F}.\hat{n}\,dS = 6V$.

ANSWERS

6. 20   7. $abc(a + b + c)$

2.12. Short Answer Questions

1. What are the values of $\nabla.\vec{r}$ and $\nabla \times \vec{r}$?

Solution: Let $\vec{r} = \vec{\imath} + y\vec{\jmath} + z\vec{k}$.

Hence $\nabla.\vec{r} = \frac{\partial}{\partial x}(x) + \frac{\partial}{\partial y}(y) + \frac{\partial}{\partial z}(z) = 1 + 1 + 1 = 3.$

Also, $\nabla \times \vec{r} = \begin{vmatrix} \vec{\imath} & \vec{\jmath} & \vec{k} \\ \frac{\partial}{\partial x} & \frac{\partial}{\partial y} & \frac{\partial}{\partial z} \\ x & y & z \end{vmatrix} = \vec{\imath}(0-0) - \vec{\jmath}(0-0) + \vec{k}(0-0) = \vec{0}.$

2. What is the greatest rate of increase of $u = xyz^2$ at the point(1,0,3)?

Solution: $u = xyz^2$

$$\nabla u = \frac{\partial u}{\partial x}\vec{\imath} + \frac{\partial u}{\partial y}\vec{\jmath} + \frac{\partial u}{\partial z}\vec{k} = yz^2\vec{\imath} + xz^2\vec{\jmath} + 2xyz\vec{k}$$

$$\nabla u_{\,at\ (1,0,3)} = 9\vec{\jmath}.$$

$\therefore$ The greatest rate of increase of $u = xyz^2$ at the point $(1,0,3) = |\Delta u|$

$$= |9i| = 9.$$

3. Calculate the *curl* of $\vec{F} = xyz\vec{\imath} + 3x^2y\vec{\jmath} + (xz^2 - y^2z)\vec{k}$ at $(1,-1,1)$.

Solution:

$$curl\ \vec{F} = \begin{vmatrix} \vec{\imath} & \vec{\jmath} & \vec{k} \\ \frac{\partial}{\partial x} & \frac{\partial}{\partial y} & \frac{\partial}{\partial z} \\ xyz & 3x^2y & xz^2 - y^2z \end{vmatrix} = \vec{\imath}(-2yz - 0) - \vec{\jmath}(z^2 - xy) + \vec{k}(6xy - xz)$$

$$curl\ \vec{F}_{at\ (1,-1,1)} = 2\vec{\imath} - 2\vec{\jmath} - 7\vec{k}.$$

4. Find curl$\vec{F}$ if $\vec{F} = \nabla(x^3 + y^3 + z^3 - 3xyz)$.

Solution: $\vec{F} = \nabla(x^3 + y^3 + z^3 - 3xyz)$

$$= (3x^2 - 3yz)\vec{\imath} + (3y^2 - 3xz)\vec{\jmath} + (3z^3 - 3xy)\vec{k}$$

$$curl\ \vec{F} = \begin{vmatrix} \vec{\imath} & \vec{\jmath} & \vec{k} \\ \frac{\partial}{\partial x} & \frac{\partial}{\partial y} & \frac{\partial}{\partial z} \\ 3x^2 - 3yz & 3y^2 - 3xz & 3z^3 - 3xy \end{vmatrix}$$

$$= \vec{\imath}(-3x + 3x) - (-3y + 3y)\vec{\jmath} + \vec{k}(-3z + 3z) = \vec{0}.$$

5. If $\vec{F} = x^3\vec{i} + y^3\vec{j} + z^3\vec{k}$, find $div\ curl\ \vec{F}$.

Solution: $$curl\,\vec{F} = \begin{vmatrix} \vec{i} & \vec{j} & \vec{k} \\ \frac{\partial}{\partial x} & \frac{\partial}{\partial y} & \frac{\partial}{\partial z} \\ x^3 & y^3 & z^3 \end{vmatrix}$$

$$= \vec{i}(0-0) - \vec{j}(0-0) + \vec{k}(0-0) = \vec{0}.$$

$\therefore div\left(curl\,\vec{F}\right) = div\left(\vec{0}\right) = 0.$

6. Prove that $curl\,(grad\,\varphi) = 0$. ***[AU, May/June 2014]***

Solution: $$grad\,\varphi = \nabla\varphi = \frac{\partial\varphi}{\partial x}\vec{i} + \frac{\partial\varphi}{\partial y}\vec{j} + \frac{\partial\varphi}{\partial z}\vec{k}$$

$$\therefore curl\ grad\,\varphi = \nabla \times (\nabla\varphi) = \begin{vmatrix} \vec{i} & \vec{j} & \vec{k} \\ \frac{\partial}{\partial x} & \frac{\partial}{\partial y} & \frac{\partial}{\partial z} \\ \frac{\partial\varphi}{\partial x} & \frac{\partial\varphi}{\partial y} & \frac{\partial\varphi}{\partial z} \end{vmatrix}$$

$$= \vec{i}\left(\frac{\partial^2\varphi}{\partial y\partial z} - \frac{\partial^2\varphi}{\partial y\partial z}\right) - \vec{j}\left(\frac{\partial^2\varphi}{\partial x\partial z} - \frac{\partial^2\varphi}{\partial z\partial x}\right) + \vec{k}\left(\frac{\partial^2\varphi}{\partial x\partial y} - \frac{\partial^2\varphi}{\partial y\partial x}\right)$$

$$= 0\vec{i} - 0\vec{j} + 0\vec{k} = \vec{0}$$

7. Show that $div\ curl\,\vec{F} = 0$.

Solution: Let $\vec{f} = f_1\vec{i} + f_2\vec{j} + f_3\vec{k}$.

$$\text{Then } curl\vec{f} = \nabla \times \vec{f} = \begin{vmatrix} \vec{i} & \vec{j} & \vec{k} \\ \frac{\partial}{\partial x} & \frac{\partial}{\partial y} & \frac{\partial}{\partial z} \\ f_1 & f_2 & f_3 \end{vmatrix}$$

$$= \left(\frac{\partial f_3}{\partial y} - \frac{\partial f_2}{\partial z}\right)\vec{i} - \left(\frac{\partial f_3}{\partial x} - \frac{\partial f_1}{\partial z}\right)\vec{j} + \left(\frac{\partial f_2}{\partial x} - \frac{\partial f_1}{\partial y}\right)\vec{k}$$

$$\therefore div\ curl\,\vec{f} = \nabla.(\nabla \times \vec{f}) = \frac{\partial}{\partial x}\left(\frac{\partial f_3}{\partial y} - \frac{\partial f_2}{\partial z}\right) - \frac{\partial}{\partial y}\left(\frac{\partial f_3}{\partial x} - \frac{\partial f_1}{\partial z}\right) + \frac{\partial}{\partial z}\left(\frac{\partial f_2}{\partial x} - \frac{\partial f_1}{\partial y}\right)$$

$$= \frac{\partial^2 f_3}{\partial x\partial y} - \frac{\partial^2 f_2}{\partial x\partial z} - \frac{\partial^2 f_3}{\partial y\partial x} + \frac{\partial^2 f_1}{\partial y\partial z} + \frac{\partial^2 f_2}{\partial z\partial x} - \frac{\partial^2 f_1}{\partial z\partial y}$$

$$= 0.$$

8. Find a unit vector normal to the surface $x^2 + y^2 - z = 10$ at (1, 1, 1).

Solution:

Let $\varphi = x^2 + y^2 - z - 10$

$$\nabla\varphi = \frac{\partial\varphi}{\partial x}\vec{\imath} + \frac{\partial\varphi}{\partial y}\vec{\jmath} + \frac{\partial\varphi}{\partial z}\vec{k} = 2x\vec{\imath} + 2y\vec{\jmath} - \vec{k}.$$

$\therefore\ \nabla\varphi\ at\ (1,1,1) = 2\vec{\imath} + 2\vec{\jmath} - \vec{k}$

$$|\nabla\varphi| = \sqrt{4+4+1} = 3.$$

Unit normal vector is $\hat{n} = \frac{\nabla\varphi}{|\nabla\varphi|} = \frac{2\vec{\imath}+2\vec{\jmath}-\vec{k}}{3}.$

9. Find the unit normal vector to the surface $x^3 + y^2 = z$ at $(1, 1, 2)$.

Solution: **[AU, *April/May 2017*]**

Let $\varphi = x^3 + y^2 - z$

$$\nabla\varphi = \frac{\partial\varphi}{\partial x}\vec{\imath} + \frac{\partial\varphi}{\partial y}\vec{\jmath} + \frac{\partial\varphi}{\partial z}\vec{k} = 3x^2\vec{\imath} + 2y\vec{\jmath} - \vec{k}.$$

$\therefore\ \nabla\varphi\ at\ (1,1,2) = 3\vec{\imath} + 2\vec{\jmath} - \vec{k}$

$$|\nabla\varphi| = \sqrt{9+4+1} = \sqrt{14}.$$

Unit normal vector is $\hat{n} = \frac{\nabla\varphi}{|\nabla\varphi|} = \frac{3\vec{\imath}+2\vec{\jmath}-\vec{k}}{\sqrt{14}}.$

10. Find the directional derivative of $\varphi = x^2yz + 4xz^2 + xyz$ at (1, 2, 3) in the direction of $2\vec{\imath} + \vec{\jmath} - \vec{k}$.

Solution: $\nabla\varphi = \frac{\partial\varphi}{\partial x}\vec{\imath} + \frac{\partial\varphi}{\partial y}\vec{\jmath} + \frac{\partial\varphi}{\partial y}\vec{k}$

$$= (2xyz + 4z^2 + yz)\vec{\imath} + (x^2z + xz)\vec{\jmath} + (x^2y + 8xz + xy)\vec{k}$$

$\nabla\varphi$ at $(1,2,3) = 54\vec{\imath} + 6\vec{\jmath} + 28\vec{k}$.

Let $\vec{a} = 2\vec{\imath} + \vec{\jmath} - \vec{k}$, then $|\vec{a}| = \sqrt{4+1+1} = \sqrt{6}$.

$\therefore$ The directional derivative $= \frac{\nabla\varphi.\vec{a}}{|\vec{a}|} = \frac{(54\vec{\imath}+6\vec{\jmath}+28\vec{k}).(2\vec{\imath}+\vec{\jmath}-\vec{k})}{\sqrt{6}}$

$$= \frac{86}{\sqrt{6}}.$$

11. Find the maximum directional derivative of $\varphi = x^2 + yz^2$ at (1, 2, 3).

Solution : $\nabla\varphi = \frac{\partial\varphi}{\partial x}\vec{i} + \frac{\partial\varphi}{\partial y}\vec{j} + \frac{\partial\varphi}{\partial z}\vec{k} = 2x\vec{i} + z^2\vec{j} + 2yz\vec{k}$

$\nabla\varphi_{at\ (1,2,3)} = 2\vec{i} + 9\vec{j} + 12\vec{k}.$

The directional derivative is maximum along the direction of $\nabla\varphi$.

i.e., D.D. is maximum along the direction of $2\vec{i} + 9\vec{j} + 12\vec{k}$.

12. In what direction from (3,1,-2) is the directional derivative of $\varphi = x^2y^2z^4$ maximum? Find also the magnitude of this maximum. ***[AU, April/May 2015]***

Solution:

Given $\varphi = x^2y^2z^4$.

$\nabla\varphi = \frac{\partial\varphi}{\partial x}\vec{i} + \frac{\partial\varphi}{\partial y}\vec{j} + \frac{\partial\varphi}{\partial z}\vec{k}$

$= 2xy^2z^4\vec{i} + 2x^2yz^4\vec{j} + 4x^2y^2z^3\vec{k}$.

$\therefore \nabla\varphi$ at $(3,1,-2) = 96\vec{i} + 288\vec{j} - 288\vec{k} = 96(\vec{i} + 3\vec{j} - 3\vec{k})$.

$\therefore$ The directional derivative of $\varphi = x^2y^2z^4$ at $(3,1,-2)$ is maximum along the direction $96(\vec{i} + 3\vec{j} - 3\vec{k})$.

The magnitude of the maximum directional derivative $= |\nabla\varphi|$

$$= 96\sqrt{1+4+9} = 96\sqrt{14}.$$

13. Find the unit normal vector to the surface $x^2 + y^2 = z$ at $(1,-2,5)$

Solution: Let $\varphi = x^2 + y^2 - z$. ***[AU, Nov/Dec 2017, May/June 2014]***

Then $\nabla\varphi = \frac{\partial\varphi}{\partial x}\vec{i} + \frac{\partial\varphi}{\partial y}\vec{j} + \frac{\partial\varphi}{\partial z}\vec{k} = 2x\vec{i} + 2y\vec{j} - \vec{k}$.

$\therefore \nabla\varphi$ at $(1,-2,5) = 2\vec{i} - 4\vec{j} - \vec{k}$

$\therefore |\nabla\varphi| = \sqrt{4+16+1} = \sqrt{21}$.

Hence unit normal vector is $\hat{n} = \frac{\nabla\varphi}{|\nabla\varphi|} = \frac{1}{\sqrt{21}}(2\vec{i} - 4\vec{j} - \vec{k})$.

14. If $\vec{F} = z\vec{i} + x\vec{j} + y\vec{k}$, prove that $curl\, curl\, \vec{F} = \vec{0}$.

Solution: $$curl\, \vec{F} = \begin{vmatrix} \vec{i} & \vec{j} & \vec{k} \\ \frac{\partial}{\partial x} & \frac{\partial}{\partial y} & \frac{\partial}{\partial z} \\ z & x & y \end{vmatrix}$$

$$= \vec{i}(1-0) - \vec{j}(0-1) + \vec{k}(1-0)$$

$$= \vec{i} + \vec{j} + \vec{k}.$$

$$curl\, curl\, \vec{F} = \begin{vmatrix} \vec{i} & \vec{j} & \vec{k} \\ \frac{\partial}{\partial x} & \frac{\partial}{\partial y} & \frac{\partial}{\partial z} \\ 1 & 1 & 1 \end{vmatrix}$$

$$= \vec{i}(0-0) - \vec{j}(0-0) + \vec{k}(0-0) = \vec{0}.$$

15. Prove that $\vec{F} = (x+3y)\vec{i} + (z-3y)\vec{j} + (x+2z)\vec{k}$ is solenoidal.

Solution: $div\, \vec{F} = \frac{\partial}{\partial x}(x+3y) + \frac{\partial}{\partial y}(z-3y) + \frac{\partial}{\partial z}(x+2z)$

$$= 1 - 3 + 2 = 0.$$

$\therefore$ $\vec{F}$ is solenoidal.

16. If $\vec{F} = (x+3y)\vec{i} + (y-2z)\vec{j} + (x+2kz)\vec{k}$ has divergence zero, find the unknown value of k. **[AU, *Nov/Dec 2017*]**

Solution: $div\, \vec{F} = \frac{\partial}{\partial x}(x+3y) + \frac{\partial}{\partial y}(y-2z) + \frac{\partial}{\partial z}(x+2k) = 1 + 1 + 2k = 2 + 2k$

Given that $div\, \vec{F} = 0$.

$\therefore$ $2 + 2k = 0$.

Hence $k = -1$.

17. Prove that $3x^2y\,\vec{i} + (yz - 3xy^2)\vec{j} - \frac{z^2}{2}\vec{k}$ is a solenoidal vector.

Solution: ***[AU, Nov/Dec 2015]***

$$div\, \vec{F} = \frac{\partial}{\partial x}(3x^2y) + \frac{\partial}{\partial y}(yz - 3xy^2) + \frac{\partial}{\partial z}\left(-\frac{z^2}{2}\right) = 6xy + z - 6xy - z = 0.$$

$\therefore$ $\vec{F}$ is solenoidal.

18. Prove that $(3x + 2y + 4z)\vec{i} + (2x + 5y + 4z)\vec{j} + (4x + 4y - 8z)\vec{k}$ is irrotational.

Solution: $curl\,\vec{F} = \begin{vmatrix} \vec{i} & \vec{j} & \vec{k} \\ \frac{\partial}{\partial x} & \frac{\partial}{\partial y} & \frac{\partial}{\partial z} \\ 3x + 2y + 4z & 2x + 5y + 4z & 4x + 4y - 8z \end{vmatrix}$

$$= \vec{i}(4 - 4) - \vec{j}(4 - 4) + \vec{k}(2 - 2) = \vec{0}.$$

$\therefore\ \vec{F}$ is irrotational.

19. If φ is a scalar point function, prove that $\nabla\varphi$ is both solenoidal and irrotational provided φ is a solution of Laplace equation.

(or) If $\nabla^2\varphi = 0$, then prove that $\nabla\varphi$ is both solenoidal and irrotational.

Solution:

Since φ is a solution of Laplace equation, $\nabla^2\varphi = 0$.

(i) $div(\nabla\varphi) = \nabla.\nabla\varphi = \left(\vec{i}\frac{\partial}{\partial x} + \vec{j}\frac{\partial}{\partial y} + \vec{k}\frac{\partial}{\partial z}\right).\left(\vec{i}\frac{\partial\varphi}{\partial x} + \vec{j}\frac{\partial\varphi}{\partial y} + \vec{k}\frac{\partial\varphi}{\partial z}\right)$

$$= \frac{\partial^2\varphi}{\partial x^2} + \frac{\partial^2\varphi}{\partial y^2} + \frac{\partial^2\varphi}{\partial z^2}$$

$$= \nabla^2\varphi = 0, \text{ using (1)}$$

$\therefore \nabla\varphi$ is solenoidal.

(ii) $curl\,(\nabla\varphi) = \begin{vmatrix} \vec{i} & \vec{j} & \vec{k} \\ \frac{\partial}{\partial x} & \frac{\partial}{\partial y} & \frac{\partial}{\partial z} \\ \frac{\partial\phi}{\partial x} & \frac{\partial\phi}{\partial y} & \frac{\partial\phi}{\partial z} \end{vmatrix}$

$$= \vec{i}\left(\frac{\partial^2\phi}{\partial y\partial z} - \frac{\partial^2\phi}{\partial z\partial y}\right) - \vec{j}\left(\frac{\partial^2\phi}{\partial x\partial z} - \frac{\partial^2\phi}{\partial z\partial x}\right) + \vec{k}\left(\frac{\partial^2\phi}{\partial x\partial y} - \frac{\partial^2\phi}{\partial y\partial x}\right) = \vec{0}.$$

$\therefore \nabla\varphi$ is irrotational.

20. Find a such that $(3x - 2y + z)\vec{i} + (4x + ay - z)\vec{j} + (x - y + 2z)\vec{k}$ is solenoidal.

Solution: ***[AU, April/May 2015]***

$\vec{F}$ is solenoidal if $div\,\vec{F} = 0$.

i.e., $\frac{\partial}{\partial x}(3x - 2y + z) + \frac{\partial}{\partial y}(4x + ay - z) + \frac{\partial}{\partial z}(x - y + 2z) = 0$

$$3 + a + 2 = 0$$

$$a = -5.$$

21. Find α, β, γ so that $\vec{A} = (x + 2y + \alpha z)\vec{\imath} + (\beta x - 3y - z)\vec{\jmath} + (4x + \gamma y + 2x)\vec{k}$ is irrotational.

Solution: $\vec{A}$ is irrotational if $curl\, \vec{A} = \vec{0}$.

$$\text{i.e.,} \begin{vmatrix} \vec{\imath} & \vec{\jmath} & \vec{k} \\ \frac{\partial}{\partial x} & \frac{\partial}{\partial y} & \frac{\partial}{\partial z} \\ x + 2y + \alpha z & \beta x - 3y - z & 4x + \gamma y + 2z \end{vmatrix} = 0$$

$$\vec{\imath}(\gamma + 1) - \vec{\jmath}(4 - \alpha) + \vec{k}(\beta - 2) = 0\vec{\imath} + 0\vec{\jmath} + 0\vec{k}$$

$$\gamma + 1 = 0, \quad \alpha - 4 = 0, \quad \beta - 2 = 0$$

$\therefore \quad \alpha = 4,\ \beta = 2$ and $\gamma = -1$.

22. If $\vec{a}$ and $\vec{b}$ are irrotational, prove that $\vec{a} \times \vec{b}$ is solenoid.

Solution: Given $curl\, \vec{a} = \vec{0}$ and $curl\, \vec{b} = \vec{0}$.

Now $div\,(\vec{a} \times \vec{b}) = \vec{b}.curl\, \vec{a} - \vec{a}.curl\, \vec{b}$

$= \vec{b}.\vec{0} - \vec{a}.\vec{0} = 0.$

$\therefore\ \vec{a} \times \vec{b}$ is solenoidal.

23. If $\nabla\varphi$ is solenoidal, find the value of $\nabla^2\varphi$.

Solution:

Since $\nabla\varphi$ is solenoidal, $div(\nabla\varphi) = 0$.

i.e., $\nabla.\nabla\varphi = 0$

i.e., $\nabla^2\varphi = 0$.

24. Evaluate $\nabla^2 \log r$. ***[AU, May/June 2016]***

Solution: We have $r = |\vec{r}| = \sqrt{x^2 + y^2 + z^2}$

$\therefore r^2 = x^2 + y^2 + z^2$.

$\therefore 2r \frac{\partial r}{\partial x} = 2x \Rightarrow \frac{\partial r}{\partial x} = \frac{x}{r}$.

Similarly, $\frac{\partial r}{\partial y} = \frac{y}{r}$ and $\frac{\partial r}{\partial z} = \frac{z}{r}$.

Now, $\frac{\partial}{\partial x}(\log r) = \frac{1}{r}\frac{\partial r}{\partial x} = \frac{1}{r}\frac{x}{r} = x\, r^{-2}$.

$$\frac{\partial^2}{\partial x^2}(\log r) = \frac{\partial}{\partial x}(x\,r^{-2})$$

$$= x(-2)r^{-3}\frac{\partial r}{\partial x} + r^{-2}$$

$$= -2xr^{-3}\frac{x}{r} + r^{-2}$$

$$= -2x^2r^{-4} + r^{-2}\,.$$

Similarly, $\dfrac{\partial^2}{\partial y^2}(\log r) = -2\,y^2\,r^{-4} + r^{-2}$.

And $\dfrac{\partial^2}{\partial z^2}(\log r) = -2\,z^2\,r^{-4} + r^{-2}$.

Now, $\nabla^2 \log r = \left(\dfrac{\partial^2}{\partial x^2} + \dfrac{\partial^2}{\partial y^2} + \dfrac{\partial^2}{\partial z^2}\right)(\log r)$

$$= -2x^2r^{-4} + r^{-2} - 2y^2r^{-4} + r^{-2} - 2z^2r^{-4} + r^{-2}$$

$$= 3r^{-2} - 2r^{-4}[x^2 + y^2 + z^2]$$

$$= 3r^{-2} - 2r^{-4}r^2$$

$$= 3r^{-2} - 2r^{-2} = r^{-2} = \frac{1}{r^2}\,.$$

25. Prove that $\nabla\left(e^{x^2+y^2+z^2}\right) = 2e^{r^2}\vec{r}$.

Solution: We know that $\vec{r} = x\vec{\imath} + y\vec{\jmath} + z\vec{k}$

$r^2 = x^2 + y^2 + z^2$

$\dfrac{\partial r}{\partial x} = \dfrac{x}{r}$, $\dfrac{\partial r}{\partial y} = \dfrac{y}{r}$ and $\dfrac{\partial r}{\partial z} = \dfrac{z}{r}$.

Now $\nabla\left(e^{x^2+y^2+z^2}\right) = \nabla\left(e^{r^2}\right)$

$$= \frac{\partial}{\partial x}(e^{r^2})\vec{\imath} + \frac{\partial}{\partial y}(e^{r^2})\vec{\jmath} + \frac{\partial}{\partial z}(e^{r^2})\vec{k} = e^{r^2}2r\frac{\partial r}{\partial x}\vec{\imath} + e^{r^2}2r\frac{\partial r}{\partial y}\vec{\jmath} + e^{r^2}2r\frac{\partial r}{\partial y}\vec{k}$$

$$= 2\,r\,e^{r^2}\left[\frac{x}{r}\vec{\imath} + \frac{y}{r}\vec{\jmath} + \frac{z}{r}\vec{k}\right]$$

$$= 2e^{r^2}(x\vec{\imath} + y\vec{\jmath} + z\vec{k}) = 2e^{r^2}\vec{r}.$$

26. Prove that $\nabla r = \frac{\vec{r}}{r}$.

Solution: $r = |\vec{r}| = \sqrt{x^2 + y^2 + z^2}$.

Then $\frac{\partial r}{\partial x} = \frac{x}{r}, \frac{\partial r}{\partial y} = \frac{y}{r}, \frac{\partial r}{\partial z} = \frac{z}{r}$.

$$\therefore \quad \nabla r = \frac{\partial r}{\partial x}\vec{i} + \frac{\partial r}{\partial y}\vec{j} + \frac{\partial r}{\partial z}\vec{k}$$

$$= \frac{x}{r}\vec{i} + \frac{y}{r}\vec{j} + \frac{z}{r}\vec{k} \quad = \frac{1}{r}\left(x\vec{i} + y\vec{j} + z\vec{k}\right) = \frac{\vec{r}}{r}.$$

27. Find $\nabla(r^n)$

Solution: $\nabla(r^n) = \frac{\partial r^n}{\partial x}\vec{i} + \frac{\partial r^n}{\partial y}\vec{j} + \frac{\partial r^n}{\partial z}\vec{k}$

$$= nr^{n-1}\frac{\partial r}{\partial x}\vec{i} + nr^{n-1}\frac{\partial r}{\partial y}\vec{j} + nr^{n-1}\frac{\partial r}{\partial z}\vec{k}$$

$$= nr^{n-1}\frac{x}{r}\vec{i} + nr^{n-1}\frac{y}{r}\vec{j} + nr^{n-1}\frac{z}{r}\vec{k}$$

$$= nr^{n-2}\left(x\vec{i} + y\vec{j} + z\vec{k}\right) = nr^{n-2}\vec{r}.$$

28. If $\vec{F} = 3xy\vec{i} - y^2\vec{j}$, evaluate $\int_C \vec{F}.d\vec{r}$ where C is the curve $y = 2x^2$ in the $xy-$ plane from (0, 0) to (1, 2).

Solution: $\vec{F}.d\vec{r} = (3xy\vec{i} - y^2\vec{j}).\left(dx\vec{i} + dy\vec{j} + dz\vec{k}\right)$

$$= 3xy\,dx - y^2 dy$$

$$= 3x(2x^2)dx - (2x^2)^2(4x\,dx) \quad [\because y = 2x^2]$$

$$= (6x^3 - 16x^5)dx$$

$$\therefore \int_C \vec{F}.d\vec{r} = \int_0^1 (6x^3 - 16x^5)\,dx$$

$$= \left[\frac{6x^4}{4} - \frac{16x^6}{6}\right]_0^1 = \frac{6}{4} - \frac{16}{6} = \frac{-7}{6}.$$

29. Find the work done by the force $\vec{F} = x\vec{i} + 2y\vec{j}$ when it moves a particle on the curve $2y = x^2$ from (0, 0) to (2 ,2).

Solution: Given curve is $y = \frac{x^2}{2}$

Then $dy = x\,dx$.

$\vec{F}.d\vec{r} = xdx + 2ydy = xdx + x^2xdx$

$= (x + x^3)dx$

$\therefore$ Work done $= \int_C \vec{F}.d\vec{r}$

$= \int_0^2 (x + x^3)dx$

$= \left[\frac{x^2}{2} + \frac{x^4}{4}\right]_0^2 = 2 + 4 = 6.$

30. Evaluate the integral $\int_C \vec{F}.d\vec{r}$ if $\vec{F} = xy^2\vec{i} + (x^2 + y^2)\vec{j}$ and C is the curve given by $y = x^2 - 4$ from (2,0) to (4,12). **[AU, *Nov/Dec 2017*]**

Solution: Given curve is $y = x^2 - 4$ or $x^2 = y + 4$.

Then $dy = 2x\,dx$.

$\vec{F}.d\vec{r} = xy^2dx + (x^2 + y^2)dy = xy^2\frac{dy}{2x} + (y + 4 + y^2)dy$

$= \left(\frac{3y^2}{2} + y + 4\right)dy$

$\therefore$ Work done $= \int_C \vec{F}.d\vec{r}$

$= \int_0^{12} \left(\frac{3y^2}{2} + y + 4\right)dx$

$= \left[\frac{3y^3}{2\times 3} + \frac{y^2}{2} + 4y\right]_0^{12} = \frac{12^3}{2} + \frac{12^2}{2} + 48 = 984.$

31. If $\vec{F} = x^2\vec{i} + xy^2\vec{j}$, evaluate the line integral $\int_C \vec{F}.d\vec{r}$ from (0, 0) to (1, 1) along the path $y = x$. **[*AU, May/June 2005, May/June 2006*]**

Solution: Given curve is $y = x$. Then $dy = dx$

$\vec{F}.d\vec{r} = x^2dx + xy^2dy = x^2dx + x^3dx = (x^2 + x^3)dx$

$$\therefore \int_C \vec{F}.d\vec{r} = \int_0^1 (x^2 + x^3)dx = \left[\frac{x^3}{3} + \frac{x^4}{4}\right]_0^1$$

$= \frac{1}{3} + \frac{1}{4} = \frac{7}{12}.$

32. Using Green's theorem, find the area of the ellipse $\frac{x^2}{a^2} + \frac{y^2}{b^2} = 1.$

Solution: Area $= \frac{1}{2}\int_C (xdy - ydx)$ (1)

On the ellipse, $x = a\cos\theta$, $y = b\sin\theta$ and θ varies from 0 to 2π.

$\therefore\ dx = -a\sin\theta\, d\theta,\quad dy = b\cos\theta\, d\theta.$

$\therefore$ (1) gives Area $= \frac{1}{2}\int_0^{2\pi} (ab\cos^2\theta + ab\sin^2\theta)d\theta$

$$= \frac{1}{2}\int_0^{2\pi} ab\, d\theta$$

$$= \frac{ab}{2}[\theta]_0^{2\pi}$$

$$= \pi ab.$$

33. Evaluate $\int_C [(2xy^2 - y)dx + (2x^2y + x)dy]$, where C is the circle of unit radius with centre at the origin, using Green's theorem.

Solution:

Let $M = 2xy^2 - y$ and $N = 2x^2y + x$.

$\frac{\partial M}{\partial y} = 4xy - 1$ and $\frac{\partial N}{\partial x} = 4xy + 1$

$\frac{\partial N}{\partial x} - \frac{\partial M}{\partial y} = (4xy + 1) - (4xy - 1) = 2.$

$\therefore$ By Green's theorem,

$$\int_C Mdy + Ndy = \iint_R \left(\frac{\partial N}{\partial x} - \frac{\partial M}{\partial y}\right) dxdy$$

$$= \iint_R 2\, dxdy$$

$$= 2 \times \iint_R dxdy$$

$= 2 \times$ Area of the circle $x^2 + y^2 = 1$

$$= 2\pi(1)^2 = 2\pi.$$

34. Evaluate $\int_C (xdy - ydx)$, where C is the circle $x^2 + y^2 = a^2$.

Solution:

By Green's theorem,

Area $= \frac{1}{2}\int_C (xdy - ydx)$

$\therefore \int_C (xdy - ydx) = 2 \times \text{Area} = 2 \times \pi a^2 = 2\pi a^2.$

CHAPTER 3

ANALYTIC FUNCTIONS

3.1. Introduction

Complex analysis, traditionally known as the theory of functions of a complex variable, is the branch of mathematical analysis that investigates functions of complex numbers It is particularly concerned with analytic functions of complex variables (or, more generally, meromorphic functions). Because the separate real and imaginary parts of any analytic function must satisfy Laplace's equation. After a brief review of complex numbers as points in the complex plane, we will first discuss analyticity and give plenty of examples of analytic functions. Then discuss the theory of conformal mappings which has many physical applications and is also used throughout analytic number theory. Complex variable techniques have been used in a wide variety of areas of engineering. This has been particularly true in areas such as electromagnetic field theory, fluid dynamics, aerodynamics and elasticity, with the rapid developments in computer technology.

3.2. Preliminaries

A complex number z is an ordered pair (x, y) of real numbers and is written as $z = x + iy$, where $i = \sqrt{-1}$ .The real numbers x and y are called the real and imaginary parts of z, denoted by $Re(z)$ and $Im(z)$.

1. **Circle:** $|z - a| = r$ represents a circle C with centre at the point a and radius r.
2. **Annulus:** The set of points which lie between two concentric circles C_1: $|z - a| = r_1$ and C_2: $|z - a| = r_2$ defines an open annulus i.e., the set of points which satisfies the inequality $r_1 < |z - a| < r_2$.

   The set of points which satisfies the inequality $r_1 \le |z - a| \le r_2$ defines a closed annulus. It is to be noted that $r_1 \le |z - a| < r_2$ is neither open nor closed.
3. **Neighbourhood:** δ-Neighbourhood of a point z_0 is the set of all points z for which $|z - z_0| < \delta$ where δ is a positive constant. If we exclude the point z_0 from the open disk $|z - z_0| < \delta$ then it is called the deleted neighbourhood of the point z_0 and is written as $0 < |z - z_0| < \delta$.

4. **Interior and exterior points:** A point z is an interior point of S if all the points in some δ - neighbourhood of z are in S and an exterior point of S if they are outside S.
5. **Boundary point:** A point z is a boundary point of S if every δ -neighbourhood of z contains at least one point of S and at least one point not in S. For example, the points on the circle $|z - z_0| = r$ are the boundary points for the disk $|z - z_0| \leq r$.
6. **Connected set:** An open set S is connected if any two points z_1 and z_2 belonging to S can be joined by a polygonal line which is totally contained in S.
7. **Domain:** An open connected set is called a domain denoted by D.
8. **Region:** A region is a domain together with all, some or none of its boundary points. Thus a domain is always a region but a region may or may not be a domain.

Function of a Complex Variable

If $z = x + iy$ is any complex number, then $\overline{z} = x - iy$ is called its *conjugate*.

Clearly, $|z| = |\bar{z}|$, $|z|^2 = z\,\bar{z}$, $Re(z) = \frac{z+\bar{z}}{2}$, $Im(z) = \frac{z-\bar{z}}{2i}$.

If $z = x + iy$ and $w = u + iv$ are two complex variables such that there exists one or more values of w, corresponding to each value of z in a certain region R of the z -plane, then w is called a *function of* z and is written as $w = f(z)$ or $w = f(x + iy)$. When $w = u + iv = f(z) = f(x + iy)$, clearly u and v are functions of the variables x and y.

For example , if $w = z^2$, then $u + iv = (x + iy)^2 = (x^2 - y^2) + i(2xy)$.

Thus $u = x^2 - y^2$ and $v = 2xy$

$\therefore \quad w = f(z) = u(x,y) + i\,v(x,y)$.

If z is expressed in the polar form, then u and v are functions of r and θ.

If for every value of z, there corresponds a unique value of w, then w is called a *single-valued function of* z. For example, $w = z^2$ and $w = \frac{1}{z}$ are single-valued functions of z.

If, for every value of z, there correspond more than one value of w, then w is called a *multiple-valued function of* z. For example, $w = z^{\frac{1}{4}}$ and $w = amp(z)$are multiple-valued functions of z. $w = z^{\frac{1}{4}}$ is four valued and $w = amp(z)$ is infinitely many valued for $z \neq 0$.

Limit of a Complex Function

A function $w = f(z)$ is said to tend to *limit* l as z approaches a point z_0, if for every real ε, we can find a positive real δ such that $|f(z) - l| < \varepsilon$ for $|(z - z_0)| < \delta$.

And we write$\lim_{z \to z_0} f(z) = l$. If D is the domain of the real-valued functions $u(x, y)$ and $v(x, y)$, then the equations $u = u(x, y)$ and $v = v(x, y)$ describe a transformation (or mapping) from D in the xy plane into the uv plane (also called the *w plane*). Hence we can also consider the function $w = f(z) = u(x, y) + iv(x, y)$ to be a transformation from the set D in the z −plane onto the range R in the w − plane.

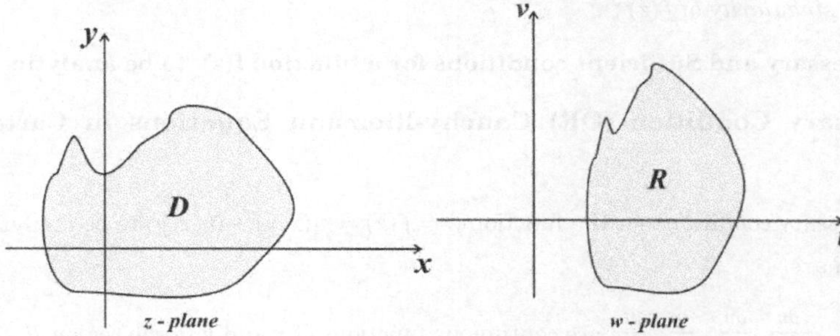

Continuity of f(z)

A function $w = f(z)$ is said to be *continuous* at $z = z_0$, if $\lim_{z \to z_0} f(z) = f(z_0)$. Further $f(z)$ is said to be continuous in any region R of the z-plane, if it is continuous at every point of that region.

Also if $w = f(z) = u(x, y) + i\,v(x, y)$ is continuous at $z = z_0$, then $u(x, y)$ and $v(x, y)$ are also continuous at $z = z_0$, i.e. at $x = x_0$ and $y = y_0$. Conversely if $u(x, y)$ and $v(x, y)$ are continuous at (x_0, y_0) , then $f(z)$ will be continuous at $z = z_0$.

Derivative of f(z)

Let $f(z)$ be a function defined in some neighbourhood of z_0. Then f is said to be *differentiable* at z_0 if $\lim\limits_{z \to 0} \frac{f(z) - f(z_0)}{z - z_0}$ exists and is finite, and this limit is written as $f'(z_0)$.

If we write $\Delta z = z - z_0$, then we have $f'(z_0) = \lim_{\Delta z \to 0} \frac{f(z_0+\Delta z)-f(z_0)}{\Delta z}$.

If $w = f(z)$, then we have $f'(z) = \frac{dw}{dz}$.

3.3. Analytic Functions

A single valued function $f(z)$ is said to be analytic at the point z_0, if it possesses a derivative at z_0 and at every point in some neighbourhood of z_0. A function $f(z)$ is said to be analytic in a region R of the z-plane, it is analytic at every point of R.

An analytic function is also called a *regular function* or a *holomorphic function*.

A function $f(z)$ which is analytic at every point of the finite complex plane is called an *entire function*. A point, at which a function $f(z)$ is not analytic is called a *singular point* or *singularity* of $f(z)$.

Necessary and Sufficient conditions for a function f(z) to be analytic

Necessary Condition (OR) Cauchy-Riemann Equations in Cartesian Form

Necessary conditions for the function $w = f(z) = u(x,y) + iv(x,y)$ to be analytic in a region R are

(i) $\frac{\partial u}{\partial x}, \frac{\partial u}{\partial y}, \frac{\partial v}{\partial x}, \frac{\partial v}{\partial y}$ are continuous functions of x and y in the region R.

(ii) $\frac{\partial u}{\partial x} = \frac{\partial v}{\partial y}, \frac{\partial u}{\partial y} = -\frac{\partial v}{\partial x}$.

Proof: Let $w = f(z) = u(x,y) + iv(x,y)$ be analytic in a region R, then $\frac{dw}{dz} = f'(z)$ exists uniquely at every point of that region.

Let Δx and Δy be the increments in x and y respectively. Let Δu , Δv and Δz be the corresponding increments in u , v and z respectively. Then

$$f'(z) = \lim_{\Delta z \to 0} \frac{f(z+\Delta z)-f(z)}{\Delta z} = \lim_{\Delta z \to 0} \frac{(u+\Delta u)+i(v+\Delta v)-(u+iv)}{\Delta z}$$

$$= \lim_{\Delta z \to 0} \left(\frac{\Delta u}{\Delta z} + i\frac{\Delta v}{\Delta z}\right) \qquad (1)$$

Since the function $w = f(z)$ is analytic in the region R, the limit (1) must exist independent of the manner in which $\Delta z \to 0$, i.e., Δx and $\Delta y \to 0$.

First, let $\Delta z \to 0$ along a line parallel to x -axis so that $\Delta y = 0$ and $\Delta z = \Delta x$.

[since $z = x + iy$, $z + \Delta z = (x + \Delta x) + i(y + \Delta y)$ and $\Delta z = \Delta x + i\Delta y$]

$\therefore$ From (1), $f'(z) = \lim_{\Delta z \to 0}\left(\frac{\Delta u}{\Delta x} + i\frac{\Delta v}{\Delta x}\right) = \frac{\partial u}{\partial x} + i\frac{\partial v}{\partial x}$ (2)

Now, let $\Delta z \to 0$ along a line parallel to y -axis so that $\Delta x = 0$ and $\Delta z = i\Delta y$.

$\therefore$ From (1), $f'(z) = \lim_{\Delta z \to 0}\left(\frac{\Delta u}{i\Delta y} + i\frac{\Delta v}{i\Delta y}\right) = \frac{1}{i}\frac{\partial u}{\partial y} + \frac{\partial v}{\partial y}$

$= \frac{\partial v}{\partial y} - i\frac{\partial u}{\partial y}$ $[\frac{1}{i} = -i]$ (3)

From (2) and (3), we have $\frac{\partial u}{\partial x} + i\frac{\partial v}{\partial x} = \frac{\partial v}{\partial y} - i\frac{\partial u}{\partial y}$

Equating the real and imaginary parts, $\frac{\partial u}{\partial x} = \frac{\partial v}{\partial y}$ and $\frac{\partial u}{\partial y} = -\frac{\partial v}{\partial x}$.

Hence the necessary condition for $f(z)$ to be analytic is that the $C.R.$ equations must be satisfied.

Sufficient Condition: Let $f(z) = u + iv$ be a single – valued function possessing partial derivatives $\frac{\partial u}{\partial x}, \frac{\partial u}{\partial y}, \frac{\partial v}{\partial x}, \frac{\partial v}{\partial y}$ at each point of a region R satisfying $C.R.$ equations. i.e., $\frac{\partial u}{\partial x} = \frac{\partial v}{\partial y}$ and $\frac{\partial u}{\partial y} = -\frac{\partial v}{\partial x}$.

We shall show that $f(z)$ is analytic, i.e., $f'(z)$ exists at every point of a region R.

By Taylor's theorem for functions of two variables, we have, on omitting second and higher degree terms of Δx and Δy

$$f(z + \Delta z) = u(x + \Delta x,\ y + \Delta y) + iv(x + \Delta x,\ y + \Delta y)$$

$$= \left[u(x,y) + \left(\frac{\partial u}{\partial x}\Delta x + \frac{\partial u}{\partial y}\Delta y\right)\right] + i\left[v(x,y) + \left(\frac{\partial v}{\partial x}\Delta x + \frac{\partial v}{\partial y}\Delta y\right)\right]$$

$$= [u(x,y) + iv(x,y)] + \left(\frac{\partial u}{\partial x} + i\frac{\partial v}{\partial x}\right)\Delta x + \left(\frac{\partial u}{\partial y} + i\frac{\partial v}{\partial y}\right)\Delta y$$

$$= f(z) + \left(\frac{\partial u}{\partial x} + i\frac{\partial v}{\partial x}\right)\Delta x + \left(\frac{\partial u}{\partial y} + i\frac{\partial v}{\partial y}\right)\Delta y$$

$$f(z + \Delta z) - f(z) = \left(\frac{\partial u}{\partial x} + i\frac{\partial v}{\partial x}\right)\Delta x + \left(\frac{\partial u}{\partial y} + i\frac{\partial v}{\partial y}\right)\Delta y$$

$$= \left(\frac{\partial u}{\partial x} + i\frac{\partial v}{\partial x}\right)\Delta x + \left(-\frac{\partial v}{\partial x} + i\frac{\partial u}{\partial x}\right)\Delta y \quad \text{[Using } C.R. \text{ equations]}$$

$$= \left(\frac{\partial u}{\partial x} + i\frac{\partial v}{\partial x}\right)\Delta x + \left(\frac{\partial u}{\partial x} + i\frac{\partial v}{\partial x}\right) i\Delta y$$

$$= \left(\frac{\partial u}{\partial x} + i\frac{\partial v}{\partial x}\right)(\Delta x + i\Delta y) = \left(\frac{\partial u}{\partial x} + i\frac{\partial v}{\partial x}\right)\Delta z$$

$$\therefore \quad \frac{f(z+\Delta z) - f(z)}{\Delta z} = \frac{\partial u}{\partial x} + i\frac{\partial v}{\partial x}$$

$$\therefore \qquad f'(z) = \lim_{\Delta z \to 0} \frac{f(z+\Delta z)-f(z)}{\Delta z} = \frac{\partial u}{\partial x} + i\frac{\partial v}{\partial x}$$

Thus $f'(z)$ exists, because $\frac{\partial u}{\partial x}, \frac{\partial v}{\partial x}$ exist.

Hence $f(z)$ is analytic.

Note :

1. When $f(z)$ is analytic, $f'(z)$ exists and is given by $f'(z) = u_x + iv_x$ or $f'(z) = v_y - i\,u_y$.
2. If $w = f(z) = u(x,y) + iv(x,y)$, then we have $\frac{dw}{dz} = f'(z) = \frac{\partial u}{\partial x} + i\frac{\partial v}{\partial x} = \frac{\partial}{\partial x}(u + iv) = \frac{\partial w}{\partial x}$. Similarly, we have
$$\frac{dw}{dz} = f'(z) = \frac{\partial v}{\partial y} - i\frac{\partial u}{\partial y} = -i\,\frac{\partial}{\partial y}(u + iv) = -i\,\frac{\partial w}{\partial y}.$$

Cauchy-Riemann Equations in Polar Co-ordinates

In the Argand's diagram, the complex number z is represented by the point $P(x, y)$. If (r, θ) are the polar coordinates of P, then $r = \sqrt{x^2 + y^2}$ is called the modulus of z and is denoted by $|z|$. Also $\theta = \tan^{-1}\frac{y}{x}$ is called the argument of z and is denoted by arg. z. Every non-zero complex number z can be expressed as $z = r(\cos\theta + i\sin\theta) = e^{i\theta}$.

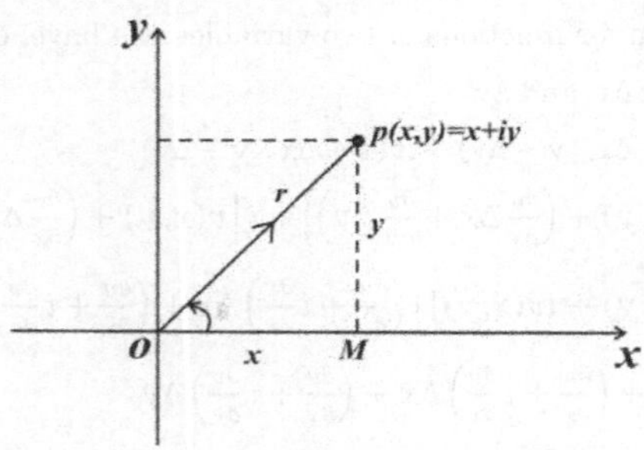

Let (r, θ) be the polar coordinates of the point whose cartesian coordinates are (x, y), then $x = r\cos\theta$, $y = r\sin\theta$.

Now, $z = x + iy = r(\cos\theta + i\sin\theta) = re^{i\theta}$

$$\therefore \; u + i\,v = f(z) = f(re^{i\theta}) \qquad (1)$$

Differentiating (1) partially w.r.t. r, we have

$$\frac{\partial u}{\partial r} + i\,\frac{\partial v}{\partial r} = f'\left(re^{i\theta}\right).e^{i\theta} \qquad (2)$$

Differentiating (1) partially w.r.t. θ, we have

$$\frac{\partial u}{\partial \theta} + i\,\frac{\partial v}{\partial \theta} = f'\left(re^{i\theta}\right).ire^{i\theta} = ir\left(\frac{\partial u}{\partial r} + i\,\frac{\partial v}{\partial r}\right) \text{ using (2)}$$

$$= -r\frac{\partial v}{\partial r} + ir\,\frac{\partial u}{\partial r}$$

Equating real and imaginary parts, we get

$$\frac{\partial u}{\partial \theta} = -r\frac{\partial v}{\partial r} \text{ and } \frac{\partial v}{\partial \theta} = r\,\frac{\partial u}{\partial r}$$

$$\frac{\partial u}{\partial r} = \frac{1}{r}\,\frac{\partial v}{\partial \theta} \text{ and } \frac{\partial v}{\partial r} = -\frac{1}{r}\frac{\partial u}{\partial \theta}$$

which is the polar form of C-R equations.

Example1: Show that the function $f(z) = e^x(\cos y + i\sin y)$ is analytic and find its derivative (or) Show that the function $w = e^z$ is analytic everywhere in the complex plane. Also find $\frac{dw}{dz}$.

Solution:

Let $f(z) = u + i\,v = e^x(\cos y + i\sin y)$.

Then $u = e^x\cos y$ and $v = e^x\sin y$.

$$\therefore\ \frac{\partial u}{\partial x} = e^x\cos y\,,\ \frac{\partial u}{\partial y} = -\,e^x\sin y\,,\ \frac{\partial v}{\partial x} = e^x\sin y \text{ and } \frac{\partial v}{\partial y} = e^x\cos y\,.$$

Hence $\frac{\partial u}{\partial x} = \frac{\partial v}{\partial y}$ and $\frac{\partial u}{\partial y} = -\,\frac{\partial v}{\partial x}$.

$\therefore$ u and v satisfy Cauchy-Riemann equations.

Further $\frac{\partial u}{\partial x}, \frac{\partial u}{\partial y}, \frac{\partial v}{\partial x}$ and $\frac{\partial v}{\partial y}$ are all continuous.

$\therefore f(z)$ is analytic.

$$\text{Now } \frac{dw}{dz} = f'(z) = \frac{\partial u}{\partial x} + i\,\frac{\partial v}{\partial x} = e^x\cos y + i\,e^x\sin y$$

$$= e^x(\cos y + i\sin y) = e^x\,.\,e^{iy} = e^{x+iy} = e^z$$

Example 2: Prove that the function $f(z) = z^2 - \bar{z}$ is not analytic.

Solution:

Let $f(z) = u + iv$.

Then $f(z) = z^2 - \bar{z} = (x+iy)^2 - \overline{(x+iy)}$

$= (x^2 - y^2 - x) + i(2xy + y).$

Hence $u = x^2 - y^2 - x$ and $v = 2xy + y$.

$\therefore \quad \frac{\partial u}{\partial x} = 2x - 1, \quad \frac{\partial u}{\partial y} = -2y,$

$\frac{\partial v}{\partial x} = 2y$ and $\frac{\partial v}{\partial y} = 2x + 1.$

Since $\frac{\partial u}{\partial x} \neq \frac{\partial v}{\partial y}$, the C - R equations are not satisfied at any point.

$\therefore \quad f(z)$ is not analytic.

Example 3: Test the analyticity of the function $w = \sin z$. Also find $\frac{dw}{dz}$.

Solution:

Let $w = f(z) = \sin z$.

i.e., $u + iv = \sin(x+iy) = \sin x \cos iy + \cos x \sin iy$

$= \sin x \cosh y + i \cos x \sinh y.$

$\therefore \quad u = \sin x \cosh y$ and $v = \cos x \sinh y$

$u_x = \cos x \cosh y$ and $v_x = -\sin x \sinh y$

$u_y = \sin x \sinh y$ and $v_y = \cos x \cosh y.$

$\therefore \quad u_x = v_y$ and $u_y = -v_x.$

Hence C - R equations are satisfied.

Also the four partial derivatives u_x, u_y, v_x, v_y are continuous.

Hence $w = \sin z$ is analytic.

Now $\frac{dw}{dz} = f'(z) = u_x + iv_x.$

$= \cos x \cosh y + i(-\sin x \sinh y)$

$= \cos x \cos iy - \sin x \sin iy$

$= \cos\,(x + iy) = \cos z.$

Example 4: Show that the function $f(z) = |z|^2$ defined on the complex plane is differentiable at the origin, and is not analytic at any point. **[*AU Apr/May 2015*]**

Solution: Let $f(z) = |z|^2 = z\bar{z}$.

Then $u + iv = (x + iy)(x - iy)$.

$= x^2 + y^2 + i0$.

Hence $u = x^2 + y^2$ and $v = 0$.

$\therefore \quad \dfrac{\partial u}{\partial x} = 2x, \quad \dfrac{\partial u}{\partial y} = 2y$ and $\dfrac{\partial v}{\partial x} = 0 = \dfrac{\partial v}{\partial y}$.

$\therefore \quad \dfrac{\partial u}{\partial x} \neq \dfrac{\partial v}{\partial y}$ and $\dfrac{\partial u}{\partial y} \neq -\dfrac{\partial v}{\partial x}$.

$\therefore$ C - R equations are not satisfied when $z \neq 0$.

$\therefore$ $f(z)$ is not analytic when $z \neq 0$.

$\therefore$ $f(z)$ is not differentiable at $z \neq 0$.

When $z = 0$, we have $x = 0$ and $y = 0$.

$\therefore$ $u_x = v_y$ and $u_y = -v_x$.

Hence the *C-R* equations are satisfied only at the origin. Therefore $f(z)$ is differentiable only at the origin.

i.e., $f(z)$ is not differentiable at the neighbourhood of origin.

$\therefore$ $f(z) = |z|^2$ is not analytic at the origin.

Example 5: Show that the function $f(z) = x^3 - 3xy^2 + i(3x^2y - y^3)$ is analytic for all z, and find its derivative.

Solution: Let $f(z) = u + iv$.

Then $u = x^3 - 3xy^2$ and $v = 3x^2y - y^3$

$u_x = 3x^2 - 3y^2, \; u_y = -6xy$,

$v_x = 6xy \, , \; v_y = 3x^2 - 3y^2$.

$\therefore \quad u_x = v_y$ and $u_y = -v_x$.

Hence *C-R* equations are satisfied. Moreover u_x, u_y, v_x and v_y are continuous everywhere.

$\therefore$ f is differentiable everywhere.

Now, $f'(z) = u_x + i\,v_x = (3x^2 - 3y^2) + i(6xy)$

$= 3\,(x^2 - y^2 + i2xy) = 3\,(x + iy)^2 = 3z^2$.

Example 6: Prove that $w = \frac{z}{z+a}$ where $a \neq 0$ is analytic whereas $w = \frac{\bar{z}}{\bar{z}+a}$ is not analytic. ***[AU, May/June 2016]***

Solution: Let $w = u + iv$ and $z = x + iy$.

Now $w = \frac{x+iy}{x+iy+a}$

$$u + iv = \frac{x+iy}{(x+a)+iy} \cdot \frac{(x+a)-iy}{(x+a)-iy}$$

$$= \frac{x(x+a)+y^2+i(ay)}{(x+a)^2+y^2} = \frac{x^2+ax+y^2+i(ay)}{x^2+2ax+a^2+y^2}$$

$$\therefore \quad u = \frac{x^2+ax+y^2}{x^2+2ax+a^2+y^2} \text{ and } v = \frac{ay}{x^2+2ax+a^2+y^2}$$

$$u_x = \frac{(x^2+2ax+a^2+y^2)(2x+a)-(x^2+ax+y^2)(2x+2a)}{(x^2+2ax+a^2+y^2)^2} = \frac{ax^2-ay^2+2a^2x+a^3}{(x^2+2ax+a^2+y^2)^2}$$

$$u_y = \frac{(x^2+2ax+a^2+y^2)(2y)-(x^2+ax+y^2)(2y)}{(x^2+2ax+a^2+y^2)^2} = \frac{2axy+2a^2y}{(x^2+2ax+a^2+y^2)^2}$$

$$v_x = \frac{-ay\,(2x+2a)}{(x^2+2ax+a^2+y^2)^2} = \frac{-2axy-2a^2y}{(x^2+2ax+a^2+y^2)^2},$$

$$v_y = \frac{(x^2+2ax+a^2+y^2)a-ay\,(2y)}{(x^2+2ax+a^2+y^2)^2} = \frac{ax^2-ay^2+2a^2x+a^3}{(x^2+2ax+a^2+y^2)^2}.$$

$$\therefore \quad u_x = v_y \text{ and } u_y = -v_x.$$

Hence *C-R* equations are satisfied. Moreover u_x, u_y, v_x and v_y are continuous everywhere. Hence w is analytic.

Now $w = \frac{\bar{z}}{\bar{z}+a}$

$$w = \frac{x-iy}{x-iy+a}$$

$$u + iv = \frac{x-iy}{(x+a)-iy} \cdot \frac{(x+a)+iy}{(x+a)+iy}$$

$$= \frac{x(x+a)+y^2-i(ay)}{(x+a)^2+y^2} = \frac{x^2+ax+y^2-i(ay)}{x^2+2ax+a^2+y^2}$$

$$\therefore \quad u = \frac{x^2+ax+y^2}{x^2+2ax+a^2+y^2} \text{ and } v = \frac{-ay}{x^2+2ax+a^2+y^2}$$

$$u_x = \frac{(x^2+2ax+a^2+y^2)(2x+a)-(x^2+ax+y^2)(2x+2a)}{(x^2+2ax+a^2+y^2)^2} = \frac{ax^2-ay^2+2a^2x+a^3}{(x^2+2ax+a^2+y^2)^2}$$

$$u_y = \frac{(x^2+2ax+a^2+y^2)(2y)-(x^2+ax+y^2)(2y)}{(x^2+2ax+a^2+y^2)^2} = \frac{2axy+2a^2y}{(x^2+2ax+a^2+y^2)^2}$$

$$v_x = \frac{ay\,(2x+2a)}{(x^2+2ax+a^2+y^2)^2} = \frac{2axy+2a^2y}{(x^2+2ax+a^2+y^2)^2}.$$

$$v_y = \frac{(x^2+2ax+a^2+y^2)(-a)+ay\,(2y)}{(x^2+2ax+a^2+y^2)^2} = \frac{-ax^2+ay^2-2a^2x-a^3}{(x^2+2ax+a^2+y^2)^2}.$$

$\therefore \quad u_x \neq v_y$ and $u_y \neq -v_x$.

Hence C - R equations are not satisfied. $\therefore$ w is not analytic.

Example 7: Verify whether the following functions are analytic, and if so find their derivatives.

(a) $f(z) = \frac{1}{z}$, (b) $f(z) = \sinh z$.

Solution:

(a) Let $f(z) = \frac{1}{z} = \frac{1}{x+iy} = \frac{(x-iy)}{(x+iy)(x-iy)}$

$$= \frac{x-iy}{x^2+y^2} = u(x,y) + iv(x,y).$$

$\therefore \; u = \frac{x}{x^2+y^2}$ and $v = \frac{-y}{x^2+y^2}$.

$$\therefore \; u_x = \frac{(x^2+y^2).\,1 - x\,.\,2x}{(x^2+y^2)^2} = \frac{y^2-x^2}{(x^2+y^2)^2}, \quad u_y = \frac{-2xy}{(x^2+y^2)^2},$$

$$v_y = \frac{(x^2+y^2).\,(-1)-(-y).\,2y}{(x^2+y^2)^2} = \frac{y^2-x^2}{(x^2+y^2)^2}, \text{ and } v_x = \frac{2xy}{(x^2+y^2)^2}.$$

$\therefore \; u_x = v_y$ and $u_y = -v_x$.

Hence the C - R equations are satisfied.

Also u_x, u_y, v_x, v_y are continuous and hence $f(z)$ is analytic.

Now $f'(z) = u_x + i\,v_x$

$$= \frac{y^2-x^2}{(x^2+y^2)^2} + i\,\frac{2xy}{(x^2+y^2)^2}$$

$$= \frac{-[x^2-2\,i\,xy-y^2]}{(x^2+y^2)^2} = \frac{-(x-iy)^2}{((x+iy)(x-iy))^2}$$

$$= -\frac{1}{(x+iy)^2} = -\frac{1}{z^2}.$$

(b)Let $f(z) = \sinh z = \frac{1}{i} \sin i\,(x+iy) = -i\, \sin(ix - y)$

$$u + i\,v \; = \sinh x\, \cos y + i\, \cosh x\, \sin y$$

Then $\quad u = \sinh x\, \cos y\,, \qquad v = \cosh x\, \sin y$

$\therefore \quad u_x = \cosh x\, \cos y\,; \quad v_x = \sinh x\, \sin y\,,$

$\quad u_y = -\sinh x\, \sin y\,; \; v_y = \cosh x\, \cos y\,.$

$\therefore \quad u_x = v_y$ and $u_y = -v_x$

Hence $C - R$ equations are satisfied.

Also u_x, u_y, v_x, v_y are continuous and hence $f(z)$ is analytic.

Also, $f'(z) = u_x + i\, v_x$

$$= \cosh x\, \cos y + i\, \sinh x\, \sin y$$

$$= \cos ix\, \cos y + \sin ix\, \sin y [\because \cos(ix) = \cosh x\,, \sin(ix) = i \sinh x]$$

$$= \cos(ix - y) = \cos i(x + iy)$$

$$= \cos(iz) = \cos hz\,.$$

Example 8: Using polar co-ordinates, prove that the following functions are analytic and also find their derivatives. **(i)** z^n (n is a positive integer) **(ii)** $\log z$.

(or) Prove that $f(z) = z^n$ is analytic for all values of n and find its derivatives.

[AU, *Nov/Dec 2017*]

Solution:

(i) Let $f(z) = u(r,\theta) + iv(r,\theta)$.

$$= z^n = \left(r\, e^{i\theta}\right)^n = r^n\, (\cos n\theta + i \sin n\theta).$$

$\therefore \quad u(r,\theta) = r^n \cos n\theta$ and $v(r,\theta) = r^n \sin n\theta$.

$\therefore \quad u_r = n\, r^{n-1} \cos n\theta, \; v_r = n\, r^{n-1} \sin n\theta\,,$

$\quad u_\theta = -n\, r^n \sin n\theta$ and $v_\theta = n\, r^n \cos n\theta$.

Hence $\quad u_r = \frac{1}{r}\, v_\theta$ and $v_r = -\frac{1}{r} u_\theta$.

Also u_r, u_θ, v_r, v_θ exist and all are continuous at all points.

Hence the C - R equations are satisfied.

Thus $f(z)$ is analytic at all points.

Now, $f'(z) = e^{-i\theta}(u_r + i\, v_r)$

$$= e^{-i\theta}\, n\, r^{n-1}(\cos n\theta + i\, \sin n\theta)$$

$$= n\, r^{n-1}\, e^{-i\theta}\, .\, e^{n\,\theta} = n\left[r\, e^{i\theta}\right]^{n-1} = n\, z^{n-1}.$$

(ii) Let $f(z) = u(r,\theta) + iv(r,\theta) = \log z = \log r\, e^{i\theta} = \log r + i\,\theta$.

$\therefore \quad u(r,\theta) = \log r \quad$ and $\quad v(r,\theta) = \theta$

$\therefore \quad u_r = \frac{1}{r}, \; u_\theta = 0$ and $\; v_r = 0\, ,\; v_\theta = 1$.

Hence $u_r, \; u_\theta, \; v_r, \, v_\theta$ all exist and are continuous.

Also $u_r = \frac{1}{r}\, v_\theta$ and $v_r = -\frac{1}{r} u_\theta$ except at $z = 0$.

Hence the C - R equations are satisfied except at $z = 0$.

Thus $f(z)$ is analytic except at $z = 0$.

Also $f'(z) = e^{-i\theta}(u_r + i\, v_r)$

$$= e^{-i\theta}\left(\frac{1}{r} + 0\right)$$

$$= \frac{1}{r\, e^{i\theta}} = \frac{1}{z} \quad (z \neq 0)\, .$$

Example 9: If a function $f(z)$ is analytic, show that it is independent of $\bar{z}$.

Solution: Let $w = f(z) = u + iv$. We have $z = x + iy$

$\therefore \quad \bar{z} = x - iy$. Hence $x = \frac{z+\bar{z}}{2}$ and $y = \frac{z-\bar{z}}{2i}$.

$$\therefore \quad \frac{\partial x}{\partial \bar{z}} = \frac{1}{2}, \quad \frac{\partial y}{\partial \bar{z}} = -\frac{1}{2i}.$$

We have, $\frac{dw}{d\bar{z}} = \frac{du}{d\bar{z}} + i\frac{dv}{d\bar{z}}$.

$$\text{Now, } \frac{du}{d\bar{z}} = \frac{\partial u}{\partial x}\frac{\partial x}{\partial \bar{z}} + \frac{\partial u}{\partial y}\frac{\partial y}{\partial \bar{z}}$$

$$= \frac{\partial u}{\partial x}\left(\frac{1}{2}\right) + \frac{\partial u}{\partial y}\left(-\frac{1}{2i}\right).$$

$$\frac{dv}{d\bar{z}} = \frac{\partial v}{\partial x}\frac{\partial x}{\partial \bar{z}} + \frac{\partial v}{\partial y}\frac{\partial y}{\partial \bar{z}}$$

$$= \frac{\partial v}{\partial x}\left(\frac{1}{2}\right) + \frac{\partial v}{\partial y}\left(-\frac{1}{2i}\right).$$

$$\therefore \quad \frac{dw}{d\bar{z}} = \frac{1}{2}\left(\frac{\partial u}{\partial x} - \frac{\partial v}{\partial y}\right) - \frac{1}{2i}\left(\frac{\partial u}{\partial y} + \frac{\partial v}{\partial x}\right)$$

$= \frac{1}{2}(0) - \frac{1}{2i}(0) = 0$, by Cauchy – Riemann equation.

Hence w is independent of $\bar{z}$.

Example 10: If $w = f(z)$ is regular, prove that $\frac{dw}{dz} = \frac{\partial w}{\partial x} = -i\frac{\partial w}{\partial y}$ where $z = x + iy$. Also prove that $\frac{\partial^2 w}{\partial z \partial \bar{z}} = 0$.

Solution: Let $w = f(z) = u + iv$.

Since $f(z)$ is analytic, we have

$u_x = v_y$ and $u_y = -v_x$.

Now $\frac{dw}{dz} = f'(z) = u_x + iv_x$

$= \frac{\partial u}{\partial x} + i\frac{\partial v}{\partial x}$

$$= \frac{\partial}{\partial x}(u + iv) = \frac{\partial w}{\partial x}. \qquad (1)$$

And $\frac{dw}{dz} = f'(z) = u_x + iv_x$

$= v_y - iu_y$, using C - R equations.

$= -i\left(\frac{\partial u}{\partial y} + i\frac{\partial v}{\partial y}\right)$

$$= -i\frac{\partial}{\partial y}(u + iv) = -i\frac{\partial w}{\partial y} \qquad (2)$$

From (1) and (2), we have

$\frac{dw}{dz} = \frac{\partial w}{\partial x} = -i\frac{\partial w}{\partial y}$

We know that $\frac{\partial w}{\partial \bar{z}} = 0$

Differentiating partially with respect to z, we get $\frac{\partial^2 w}{\partial z \partial \bar{z}} = 0$.

Example 11: For what values of a, b, c the function $f(z) = x - 2ay + i(bx - cy)$ is analytic?

Solution: Let $f(z) = x - 2ay + i(bx - cy) = u + iv$.

Then $u = x - 2ay$ and $v = bx - cy$.

$\therefore \quad u_x = 1;\ u_y = -2a;\ v_x = b;\ v_y = -c$.

Since $f(z)$ is analytic, $\frac{\partial u}{\partial x} = \frac{\partial v}{\partial y}$ and $\frac{\partial u}{\partial y} = -\frac{\partial v}{\partial x}$.

$\therefore \quad 1 = -c$ and $-2a = -b$.

$\therefore \quad c = -1$ and $b = 2a$.

Example 12: Find a, b, c if $f(z)$ is analytic, where $f(z) = ax^2 - by^2 + i\,cxy$. Express $f(z)$ in terms of z.

Solution: Let $f(z) = ax^2 - by^2 + i\,cxy = u(x,y) + iv(x,y)$.

Then $u(x,y) = ax^2 - by^2$ and $v(x,y) = cxy$.

Since $f(z)$ is analytic, $\frac{\partial u}{\partial x} = \frac{\partial v}{\partial y}$ and $\frac{\partial u}{\partial y} = -\frac{\partial v}{\partial x}$.

$\therefore\ 2ax = cx$ and $-2by = -cy$.

$\therefore\quad c = 2a$ and $c = 2b$.

i.e., $a = \frac{c}{2}$ and $b = \frac{c}{2} = \frac{2a}{2} = a$.

$\therefore\quad b = a,\ c = 2a$ and $f(z) = ax^2 - ay^2 + i\,2axy$ (1)

Putting $x = z$ and $y = 0$ in (1), we get $f(z) = az^2$.

Example 13: For what values of z do the following functions cease to be analytic?

(i) $\frac{z+i}{(z-i)^2}$ (ii) $\tan^2 z$.

Solution:

(i) Let $f(z) = \frac{z+i}{(z-i)^2}$.

Then $f'(z) = \frac{(z-i)^2 - (z+i)\ 2(z-i)}{(z-i)^4} = -\frac{(z+3i)}{(z-i)^3}$

$\therefore\ f'(z) \to \infty$ at $z = i$.

Hence $f(z)$ is not analytic at $z = i$.

(ii) Let $f(z) = \tan^2 z$.

Then $f'(z) = 2\tan z\ \sec^2 z = 2\frac{\sin z}{\cos^3 z}$

$\therefore\ f'(z) \to \infty$ when $\cos^3 z = 0$.

i.e., when $z = (2n-1)\frac{\pi}{2},\ n = 1, 2, 3, \cdots$

Thus $f(z)$ is not analytic at $z = (2n-1)\frac{\pi}{2},\ n = 1, 2, 3, \cdots$

Example 14: Show that the function $f(z) = \sqrt{|xy|}$ is not analytic at the origin, although the Cauchy-Riemann equations are satisfied at that point.

Solution: Let $f(z) = u(x,y) + iv(x,y) = \sqrt{|xy|}$

$\therefore \quad u(x,y) = \sqrt{|xy|}$ and $v(x,y) = 0.$

Hence, at the origin, we have

$$u_x = \lim_{\Delta x \to 0} \frac{u(\Delta x,0)-u(0,0)}{\Delta x} = \lim_{\Delta x \to 0} \frac{0-0}{\Delta x} = 0.$$

Similarly, $u_y = \lim_{\Delta y \to 0} \frac{u(0,\Delta y)-u(0,0)}{\Delta y} = \lim_{\Delta y \to 0} \frac{0-0}{\Delta y} = 0.$

$$v_x = \lim_{\Delta x \to 0} \frac{v(\Delta x,0)-v(0,0)}{\Delta x} = \lim_{\Delta x \to 0} \frac{0-0}{\Delta x} = 0,$$

and $\quad v_y = \lim_{\Delta y \to 0} \frac{v(0,y)-v(0,0)}{\Delta y} = \lim_{\Delta y \to 0} \frac{0-0}{\Delta y} = 0.$

$\therefore \quad u_x = v_y$ and $u_y = -v_x$.

Hence Cauchy-Riemann equations are satisfied at the origin.

Now, $\quad f'(0) = \lim_{z \to 0} \frac{f(z)-f(0)}{z} = \lim_{z \to 0} \frac{\sqrt{|xy|}-0}{x+iy}$

Letting $z \to 0$ along the straight line $y = mx$, we see that

$f'(0) = \lim_{x \to 0} \frac{\sqrt{|m\,x^2|}-0}{(1+im)x} = \lim_{x \to 0} \frac{\sqrt{|m|}}{(1+im)} = \frac{\sqrt{m}}{1+im}$, which is not unique. Hence $f'(0)$ does not exist and so $f(z)$ is not analytic at the origin.

3.4. Properties of Analytic Functions

Harmonic Function: The Laplace equation in two dimensions is $\frac{\partial^2 \varphi}{\partial x^2} + \frac{\partial^2 \varphi}{\partial y^2} = 0.$

Here the operator $\frac{\partial^2}{\partial x^2} + \frac{\partial^2}{\partial y^2}$ is called the *Laplacian operator* and is denoted by ∇^2. Using this operator, the Laplace equation is usually written as $\nabla^2 \varphi = 0$. The Laplace equation in polar coordinates is defined as $\frac{\partial^2 \varphi}{\partial r^2} + \frac{1}{r}\frac{\partial \varphi}{\partial r} + \frac{1}{r^2}\frac{\partial^2 \varphi}{\partial \theta^2} = 0$. A function of x, y which possesses continuous partial derivatives of the first and second orders and satisfies Laplace's equation is called a *harmonic function*.

Property 1: The real and imaginary parts of an analytic function $f(z) = u + iv$ satisfy the Laplace equation. ie, $\nabla^2 u = 0$ and $\nabla^2 v = 0$.

(**or**) Prove that the real and imaginary of an analytic function are harmonic functions.

[AU, May/June 2014]

Proof: Let $f(z) = u(x,y) + iv(x,y)$ be an analytic function in a domain D. Then by Cauchy-Riemann equations,

$$\frac{\partial u}{\partial x} = \frac{\partial v}{\partial y} \text{ and } \frac{\partial u}{\partial y} = -\frac{\partial v}{\partial x}. \quad (1)$$

Now, assume that the second order partial derivatives of u and v exist and are continuous functions of x, y, so that $\frac{\partial^2 u}{\partial x \partial y} = \frac{\partial^2 u}{\partial y \partial x}$ (2)

and $$\frac{\partial^2 v}{\partial x \partial y} = \frac{\partial^2 v}{\partial y \partial x} \quad (3)$$

Differentiating both sides of (1) partially with respect to x and y, we get

$$\frac{\partial^2 u}{\partial x^2} = \frac{\partial^2 v}{\partial x\, \partial y} \text{ and } \frac{\partial^2 u}{\partial y^2} = -\frac{\partial^2 v}{\partial y \partial x}.$$

Hence $$\frac{\partial^2 u}{\partial x^2} + \frac{\partial^2 u}{\partial y^2} = 0, \text{ using (2) and (3)} \quad (4)$$

Similarly, $$\frac{\partial^2 v}{\partial x^2} + \frac{\partial^2 v}{\partial y^2} = 0 \quad (5)$$

The equations (4) and (5) show that the functions u and v satisfy the Laplace equations $\nabla^2 u = 0$ and $\nabla^2 v = 0$.

Note: If u and v are harmonic functions such that $u + iv$ is analytic, then each is called the *conjugate harmonic* function of each other.

Property 2: Real and imaginary parts of an analytic function $w = u(r,\theta) + iv(r,\theta)$ satisfy the Laplace equation in polar coordinates.

Proof: Since $w = u(r,\theta) + iv(r,\theta)$ is analytic in some region of the complex plane, the Cauchy-Riemann equations are satisfied.

$$\therefore \quad \frac{\partial u}{\partial r} = \frac{1}{r}\frac{\partial v}{\partial \theta} \quad (1)$$

and $$\frac{\partial v}{\partial r} = -\frac{1}{r}\frac{\partial u}{\partial \theta} \quad (2)$$

Differencing (1) partially with respect to r, and (2) with respect to θ, we have

$$\frac{\partial^2 u}{\partial r^2} = \frac{1}{r}\frac{\partial^2 v}{\partial r\, \partial \theta} - \frac{1}{r^2}\frac{\partial v}{\partial \theta} \quad (3)$$

and $\frac{1}{r}\frac{\partial^2 u}{\partial \theta^2} = -\frac{\partial^2 v}{\partial \theta \partial r}$ (4)

Assume that the mixed derivatives are continuous so that

$$\frac{\partial^2 v}{\partial r \partial \theta} = \frac{\partial^2 v}{\partial \theta \partial r}. \quad (5)$$

Then (1), (3), (4) and (5) give

$$\frac{\partial^2 u}{\partial r^2} + \frac{1}{r}\frac{\partial u}{\partial r} + \frac{1}{r^2}\frac{\partial^2 u}{\partial \theta^2} = \frac{1}{r}\left(\frac{\partial^2 v}{\partial r \partial \theta} - \frac{\partial^2 v}{\partial \theta \partial r}\right) = 0.$$

Thus u satisfies the Laplace equation in polar co-ordinates. Similarly, we can prove that v also satisfies the Laplace equations in polar coordinates.

Orthogonal system: Two families of curves $u(x,y) = c_1$ and $v(x,y) = c_2$ are said to form an *orthogonal system* if they intersect at right angles at each of their point of intersection.

Property 3: If $w = u(x,y) + iv(x,y)$ is an analytic function, the curves of the family $u(x,y) = a$ and the curves of the family $v(x,y) = b$, cut orthogonally, where a and b are arbitrary constants.

Proof: Consider two families of curves

$u(x,y) = a$ and $v(x,y) = b$

Taking differentials on both sides, we get $du = 0$ and $dv = 0$.

i.e., $\frac{\partial u}{\partial x}dx + \frac{\partial u}{\partial y}dy = 0.$

$\therefore \quad \frac{dy}{dx} = -\left(\frac{\partial u}{\partial x}\right) / \left(\frac{\partial u}{\partial y}\right) = m_1$, where m_1 is the slope of the curve $u(x,y) = a$ at (x,y).

Similarly, we have

$m_2 = \frac{dy}{dx} = -\left(\frac{\partial v}{\partial x}\right) / \left(\frac{\partial v}{\partial y}\right)$, where m_2 is the slope of the curve $v(x,y) = b$ at (x,y).

$$\therefore m_1 m_2 = \frac{\left(\frac{\partial u}{\partial x}\right)\left(\frac{\partial v}{\partial x}\right)}{\left(\frac{\partial u}{\partial y}\right)\left(\frac{\partial v}{\partial y}\right)} \quad (1)$$

Since $w = u(x,y) + iv(x,y)$ is analytic, we have the Cauchy-Riemann equations $\frac{\partial u}{\partial x} = \frac{\partial v}{\partial y}$ and $\frac{\partial u}{\partial y} = -\frac{\partial v}{\partial x}$.

$\therefore$ (1) gives $m_1 m_2 = -1$.

Thus the product of the slope of the curves is -1.

Hence the curves intersect at right angles, i.e., they form an orthogonal system.

We conclude that the two families of the curves $u(x,y) = a$ and $v(x,y) = b$ are orthogonal.

Property 4: An analytic function with constant modulus is constant.

Proof: Let $f(z) = u + iv$ be an analytic function with constant modulus.

Thus $|f(z)| = |u + iv| = \text{constant}$.

i.e., $\sqrt{u^2 + v^2} = c$ (say)

Squaring on both sides, we get $u^2 + v^2 = c^2$ (1)

Differentiating (1) partially with respect to x and y, we get

$$2u\frac{\partial u}{\partial x} + 2v\frac{\partial v}{\partial x} = 0 \text{ and } 2u\frac{\partial u}{\partial y} + 2v\frac{\partial v}{\partial y} = 0.$$

$$u\frac{\partial u}{\partial x} + v\frac{\partial v}{\partial x} = 0 \quad (2)$$

and $$u\frac{\partial u}{\partial y} + v\frac{\partial v}{\partial y} = 0 \quad (3)$$

Using Cauchy-Riemann equations, (3) becomes

$$-u\frac{\partial v}{\partial x} + v\frac{\partial u}{\partial x} = 0 \quad (4)$$

Squaring and adding (2) and (4), we get $(u^2 + v^2)\left[\left(\frac{\partial u}{\partial x}\right)^2 + \left(\frac{\partial v}{\partial x}\right)^2\right] = 0.$

Thus $\left(\frac{\partial u}{\partial x}\right)^2 + \left(\frac{\partial v}{\partial x}\right)^2 = 0$ provided $u^2 + v^2 \neq 0$.

i.e., $|f'(z)|^2 = 0$

i.e, $|f'(z)| = 0$

$\therefore f(z)$ is constant.

Example 1: Let $f(z)$ be an analytic function in the domain D. Show that if $Re[f(z)] = 0$ at all points in D, then f is constant in D.

Solution: Let $f(z) = u(x, y) + iv(x, y)$.

Given that $u = 0$.

$\therefore u_x = 0$ and $u_y = 0$.

Since $f(z)$ is an analytic function, the C - R equations are satisfied.

$\Rightarrow v_y = 0$ and $v_x = 0$ on D.

Now $f'(z) = u_x + iv_x = 0$ on D.

Hence $f(z)$ is a constant on D.

Example 2: Let $f(z)$ be an analytic function defined in a domain D. Show that $f(z)$ is constant if $Re\,[f(z)] = [Im\, f(z)]^2$ is true for all $z \in D$.

Solution:

Let $f(z) = u(x,y) + iv(x,y)$.

Given that $u = v^2$.

$\therefore \quad u_y = 2\,v\,v_y \qquad (1)$

$u_x = 2\,v\,v_x \qquad (2)$

Since $f(z)$ is analytic, we have

$$u_x = v_y \text{ and } u_y = -\,v_x\,.$$

From (1) and (2), we have $u_y = 2v\;\; u_x = 2v\,(2v\,v_x)$.

From (2), we have

$$-\,v_x = 4\,v^2 v_x\,.$$

$\therefore\; v_x\,(1 + 4v^2) = 0$

$\therefore\; v_x = 0,$ since $1 + 4\,v^2 \neq 0$

$\therefore\; u_y = 0,\, v_y = 0$ and $u_x = 0$.

$\therefore\; f'(z) = u_x + i\,v_x = 0$ for all $z \in D$

Hence $f(z)$ is a constant on D.

Example 3: Let $f(z) = u(x,y) + iv(x,y)$ be an analytic function in a domain D. Prove that $f(z)$ is constant in D if $f'(z)$ vanishes identically in D.

Solution: Since $f(z)$ is analytic, we have

$$f'(z) = u_x + i\,v_x = v_y - i\,u_y, \text{ using } C-R \text{ equations.}$$

So, if $f'(z) = 0$, then

$u_x + i\,v_x = 0$ and $v_y - i\,u_y = 0$.

$\therefore \quad u_x = 0\,,\; v_x = 0\,,\; v_y = 0$ and $u_y = 0$.

It follows that u and v are constants so that $f(z)$ is constant.

Example 4: Let $f(z) = u(x,y) + iv(x,y)$ be an analytic function in a domain D. Prove that $f(z)$ is constant in D if $Re[f(z)] = u$ is constant.

Solution: If u is a constant, then $u_x = 0 = u_y$.

$\Rightarrow v_x = 0 = v_y$. [Using $C-R$ equations]

Hence v is a constant.

$\therefore f(z) = u + iv$ is a constant.

Example 5: Let $f(z) = u(x,y) + iv(x,y)$ be an analytic function in a domain D. Prove that $f(z)$ is constant in D if $Im[f(z)] = v$ is constant.

Solution: If v is a constant, then $v_x = 0 = v_y$.

$\Rightarrow u_x = 0 = u_y$. [Using $C-R$ equations]

Hence u is a constant. $\therefore f(z) = u + iv$ is a constant.

Example 6: Let $f(z) = u(x,y) + iv(x,y)$ be an analytic function in a domain D. Prove that $f(z)$ is constant in D if $\arg[f(z)]$ is constant.

Solution:

We have $\arg f(z) = \tan^{-1}\frac{v}{u}$.

If $\arg(f(z)) = c$, a constant, then $\frac{v}{u} = \tan c$ or $u = v \cot c$.

Putting $\cot c = k$, we have $u - kv = 0$.

Now, $(1 + ik)f = (1 + ik)(u + iv) = (u - kv) + i(ku + v) = i(ku + v)$, which shows that $Re[(1 + ik)f] = 0$.

Hence $(1 + ik)f$ is constant.

Since $1 + ik$ is a constant, we see that $f(z)$ is a constant.

Example 7: If $f(z) = u + iv$ is an analytic function, then prove that $\left(\frac{\partial^2}{\partial x^2} + \frac{\partial^2}{\partial y^2}\right) |f(z)|^2 = 4|f'(z)|^2$. **[AU, Nov/Dec 2014]**

Solution: Let $z = x + iy$.

Then $\bar{z} = x - iy$, so that

$$x = \frac{z+\bar{z}}{2} \text{ and } y = \frac{z-\bar{z}}{2i}.$$

$\therefore \quad \frac{\partial x}{\partial z} = \frac{1}{2}; \ \frac{\partial x}{\partial \bar{z}} = \frac{1}{2}; \quad \frac{\partial y}{\partial z} = -\frac{i}{2}; \ \frac{\partial y}{\partial \bar{z}} = \frac{i}{2}.$

Now, $\frac{\partial}{\partial z} = \frac{\partial}{\partial x}\frac{\partial x}{\partial z} + \frac{\partial}{\partial y}\frac{\partial y}{\partial z} = \frac{1}{2}\left(\frac{\partial}{\partial x} - i\frac{\partial}{\partial y}\right)$

and $\frac{\partial}{\partial \bar{z}} = \frac{\partial}{\partial x}\frac{\partial x}{\partial \bar{z}} + \frac{\partial}{\partial y}\frac{\partial y}{\partial \bar{z}} = \frac{1}{2}\left(\frac{\partial}{\partial x} + i\frac{\partial}{\partial y}\right)$

$\therefore \quad \frac{\partial^2}{\partial z\,\partial \bar{z}} = \frac{1}{4}\left(\frac{\partial}{\partial x} - i\frac{\partial}{\partial y}\right)\left(\frac{\partial}{\partial x} + i\frac{\partial}{\partial y}\right)$

$= \frac{1}{4}\left(\frac{\partial^2}{\partial x^2} + \frac{\partial^2}{\partial y^2}\right) \left[\because \frac{\partial^2}{\partial y \partial x} = \frac{\partial^2}{\partial x \partial y}\right].$

$\therefore \quad \frac{\partial^2}{\partial x^2} + \frac{\partial^2}{\partial y^2} = 4\,\frac{\partial^2}{\partial z\,\partial \bar{z}}.$ (1)

Let $f(z) = u + iv$ be analytic.

$\therefore \ \overline{f(z)} = f(\bar{z})$

$|f(z)|^2 = f(z)\overline{f(z)} = f(z)f(\bar{z})$ (2)

$\therefore \left(\frac{\partial^2}{\partial x^2} + \frac{\partial^2}{\partial y^2}\right)|f(z)|^2 = 4\,\frac{\partial}{\partial z}\frac{\partial}{\partial \bar{z}}[|f(z)|^2]$ by (1)

$= 4\,\frac{\partial}{\partial z}\frac{\partial}{\partial \bar{z}}[f(z)f(\bar{z})]$ by(2)

$= 4\,\frac{\partial}{\partial z}[f(z)]\frac{\partial}{\partial \bar{z}}[f(\bar{z})]$

$= 4f'(z)f'(\bar{z}) \quad = 4\,|f'(z)|^2.$

Example 8: Show that a harmonic function 'u' satisfies the formal differential equation $\frac{\partial^2 u}{\partial z \partial \bar{z}} = 0$ and hence prove that $log|f'(z)|$ is harmonic, where $f(z)$ is a regular function.

Solution: ***[AU, April/May 2017, April/May 2015]***

Let $z = x + iy.$ (1)

Then $\bar{z} = x - iy.$ (2)

So that $x = \frac{z+\bar{z}}{2}$ and $y = \frac{z-\bar{z}}{2i}.$

$\therefore \quad \frac{\partial x}{\partial z} = \frac{1}{2}; \ \frac{\partial x}{\partial \bar{z}} = \frac{1}{2}; \quad \frac{\partial y}{\partial z} = -\frac{i}{2}; \ \frac{\partial y}{\partial \bar{z}} = \frac{i}{2}.$

Now $\frac{\partial}{\partial z} = \frac{\partial}{\partial x}\frac{\partial x}{\partial z} + \frac{\partial}{\partial y}\frac{\partial y}{\partial z} = \frac{1}{2}\left(\frac{\partial}{\partial x} - i\frac{\partial}{\partial y}\right)$

and $\frac{\partial}{\partial \bar{z}} = \frac{\partial}{\partial x}\frac{\partial x}{\partial \bar{z}} + \frac{\partial}{\partial y}\frac{\partial y}{\partial \bar{z}} = \frac{1}{2}\left(\frac{\partial}{\partial x} + i\frac{\partial}{\partial y}\right)$

$$\therefore \quad \frac{\partial^2}{\partial z\,\partial \bar{z}} = \frac{1}{4}\left(\frac{\partial}{\partial x} - i\frac{\partial}{\partial y}\right)\left(\frac{\partial}{\partial x} + i\frac{\partial}{\partial y}\right)$$

$$= \frac{1}{4}\left(\frac{\partial^2}{\partial x^2} + \frac{\partial^2}{\partial y^2}\right) \quad \left[\because \ \frac{\partial^2}{\partial y\partial x} = \frac{\partial^2}{\partial x\partial y}\right].$$

$$\therefore \frac{\partial^2}{\partial x^2} + \frac{\partial^2}{\partial y^2} = 4\,\frac{\partial^2}{\partial z\,\partial \bar{z}}.$$

Now $\dfrac{\partial^2 u}{\partial z\partial \bar{z}} = \dfrac{1}{4}\left(\dfrac{\partial^2 u}{\partial x^2} + \dfrac{\partial^2 u}{\partial y^2}\right) = 0$ [$\because u$ is harmonic]

To prove: $\left(\dfrac{\partial^2}{\partial x^2} + \dfrac{\partial^2}{\partial y^2}\right)(\log|f'(z)|) = 0.$

Let $f(z) = u + iv$ be analytic.

Now , $|f'(z)|^2 = f'(z)\overline{f'(z)} = f'(z)f'(\bar{z}).$

$$\Rightarrow \quad |f'(z)| = [f'(z)f'(\bar{z})]^{\frac{1}{2}}$$

$$\therefore \quad \log|f'(z)| = \frac{1}{2}\,[\log f'(z) + \log f'(\bar{z})]$$

$$\left(\frac{\partial^2}{\partial x^2} + \frac{\partial^2}{\partial y^2}\right)(\log|f'(z)|) = 4\,\frac{\partial^2}{\partial z\,\partial \bar{z}}(\log|f'(z)|)$$

$$\therefore \quad \frac{\partial}{\partial \bar{z}}(\log|f'(z)|) = \frac{1}{2}\,\frac{\partial}{\partial \bar{z}}[\log f'(z) + \log f'(\bar{z})] = \frac{1}{2}\,\frac{f''(\bar{z})}{f'(\bar{z})}$$

[$\because f'(\bar{z})$ is independent of $\bar{z}$]

$$\therefore \quad \frac{\partial^2}{\partial z\,\partial \bar{z}}(\log|f'(z)|) = \frac{1}{2}\,\frac{\partial}{\partial z}\left(\frac{f''(\bar{z})}{f'(\bar{z})}\right)$$

$= 0$ [$\because f''(\bar{z})$ and $f'(\bar{z})$ are independent of z.]

$$\therefore \left(\frac{\partial^2}{\partial x^2} + \frac{\partial^2}{\partial y^2}\right)(\log|f'(z)|) = 0.$$

Example 9: If $f(z) = u + iv$ is an analytic function, then prove that $\left(\dfrac{\partial^2}{\partial x^2} + \dfrac{\partial^2}{\partial y^2}\right)|f(z)|^p = p^2\,|f'(z)|^2\,|f(z)|^{p-2}.$ ***[AU, Nov/Dec 2015]***

Solution: Let $z = x + iy$

Then $\bar{z} = x - iy$

So that $x = \dfrac{z+\bar{z}}{2}$ and $y = \dfrac{z-\bar{z}}{2i}.$

$$\therefore \frac{\partial x}{\partial z} = \frac{1}{2};\ \frac{\partial x}{\partial \bar{z}} = \frac{1}{2};\quad \frac{\partial y}{\partial z} = -\frac{i}{2};\ \frac{\partial y}{\partial \bar{z}} = \frac{i}{2}.$$

Now, $\dfrac{\partial}{\partial z} = \dfrac{\partial}{\partial x}\dfrac{\partial x}{\partial z} + \dfrac{\partial}{\partial y}\dfrac{\partial y}{\partial z} = \dfrac{1}{2}\left(\dfrac{\partial}{\partial x} - i\dfrac{\partial}{\partial y}\right)$

And $\quad \dfrac{\partial}{\partial \bar{z}} = \dfrac{\partial}{\partial x}\dfrac{\partial x}{\partial \bar{z}} + \dfrac{\partial}{\partial y}\dfrac{\partial y}{\partial \bar{z}} = \dfrac{1}{2}\left(\dfrac{\partial}{\partial x} + i\dfrac{\partial}{\partial y}\right)$

$$\therefore \quad \frac{\partial^2}{\partial z\,\partial \bar{z}} = \frac{1}{4}\left(\frac{\partial}{\partial x} - i\frac{\partial}{\partial y}\right)\left(\frac{\partial}{\partial x} + i\frac{\partial}{\partial y}\right)$$

$$= \frac{1}{4}\left(\frac{\partial^2}{\partial x^2} + \frac{\partial^2}{\partial y^2}\right) \quad \left[\because \frac{\partial^2}{\partial y \partial x} = \frac{\partial^2}{\partial x \partial y}\right].$$

$$\therefore \quad \frac{\partial^2}{\partial x^2} + \frac{\partial^2}{\partial y^2} = 4\,\frac{\partial^2}{\partial z\,\partial \bar{z}}. \qquad (1)$$

Let $f(z) = u + iv$ be analytic.

$\therefore \quad \overline{f(z)} = f(\bar{z})$

$|f(z)|^2 = f(z)\overline{f(z)} = f(z)f(\bar{z})$

$|f(z)| = [f(z)f(\bar{z})]^{\frac{1}{2}}$

$$\therefore \quad |f(z)|^p = [f(z)f(\bar{z})]^{\frac{p}{2}} = [f(z)]^{\frac{p}{2}}\,[f(\bar{z})]^{\frac{p}{2}} \qquad (2)$$

$$\therefore \left(\frac{\partial^2}{\partial x^2} + \frac{\partial^2}{\partial y^2}\right)|f(z)|^p = 4\,\frac{\partial}{\partial z}\frac{\partial}{\partial \bar{z}}[|f(z)|^p] \quad \text{by (1)}$$

$$= 4\,\frac{\partial}{\partial z}\frac{\partial}{\partial \bar{z}}[f(z)]^{\frac{p}{2}}\,[f(\bar{z})]^{\frac{p}{2}} \quad \text{by(2)}$$

$$= 4\,\frac{\partial}{\partial z}[f(z)]^{\frac{p}{2}}\frac{\partial}{\partial \bar{z}}\,[f(\bar{z})]^{\frac{p}{2}}$$

$$= 4\,\frac{p}{2}[f(z)]^{\frac{p}{2}-1}f'(z)\frac{p}{2}\,[f(\bar{z})]^{\frac{p}{2}-1}f'(\bar{z})$$

$$= p^2[f(z)f(\bar{z})]^{\frac{p}{2}-1}f'(z)f'(\bar{z})$$

$$= p^2[|f(z)|^2]^{\frac{p}{2}-1}|f'(z)|^2 = p^2\,|f(z)|^{p-2}\,|f'(z)|^2.$$

Example 10: If $u(x, y)$ and $v(x, y)$ are harmonic functions in a region R, prove that the function $\left(\dfrac{\partial u}{\partial y} - \dfrac{\partial v}{\partial x}\right) + i\left(\dfrac{\partial u}{\partial x} + \dfrac{\partial v}{\partial y}\right)$ is analytic function of $z = x + iy$.

Solution: Let $U = \dfrac{\partial u}{\partial y} - \dfrac{\partial v}{\partial x}$ and $V = \dfrac{\partial u}{\partial x} + \dfrac{\partial v}{\partial y}$.

We prove that $U + iV$ is analytic.

Since u and v are harmonic, $\dfrac{\partial^2 u}{\partial x^2} + \dfrac{\partial^2 u}{\partial y^2} = 0$ and $\dfrac{\partial^2 v}{\partial x^2} + \dfrac{\partial^2 v}{\partial y^2} = 0.$ (1)

Now, $\dfrac{\partial U}{\partial x} = \dfrac{\partial}{\partial x}\left(\dfrac{\partial u}{\partial y} - \dfrac{\partial v}{\partial x}\right) = \dfrac{\partial^2 u}{\partial x\,\partial y} - \dfrac{\partial^2 v}{\partial x^2}$,

$\dfrac{\partial U}{\partial y} = \dfrac{\partial}{\partial y}\left(\dfrac{\partial u}{\partial y} - \dfrac{\partial v}{\partial x}\right) = \dfrac{\partial^2 u}{\partial y^2} - \dfrac{\partial^2 v}{\partial x\,\partial y}$,

$$\frac{\partial V}{\partial x} = \frac{\partial}{\partial x}\left(\frac{\partial u}{\partial x} + \frac{\partial v}{\partial y}\right) = \frac{\partial^2 u}{\partial x^2} + \frac{\partial^2 v}{\partial x\, \partial y},$$

$$\frac{\partial V}{\partial y} = \frac{\partial}{\partial y}\left(\frac{\partial u}{\partial x} + \frac{\partial v}{\partial y}\right) = \frac{\partial^2 u}{\partial x\, \partial y} + \frac{\partial^2 v}{\partial y^2}.$$

$$\therefore \quad \frac{\partial U}{\partial x} = \frac{\partial^2 u}{\partial x\, \partial y} - \frac{\partial^2 v}{\partial x^2}$$

$$= \frac{\partial^2 u}{\partial x\, \partial y} + \frac{\partial^2 v}{\partial y^2}, \text{ using (1)}$$

$$= \frac{\partial V}{\partial y}.$$

And $\dfrac{\partial U}{\partial y} = \dfrac{\partial^2 u}{\partial y^2} - \dfrac{\partial^2 v}{\partial x\, \partial y} = -\dfrac{\partial^2 u}{\partial x^2} - \dfrac{\partial^2 v}{\partial x\, \partial y}$, using (1)

$$= -\frac{\partial V}{\partial x}.$$

$\therefore \dfrac{\partial U}{\partial x} = \dfrac{\partial V}{\partial y}$ and $\dfrac{\partial U}{\partial y} = -\dfrac{\partial V}{\partial x}$.

Thus the functions U and V satisfies the C-R equations.

Since U and V are harmonic, all partial derivatives of U and V are continuous.

Hence $U + iV$ is analytic.

Example 11: Prove that $u = e^{-y} \cos x$ and $v = e^{-x} \sin y$ satisfy Laplace equations, but that $(u + iv)$ is not an analytic function of z.

Solution: $u = e^{-y} \cos x$

$$\therefore \frac{\partial u}{\partial x} = -e^{-y} \sin x \Rightarrow \frac{\partial^2 u}{\partial x^2} = -e^{-y} \cos x$$

$$\frac{\partial u}{\partial y} = -e^{-y} \cos x \Rightarrow \frac{\partial^2 u}{\partial y^2} = e^{-y} \cos x$$

$\therefore \dfrac{\partial^2 u}{\partial x^2} + \dfrac{\partial^2 u}{\partial y^2} = 0$ and so u satisfies Laplace equation.

$v = e^{-x} \sin y$

$$\therefore \frac{\partial v}{\partial x} = -e^{-x} \sin y \Rightarrow \frac{\partial^2 v}{\partial x^2} = e^{-x} \sin y$$

$$\frac{\partial v}{\partial y} = e^{-x} \cos y \Rightarrow \frac{\partial^2 v}{\partial y^2} = -e^{-x} \sin y$$

$\therefore \dfrac{\partial^2 v}{\partial x^2} + \dfrac{\partial^2 v}{\partial y^2} = 0$ and so v satisfies Laplace equation.

Since $\dfrac{\partial u}{\partial x} \neq \dfrac{\partial v}{\partial y}$ and $\dfrac{\partial u}{\partial y} \neq -\dfrac{\partial v}{\partial x}$, the Cauchy - Riemann equations are not satisfied for the function $u + iv$.

Hence $u + iv$ is not an analytic function of z.

Example 12: If $u = x^2 - y^2$ and $v = \frac{-y}{x^2+y^2}$, prove that u and v satisfies Laplace equation but is not an analytic function (u and v are not harmonic conjugates).

Solution: **[*AU, Nov/Dec 2014*]**

Given $u = x^2 - y^2$

$$\therefore \frac{\partial u}{\partial x} = 2x \quad \Rightarrow \quad \frac{\partial^2 u}{\partial x^2} = 2$$

$$\frac{\partial u}{\partial y} = -2y \quad \Rightarrow \quad \frac{\partial^2 u}{\partial y^2} = -2$$

$\therefore \frac{\partial^2 u}{\partial x^2} + \frac{\partial^2 u}{\partial y^2} = 0$ and so u satisfies Laplace equation.

Also given $v = \frac{-y}{x^2+y^2}$

$$\therefore \frac{\partial v}{\partial x} = \frac{2xy}{(x^2+y^2)^2} \Rightarrow \frac{\partial^2 v}{\partial x^2} = \frac{2y^3-6x^2y}{(x^2+y^2)^3}$$

$$\frac{\partial v}{\partial y} = \frac{y^2-x^2}{(x^2+y^2)^2} \Rightarrow \frac{\partial^2 v}{\partial y^2} = \frac{6x^2y-2y^3}{(x^2+y^2)^3}$$

$\therefore \frac{\partial^2 v}{\partial x^2} + \frac{\partial^2 v}{\partial y^2} = 0$ and so v satisfies Laplace equation.

Since $\frac{\partial u}{\partial x} \neq \frac{\partial v}{\partial y}$ and $\frac{\partial u}{\partial y} \neq -\frac{\partial v}{\partial x}$, the Cauchy - Riemann equations are not satisfied for the function $u + iv$.

Hence $u + iv$ is not an analytic function of z.

Example 13: If $f(z) = u + iv$ is analytic function of z in a region R, prove that $\nabla^2\{Re\, f(z)\}^2 = \nabla^2\{Im\, f(z)\}^2 = 2\,|f'(z)|^2$

(or) If $f(z) = u + iv$ is an analytic function in $z = x + iy$ then prove that

$\left(\frac{\partial^2}{\partial x^2} + \frac{\partial^2}{\partial y^2}\right)|u|^2 = 2|f'(z)|^2$. **[*AU, May/June 2016*]**

Solution: Let $f(z) = u + iv$ be analytic.

Then $f'(z) = u_x + iv_x = u_x - iu_y = v_y + iv_x$.

We have $\nabla^2\{Re\, f(z)\}^2 = \left(\frac{\partial^2}{\partial x^2} + \frac{\partial^2}{\partial y^2}\right)(u^2)$.

Now, $\frac{\partial}{\partial x}(u^2) = 2u\,\frac{\partial u}{\partial x}$, $\frac{\partial}{\partial y}(u^2) = 2u\,\frac{\partial u}{\partial y}$.

$$\therefore \quad \frac{\partial^2}{\partial x^2}(u^2) = 2\left[u\,\frac{\partial^2 u}{\partial x^2} + \left(\frac{\partial u}{\partial x}\right)^2\right],$$

$$\frac{\partial^2}{\partial y^2}(u^2) = 2\left[u\,\frac{\partial^2 u}{\partial y^2} + \left(\frac{\partial u}{\partial y}\right)^2\right].$$

$$\therefore \nabla^2\{Re\, f(z)\}^2 = 2\left[u\left(\frac{\partial^2 u}{\partial x^2} + \frac{\partial^2 u}{\partial y^2}\right) + \left(\frac{\partial u}{\partial x}\right)^2 + \left(\frac{\partial u}{\partial y}\right)^2\right]$$

$$= 2\left(u_x^2 + u_y^2\right) = 2\,|f'(z)|^2 . \left[\because\ \frac{\partial^2 u}{\partial x^2} + \frac{\partial^2 u}{\partial y^2} = 0\right]$$

Similarly, $\nabla^2\{Im\, f(z)\}^2 = \nabla^2(v^2) = 2\,|f'(z)|^2$.

EXERCISES

1. Show that the function $f(z)$ defined by $f(z) = \frac{xy\,(y-ix)}{x^2+y^2}$, $z \neq 0$ and $f(0) = 0$ is not analytic at the origin, though it satisfies C - R equations at the origin.
2. Show that the function $f(z) = e^{-y}(\cos x + i \sin x)$ is analytic and find its derivative.
3. Show that $f(z) = e^{-z}$ is analytic everywhere in the complex plane and find $f'(z)$.
4. Verify whether the following functions are analytic, and if so find their derivatives.

   (*i*) $\cos z$ (*ii*) $\cosh z$ (*iii*) $\sec z$ (*iv*) z^2
5. Determine the function $2xy + i(x^2 - y^2)$ is analytic or not. **[*AU, May 2002*]**
6. Test whether the function $f(z) = \frac{1}{2}\log(x^2 + y^2) + i \tan^{-1}\left(\frac{y}{x}\right)$ is analytic or not.
7. Prove that $f(z) = z + \bar{z}$ is nowhere differentiable.
8. Find the constants a, b, c if $f(z) = x + ay + i(bx + cy)$ is analytic.
9. Find where each of the following functions ceases to be analytic.

   (*i*) $\frac{z}{z^2+1}$ (*ii*) $\frac{z^2-4}{z^2+1}$ (*iii*) $z^3 - 4z - 1$
10. Determine the constants a and b so that the function $w = (x^2 + ay^2 - 2xy) + i(bx^2 - y^2 + 2xy)$ is analytic. Also find the derivative of w.
11. If $f(z) = \log r + i\theta$ is single - valued and continuous in a domain, show that $f(z)$ is analytic and find $f'(z)$.
12. Prove that the function $f(z) = xy + iy$ is continuous everywhere but is not analytic.

13. Let $f(z) = f(re^{i\theta}) = u(r,\theta) + iv(r,\theta)$ be analytic on a domain D. Establish the polar form of Laplace's equation: $r^2\frac{\partial^2 u}{\partial r^2} + r\frac{\partial u}{\partial r} + \frac{\partial^2 u}{\partial \theta^2} = 0$.
14. If $f(z) = u + iv$ is an analytic function, prove that $\nabla^2(\log|f(z)|) = 0$. [Hint : $\log|f(z)| = \frac{1}{2}\log(u^2 + v^2)$].
15. If $u + iv$ is an analytic function, find the conditions under which $v + iu$ is analytic.
16. If $f(z)$ is analytic in a domain D, prove the following:
    (i) $f'(z) = 0$ everywhere implies $f(z)$ is constant.
    (ii) $f'(z)$ is also an analytic function.
17. Show that the families of curves $u = c_1$ and $v = c_2$ cut orthogonally, where $w = u + iv = \frac{1}{z}$.
18. Prove that $\left(\frac{\partial}{\partial x}|f(z)|\right)^2 + \left(\frac{\partial}{\partial y}|f(z)|\right)^2 = |f'(z)|^2$.
19. If u is the real part of an analytic function $f(z)$, prove that $\nabla^2(u^p) = p(p-1)u^{p-2}|f'(z)|^2$.
20. If $f(z) = e^z$, then show that u and v are harmonic functions.

ANSWERS

2. $f'(z) = -e^{-y}(\sin x - i\cos x)$

4. (*i*) $-\sin z$ (*ii*) $\sinh z$ (*iii*) $\sec z \tan z$ (*iv*) $2z$

5. not analytic   6. Analytic.   8. $a = -b,\ c = 1$.

9. (*i*) $z = \pm 1$ (*ii*) $z = \pm i$ (*iii*) $f(z)$ is analytic.

10. $a = -1,\ b = 1;\ (1 + i)z^2$ 11. $f'(z) = \frac{1}{z}$.   15. u and v are constants.

3.5. Construction of Analytic Functions

We next develop methods of finding an analytic function $f(z) = u + iv$ of which either real part or imaginary part is known.

Method I

Let $u(x,y)$, the real part of the analytic function $f(z) = u(x,y) + iv(x,y)$ be known. In this method, we first find $v(x,y)$ and then find $u(x,y) + iv(x,y)$ as a function of z.

Since $u(x,y)$ is given, $\frac{\partial u}{\partial x}$ and $\frac{\partial u}{\partial y}$ can be found out.

We have, $dv = \frac{\partial v}{\partial x}\, dx + \frac{\partial v}{\partial y} dy$

$$dv = -\frac{\partial u}{\partial y}\, dx + \frac{\partial u}{\partial x} dy \text{ , by Cauchy-Riemann equations} \qquad (1)$$

Let $M = -\frac{\partial u}{\partial y}$ and $N = \frac{\partial u}{\partial x}$.

Then (1) becomes

$dv = M\, dx + N\, dy.$

Now, $\frac{\partial M}{\partial y} = -\frac{\partial^2 u}{\partial y^2}$, $\frac{\partial N}{\partial x} = \frac{\partial^2 u}{\partial x^2}$. Since $u(x, y)$ is the real part of an analytic function, we have u is harmonic and so it satisfies the Laplace equation

$$\frac{\partial^2 u}{\partial x^2} + \frac{\partial^2 u}{\partial y^2} = 0.$$

$\therefore\ \frac{\partial^2 u}{\partial x^2} = -\frac{\partial^2 u}{\partial y^2}$. Hence $\frac{\partial M}{\partial y} = \frac{\partial N}{\partial x}$.

$\therefore\ M\, dx + N\, dy$ is exact.

Hence equation (1) can be integrated and v can be found. Using this v, we can determine $f(z) = u + iv$.

Method II: Milne –Thomson Method

With the help of this method, we can directly construct $f(z)$ in terms of z without first finding out v when u is given or u when v is given.

We have $z = x + iy$.

$\therefore\ \bar{z} = x - iy$.

Hence $x = \frac{z+\bar{z}}{2}$, $y = \frac{z-\bar{z}}{2i}$. Assume that the real part $u(x, y)$ of the analytic function $f(z)$ is given.

Now $f(z) = u(x, y) + iv(x, y) = u\left(\frac{z+\bar{z}}{2}, \frac{z-\bar{z}}{2i}\right) + i\, v\left(\frac{z+\bar{z}}{2}, \frac{z-\bar{z}}{2i}\right)$.

Setting $z = \bar{z}$, we get $f(z) = u(z, 0) + iv(z, 0)$.

Since $f(z)$ is analytic, we have

$$f'(z) = u_x + i\, v_x = u_x - iu_y, \text{ by Cauchy - Riemann equation.}$$

Choose $\quad \frac{\partial u}{\partial x} = \varphi_1(x, y) = \varphi_1(z, 0)$ and $\frac{\partial u}{\partial y} = \varphi_2(x, y) = \varphi_2(z, 0)$.

Then $\quad f'(z) = \varphi_1(z, 0) - i\, \varphi_2(z, 0)$.

Integrating, $f(z) = \int[\varphi_1(z, 0) - i\, \varphi_2(z, 0)]\ dz + c$, where c is an arbitrary constant.

Similarly, if v is given, then $f'(z) = v_y + i\, v_x$.

Choosing $\frac{\partial v}{\partial y} = \psi_1(x,y) = \psi_1(z,0)$ and $\frac{\partial v}{\partial x} = \psi_2(x,y) = \psi_2(z,0)$, we have

$$f'(z) = \psi_1(z,0) + i\,\psi_2(z,0).$$

Integrating, $f(z) = \int[\psi_1(z,0) + i\,\psi_2(z,0)]\, dz + c$.

Steps Involved in Milne's Thomson Method

Case 1: When only real part $u(x,y)$ is given

1. Find $\frac{\partial u}{\partial x}$ write it as equal to $\varphi_1(x,y)$
2. Find $\frac{\partial u}{\partial y}$ and write it as equal to $\varphi_2(x,y)$
3. Find $\varphi_1(z,0)$ and $\varphi_2(z,0)$ by replacing x by z and y by 0.
4. $f(z)$ is obtained by the formula $f(z) = \int[\varphi_1(z,0) - i\,\varphi_2(z,0)]\, dz + c$, directly in terms of z.

Case 2: When only imaginary part $v(x,y)$ is given

1. Find $\frac{\partial v}{\partial y}$ write it as equal to $\psi_1(x,y)$
2. Find $\frac{\partial v}{\partial x}$ and write it as equal to $\psi_2(x,y)$
3. Find $\psi_1(z,0)$ and $\psi_2(z,0)$ by replacing x by z and y by 0.
4. $f(z)$ is obtained by the formula $f(z) = \int[\psi_1(z,0) + i\,\psi_2(z,0)]\, dz + c$, directly in terms of z.

Case 3: When $u - v$ is given

1. $f(z) = u + iv$ (1)
2. $if(z) = iu - v$ (2)
3. Add (1) and (2) to get $(1+i)f(z) = (u - v) + i(u + v)$

   (or) $F(z) = U + iV$ where $F(z) = (1+i)f(z)$, $U = u - v$, $V = u + v$
4. Since $u - v$ is given, $U(x,y)$ is given
5. Find $\frac{\partial U}{\partial x}$ and write it as equal to $\varphi_1(x,y)$
6. Find $\frac{\partial U}{\partial y}$ and write it as equal to $\varphi_2(x,y)$
7. Find $\varphi_1(z,0)$ and $\varphi_2(z,0)$ by replacing x by z and y by 0.
8. $F(z)$ is obtained by the formula $F(z) = \int[\varphi_1(z,0) - i\,\varphi_2(z,0)]\, dz + c$.
9. $f(z)$ is determined by $f(z) = \frac{F(z)}{1+i}$, directly in terms of z.

Example 1: Find an analytic function whose real part is $u(x,y) = x^3 - 3xy^2$.

Solution: Let $f(z)$ be an analytic function whose real part is

$u(x,y) = x^3 - 3xy^2$.

Method I

$$u(x,y) = x^3 - 3xy^2.$$

$$\therefore \quad \frac{\partial u}{\partial x} = 3x^2 - 3y^2$$

and $\frac{\partial u}{\partial y} = -6xy.$

Let $v(x,y)$ be the harmonic conjugate of u.

$$\frac{\partial v}{\partial x} = -\frac{\partial u}{\partial y} = 6xy \text{ and } \frac{\partial v}{\partial y} = \frac{\partial u}{\partial x} = 3x^2 - 3y^2.$$

Now $dv = \frac{\partial v}{\partial x}\,dx + \frac{\partial v}{\partial y}\,dy$

$$= 6xy\,dx + (3x^2 - 3y^2)\,dy$$

$$= (6xy\,dx + 3x^2dy) - 3y^2dy = 3\,d(x^2y) - d(y^3).$$

Integrating, $\int dv = 3\int d(x^2y) - \int d(y^3)$

$$\therefore \quad v = 3x^2y - y^3 + c.$$

$\therefore$ The corresponding analytic function is $f(z) = (x^3 - 3xy^2) + i(3x^2y - y^3 + c)$.

Method II

Let $\frac{\partial u}{\partial x} = \varphi_1(x,y)$ and $\frac{\partial u}{\partial y} = \varphi_2(x,y)$.

$\therefore \quad \varphi_1(z,0) = 3z^2$ and $\varphi_2(z,0) = 0$.

Then by Milne Thomson method,

$$f(z) = \int[\varphi_1(z,0) - i\,\varphi_2(z,0)]\,dz + c$$

$$= \int 3z^2\,dz + c = \frac{3z^3}{3} + c$$

$$\therefore \quad f(z) = z^3 + c.$$

Example 2: Show that the function $u(x,y) = -x^3 + 3xy^2 + 2y + 1$ is harmonic. Find the harmonic conjugate of $u(x,y)$ and construct the corresponding analytic function.

Solution: We have

$$u(x,y) = -x^3 + 3xy^2 + 2y + 1.$$

$$\therefore \quad \frac{\partial u}{\partial x} = -3x^2 + 3y^2,$$

$$\frac{\partial^2 u}{\partial x^2} = -6x,$$

$$\frac{\partial u}{\partial y} = 6xy + 2,$$

and $\dfrac{\partial^2 u}{\partial y^2} = 6x.$

Hence $\dfrac{\partial^2 u}{\partial x^2} + \dfrac{\partial^2 u}{\partial y^2} = -6x + 6x = 0.$

$\therefore$ $u(x,y)$ is harmonic.

Now $\varphi_1(z,0) = \dfrac{\partial u}{\partial x}(z,0) = -3z^2$

$\varphi_2(z,0) = \dfrac{\partial u}{\partial y}(z,0) = 2$

By Milne Thomson Method,

$$f(z) = \int [\varphi_1(z,0) - i\,\varphi_2(z,0)]\, dz + c$$

$$= \int (-3z^2 - 2i)\, dz + c = -z^3 - 2iz + c$$

Now, let $f(z) = u + iv$.

Then
$$u + iv = -z^3 - 2iz + c$$
$$= -(x+iy)^3 - 2i(x+iy) + c$$
$$= -[x^3 - iy^3 + 3ix^2y - 3xy^2] - 2ix + 2y + c$$
$$= -x^3 + 3xy^2 + 2y + i\,(-y^3 + 3x^2y - 2x) + (l + ik).$$

$\therefore$ $v(x,y) = -y^3 + 3x^2y - 2x + k$ is the harmonic conjugate of u.

Example 3: Show that $u(x,y) = \frac{1}{2}\log(x^2 + y^2)$ is harmonic and also find the corresponding analytic function $f(z) = u(x,y) + iv(x,y)$ and the harmonic conjugate of $u(x,y)$. **[AU, *Nov/Dec 2017*]**

Solution: We have $u(x,y) = \log(x^2 + y^2)$.

$\therefore \dfrac{\partial u}{\partial x} = \dfrac{1}{2}\dfrac{2x}{x^2+y^2} = \dfrac{x}{x^2+y^2}, \quad \dfrac{\partial u}{\partial y} = \dfrac{1}{2}\dfrac{2y}{x^2+y^2} = \dfrac{y}{x^2+y^2},$

$\dfrac{\partial^2 u}{\partial x^2} = \dfrac{(x^2+y^2)-2x^2}{(x^2+y^2)^2} = \dfrac{y^2-x^2}{(x^2+y^2)^2},$

and $\dfrac{\partial^2 u}{\partial y^2} = \dfrac{(x^2+y^2)-2y^2}{(x^2+y^2)^2} = \dfrac{x^2-y^2}{(x^2+y^2)^2}.$

$\therefore \dfrac{\partial^2 u}{\partial x^2} + \dfrac{\partial^2 u}{\partial y^2} = 0.$

Hence u is harmonic.

To find the analytic function:

$\varphi_1(z,0) = \dfrac{\partial u}{\partial x}(z,0) = \dfrac{z}{z^2} = \dfrac{1}{z}$

$\varphi_2(z,0) = \dfrac{\partial u}{\partial y}(z,0) = 0.$

By Milne Thomson Method,

$f(z) = \int [\varphi_1(z,0) - i\,\varphi_2(z,0)]\, dz + c$

$= \int \frac{1}{z}\, dz + c$

$f(z) = \log z + c.$

To find the harmonic conjugate v :

Now, put $z = x + iy$ and $f(z) = u + iv$.

$u + iv = \log(x + iy) + c$

$= \frac{1}{2}\log(x^2 + y^2) + i\tan^{-1}(y/x) + c$

$\therefore v = \tan^{-1}(y/x) + c.$

Example 4: Can $v = \tan^{-1}(y/x)$ be the imaginary part of an analytic function? If so construct an analytic $f(z) = u + iv$, taking v as the imaginary part and hence find u.

Solution: ***[AU, May/June 2016]***

Given $v = \tan^{-1}(y/x)$

$\psi_1(x,y) = \dfrac{\partial v}{\partial y} = \dfrac{1}{1+\left(\frac{y}{x}\right)^2}\dfrac{1}{x} = \dfrac{x}{x^2+y^2}$

$$\frac{\partial^2 v}{\partial y^2} = \frac{-2xy}{(x^2+y^2)^2}$$

$$\psi_2(x,y) = \frac{\partial v}{\partial x} = \frac{1}{1+\left(\frac{y}{x}\right)^2}\left(\frac{-y}{x^2}\right) = \frac{-y}{x^2+y^2}$$

$$\frac{\partial^2 v}{\partial x^2} = \frac{2xy}{(x^2+y^2)^2}$$

$\therefore \quad \frac{\partial^2 v}{\partial x^2} + \frac{\partial^2 v}{\partial y^2} = 0$. Hence v is harmonic.

$\therefore \quad v = \tan^{-1}\left({}^{y}/_{x}\right)$ is the imaginary part of an analytic function.

To find the analytic function:

$\psi_1(z,0) = \frac{1}{z}$ and $\psi_2(z,0) = 0$.

By Milne's Thomson method,

$$f(z) = \int [\psi_1(z,0) + i\,\psi_2(z,0)]dz + c$$

$$= \int \frac{1}{z} dz + c$$

$$= \log z + c.$$

Example 5: Find the analytic function whose real part is $e^{2x} \sin 2y$.

Solution:

Given $\quad u = e^x \sin 2y$.

Then $\varphi_1(x,y) = \frac{\partial u}{\partial x} = 2e^{2x} \sin 2y$.

$\therefore \quad \varphi_1(z,0) = 2e^{2z} \sin 0 = 0$

And $\varphi_2(x,y) = \frac{\partial u}{\partial y} = 2e^{2x} \cos 2y$.

$\therefore \quad \varphi_2(z,0) = 2e^{2z} \cos 0 = 2e^{2z}$

By Milne's Thomson method,

$$f(z) = \int [\varphi_1(z,0)dz - i\varphi_2(z,0)]dz$$

$$= 0 - i \int 2e^{2z}\, dz$$

$$= -i\, 2\frac{e^{2z}}{2} + c = -ie^{2z} + c.$$

Example 6: Find the analytic function $w = u + iv$, given $v = e^{-2xy}\sin(x^2 - y^2)$.

Solution:

Given $v = e^{-2xy}\sin(x^2 - y^2)$.

$$\psi_1(x,y) = \frac{\partial v}{\partial y} = e^{-2xy}\cos(x^2 - y^2)(-2y) + e^{-2xy}(-2x)\sin(x^2 - y^2).$$

$\therefore \quad \psi_1(z,0) = 0 - 2z\sin(z^2) = -2z\sin(z^2)$

$$\psi_2(x,y) = \frac{\partial v}{\partial x} = e^{-2xy}\cos(x^2 - y^2)(2x) + e^{-2xy}(-2y)\sin(x^2 - y^2)$$

$\therefore \quad \psi_2(z,0) = 2z\cos(z^2) - 0 = 2z\cos(z^2)$.

By Milne's Thomson method,

$$f(z) = \int[\psi_1(z,0) + i\,\psi_2(z,0)]dz + c$$
$$= \int[-2z\sin(z^2) + i\,2z\cos(z^2)]dz + c$$
$$= \int[-\sin(z^2) + i\cos(z^2)]\,2zdz + c$$
$$= \int[-\sin t + i\cos t]dt + c, \text{ by putting } z^2 = t.$$
$$= \cos t + i\sin t + c$$
$$= \cos(z^2) + i\sin(z^2) + c = e^{iz^2} + c.$$

Example 7: Find the analytic function $f(z) = u + iv$ whose real part is $y + e^x\cos y$.

Solution:

Given $u = y + e^x\cos y$.

$$\varphi_1(x,y) = \frac{\partial u}{\partial x} = e^x\cos y.$$

$\therefore \quad \varphi_1(z,0) = e^z$

$$\varphi_2(x,y) = \frac{\partial u}{\partial y} = 1 + e^x(-\sin y).$$

$\therefore \quad \varphi_2(z,0) = 1 + 0 = 1.$

By Milne's Thomson method,

$$f(z) = \int[\varphi_1(z,0) - i\,\varphi_2(z,0)]dz + c$$
$$= \int[e^z - i]dz + c = e^z - iz + c.$$

Example 8: An electrostatic field in the *xy*-plane is given by the potential function $\varphi = 3x^2y - y^3$, find the stream function. ***[AU, Nov/Dec 2017]***

Solution: $f(z) = \varphi + i\psi = u + iv$.

Given $\varphi = u = 3x^2y - y^3$.

To find $\psi = v$ by Milne's Thomson method.

Now, $\varphi_1(x,y) = \dfrac{\partial u}{\partial x} = 6xy.$

$\therefore \quad \varphi_1(z,0) = 0$

$$\varphi_2(x,y) = \frac{\partial u}{\partial y} = 3x^2 - 2y.$$

$\therefore \quad \varphi_2(z,0) = 3z^2.$

By Milne's Thomson method,

$$f(z) = \int[\varphi_1(z,0) - i\,\varphi_2(z,0)]dz + c$$

$$= \int[0 - i3z^2]dz + c \;=\; -iz^3 + c.$$

Example 9: Show that $v = e^{-x}(x\cos y + y\sin y)$ is harmonic function. Hence find the analytic function $f(z) = u + iv$. ***[AU, Nov/Dec 2017, May/June 2014]***

Solution: Given $v = e^{-x}(x\cos y + y\sin y)$

$$\psi_1(x,y) = \frac{\partial v}{\partial y} = e^{-x}(-x\sin y + y\cos y + \sin y)$$

$$\frac{\partial^2 v}{\partial y^2} = e^{-x}(-x\cos y + \cos y - y\sin y + \cos y)$$

$$\psi_2(x,y) = \frac{\partial v}{\partial x} = e^{-x}\cos y - e^{-x}(x\cos y + y\sin y)$$

$$\frac{\partial^2 v}{\partial x^2} = -e^{-x}\cos y - e^{-x}\cos y + e^{-x}(x\cos y + y\sin y)$$

$$\therefore \frac{\partial^2 v}{\partial x^2} + \frac{\partial^2 v}{\partial y^2} = 0.$$

Hence v is harmonic.

$\therefore \psi_1(z,0) = 0$ and $\psi_2(z,0) = (1-z)e^{-z}$.

By Milne's Thomson method,

$$f(z) = \int[\psi_1(z,0) + i\,\psi_2(z,0)]dz + c$$

$$= i\int(1-z)e^{-z}dz + c$$

$$= i[-(1-z)e^{-z} + e^{-z}] + c$$

$$f(z) = ize^{-z} + c.$$

Example 10: Determine the analytic function $w = u + iv$ if $u = e^{2x}(x\cos 2y - y\sin 2y)$.

Solution: ***[AU, April/May 2015]***

Given $u = e^{2x}(x\cos 2y - y\sin 2y)$.

$$\varphi_1(x,y) = \frac{\partial u}{\partial x} = e^{2x}\cos 2y + 2e^{2x}(x\cos 2y - y\sin 2y)$$

$$\varphi_2(x,y) = \frac{\partial u}{\partial y} = e^{2x}(-2x\sin y - 2y\cos 2y - \sin 2y)$$

$\therefore \varphi_1(z,0) = (1+2z)e^{2z}$

$\varphi_2(z,0) = 0.$

By Milne's Thomson method,

$f(z) = \int [\varphi_1(z,0) - i\,\varphi_2(z,0)]dz + c$

$= \int (1+2z)e^{2z}\,dz + c$

$= \left[(1+2z)\frac{e^{2z}}{2} - 2\frac{e^{2z}}{4}\right] + c$

$= ze^{2z} + c.$

Example 11: Given that $u = \frac{\sin 2x}{\cosh 2y - \cos 2x}$. Find the analytic function whose real part is u. ***[AU, Nov/Dec 2014]***

Solution: Given $u = \frac{\sin 2x}{\cosh 2y - \cos 2x}$

$$\varphi_1(x,y) = \frac{\partial u}{\partial x} = \frac{(\cosh 2y - \cos 2x)(2\cos 2x) - \sin 2x\,(2\sin 2x)}{(\cosh 2y - \cos 2x)^2}$$

$$\therefore \varphi_1(z,0) = \frac{(1-\cos 2z)(2\cos 2z) - \sin 2z\,(2\sin 2z)}{(1-\cos 2z)^2}$$

$$= \frac{2\cos 2z(1-\cos 2z) - 2\sin^2 2z}{(1-\cos 2z)^2}$$

$$= \frac{2\cos 2z - 2\cos^2 2z - 2\sin^2 2z}{(1-\cos 2z)^2}$$

$$= \frac{2\cos 2z - 2(\cos^2 2z + \sin^2 2z)}{(1-\cos 2z)^2}$$

$$= \frac{2\cos 2z - 2}{(1-\cos 2z)^2} = \frac{2(\cos 2z - 1)}{(\cos 2z - 1)^2}$$

$$= \frac{2}{\cos 2z - 1} = \frac{2}{-2\sin^2 z}$$

$$= -\text{cosec}^2 z.$$

$$\varphi_2(x,y) = \frac{\partial u}{\partial y} = \frac{(\cosh 2y - \cos 2x)(0) - \sin 2x\,(2\sinh 2y)}{(\cosh 2y - \cos 2x)^2}$$

$\varphi_2(z,0) = 0 - 0 = 0.$

By Milne's – Thomson method,

$f(z) = \int [\varphi_1(z,0) - i\varphi_2(z,0)]dz + c$

$= \int (-\text{cosec}^2 z - i \times 0)dz + c = \cot z + c.$

Example 12: Find the analytic function $f(z) = u + iv$, given that $2u - 3v = 3y^2 - 2xy - 3x^2 + 3y - x$ and $f(0) = 0$.

Solution: Let $f(z) = u + iv$.

Then $\quad if(z) = iu - v.$

$\therefore \quad 2f(z) + 3if(z) = 2u - 3v + i(2v + 3u)$

i.e., $(2 + 3i)f(z) = (2u - 3v) + i(2v + 3u) \qquad (1)$

Let $F(z) = (2 + 3i)\, f(z)$, $U = 2u - 3v$ and $V = 2v + 3u$.

Then (1) shows that $F(z) = U + iV$ is an analytic function.

Now, $\dfrac{\partial U}{\partial x} = \dfrac{\partial}{\partial x}(2u - 3v) = -2y - 6x - 1 = \varphi_1(x, y)$,

and $\dfrac{\partial U}{\partial y} = \dfrac{\partial}{\partial y}(2u - 3v) = 6y - 2x + 3 \quad = \varphi_2(x, y)$.

$$\varphi_1(z, 0) = \frac{\partial U}{\partial x}(z, 0) = -6z - 1 \text{ and } \varphi_2(z, 0) = \frac{\partial U}{\partial y}(z, 0) = -2z + 3.$$

By Milne Thomson Method,

$$\begin{aligned} F(z) &= \int [\varphi_1(z, 0) - i\varphi_2(z, 0)]dz + c \\ &= \int [-6z - 1 - i(-2z + 3)]dz + c \\ &= -\frac{6\,z^2}{2} - z - i\left(-\frac{2\,z^2}{2} + 3z\right) + c = -3z^2 - z + i(z^2 - 3z) + c. \end{aligned}$$

$$\therefore \quad f(z) = \frac{1}{2+3i}[-3z^2 - z + i(z^2 - 3z) + c].$$

Also $f(0) = 0$ gives $c = 0$.

$$\therefore \quad f(z) = \left(\frac{-3+11i}{13}\right) z^2 - \left(\frac{11+3i}{13}\right) z \quad \text{is the required analytic function.}$$

Example 13: Find the analytic function $f(z) = u + iv$, given that $2u + 3v = e^x(\cos y - \sin y)$. ***[AU, April/May 2017]***

Solution: Let $f(z) = u + iv$.

Then $\quad if(z) = iu - v.$

$\therefore \quad 2f(z) - 3if(z) = 2u + 3v + i(2v - 3u)$

i.e., $(2 - 3i)f(z) = (2u + 3v) + i(2v - 3u) \qquad (1)$

Let $F(z) = (2 - 3i)\, f(z)$, $U = 2u + 3v$ and $V = 2v - 3u$.

Then (1) shows that $F(z) = U + iV$ is an analytic function.

Now, $\frac{\partial U}{\partial x} = \frac{\partial}{\partial x}(2u + 3v) = e^x(\cos y - \sin y) = \varphi_1(x, y)$,

and $\frac{\partial U}{\partial y} = \frac{\partial}{\partial y}(2u + 3v) = e^x(-\sin y - \cos y) = \varphi_2(x, y)$.

$\varphi_1(z, 0) = \frac{\partial U}{\partial x}(z, 0) = e^z$ and $\varphi_2(z, 0) = \frac{\partial U}{\partial y}(z, 0) = -e^z$.

By Milne Thomson Method,

$$F(z) = \int[\varphi_1(z, 0) - i\varphi_2(z, 0)]dz + c$$

$$= \int[e^z - i(-e^z)]dz + c$$

$$= \int(1 + i)e^z dz + c = (1 + i)e^z + c.$$

$$\therefore \quad f(z) = \frac{1}{2-3i}[(1 + i)e^z + c].$$

$$\therefore \quad f(z) = \frac{(1+i)(2+3i)}{(2-3i)(2+3i)}e^z + c_1$$

$= \frac{-1+5i}{13}e^z + c_1$ is the required analytic function.

Example 14: If $f(z) = u + iv$ is an analytic function of $z = x + iy$ and $u - v = e^x(\cos y - \sin y)$, find $f(z)$ in terms of z.

Solution: $f(z) = u + iv$

$\therefore \quad if(z) = iu - v$

$\therefore \quad (u - v) + i(u + v) = (1 + i)f(z)$. (1)

Let $F(z) = (1 + i)f(z)$, $U = u - v$ and $V = u + v$.

Then (1) shows that $F(z) = U + iV$ is an analytic function.

Now, $U = e^x(\cos y - \sin y)$

$\therefore \quad \frac{\partial U}{\partial x} = e^x(\cos y - \sin y) = \varphi_1(x, y)$,

and $\frac{\partial U}{\partial y} = e^x(-\sin y - \cos y) = \varphi_2(x, y)$.

$\varphi_1(z, 0) = \frac{\partial U}{\partial x}(z, 0) = e^z$ and $\varphi_2(z, 0) = \frac{\partial U}{\partial y}(z, 0) = -e^z$.

By Milne Thomson Method,

$$F(z) = \int [\varphi_1(z,0) - i\varphi_2(z,0)]dz + c$$

$$= \int (1+i)e^z dz + c$$

$$F(z) = (1+i)e^z + c$$

$$\therefore\ (1+i)\, f(z) = (1+i)e^z + c$$

i.e. $\quad f(z) = e^z + c_1.$

Example 15: Determine the analytical function $f(z) = u + iv$, given $u - v = \frac{\cos x + \sin x - e^{-y}}{2(\cos x - \cosh y)}$, and $f\left(\frac{\pi}{2}\right) = 0.$ ***[AU, Nov/Dec 2015]***

Solution: $u + iv = f(z)$

$$\therefore\quad iu - v = if(z)$$

$$\therefore\ (u - v) + i(u + v) = (1+i)f(z)$$

Let $F(z) = (1+i)f(z)$, $U = u - v$ and $V = u + v$.

Then $F(z) = U + iV$ is an analytic function.

Now, $U = \frac{\cos x + \sin x - e^{-y}}{2(\cos x - \cosh y)}$

$$= \frac{\cos x + \sin x - \cosh y + \sinh y}{2(\cos x - \cosh y)} \qquad [\because e^{-y} = \cosh y - \sinh y]$$

$$= \frac{1}{2} + \frac{\sin x + \sinh y}{2(\cos x - \cosh y)}$$

$$\therefore\quad \frac{\partial U}{\partial x} = \frac{1}{2}\,\frac{(\cos x - \cosh y)\cos x - (\sin x + \sinh y)(-\sin x)}{(\cos x - \cosh y)^2}$$

$$= \frac{1}{2}\,\frac{1 - \cos x \cosh y + \sin x \sinh y}{(\cos x - \cosh y)^2} = \varphi_1(x,y),$$

$$\varphi_1(z,0) = \frac{1}{2}\,\frac{1 - \cos z}{(\cos z - 1)^2} = \frac{1}{2(1 - \cos z)}$$

And $\frac{\partial U}{\partial y} = \frac{1}{2}\,\frac{(\cos x - \cosh y)\cosh y - (\sin x + \sinh y)(-\sinh y)}{(\cos x - \cosh y)^2}$

$$= \frac{1}{2}\,\frac{\cos x \cosh y + \sin x \sinh y - 1}{(\cos x - \cosh y)^2} = \varphi_2(x,y).$$

$$\varphi_2(z,0) = \frac{1}{2}\frac{\cos z-1}{(\cos z-1)^2} = \frac{-1}{2(1-\cos z)}$$

By Milne Thomson Method,

$$F(z) = \int [\varphi_1(z,0) - i\varphi_2(z,0)]dz + c$$

$$= \int \left[\frac{1}{2(1-\cos z)} + i\frac{1}{2(1-\cos z)}\right] dz + c$$

$$= \frac{1+i}{2}\int \frac{1}{2\sin^2 z/2} dz + c = \frac{1+i}{4}\int cosec^2\,(z/2)\, dz + c$$

$$= -\left(\frac{1+i}{2}\right)\cot\frac{z}{2} + c$$

$$\therefore\ (1+i)f(z) = -\left(\frac{1+i}{2}\right)\cot\frac{z}{2} + c$$

$$\Rightarrow \quad f(z) = \frac{-1}{2}\cot\frac{z}{2} + \frac{c}{1+i} \qquad (1)$$

Now $f\left(\frac{\pi}{4}\right) = \frac{-1}{2}\cot\frac{\pi}{4} + \frac{c}{1+i}$

$$0 = \frac{-1}{2} + \frac{c}{1+i} \Rightarrow \frac{c}{1+i} = \frac{1}{2}$$

$\therefore$ From (1), $f(z) = \frac{-1}{2}\cot\frac{z}{2} + \frac{1}{2} = \frac{1}{2}\left(1 - \cot\frac{z}{2}\right)$.

Example 16: Find the analytic function $f(z) = u + iv$ if $u + v = \frac{\sin 2x}{\cosh 2y - \cos 2x}$.

Solution:

Let $f(z) = u + iv$. (1)

$\therefore\ if(z) = iu - v$ (2)

(1) + (2) gives $(1+i)f(z) = (u-v) + i(u+v)$

Let $F(z) = U + iV$, $U = u - v$ and $V = u + v$.

Then $F(z) = U + iV$ is an analytic function.

Given $V = \frac{\sin 2x}{\cosh 2y - \cos 2x}$

$$\psi_1(x,y) = \frac{\partial V}{\partial y} = \frac{(\cosh 2y - \cos 2x)(0) - \sin 2x(2\sinh 2y)}{(\cosh 2y - \cos 2x)^2}$$

$\therefore \psi_1(z,0) = 0$

$$\psi_2(x,y) = \frac{\partial V}{\partial x} = \frac{(\cosh 2y - \cos 2x)(2\cos 2x) - \sin 2x(2\sin 2x)}{(\cosh 2y - \cos 2x)^2}$$

$$\psi_2(z,0) = \frac{(1-\cos 2z)(2\cos 2z) - \sin 2z(2\sin 2z)}{(1-\cos 2z)^2}$$

$$= \frac{2\cos 2z - 2(\cos^2 2z + \sin^2 2z)}{(1-\cos 2z)^2} = \frac{2}{\cos 2z - 1} = -\text{cosec}^2\, z.$$

By Milne's – Thomson method,

$$F(z) = \int[\psi_1(z,0) + i\,\psi_2(z,0)]dz + c = \int(0 - i\,\text{cosec}^2 z)dz + c = i\cot z + c.$$

i.e., $(1+i)f(z) = i\cot z + c$

$$f(z) = \frac{i}{1+i}\cot z + \frac{c}{1+i}$$

$$f(z) = \left(\frac{1+i}{2}\right)\cot z + c_1.$$

EXERCISES

1. If $u = 3x^2y - y^3$, find v where $w = u + iv$ is an analytic function.
2. Determine the analytic function whose real part is $u = x^3 - 3xy^2 + 3x^2 - 3y^2 + 1$
3. Construct the analytic function $f(z)$ for which the real part is $e^x \cos y$.
4. Find the regular function whose imaginary part is $e^{-x}(x\cos y + y\sin y)$.
5. If $u = (x-1)^3 - xy^2 + 3y^2$, find v so that $u + iv$ is an analytic function of $z = x + iy$.
6. Does an analytic function $f(z) = u(x,y) + iv(x,y)$ exist for which $v(x,y) = x^3y - xy^3 + xy + x + y$ why or why not?
7. Find whether $y + e^x\cos y$ can be real part of an analytic function of z. If so, determine the function.
8. If $u + v = \frac{2\sin 2x}{e^{2y} + e^{-2y} - 2\cos x}$, and $f(z) = u + iv$ is an analytic function of $z = x + iy$, find $f(z)$ in terms of z.
9. Determine the analytic function $f(z) = u + iv$ given that $3u + 2v = y^2 - x^2 + 16xy$.

ANSWERS

1. $v = -x^3 + 3y^2x + c$
2. $f(z) = z^3 + 3z^2 + c$
3. $f(z) = e^z + c$
4. $f(z) = i\,z\,e^{-z} + c$
5. $v(x,y) = -y^3 + 6x^2y - 12xy + 3y$
6. Yes
7. $f(z) = e^z - iz + c$
8. $f'(z) = \frac{1}{2}(1+i)\cot z + c$
9. $f(z) = (1-2i)z^2 + c$.

Mapping

In the complex domain, the function $w = f(z)$ i.e., $u + iv = f(x + iy)$ (1)

involves four real variables x, y, u, v. Hence a four dimensional region is required to plot (1) in the cartesian region. As it is not possible to have 4-dimensional graph papers, we make use of two complex planes, one for the variable $z = x + iy$, and the other for the variable $w = u + iv$. If the point z describes some curve C in the z-plane, the point w will move along a corresponding curve C' in the w-plane, since to each point (x, y), there corresponds a point (u, v). We then, say that a curve C in the z-plane is mapped into the corresponding curve C' in the w-plane by the function $w = f(z)$ which defines a *mapping* or *transformation* of the z-plane into the w-plane.

3.6. Conformal Mapping

Suppose two curves C_1 and C_2 in the z -plane intersect at the point P and the corresponding curves C_1' and C_2' in the w –plane intersect at P' under the transformation $w = f(z)$. If the angle of intersection of the curves at P is the same as the angle of intersection of the curves at P', both in magnitude and direction, then the transformation is said to be *conformal* at P. A transformation which preserves angles both in magnitude and direction between every pair of curves through a point is said to be *conformal* at the point.

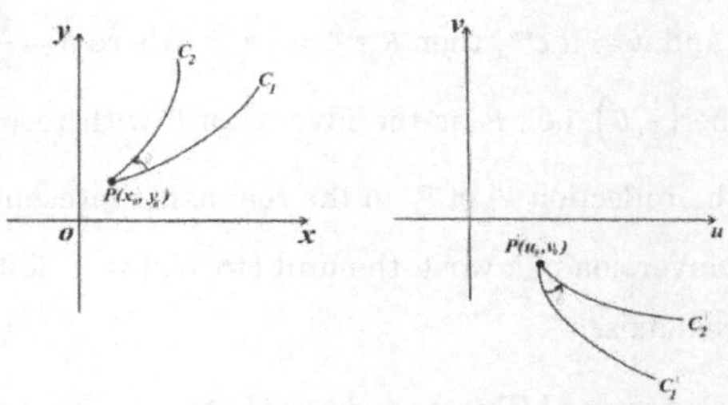

Remark: If $f(z)$ is analytic and $f'(z) \neq 0$ in a region R in the z -plane, then the mapping $w = f(z)$ is conformal at all points of R.

Critical Point: A point at which $f'(z) = 0$ is called a *critical point* of the transformation $w = f(z)$.

Note:

1. The mapping is not conformal at a critical point.
2. A harmonic function remains harmonic under a conformal transformation.
3. Coefficient of magnification for the conformal transformation $w = f(z)$ at $z = z_0$ is given by $|f'(z_0)|$.
4. The *angle of rotation* of the map $w = f(z)$ at a point $z = z_0$ is given by $\arg f'(z_0)$ [or amp $f'(z_0)$].

3.7. Some Standard Transformations

1. Translation: $w = z + c$, where c is a complex constant.

If $z = x + iy$, $c = a + ib$ and $w = u + iv$ then the transformation becomes $u + iv = (x + a) + i(y + b)$. i.e., The point $P(x, y)$ in the z -plane is mapped onto the point $P'(x, y)$ in the w –plane. Every point in the z -plane is mapped onto the w –plane in the same way. Accordingly, this transformation maps a figure in the z -plane into a figure in the w -plane of the same shape and size. In particular, this transformation changes circles into circles.

2. Magnification and Rotation: $w = cz$, where $c \neq 0$ is a complex constant.

If $c = |c|e^{i\alpha}$ and $z = re^{i\theta}$, then $w = |c|r\, e^{i(\theta+\alpha)}$. The two figures in z-plane and w -plane are similar and similarly situated about their respective origins but the figure in w-plane is $|c|$ times the figure in the z –plane and rotates it through an angle $\arg(c)$.

3. Inversion and Reflection: $w = \frac{1}{z}$.

If $z = re^{i\theta}$, $r > 0$, and $w = R\, e^{i\varphi}$, then $R\, e^{i\varphi} = \frac{1}{r}e^{-i\theta}$ where $R = \frac{1}{r}$ and $\varphi = -\theta$. Thus, if P be (r, θ) and P_1 be $\left(\frac{1}{r}, \theta\right)$, i.e., P_1 is the inverse of P with respect to the unit circle with centre O, then the reflection P' of P_1 in the real axis represents $w = \frac{1}{z}$. Hence this transformation is an inversion of z w.r.t. the unit circle $|z| = 1$ followed by reflection of the inverse into the real axis.

3.8. Special Conformal Transformation

Transformation $w = z^2$:

We have $u + iv = (x + iy)^2 = x^2 - y^2 + 2ixy$.

$\therefore\ u = x^2 - y^2$ and $v = 2xy$ (1)

If u is constant (say, a), then $x^2 - y^2 = a$ which is a rectangular hyperbola. Similarly, if v is constant (say, b) then $xy = \frac{b}{2}$ which also represents a rectangular hyperbola. Hence a pair of lines $u = a$, $v = b$ parallel to the axes in the w -plane, map into pair of orthogonal rectangular hyperbolae in the z –plane.

Again, if x is constant (say, c), then $y = \frac{v}{2c}$ and $y^2 = c^2 - u$. Elimination of y from these equations gives $v^2 = 4c^2(c^2 - u)$, which represents a parabola. Similarly if y is constant (say, d), then elimination of x from the equations (1) gives $v^2 = 4d^2(d^2 + u)$, which is also a parabola. Hence the pair of lines $x = c$ and $y = d$ parallel to the axes in the z -plane, map into orthogonal parabolas in the w –plane. Also since $\frac{dw}{dz} = 2z = 0$ for $z = 0$, therefore, it is a critical point of a mapping.

Taking $z = re^{i\theta}$ and $w = R\,e^{i\varphi}$ then in polar form $w = z^2$ becomes $R\,e^{i\varphi} = r^2 e^{2i\theta}$.

This shows that upper half of the z –plane $0 < \theta < \pi$ transforms into the entire w –plane $0 \leq \varphi < 2\pi$. The same is true of the lower half.

Example 1: Find the critical points of $w = \sin z$.

Solution:

Let $f(z) = \sin z$.

$\therefore\ f'(z) = \cos z$.

Then $\cos z = 0$ gives $z = \frac{n\pi}{2}$, where n is an odd integer.

Hence the critical points are given by $z = \pm\frac{\pi}{2},\ \pm\frac{3\pi}{2},\ \pm\frac{5\pi}{2}, \cdots$

Example 2: Find the critical points of the transformation $w^2 = (z-\alpha)(z-\beta)$

Solution: ***[AU, Nov/Dec 2014, May/June 2016]***

Given $w^2 = (z-\alpha)(z-\beta)$ (1)

Differentiating (1) with respect to z, we get

$$2w\frac{dw}{dz} = (z-\alpha) + (z-\beta)$$

$$= 2z - (\alpha+\beta)$$

$$\frac{dw}{dz} = \frac{2z-(\alpha+\beta)}{2w}.$$

The critical points occur at $\frac{dw}{dz} = 0$ and $\frac{dz}{dw} = 0$

i.e., $\frac{2z-(\alpha+\beta)}{2w} = 0$ and $\frac{2w}{2z-(\alpha+\beta)} = 0$

$2z - (\alpha+\beta) = 0$ and $w = 0$

$$2z = \alpha+\beta \text{ and } \sqrt{(z-\alpha)(z-\beta)} = 0$$

i.e., $z = \frac{\alpha+\beta}{2}$ and $(z-\alpha) = 0, (z-\beta) = 0$.

$\therefore\ z = \alpha, \beta, \frac{\alpha+\beta}{2}$.

$\therefore$ The critical points are $\alpha,\ \beta,\ \frac{\alpha+\beta}{2}$.

Example 3: What is the region of the w −plane into which the rectangular region in the z −plane bounded by the lines $x = 0, y = 0, x = 1, y = 2$ is mapped under the transformation $w = z + (2 - i)$.

Solution:

Let $z = x + iy$ and $w = u + iv$.

Now $w = z + (2 - i)$

$\Rightarrow u + iv = (x + iy) + (2 - i) = (x + 2) + i(y - 1)$

Hence $u = x + 2,\ v = y - 1$.

When $x = 0, u = 2$, when $y = 0, v = -1$, when $x = 1, u = 3$ and when $y = 2, v = 1$.

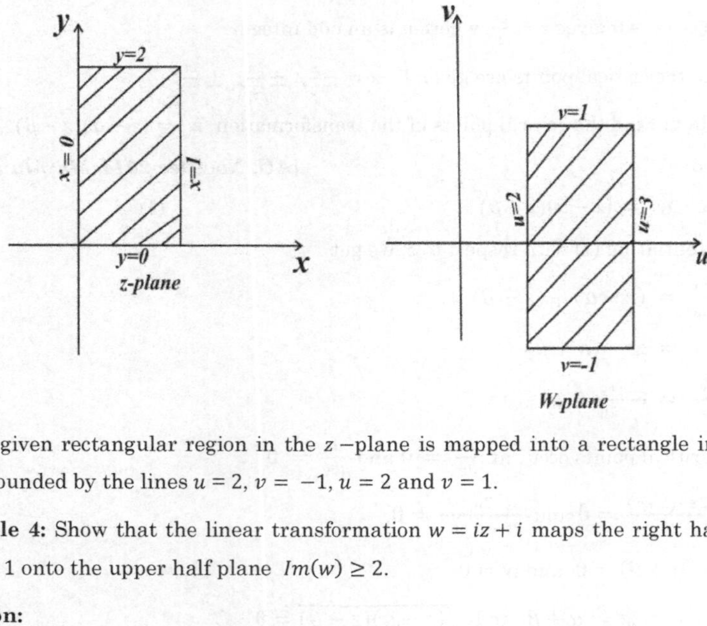

The given rectangular region in the z −plane is mapped into a rectangle in the w-plane bounded by the lines $u = 2$, $v = -1$, $u = 2$ and $v = 1$.

Example 4: Show that the linear transformation $w = iz + i$ maps the right half-plane $Re(z) \geq 1$ onto the upper half plane $Im(w) \geq 2$.

Solution:

Let $z = x + iy$ and $w = u + iv$.

Then $w = iz + i$

$\Rightarrow u + iv = i(x + iy) + i = -y + i(x + 1)$

Hence $u = -y$ and $v = x + 1$ are the transformation equations.

$\therefore \quad y = -u$ and $x = v - 1$.

The given region is $Re(z) \geq 1$

i.e., $x \geq 1$

i.e., $v - 1 \geq 1$

i.e., $v \geq 2$

Hence $Im(w) \geq 2$.

Example 5: Let $w = (2 + i)z - 2i$. Find the triangle onto which the triangle with vertices $z_1 = -2 + i$, $z_2 = -2 + 2i$ and $z_3 = 2 + i$ is mapped.

Solution: The given transformation is

$$w = (2 + i)z - 2i.$$

$$\therefore u + iv = (2 + i)(x + iy) - 2i = (2x - y) + i(x + 2y - 2).$$

Hence the transformation equations are

$$u = 2x - y \text{ and } v = x + 2y - 2.$$

When $z_1 = -2 + i$, we have $x = -2$ and $y = 1$.

$\therefore \quad u = -5, \quad v = -2.$

$\therefore \quad z_1 = (-2, 1)$ is mapped into $(-5, -2)$.

Similarly, when $z_2 = -2 + 2i$, $u = -6$, $v = 0$, and so z_2 is mapped into $(-6, 0)$, and when $z_3 = 2 + i$, $u = 3$, $v = 2$ and so z_3 is mapped into $(3, 2)$.

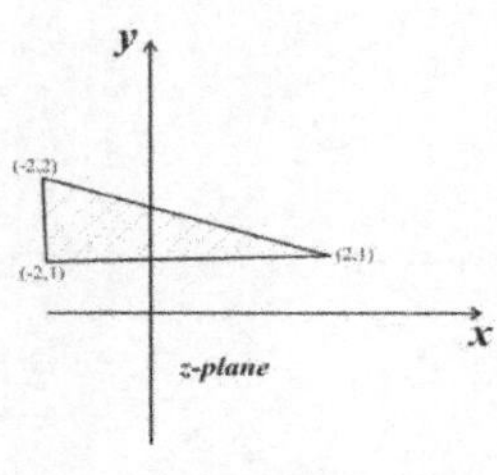

Figure 1

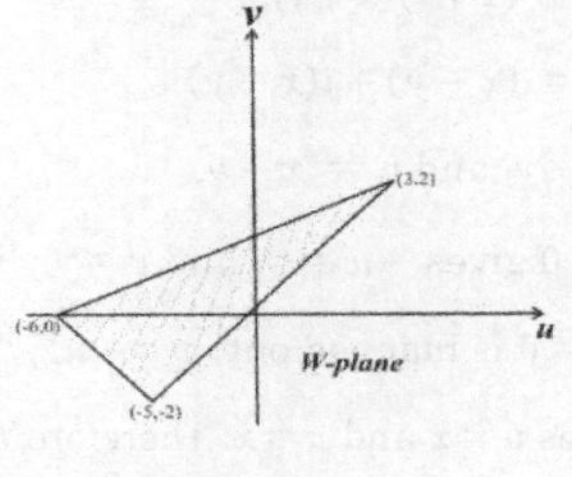

Figure 2

The given triangle is mapped onto the triangle with vertices $(-5, -2)$, $(-6, 0)$ and $(3, 2)$ in the w-plane. The given triangle and the transformed triangle are given in Figure 1 and Figure 2, respectively.

Example 6: Show that the image of $|z+1+i|<1$ under the transformation $w = (3-4i)z+6+2i$ is the open disc $|w+1-3i|<5$.

Solution: The given transformation is $w = (3-4i)z+6+2i$.

$\therefore\ (3-4i)\,z = w-6-2i.$

$\therefore\qquad z = \frac{w-6-2i}{3-4i}.$

The given region is $|z+1+i|<1$.

$\therefore\ \left|\frac{w-6-2i}{3-4i}+1+i\right|<1.$

i.e., $\left|\frac{w-6-2i+(1+i)(3-4i)}{3-4i}\right|<1.$

$\therefore\qquad |w-6-2i+7-i|<|3-4i|$

i.e., $|w+1-3i|<\sqrt{9+16}=\sqrt{25}.$

i.e., $|w+1-3i|<5$ which is the required image.

Example 7: Determine the region of the w –plane in to which the rectangular region in the z –plane bounded by $x=0,\ y=0,\ x=2,\ y=1$ is mapped under the transformation $w = \sqrt{2}\,e^{i\frac{\pi}{4}z}$.

Solution: The given transformation is $w=\sqrt{2}\,e^{i\frac{\pi}{4}z}$.

$\therefore u+iv = \sqrt{2}\left(\cos\frac{\pi}{4}+i\sin\frac{\pi}{4}\right)(x+iy).$

$= \sqrt{2}\left(\frac{1}{\sqrt{2}}+\frac{i}{\sqrt{2}}\right)(x+iy)$

$= (1+i)(x+iy)$

$= (x-y)+i(x+y).$

$\therefore u = x-y$ and $v = x+y$.

Now, $x=0$ gives $-u = y$ and $v=y$. Therefore, $v=-u$.

Hence $x=0$ is mapped onto $v=-u$.

$y=0$ gives $u=x$ and $v=x$. Therefore, $u=v$.

Hence $y=0$ is mapped onto $u=v$.

$x=2$ gives $u=2-y$ and $v=2+y$.

Hence $y=2-u$ and so $v=2+2-u$. Therefore, $u+v=4$.

Hence $x=2$ is mapped onto $u+v=4$.

$y = 1$ gives $u = x - 1$ and $v = x + 1$.

Hence $x = u + 1$ and so $v = u + 2$.

$\therefore$ $y = 1$ is mapped onto $v = u + 2$.

Thus the required image in the w-plane is the rectangle bounded by the lines $v = -u$, $v = u$, $u + v = 4$ and $v = u + 2$. (Note that the region is magnified and rotated under the given transformation.)

Example 8: Find the image of the region $y > 1$ under the transformation $w = (1 - i)z$.

Solution:

The given transformation is $w = (1 - i)z$.

i.e., $u + iv = (1 - i)(x + iy)$

$= (x + y) + i(y - x)$

$\therefore$ $u = x + y$ and $v = y - x$.

Now $u + v = 2y$ and $u - v = 2x$.

$\therefore$ $x = \dfrac{u - v}{2}$ and $y = \dfrac{u + v}{2}$.

The image of $y > 1$ under the given transformation is $\dfrac{u + v}{2} > 1$.

i.e., $u + v > 2$.

$\therefore$ The image of $y > 1$ under the given transformation is $u + v > 2$.

Example 9: Find the image of the circle $|z| = \lambda$ under the transformation $w = 5z$.

Solution:

The given transformation is $w = 5z$.

$\therefore$ $z = \dfrac{w}{5}$.

The image of $|z| = \lambda$ is $\left|\dfrac{w}{5}\right| = \lambda$

$$|w| = 5\lambda$$

$$|u + iv|^2 = 25\,\lambda^2$$

$$u^2 + v^2 = 25\,\lambda^2$$

$\therefore$ The image of $|z| = \lambda$ under the given transformation is the circle $u^2 + v^2 = (5\lambda)^2$.

Example 10: Find the image in the w-plane of the infinite strip $\frac{1}{4} \leq y \leq \frac{1}{2}$ under the transformation $w = \frac{1}{z}$. **[*AU, April/May 2015*]**

Solution: Given transformation is $w = \frac{1}{z}$.

Then $z = \frac{1}{w}$.

i.e., $x + iy = \frac{1}{u+iv}$

$$= \frac{u-iv}{u^2+v^2}$$

$\therefore x = \frac{u}{u^2+v^2}$ and $y = \frac{-v}{u^2+v^2}$.

Now $y \leq \frac{1}{2} \Rightarrow \frac{-v}{u^2+v^2} \leq \frac{1}{2} \Rightarrow -2v \leq u^2 + v^2 \Rightarrow u^2 + (v+1)^2 \geq 1$.

Also $y \geq \frac{1}{4} \Rightarrow \frac{-v}{u^2+v^2} \geq \frac{1}{4} \Rightarrow -4v \geq u^2 + v^2 \Rightarrow u^2 + (v+2)^2 \leq 4$.

Finally $\frac{1}{4} \leq y \leq \frac{1}{2} \Rightarrow u^2 + (v+2)^2 \leq 4$ and $u^2 + (v+1)^2 \geq 1$.

i.e., the region is exterior to the circle $u^2 + (v+1)^2 = 1$ and interior to the circle $u^2 + (v+2)^2 = 4$.

Example 11: Find the image of $|z+1| = 1$ under the map $w = \frac{1}{z}$.

Solution: **[*AU, Nov/Dec 2017, May/June 2014*]**

Given transformation is $w = \frac{1}{z}$.

Then $z = \frac{1}{w}$.

$\therefore$ The image of $|z+1| = 1$ is $\left|\frac{1}{w} + 1\right| = 1$.

$$\text{i.e.,} \quad \left|\frac{1+w}{w}\right| = 1$$

$$\text{i.e.,} \quad |1+w| = |w|$$

$$\text{i.e.,} \quad |1+(u+iv)| = |u+iv|$$

$$\text{i.e.,} \quad |(1+u)+iv|^2 = |u+iv|^2$$

$$(1+u)^2 + v^2 = u^2 + v^2$$

$$u^2 + 2u + 1 + v^2 = u^2 + v^2$$

$$2u + 1 = 0.$$

$\therefore$ The image of $|z+1| = 1$ under the transformation $w = \frac{1}{z}$ is a straight line $2u + 1 = 0$ in the $w-$ plane.

Example 12: Find the image of $|z-2i| = 2$ under the transformation $w = \frac{1}{z}$.

Solution: **[AU, *Nov/Dec 2017*]**

Given transformation is $w = \frac{1}{z}$.

Then $z = \frac{1}{w}$.

$\therefore$ The image of $|z-2i| = 2$ is $\left|\frac{1}{w} - 2i\right| = 2$.

i.e., $\left|\frac{1-2iw}{w}\right| = 2$

$$\begin{aligned}
\text{i.e.,} \quad |1-2iw| &= 2|w| \\
\text{i.e., } |1-2i(u+iv)| &= 2|u+iv| \\
\text{i.e., } |(1+2v) - i2u|^2 &= 4|u+iv|^2 \\
(1+2v)^2 + 4u^2 &= 4(u^2+v^2) \\
4u^2 + 4v^2 + 4v + 1 &= 4(u^2+v^2) \\
4v + 1 &= 0.
\end{aligned}$$

$\therefore$ The image of $|z-2i| = 2$ under the transformation $w = \frac{1}{z}$ is a straight line $4v+1=0$ in the w-plane.

Example 13: Find the image of $x = 2$ under the transformation $w = \frac{1}{z}$.

Solution: Given transformation is $w = \frac{1}{z}$.

Then $z = \frac{1}{w}$.

$$\begin{aligned}
\text{i.e., } x + iy &= \frac{1}{u+iv} \\
&= \frac{u-iv}{u^2+v^2}
\end{aligned}$$

$\therefore x = \frac{u}{u^2+v^2}$ and $y = \frac{-v}{u^2+v^2}$.

The image of $x = 2$ is $\frac{u}{u^2+v^2} = 2$.

$$\begin{aligned}
\text{i.e.,} \quad u &= 2(u^2+v^2) \\
\text{i.e., } 2(u^2+v^2) - u &= 0 \\
\text{i.e.,} \quad u^2 + v^2 - \frac{u}{2} &= 0 \\
\left(u - \frac{1}{4}\right)^2 + v^2 &= \left(\frac{1}{4}\right)^2
\end{aligned}$$

$\therefore$ The image of $x = 2$ under the given transformation is the circle $\left(u-\frac{1}{4}\right)^2 + v^2 = \left(\frac{1}{4}\right)^2$ in the w-plane.

Example 14: Find the image of the infinite strip $0 < y < \frac{c}{2}$ under the transformation $w = \frac{1}{z}$.

Solution:

The given transformation is $w = \frac{1}{z}$.

Then $z = \frac{1}{w}$.

i.e., $x + iy = \frac{1}{u+iv}$

$= \frac{u-iv}{u^2+v^2}$

$\therefore x = \frac{u}{u^2+v^2}$ and $y = \frac{-v}{u^2+v^2}$.

Now $y > 0$ gives $\frac{-v}{u^2+v^2} > 0$ and so $v < 0$. (1)

$y < \frac{c}{2}$ gives $\frac{-v}{u^2+v^2} < \frac{c}{2}$ and so $c(u^2 + v^2) + 2v > 0$.

i.e., $u^2 + v^2 + \frac{2v}{c} > 0$ $(\because c > 0)$

or $u^2 + \left(v + \frac{1}{c}\right)^2 - \frac{1}{c^2} > 0$

or $u^2 + \left(v + \frac{1}{c}\right)^2 > \frac{1}{c^2}$, which is the exterior of the circle with centre $\left(0, \frac{-1}{c}\right)$ and radius $\frac{1}{c}$ in the w –plane. (2)

From (1) and (2), we see that the image of the infinite strip $0 < y < \frac{c}{2}$ is the region in the w –plane which lies below the v –axis and is the exterior of the circle with centre $\left(0, \frac{-1}{c}\right)$ and radius $\frac{1}{c}$.

Example 15: Show that by means of the transformation $w = \frac{1}{z}$ the circle given by $|z - 3| = 5$ is mapped into the circle $\left|w + \frac{3}{16}\right| = \frac{5}{16}$.

Solution:

The given transformation is $w = \frac{1}{z}$.

Then $z = \frac{1}{w}$.

i.e., $x + iy = \frac{1}{u+iv}$

$= \frac{u-iv}{u^2+v^2}$

$\therefore x = \frac{u}{u^2+v^2}$ and $y = \frac{-v}{u^2+v^2}$.

The given region is the circle $|z-3| = 5$.

$\therefore |z-3|^2 = 5^2$.

i.e. , $|(x-3)+iy|^2 = 25$.

$\Rightarrow (x-3)^2 + y^2 = 25$.

i.e., $x^2 + y^2 - 6x = 16$.

$\therefore \frac{1}{u^2+v^2} - \frac{6u}{u^2+v^2} = 16$.

i.e., $u^2 + v^2 + \frac{6u}{16} = \frac{1}{16}$.

$\left(u+\frac{3}{16}\right)^2 + v^2 - \frac{9}{16^2} = \frac{1}{16}$.

$\left(u+\frac{3}{16}\right)^2 + v^2 = \frac{25}{256}$

$\Rightarrow \left|u+\frac{3}{16}+iv\right| = \frac{5}{16}$

i.e., $\left|w+\frac{3}{16}\right| = \frac{5}{16}$, which is the required image.

Example 16: Find the image of the region $2 < x < 3$ under $w = \frac{1}{z}$.

Solution: The given transformation is $w = \frac{1}{z}$.

Then $z = \frac{1}{w}$.

i.e., $x + iy = \frac{1}{u+iv}$

$= \frac{u-iv}{u^2+v^2}$

$\therefore x = \frac{u}{u^2+v^2}$ and $y = \frac{-v}{u^2+v^2}$.

The given region is $2 < x < 3$.

Now, $x < 3$ gives $\frac{u}{u^2+v^2} < 3$ so that $u^2 + v^2 > \frac{u}{3}$.

i.e., $u^2 + v^2 - \frac{u}{3} > 0$

i.e., $\left(u - \frac{1}{6}\right)^2 + v^2 > \frac{1}{6^2}$.

Similarly, $x > 2$ gives $\frac{u}{u^2+v^2} > 2$.

i.e., $u^2 - \frac{u}{2} + v^2 < 0$.

or $\left(u - \frac{1}{4}\right)^2 + v^2 < \frac{1}{4^2}$.

Thus the image of $2 < x < 3$ is the region bounded by the circles $\left(u - \frac{1}{6}\right)^2 + v^2 = \frac{1}{6^2}$ and $\left(u - \frac{1}{4}\right)^2 + v^2 = \frac{1}{4^2}$.

Example 17: Show that the transformation $w = \frac{1}{z}$ transforms all circles and straight lines in the z −plane into circles and/or straight lines in the w −plane.

Solution: ***[AU, April/May2017, Nov/Dec 2015]***

Given transformation is $w = \frac{1}{z}$.

Then $z = \frac{1}{w}$.

i.e., $x + iy = \frac{1}{u+iv} = \frac{u-iv}{u^2+v^2}$.

$\therefore\ x = \frac{u}{u^2+v^2}$ and $y = \frac{-v}{u^2+v^2}$.

The general equation of circles and straight lines is

$$a(x^2 + y^2) + 2gx + 2fy + d = 0 \qquad (1)$$

$\therefore$ The image of (1) is

$$a\left(\frac{u^2+v^2}{(u^2+v^2)^2}\right) + 2g\left(\frac{u}{u^2+v^2}\right) + 2f\left(\frac{-v}{u^2+v^2}\right) + d = 0$$

$$\frac{a + 2gu - 2fv + d(u^2+v^2)}{u^2 + v^2} = 0$$

i.e., $a + 2gu - 2fv + d(u^2 + v^2) = 0$ (2)

Thus (2) is the image of (1) under the given transformation.

Case(i) $a \neq 0,\ d \neq 0$.

Circles not passing through the origin in the z −plane mapped onto circles not passing through the origin in the w −plane.

Case(ii) $a = 0,\ d \neq 0$.

Straight lines not passing through the origin mapped onto circles passing through the origin.

Case(iii) $a \neq 0,\ d = 0$.

Circles passing through the origin is mapped onto straight lines not passing through the origin.

Case(iv) $a = 0,\ d = 0$.

Straight lines passing through the origin is mapped onto straight lines through the origin.

Hence the transformation $w = \frac{1}{z}$ transforms all circles and straight lines in the z −plane into circles and straight lines in the w −plane.

Example 18: Show that the image of the hyperbola $x^2 - y^2 = 1$ under the transformation $w = \frac{1}{z}$ is the lemniscates $R^2 = \cos 2\varphi$.

Solution: The given transformation is $w = \frac{1}{z}$. Hence $z = \frac{1}{w}$.

Now, $x + iy = \frac{1}{R\, e^{i\varphi}} = \frac{1}{R}\,(\cos\varphi - i\sin\varphi)$

$\therefore\ x = \frac{1}{R}\cos\varphi$ and $y = \frac{-1}{R}\sin\varphi$.

The given hyperbola is $x^2 - y^2 = 1$.

This gives $\frac{1}{R^2}\cos^2\varphi - \frac{1}{R^2}\sin^2\varphi = 1$.

$$\frac{1}{R^2}(\cos^2\varphi - \sin^2\varphi) = 1.$$

i.e.,
$$\frac{1}{R^2}\cos 2\varphi = 1$$

$\Rightarrow\ R^2 = \cos 2\varphi$ which is a lemniscate.

EXERCISES

1. Find the image of the circle $|z| = 1$ under the map $w = (1 + 2i)z + (1 + i)$.
2. Find the image of the half-plane $x > c$ when $c > 0$ under the map $w = \frac{1}{z}$.
3. Show that by means of $w = \frac{1}{z}$ the circle given by $|z - 2| = 7$ is mapped into the circle $\left|w + \frac{2}{45}\right| = \frac{7}{45}$.
4. Find the image of the semi infinite strip $x > 0;\ 0 < y < 2$ under the transformation $w = iz + 1$.
5. Find the image of the strip $2 < x < 4$ under the map $w = \frac{1}{z}$.
6. Find the image of the strip $0 < y < \frac{1}{2}$ under the transformation $w = \frac{1}{z}$.
7. Find the image of the region $y > 1$ under the transformation $w = (1 - i)z$.

8. Find the image of the infinite strip $0 \leq x \leq 2$ under the transformation $w = iz$.
9. Find the image of the hyperbola $x^2 - y^2 = 25$ under the transformation $w = (1+i)z$.
10. Find the image of the hyperbola $xy = 1$ under the transformation $w = 3z$.
11. Find the image of the circle $|z-1| = 1$ in the complex plane under the mapping $w = \frac{1}{z}$.

ANSWERS

1. $(u-1)^2 + (v-1)^2 = 5$ 2. Interior of the circle $\left(u - \frac{1}{2c}\right)^2 + v^2 = \left(\frac{1}{2c}\right)^2$

4. $-1 < u < 1, u > 0$ 5. Region between $\left(u - \frac{1}{8}\right)^2 + v^2 > \frac{1}{64}$ and $\left(u - \frac{1}{4}\right)^2 + v^2 < \frac{1}{16}$

6. $u^2 + (v+1)^2 > 1,\ r < 0$ 7. $u + v > 2$ 8. $0 < v < 2$

9. $uv = 25$ 10. $uv = 9$ 11. $2u - 1 = 0$.

3.9. Bilinear Transformation

A transformation of the form $w = \frac{az+b}{cz+d}$, ...(1) where a, b, c, d are complex constants and $ad - bc \neq 0$ is called a *bilinear transformation* or *linear fractional transformation* or a *mobius transformation*.

The transformation given by (1) is conformal, since $\frac{dw}{dz} = \frac{ad-bc}{(cz+d)^2} \neq 0$. The inverse mapping of (1) is $z = \frac{-dw+b}{cw-a}$ which is also a bilinear transformation.

Note

1. If $ad - bc = 0$, every point of z -plane is a critical point.
2. The expression $ad - bc$ is called the *determinant* of the bilinear transformation.
3. A bilinear transformation $w = \frac{az+b}{cz+d}$ is said to be *normalized* if $ad - bc = 1$.
4. A bilinear transformation maps circles into circles.

Invariant or Fixed points

The points which coincide with their transformation are called *invariant points* of the transformation. Fixed points of a transformation $w = f(z)$ are obtained by the equation $z = f(z)$.

Note: Every bilinear transformation has at most two fixed points.

Cross Ratio

If there are four points $z_1,\ z_2,\ z_3,\ z_4$ taken in order, then the ratio $\frac{(z_1-z_2)(z_3-z_4)}{(z_1-z_4)(z_3-z_2)}$ is called the *cross ratio* of $z_1,\ z_2,\ z_3,\ z_4$ and is denoted by $(z_1,\ z_2,\ z_3,\ z_4)$.

Theorem: A bilinear transformation preserves cross ratio of four points.

Proof:

Let $w_i \ = \ \frac{az_i+b}{cz_i+d},\ i = 1,2,3,4.$

Then $w_1 - w_2 = \frac{az_1+b}{cz_1+d} - \frac{az_2+b}{cz_2+d} = \frac{(ad-bc)(z_1-z_2)}{(cz_1+d)(cz_2+d)}$

Similarly, $w_3 - w_4 \ = \ \frac{(ad-bc)(z_3-z_4)}{(cz_3+d)(cz_4+d)}.$

$\therefore\ (w_1 - w_2)(w_3 - w_4) \ = \ \frac{(ad-bc)^2\,(z_1-z_2)\,(z_3-z_4)}{(cz_1+d)\,(cz_2+d)\,(cz_3+d)\,(cz_4+d)}.$

Similarly, $(w_2 - w_3)(w_4 - w_1) \ = \ \frac{(ad-bc)^2\,(z_2-z_3)(z_4-z_1)}{(cz_1+d)\,(cz_2+d)\,(cz_3+d)\,(cz_4+d)}.$

$\therefore \quad \frac{(w_1-w_2)\,(w_3-w_4)}{(w_2-w_3)\,(w_4-w_1)} \ = \ \frac{(z_1-z_2)\,(z_3-z_4)}{(z_2-z_3)\,(z_4-z_1)}.$

$\therefore \quad (w_1,\ w_2,\ w_3,\ w_4) \ = \ (z_1,\ z_2,\ z_3,\ z_4).$

Note

1. In the bilinear transformation $w = \frac{az+b}{cz+d},\ ad - bc \neq 0$ dividing the numerator and denominator of the right hand side by one of the four constants, we observe that there are only three independent constant. Hence three independent constants are required to determine a bilinear transformation.
2. The bilinear transformation that maps the three points $z_1,\ z_2,\ z_3$ to the three points $w_1,\ w_2,\ w_3$ is given by $(z,\ z_1,\ z_2,\ z_3) = (w,\ w_1,\ w_2,\ w_3)$.

Example 1: Find the image of the region $Re(z) \leq 0$ under the transformation

$w \ = \ \frac{2z+i}{iz-1}.$

Solution:

We have $w \ = \ \frac{2z+i}{iz-1}$.

Solving for z, we get

$(iz-1)w = 2z+i.$

i.e., $z(iw-2) = i+w$

or $\qquad z = \frac{i+w}{iw-2}.$

Let $z = x+iy$ and $w = u+iv$.

Hence $x+iy = \frac{i+(u+iv)}{i(u+iv)-2} = \frac{u+i(v+1)}{-2-v+iu} = \frac{-u-i(u^2+v^2+3v)}{(2+v)^2+u^2}.$

$\therefore\ x = \frac{-u}{(2+v)^2+u^2}$ and $y = \frac{-(u^2+v^2+3v)}{(2+v)^2+u^2}.$

The given region is $Re(z) \le 0$. i.e., $x \le 0$.

This gives $\frac{-u}{(2+v)^2+u^2} \le 0$ so that $-u \le 0$.

Hence $u \ge 0$. i.e., $Re(w) \ge 0$.

$\therefore$ The image of the region is $Re(w) \ge 0$.

Example 2: Prove that $w = \frac{z}{1-z}$ maps the upper half of the z-plane onto the upper half of the w −plane. What is the image of the circle $|z| = 1$ under this transformation?

Solution:

We have $w = \frac{z}{1-z}$.

Solving for z, $(1-z)w = z$

$$z(1+w) = w$$

i.e., $\qquad z = \frac{w}{1+w}.$

The given region is $Im(z) \ge 0$.

i.e., $\qquad \frac{z-\bar{z}}{2i} \ge 0.$

$\therefore \qquad \frac{1}{2i}\left[\frac{w}{1+w} - \frac{\bar{w}}{1+\bar{w}}\right] \ge 0.$

i.e., $\frac{1}{2i}[w(1+\bar{w}) - \bar{w}(1+w)] \ge 0.$

i.e., $\qquad \frac{1}{2i}[w + w\bar{w} - \bar{w} - w\bar{w}] \ge 0.$

i.e., $\qquad \frac{1}{2i}(w-\bar{w}) \ge 0.$

Let $w = u + iv$.

Hence $v \geq 0$, and so $Im(w) \geq 0$.

Thus the upper half of the w –plane is mapped onto the upper half of the w –plane.

If $|z| = 1$, then $\left|\frac{w}{1+w}\right| = 1$.

This gives $|w| = |1 + w|$.

$$\therefore \quad |u + iv| = |1 + u + iv|$$

$$\therefore \quad u^2 + v^2 = (1 + u)^2 + v^2$$

i.e., $u^2 + v^2 = 1 + 2u + u^2 + v^2$

$$\therefore \quad u = \frac{-1}{2}.$$

Hence the image of $|z| = 1$ is the straight line $u = -\frac{1}{2}$.

Example 3: Show that the transformation $w = \frac{i - iz}{1 + z}$ maps the unit circle $|z| = 1$ into the real axis of the w –plane.

Solution: We have $w = \frac{i - iz}{1 + z}$.

$$\therefore \quad w(1 + z) = i - iz$$

Hence $z = \frac{i-w}{w+i}$.

Let $w = u + iv$.

The given region is $|z| = 1$.

i.e., $\left|\frac{i-w}{w+i}\right| = 1$.

$$\therefore \quad |i - w| = |w + i|.$$

i.e., $|i - (u + iv)|^2 = |u + iv + i|^2$

i.e., $|-u + i(-v + 1)|^2 = |u + i(v + 1)|^2$

or $u^2 + (1 - v)^2 = u^2 + (v + 1)^2$

or $u^2 + 1 + v^2 - 2v = u^2 + v^2 + 2v + 1$

or $2v + 2v = 0$.

$\therefore$ $v = 0$ which is the real axis in the w –plane. Thus the unit circle $|z| = 1$ is mapped into the real axis in the w –plane.

Example 4: Prove that the transformation $w = \frac{z-1}{z+1}$ maps the imaginary axis in the z −plane into the unit circle in the w −plane.

Solution: We have $w = \frac{z-1}{z+1}$.

$\therefore \quad w(z+1) = z-1.$

$\therefore \quad z = \frac{-1-w}{w-1}.$

Let $w = u + iv$.

The given region is $x = 0$.

i.e., $Re(z) = 0.$

i.e., $Re\left(\frac{-1-w}{w-1}\right) = 0.$

i.e., $Re\left(\frac{(-1-u)-iv}{(u-1)+iv}\right) = 0.$

i.e., $Re\left(\frac{[(-1-u)-iv][(u-1)-iv]}{[(u-1)+iv][(u-1)-iv]}\right) = 0.$

i.e., $Re\left(\frac{[-u^2-v^2+1]+i2v}{(u-1)^2+v^2}\right) = 0.$

i.e., $-\left[\frac{u^2+v^2-1}{(u-1)^2+v^2}\right] = 0.$

$\therefore\ u^2 + v^2 = 1$, which is the unit circle $|w| = 1$.

Thus the imaginary axis in the z −plane is mapped into the circle $|w| = 1$ in the w −plane.

Example 5: Prove that under the transformation $w = \frac{z-i}{iz-1}$ the region $Im(z) \geq 0$ is mapped onto the region $|w| \leq 1$. Into what region $Im(z) \leq 0$ mapped?

Solution: We have $w = \frac{z-i}{iz-1}$.

$\therefore \quad w(iz-1) = z-i.$

$\therefore \quad z = \frac{w-i}{wi-1}$

Given region is $Im(z) \geq 0$.

i.e., $\frac{z-\bar{z}}{2i} \geq 0.$

$\therefore \quad z - \bar{z} \geq 0.$

Hence $\frac{w-i}{wi-1} - \frac{\bar{w}+i}{-\bar{w}i-1} \geq 0.$

$\therefore \quad (w-i)(-\bar{w}i-1)-(wi-1)(\bar{w}+i) \geq 0.$

i.e., $-w\bar{w}i-\bar{w}-w+i-w\bar{w}i+\bar{w}+w+i \geq 0.$

i.e., $\quad -2w\bar{w}i+2i \geq 0.$

i.e., $\quad w\bar{w} \leq 1.$

i.e., $\quad |w|^2 \leq 1$

$\therefore \quad |w| \leq 1$

$\therefore$ $Im(z) \geq 0$ is mapped onto $|w| \leq 1$.

Now $Im(z) \leq 0$ implies $-w\bar{w} \leq -1$ and so $w\bar{w} \geq 1$.

i.e., $\quad |w|^2 \geq 1.$

i.e., $\quad |w| \geq 1.$

$\therefore$ $Im(z) \leq 0$ is mapped into the exterior of the unit circle $|w|=1$ in the w −plane.

Example 6: Find the bilinear transformation that maps the points $z = 0, -1, i$ into the points $w = i, 0, \infty$ respectively. ***[AU, April/May 2015]***

Solution:

Given $z_1 = 0,\ z_2 = -1,\ z_3 = i$

$w_1 = i, w_2 = 0, w_3 = \infty \quad \left[\therefore \frac{1}{w_3} = 0\right]$

$$\frac{(w-w_1)(w_2-w_3)}{(w-w_3)(w_2-w_1)} = \frac{(z-z_1)(z_2-z_3)}{(z-z_3)(z_2-z_1)}$$

$$\Rightarrow \quad \frac{(w-w_1)\,w_3\left(\frac{w_2}{w_3}-1\right)}{w_3\left(\frac{w}{w_3}-1\right)(w_2-w_1)} = \frac{(z-z_1)(z_2-z_3)}{(z-z_3)(z_2-z_1)}$$

$$\Rightarrow \quad \frac{(w-i)(-1)}{(-1)(0-i)} = \frac{(z-0)(-1-i)}{(z-i)(-1-0)}$$

$$\Rightarrow \quad \frac{w-i}{-i} = \frac{z(1+i)}{z-i}$$

$$\Rightarrow \quad w-i = \frac{z(-i+1)}{z-i}$$

$$\Rightarrow \quad w = \frac{z(-i+1)}{z-i} + i$$

$$\Rightarrow \quad w = \frac{z(-i+1)+i(z-i)}{z-i}$$

$$\therefore \quad w = \frac{z+1}{z-i}.$$

Example 7: Find the bilinear transformation that maps 1, i and -1 of the z – plane onto 0, 1 and ∞ of the w – plane. ***[AU, Nov/Dec 2017, May/June 2014]***

Solution:

Given $z_1 = 1, z_2 = i, z_3 = -1$

$$w_1 = 0,\ w_2 = 1, w_3 = \infty \quad \left[\therefore \frac{1}{w} = 0\right]$$

$$\frac{(w - w_1)(w_2 - w_3)}{(w - w_3)(w_2 - w_1)} = \frac{(z - z_1)(z_2 - z_3)}{(z - z_3)(z_2 - z_1)}$$

$$\Rightarrow \frac{(w - w_1)\, w_3 \left(\frac{w_2}{w_3} - 1\right)}{w_3\left(\frac{w}{w_3} - 1\right)(w_2 - w_1)} = \frac{(z - z_1)(z_2 - z_3)}{(z - z_3)(z_2 - z_1)}$$

$$\Rightarrow \frac{(w - 0)(-1)}{(-1)(1 - 0)} = \frac{(z - 1)(i + 1)}{(z + 1)(i - 1)}$$

$$\Rightarrow w = \frac{z - 1}{z + 1}\,\frac{i + 1}{i - 1}$$

$$\Rightarrow w = \frac{z - 1}{z + 1}\,\frac{i + 1}{i - 1}\cdot\frac{i - 1}{i - 1}$$

$$\Rightarrow w = \frac{z - 1}{z + 1}(-i)$$

$$\therefore w = i\,\frac{1 - z}{1 + z}.$$

Example 8: Find the bilinear transformation that transforms the points $z = 1, i, -1$ of the z – plane into the points $w = 2, i, -2$ of the w – plane. ***[AU, May/June 2016]***

Solution:

Given $z_1 = 1, z_2 = i, z_3 = -1$

$$w_1 = 2,\ w_2 = i,\ w_3 = -2$$

$$\frac{(w - w_1)(w_2 - w_3)}{(w - w_3)(w_2 - w_1)} = \frac{(z - z_1)(z_2 - z_3)}{(z - z_3)(z_2 - z_1)}$$

$$\Rightarrow \frac{(w - 2)(i + 2)}{(w + 2)(i - 2)} = \frac{(z - 1)(i + 1)}{(z + 1)(i - 1)}$$

$$\Rightarrow \frac{(w - 2)(i + 2)}{(w + 2)(i - 2)} = \frac{z - 1}{z + 1}(-i)$$

$$\Rightarrow \frac{(w - 2)(i + 2)}{(w + 2)(i - 2)} = \frac{i - iz}{1 + z}$$

$$\Rightarrow (w - 2)(i + 2)(1 + z) = (w + 2)(i - 2)(i - iz)$$

$$\Rightarrow w[(i + 2)(1 + z) - (i - 2)(i - iz)] = 2(i + 2)(1 + z) + 2(i - 2)(i - iz)$$

$$\Rightarrow w[(i + iz + 2 + 2z) - (-1 + z - 2i + 2iz)] = 2(i + 2 + \text{iz} + 2\text{z} - 1 + \text{z} - 2\text{i} + 2\text{iz})$$

$$\Rightarrow \qquad w(3 + 3i + z - iz) = 2(1 - i + 3z + 3iz)$$

$$\Rightarrow w = 2\,\frac{(1 - i + 3z + 3iz)}{(3 + 3i + z - iz)}$$

$$= 2\,\frac{[1 - i + 3z\,(1 + i)]}{[3(1 + i) + z\,(1 - i)]}$$

$$= 2\,\frac{\frac{1-i}{1+i} + 3z}{3 + z\frac{1-i}{1+i}} = 2\,\frac{-i + 3z}{3 - iz}$$

$$\Rightarrow w = \frac{6z - 2i}{3 - iz}.$$

Example 9: Find the bilinear transformation that maps the points $z = 1,\ i$ and -1 onto the planes $w = i\,, 0,\ -i$. Hence find the image of $|z| < 1$.

Solution : ***[AU, Nov/Dec 2017, Nov/Dec 2014]***

Given $z_1 = 1,\ z_2 = i,\ z_3 = -1$

$$w_1 = i,\ w_2 = 0,\ w_3 = -i$$

$$\frac{(w - w_1)\,(w_2 - w_3)}{(w - w_3)\,(w_2 - w_1)} = \frac{(z - z_1)\,(z_2 - z_3)}{(z - z_3)\,(z_2 - z_1)}$$

$$\therefore \qquad \frac{(z - 1)\,(i + 1)}{(z + 1)\,(i - 1)} = \frac{(w - i)\,(0 + i)}{(w + i)\,(0 - i)}.$$

i.e., $$\frac{(z - 1)\,(i + 1)}{(z + 1)\,(i - 1)} = \frac{(w - i)\,i}{(w + i)\,(-i)}$$

i.e., $$(z - 1)(i + 1)(w + i) = -(w - i)(z + 1)(i - 1)$$

i.e., $$w\,[(z - 1)(i + 1) + (z + 1)(i - 1)] = i(z + 1)(i - 1) - i(z - 1)(i + 1)$$

i.e., $$w\,(2iz - 2) = i\,(2i - 2z).$$

Hence $w = \frac{i\,(i - z)}{iz - 1} = \frac{i - z}{i + z} = \frac{1 + iz}{1 - iz}$ is the required bilinear transformation.

To find z: $(1 - iz)w = 1 + iz$

i.e., $w - izw = 1 + iz$

i.e., $izw + iz = w - 1$

i.e., $z = \frac{w-1}{i\,(w+1)}$.

The image of $|z| < 1$ is $\left|\frac{w-1}{i\,(w+1)}\right| < 1$

$$\Rightarrow \quad |w - 1| < |i(w + 1)|$$

$$\Rightarrow \quad |u + iv - 1| < |i(u + iv + 1)|$$

$$\Rightarrow \quad |(u - 1) + iv| < |-v + i(u + 1)|$$

$\Rightarrow \quad \sqrt{(u-1)^2+v^2} = \sqrt{v^2+(u+1)^2}$

$\Rightarrow \quad u^2-2u+1+v^2 = v^2+u^2+2u+1.$

$\Rightarrow \quad 4u = 0 \qquad \Rightarrow u = 0.$

$\therefore$ The image of $|z| < 1$ is $u = 0$.

Example 10: Find the bilinear transformation that maps the points $z = 0,\ 1,\ \infty$ onto the planes $w = i,\ -1,\ -i$.

Solution : Given $z_1 = 0,\ z_2 = 1,\ z_3 = \infty$

$$w_1 = i,\ w_2 = -1,\ w_3 = -i$$

$$\frac{(w-w_1)(w_2-w_3)}{(w-w_3)(w_2-w_1)} = \frac{(z-z_1)(z_2-z_3)}{(z-z_3)(z_2-z_1)}$$

i.e., $$\frac{(w-w_1)(w_2-w_3)}{(w-w_3)(w_2-w_1)} = \frac{(z-z_1)\left(\frac{z_2}{z_3}-1\right)}{\left(\frac{z}{z_3}-1\right)(z_2-z_1)}$$

i.e., $$\frac{(w-i)(-1+i)}{(w+i)(-1-i)} = \frac{(z-0)(-1)}{(-1)(1-0)}$$

$$\frac{(w-i)(-1+i)}{(w+i)(-1-i)} = \frac{(z-0)(-1)}{(-1)(1-0)}$$

$$\frac{w-i}{w+i} = z\ \frac{(1+i)}{(1-i)} = z\ \frac{(1+i)}{(1-i)}\cdot\frac{(1+i)}{(1+i)}$$

$$\frac{w-i}{w+i} = iz$$

$$\frac{(w-i)+(w+i)}{(w-i)-(w+i)} = \frac{iz+1}{iz-1} \quad \text{[Applying C and D formula]}$$

$$\frac{2w}{2i} = \frac{iz+1}{iz-1}$$

$$w = i\ \frac{iz+1}{iz-1}$$

Hence $w = \frac{i-z}{iz-1}$ is the required bilinear transformation.

Example 11: Find the bilinear transformation that maps the points $z = \infty,\ i,\ 0$ onto the planes $w = 0,\ i,\ \infty$.

Solution: Given $z_1 = \infty,\ z_2 = i,\ z_3 = 0$

$$w_1 = 0,\ w_2 = i,\ w_3 = \infty$$

$$\frac{(w-w_1)(w_2-w_3)}{(w-w_3)(w_2-w_1)} = \frac{(z-z_1)(z_2-z_3)}{(z-z_3)(z_2-z_1)}$$

i.e., $$\frac{(w-w_1)\,w_3\left(\frac{w_2}{w_3}-1\right)}{w_3\left(\frac{w}{w_3}-1\right)(w_2-w_1)} = \frac{z_1\left(\frac{z}{z_1}-1\right)(z_2-z_3)}{(z-z_3)\,z_1\left(\frac{z_2}{z_1}-1\right)}$$

$\therefore$ $$\frac{(w-0)\,(0-1)}{(0-1)(i-0)} = \frac{(0-1)\,(i-0)}{(z-0)\,(0-1)}$$

i.e., $$\frac{w}{i} = \frac{i}{z}.$$

Hence $w = \frac{-1}{z}$ is the required bilinear transformation.

Example 12: Find the bilinear transformation which maps -1, 0, 1 of the z-plane onto $-1, -i$, 1 of the w-plane. Show that under this transformation the upper half of the z -plane maps onto the interior of the unit circle $|w| = 1$. ***[AU, April/May 2017]***

Solution : Given $z_1 = -1,\ z_2 = 0,\ z_3 = 1$

$w_1 = -1,\ w_2 = -i,\ w_3 = 1$

$$\frac{(w-w_1)(w_2-w_3)}{(w-w_3)(w_2-w_1)} = \frac{(z-z_1)(z_2-z_3)}{(z-z_3)(z_2-z_1)}$$

$\therefore$ $$\frac{(w+1)\,(-i-1)}{(w-1)\,(-i+1)} = \frac{(z+1)\,(0-1)}{(z-1)\,(0+1)}$$

i.e., $$\frac{-(w+1)\,(i+1)}{(w-1)\,(1-i)} = \frac{-(z+1)}{z-1}$$

i.e., $$(w+1)(i+1)(z-1) = (w-1)(1-i)(z+1)$$

i.e., $$w[(i+1)(z-1)-(1-i)(z+1)] = -(i+1)(z-1)-(1-i)(z+1)$$

$$w[(iz+z-i-1)-(z-iz+1-i)] = -(iz+z-i-1+z-iz+1-i)$$

$$w(2iz-2) = -(2z-2i)$$

$$\therefore\ w = \frac{z-i}{1-iz} = \frac{z-i}{-i^2-iz} = \frac{z-i}{-i(i+z)} = i\left(\frac{z-i}{i+z}\right).$$

Hence $w = \frac{iz+1}{i+z}$ is the required bilinear transformation.

Now, $w = \frac{z-i}{1-iz}$ gives $(1-iz)w = z-i$.

i.e., $w - izw = z - i$

i.e., $z(1+iw) = w+i$

$\therefore\ z = \frac{w+i}{1+iw}$ is the required bilinear transformation.

For the second part, let $z = x+iy$ and $w = u+iv$.

Hence $z = \frac{u+i(1+v)}{(1-v)+iu} = \frac{[u+i(1+v)]\,[(1-v)-iu]}{(1-v)^2-u^2}$

$$= \frac{2u+i(1-u^2-v^2)}{(1-v)^2-u^2}.$$

The given region is $y > 0$.

Hence $\frac{1-u^2-v^2}{(1-v)^2-u^2} > 0$.

i.e., $1 - u^2 - v^2 > 0$.

$u^2 + v^2 < 1$, which is the interior of the circle $|w| = 1$.

$\therefore$ The upper half of the z −plane is mapped onto the interior of the unit circle $|z| = 1$.

Example 13: Find the bilinear transformation which maps the points $-i,\ 0, i$ onto the point $-1,\ i,\ 1$ respectively. Into what curve the y −axis is transformed under this transformation? ***[AU, Nov/Dec 2015]***

Solution. Given $z_1 = -i,\ z_2 = 0,\ z_3 = i$

$$w_1 = -1,\ w_2 = i,\ w_3 = 1$$

$$\frac{(w-w_1)(w_2-w_3)}{(w-w_3)(w_2-w_1)} = \frac{(z-z_1)(z_2-z_3)}{(z-z_3)(z_2-z_1)}$$

$$\therefore \quad \frac{(w+1)(i-1)}{(w-1)(i+1)} = \frac{(z+i)(0-i)}{(z-i)(0+i)}$$

i.e., $$\frac{(w+1)(i-1)}{(w-1)(i+1)} = \frac{-(z+i)}{z-i}$$

$$\Rightarrow \quad \frac{w+1}{w-1} = \frac{-(z+i)(i+1)}{(z-i)(i-1)} \cdot \frac{(i+1)}{(i+1)}$$

$$\Rightarrow \quad \frac{w+1}{w-1} = \frac{(z+i)\,i}{(z-i)} = \frac{iz-1}{z-i}$$

$$\Rightarrow \quad \frac{w+1+w-1}{w+1-(w-1)} = \frac{iz-1+z-i}{iz-1-(z-i)}$$

$$\Rightarrow \quad \frac{2w}{2} = \frac{z(i+1)-(i+1)}{z(i-1)+(i-1)} = \frac{(i+1)(z-1)}{(i-1)(z+1)}$$

$$\Rightarrow \quad w = \frac{(z-1)(i+1)}{(z+1)(i-1)} \cdot \frac{(i+1)}{(i+1)}$$

$$\Rightarrow \quad w = (-i)\frac{(z-1)}{(z+1)}$$

$\therefore w = \frac{-iz+i}{z+1}$ is the required bilinear transformation.

The given curve is the y −axis, i.e., $x = 0$.

Now, $w = \frac{-iz+i}{z+1}$ gives $w(z+1) = -iz+i \Rightarrow z = \frac{i-w}{i+w}$.

Hence $x = 0$ gives $\frac{z+\bar{z}}{2} = 0$ and so $z+\bar{z} = 0$.

$$\therefore \quad \frac{i-w}{w+i} + \left(\frac{-i-\bar{w}}{\bar{w}-i}\right) = 0$$

i.e., $(i-w)(\bar{w}-i) + (-i-\bar{w})(w+i) = 0$

i.e., $-2w\bar{w} + 2 = 0 \Rightarrow w\bar{w} = 1$

$$\therefore \quad |w| = 1.$$

Thus the y −axis is mapped onto the circle $|w| = 1$ in the w −plane.

Exercises

1. Show that the bilinear transformation $w = \frac{2z+3}{z-4}$ maps the circle $x^2 + y^2 - 4x = 0$ into the line $4u + 3 = 0$.
2. Show that $w = \frac{5-4z}{4z-2}$ transforms the circle $|z| = 1$ into a circle of radius unity in the w −plane and find the centre of this circle.
3. Find the image of the quadrant $x > 0,\ y > 0$ under $w = \frac{z-1}{z+1}$.
4. Find the image of the horizontal strip $0 < y < 2$ under $w = \frac{z}{z-i}$.
5. Show that the transformation $w = \frac{i+z}{i-z}$ maps the unit disc $|z| < 1$ onto the right half-plane $Re(w) > 0$. Find the image of the lower half-plane $Im(z) < 0$ under $w = \frac{i+z}{i-z}$.
6. Show that both the transformations, $w = \frac{1+z}{1-z}$, $w = \frac{z+1}{z-1}$ map the left half plane $Re(z) \le 0$ onto $|w| \le 1$.
7. Determine the bilinear transformation that maps the points $1-2i, 2+i, 2+3i$ respectively into $2+2i, 1+3i, 4$.

Answers

3. The portion of $|w| < 1$ in the upper half-plane $Im(w) > 0$.

4. The region that lies exterior to both the circles $\left|w - \frac{1}{2}\right| = \frac{1}{2}$ and $\left|w - \frac{3}{2}\right| = \frac{1}{2}$.

7. $w = \frac{(20+18i)-(32+12i)z}{(29+17i)-(11-3i)z}$

3.10. Short Answer Questions

1. Is the function $f(z) = \bar{z}$ analytic? (or) Show that $f(z) = \bar{z}$ is not analytic at any point.

Solution: ***[AU, May/June 2014]***

Let $f(z) = u + iv$.

Now, $f(z) = \bar{z} = x - iy$.

$\therefore \quad u = x$ and $v = -y$

$\therefore \; u_x = 1,\, u_y = 0$ and $v_x = 0,\, v_y = -1$.

Hence $u_x \neq v_y$ at any point.

$\therefore$ C - R equations are not satisfied.

$\therefore$ $f(z)$ is not an analytic function.

2. Prove that $w = \sin 2z$ is an analytic function. ***[AU, Nov/Dec 2015]***

Solution: Given $w = sin\,2z$

i.e.,
$$u + iv = \sin\, 2(x + iy) = \sin 2x \cos i2y + \cos x \sin i2y$$
$$= \sin x\, \cosh 2y + i \cos x \sinh 2y.$$

$\therefore\; u = \sin 2x \cosh 2y$ and $v = \cos 2x \sinh 2y$

$u_x = 2 \cos x \cosh 2y$ and $v_x = -2\sin 2x \sinh 2y$

$u_y = 2\sin 2x \sinh 2y$ and $v_y = 2\cos 2x \cosh 2y$.

$\therefore\; u_x = v_y$ and $u_y = -v_x$.

Hence C-R equations are satisfied.

Also the four partial derivatives $u_x,\ u_{y,}\ v_x,\ v_y$ are continuous.

Hence $w = sin\, 2z$ is analytic.

3. Is the function $f(z) = z^2$ is analytic? Justify the claim. ***[AU, Nov/Dec 2017]***

Solution: Let $f(z) = u + iv$ and $z = x + iy$

Given $f(z) = z^2$

$$\Rightarrow \quad u + iv = (x + iy)^2 = x^2 + 2xiy + (iy)^2$$
$$= x^2 + 2ixy - y^2$$

$= (x^2 - y^2) + 2ixy$

$\therefore\ u = x^2 - y^2$ and $v = 2ixy$

$u_x = 2x$; $v_x = 2iy$

$u_y = -2y$; $v_y = 2ix$.

$\therefore\ u_x \neq v_y$ and $u_y \neq -v_x$.

Hence C - R equations are not satisfied.

Hence $w = z^2$ is not analytic.

4. Verify $f(z) = z^3$ is analytic or not. ***[AU, Nov/Dec 2014]***

Solution: Let $f(z) = u + iv$ and $z = x + iy$

Given $f(z) = z^3$

$\Rightarrow\ u + iv = (x + iy)^3 = x^3 + 3x^2 iy + 3x(iy)^2 + (iy)^3$

$= x^3 + 3x^2 iy - 3xy^2 - iy^3$

$= x^3 - 3xy^2 + i(3x^2 y - y^3)$

$\therefore\ u = x^3 - 3xy^2$ and $v = 3x^2 y - y^3$

$u_x = 3x^2 - 3y^2$; $v_x = 6xy$

$u_y = -6xy$; $v_y = 3x^2 - 3y^2$.

$\therefore\ u_x = v_y$ and $u_y = -v_x$.

Hence C - R equations are satisfied.

Also the four partial derivatives u_x, u_y, v_x, v_y are continuous.

Hence $w = z^3$ is analytic.

5. Prove that $f(z) = z$ is analytic everywhere.

Solution: Let $f(z) = z$.

i.e., $u + iv = x + iy$

Then $u = x$ and $v = y$.

$u_x = 1,\ v_x = 0$ and $u_y = 0,\ v_y = 1$.

Clearly $u_x = v_y$ and $u_y = -v_x$. Also u_x, u_y, v_x and v_y are continuous at any point.

$\therefore f(z) = z$ is analytic everywhere.

6. The real part of an analytic function $f(z)$ is constant, prove that $f(z)$ is a constant function. **[AU, *April/May 2017*]**

Solution: Let $f(z) = u + iv$ be an analytic function.

$\therefore u_x = v_y$ and $u_y = -v_x$ ...(1)

Given real part of $f(z)$ is constant.

$\therefore u = c$, c is a constant.

$\Rightarrow u_x = 0,\ v_x = 0$ and $u_y = 0,\ v_y = 0$, by (1)

We know, $f'(z) = u_x + iv_x = 0 + i0 = 0$.

Integrating with respect to z, we get $f(z) = c$, a constant.

7. Give an example of a function where u and v are harmonic but $u + iv$ is not analytic.

Solution: **[*AU, May/June 2016*]**

Let $u = x$ and $v = -y$.

Then $u_x = 1,\ u_{xx} = 0,\ v_x = 0,\ v_{xx} = 0,$

$u_y = 0,\ u_{yy} = 0,\ v_y = -1$ and $v_{yy} = 0$.

Clearly $u_{xx} + u_{yy} = 0$ and $v_{xx} + v_{yy} = 0$.

$\therefore u$ and v are harmonic functions.

Since $u_x \neq v_y$, $f(z) = u + iv = x - iy$ is not an analytic function.

8. Check whether $u = 2xy + 3xy^2 - 2y^3$ is a harmonic function.

Solution: Given $u = 2xy + 3xy^2 - 2y^3$

$$u_x = 2y + 3y^2,\ u_y = 2x + 6xy - 6y^2$$

$$u_{xx} = 0,\ u_{yy} = 6x - 12y.$$

Clearly $u_{xx} + u_{yy} \neq 0$.

$\therefore u$ is not a harmonic function.

9. Examine whether $y + e^x \cos y$ is harmonic. **[AU, *Nov/Dec 2017*]**

Solution: Let $u = y + e^x \cos y$.

$\therefore\ u_x = e^x \cos y,\ u_{xx} = e^x \cos y$, and $u_y = 1 - e^x \sin y,\ u_{yy} = -e^x \cos y$.

$\Rightarrow u_{xx} + u_{yy} = e^x \cos y - e^x \cos y = 0.$

$\Rightarrow$ u is harmonic.

10. Can $u = 2xy + 2x$ be the real part of an analytic function.

Solution:

Given $u = 2xy + 2x$.

$$u_x = 2y + 2, \; u_y = 2x, \; u_{xx} = 0, \; u_{yy} = 0.$$

Clearly $u_{xx} + u_{yy} = 0$.

$\therefore u$ satisfies the Laplace equation.

Hence u is the real part of an analytic function.

11. Find the value of $'a'$ so that $u(x,y) = ax^2 - y^2 + 2xy$ is harmonic.

Solution: Given $u = ax^2 - y^2 + 2xy$

$$u_x = 2ax + 2y, \; u_y = -2y + 2x$$

$$u_{xx} = 2a, \quad u_{yy} = -2.$$

Since u is harmonic, $u_{xx} + u_{yy} = 0$

$$2a - 2 = 0$$

$$a = 1.$$

12. Define conformal mapping. ***[AU, Nov/Dec 2015]***

A transformation which preserves angles both in magnitude and direction between every pair of curves through a point is said to be *conformal* at the point.

13. Find the points where $f(z) = \cos z$ is not conformal.

Solution: $f(z) = \cos z$

$$f'(z) = -\sin z.$$

A transformation is not conformal if $f'(z) = 0$

i.e., $-\sin z = 0$

$$z = n\pi, \; n \text{ is an integer.}$$

$\therefore \cos z$ is not conformal when $z = n\pi$, n is an integer.

14. Find the critical points of the transformation $w = z^2 + \frac{1}{z^2}$. **[AU, *April/May 2017*]**

Solution: Let $f(z) = z^2 + \frac{1}{z^2}$

$$f'(z) = 2z - \frac{2}{z^3}.$$

The critical points of $f(z)$ are given by $f'(z) = 0$

i.e.,
$$2z - \frac{2}{z^3} = 0$$

$$z^4 - 1 = 0$$

$$(z^2 + 1)(z^2 - 1) = 0$$

$$z = \pm 1, \pm i.$$

$\therefore$ The critical points are $z = \pm 1, \pm i$.

15. Find all points at which the mapping $w = z^2 + \frac{1}{z^2}$ is not conformal.

Solution: Let $f(z) = z^2 + \frac{1}{z^2}$

$$f'(z) = 2z - \frac{2}{z^3}.$$

The transformation $w = f(z)$ is not conformal if $f'(z) = 0$

i.e.,
$$2z - \frac{2}{z^3} = 0$$

$$z^4 - 1 = 0$$

$$(z^2 + 1)(z^2 - 1) = 0$$

$$z = \pm 1, \pm i.$$

$\therefore w = z^2 + \frac{1}{z^2}$ is not conformal at $z = \pm 1, \pm i$.

16. Find the image of the circle $|z - 1| = 1$ in the complex plane under the mapping $w = \frac{1}{z}$.

Solution: The given transformation is $w = \frac{1}{z}$.

$$\therefore \qquad z = \frac{1}{w}.$$

$\therefore$ The image of $|z-1| = 1$ is $\left|\frac{1}{w}-1\right| = 1$

i.e., $$\left|\frac{1-w}{w}\right| = 1$$

$$|1-w| = |w|$$

$$|1-(u+iv)|^2 = |u+iv|^2$$

$$(1-u)^2+v^2 = u^2+v^2$$

$$1+u^2-2u+v^2 = u^2+v^2$$

$\therefore$ $$2u-1 = 0.$$

$\therefore$ $$u = \frac{1}{2}.$$

17. Find the image of the region $|z| > 1$ under the given mapping $w = 4z$.

Solution:

The given mapping is $w = 4z$.

$\therefore z = \frac{w}{4}$.

The image of $|z| > 1$ is $\left|\frac{w}{4}\right| > 1$

$$|w| > 4$$

i.e., $$\sqrt{u^2+v^2} > 4$$

$$u^2+v^2 > 16.$$

18. Find the fixed points of $w = z^3$.

Solution:

The fixed points of $w = z^3$ are the solutions of $z = z^3$.

i.e., $z^3 - z = 0$.

i.e., $z(z^2-1) = 0$

i.e., $z = 0$ or $z = \pm 1$.

$\therefore$ The fixed points are $0, 1, -1$.

19. Find the invariant points of $f(z) = z^2$. **[AU, May/June 2014]**

Solution: The invariant points are solutions of $z = z^2$

$\Rightarrow \quad z^2 - z = 0$

$\Rightarrow \quad z(z-1) = 0$

i.e., $z = 0$ or $z = 1$.

$\therefore$ The fixed points are $0, 1$.

20. Find the invariant points of $w = \frac{z-1}{z+1}$. **[AU, April/May 2015]**

Solution: The invariant points are solutions of $z = \frac{z-1}{z+1}$

$\Rightarrow z^2 + z = z - 1$

$\Rightarrow z^2 - 1 = 0$

$\Rightarrow \quad z^2 = 1$

$\Rightarrow \quad z = 1, -1.$

$\therefore$ The invariant points are $z = 1, -1$.

21. Find the invariant points of $w = \frac{2z+6}{z+7}$.

Solution: The invariant points are solutions of $z = \frac{2z+6}{z+7}$

$\Rightarrow \quad z^2 + 7z = 2z + 6$

$\Rightarrow \quad z^2 + 5z - 6 = 0$

$\Rightarrow (z-1)(z+6) = 0$

$\Rightarrow \quad z = 1, -6.$

$\therefore$ The invariant points are $z = 1, -6$.

22. If $z = i$ is the fixed point of the bilinear transformation $w = \frac{1}{z+c}$, then find 'c'.

Solution: **[AU, Nov/Dec 2017]**

Given transformation is $w = \frac{1}{z+c}$.

The fixed points are given by $z = \frac{1}{z+c}$.

Since $z = i$ is the fixed point, $i = \frac{1}{i+c} \quad \Rightarrow \quad \frac{1}{i} = i + c$

$\Rightarrow \quad -i = i + c$

$\Rightarrow \quad c = -2i$

23. Find the image of the line $x = 1$ under the transformation $w = z^2$.

Solution: **[AU, *Nov/Dec 2017*]**

Let $f(z) = w = u + iv$ and $z = x + iy$

Given $w = z^2$

$$\Rightarrow \quad u + iv = (x + iy)^2 = x^2 + 2xiy + (iy)^2$$

$$= x^2 + 2ixy - y^2$$

$$= (x^2 - y^2) + 2ixy$$

Put $x = 1$, $u + iv = (1 - y^2) + 2iy$

$\Rightarrow u = 1 - y^2$ and $v = 2y$

$\Rightarrow u = 1 - \left(\frac{v}{2}\right)^2 \quad \Rightarrow v^2 = 4(1 - u)$ which is a parabola in w plane.

CHAPTER 4

COMPLEX INTEGRATION

4.1. Introduction

Complex analysis is an extremely useful and beautiful part of mathematics and forms the basis of many techniques employed in many branches of mathematics and physics. In fact, to a large extent complex analysis is the study of analytic functions which we have discussed in the previous chapter. In this chapter, we will extend the notions of derivatives and integrals, familiar from calculus, to the case of complex functions of a complex variable. In doing so we will come across analytic functions, which form the centerpiece of this part of the course. We will then discuss complex integration, culminating with the generalised Cauchy Integral Formula, and some of its applications. We then go on to discuss the power series representations of analytic functions and the residue calculus, which will allow us to compute many real integrals and in finite sums very easily via complex integration. Throughout engineering, transforms defined as complex integrals in one form or another play a major role in analysis and design. The use of complex variable techniques allows us to develop criteria for the stability of systems.

The concept of a real line integral, with which the reader is familiar is extended to that of a complex line integral as given below in Fig.1.

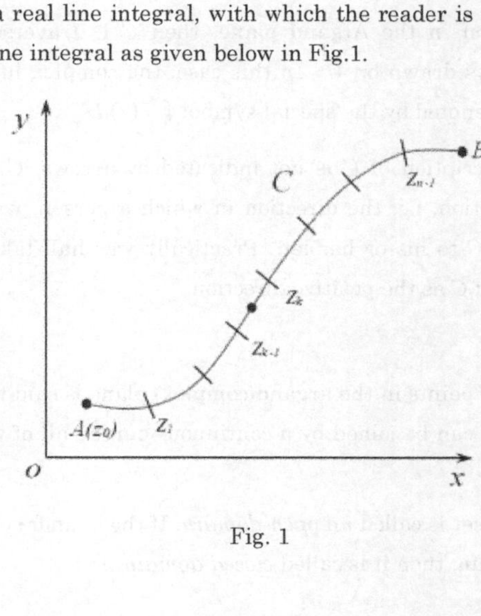

Fig. 1

Let $f(z)$ be a continuous function of the complex variable $z(=x+iy)$ and let C be any continuous curve connecting two points $A(z_0)$ and $B(z_n)$ on the z-plane. Let C be divided into n parts by means of the points $z_1, z_2, \ldots z_{n-1}$. Let $z_k - z_{k-1} = \Delta z_k$. Let α_k be an arbitrary point in the arc $z_{k-1}\, z_k$ $(k = 1,2, \ldots, n)$.

$$\text{Let } S_n = \sum_{k=1}^{n} f(\alpha_k)(z_k - z_{k-1}) \text{ or } \sum_{k=1}^{n} f(\alpha_k)\, \Delta z_k .$$

If the limit of S_n exists as $n \to \infty$ in such a way that each $\Delta z_k \to 0$ and if the limit is independent of the mode of subdivision of C and the choice of the points α_k, then it is called the complex line integral of $f(z)$ along C from A to B and denoted as $\int_C f(z)\, dz$. Practically a complex line integral is expressed in terms of two real line integrals and evaluated.

i.e If $f(z) = u(x,y) + iv(x,y)$ then

$$\int_C f(z)\, dz = \int_C (u + iv)\,(dx + idy)$$

$$= \int_C (udx - vdy) + i \int_C (vdx + udy)$$

Note: When C is an open are (of finite length) of a continuous curve, the sense of description of C is obvious, as it is traversed from $A(z_0)$ to $B(z_n)$. When C is a simple closed curve, i.e. a continuous closed curve which does not intersect itself and which encloses a finite region in the Argand plane, then C is traversed in the direction indicated by the arrows drawn on C. In this case, the complex line integral is called contour integral and denoted by the special symbol $\oint_C f(z)dz$.

If the sense of description of C is not indicated by arrows, C is traversed in the positive sense or direction, i.e. the direction in which a person, walking along C, has the interior region of C to his or her left. Practically we shall take the anticlockwise direction of traversal of C as the positive direction.

Basic Concepts

Domain: A set S of points in the argand(complex) plane is said to be a *connected set* if any two of its points can be joined by a continuous curve , all of whose points belong to S.

An open connected set is called an *open domain.* If the boundary points of S are also added to an open domain, then it is called *closed domain.*

Simply and Multiply Connected Regions

A curve is called simple closed curve if it does not cross itself [Fig.2(a)]. A curve which crosses itself is called a multiple curve [Fig.2(b)].

A region is called simply connected if every closed curve in the region encloses points of the region only, i.e., every closed curve lying in it can be contracted indefinitely without passing out of it. A region which is not simply connected is called a multiply connected region. In plain terms, a simply connected region is one which has no holes. A multiply connected region can be converted into a simply connected region by introducing cross–cuts as shown in the following figure [Fig.2(c)].

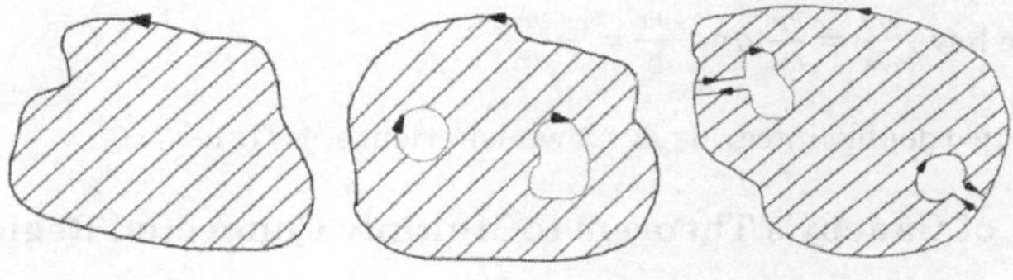

Fig. 2(a) Fig. 2(b) Fig.2(c)

Remark 1: Jordan arc is a continuous arc without multiple points.

Remark 2: Contour is a Jordan curve consisting of continuous chain of a finite number of regular arcs.

4.2. Cauchy's Integral Theorem OR Cauchy's Fundamental Theorem

If $f(z)$ is an analytic function and $f'(z)$ is continuous at each point within and on a simple closed curve C, then $\oint_C f(z)dz = 0$.

Proof: Let R be the region bounded by the curve C.

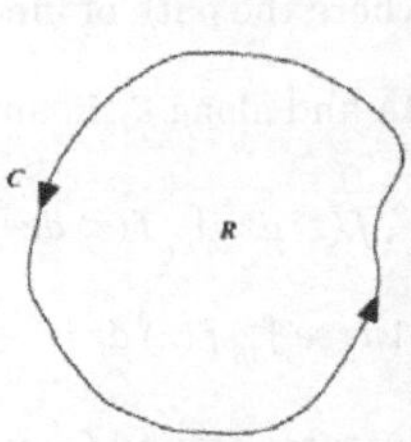

Fig. 3

Let $\quad f(z) = u(x, y) + iv(x, y)$, then

$$\oint_C f(z)dz = \oint_C (u + iv)(dx + idy)$$

$$= \oint_C (udx - vdy) + i \oint_C (vdx + udy) \quad (1)$$

Since $f'(z)$ is continuous, the partial derivatives $\frac{\partial u}{\partial x}, \frac{\partial u}{\partial y}, \frac{\partial v}{\partial x}, \frac{\partial v}{\partial y}$ are also continuous in R. Hence by Green's Theorem, we have

$$\oint_C f(z)dz = \iint_R \left(-\frac{\partial v}{\partial x} - \frac{\partial u}{\partial y}\right) dx\, dy + i \iint_R \left(\frac{\partial u}{\partial x} - \frac{\partial v}{\partial y}\right) dx\, dy \quad (2)$$

Now $f(z)$ being analytic at each point of the region R, by Cauchy-Riemann equations, we have $\frac{\partial u}{\partial x} = \frac{\partial v}{\partial y}$ *and* $\frac{\partial u}{\partial y} = -\frac{\partial v}{\partial x}$.

Thus, the two double integrals in (2) vanish. Hence $\oint_C f(z)dz = 0$.

Extension of Cauchy's Theorem to Multiply Connected Region

If $f(z)$ is analytic in the region R between two simple closed curves C_1 and C_2 then $\int_{C_1} f(z)\, dz = \int_{C_2} f(z)\, dz$ when integral along each curve is taken in anti-clockwise direction.

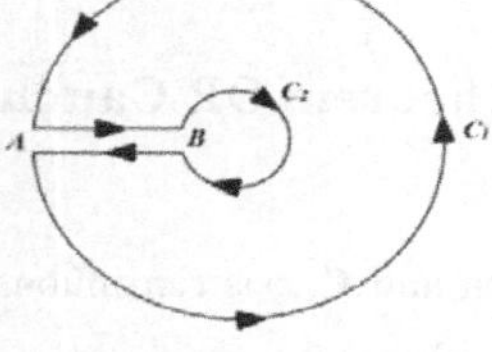

Fig. 4

Proof: We know, $\int f(z)dz = 0$, where the path of integration is along AB and curve C_2 in clockwise direction and along BA and along C_1 in anti-clockwise direction.

ie., $\int_{AB} f(z)\, dz + \int_{C_2} f(z)\, dz + \int_{BA} f(z)\, dz \ \int_{C_1} f(z)\, dz = 0$

or $\int_{C_2} f(z)\, dz + \int_{C_1} f(z)\, dz = 0$ {*since* $\int_{AB} f(z)\, dz = -\int_{BA} f(z)\, dz$ }

Reversing the direction of the integral around C_2, we get

$$\int_{C_1} f(z)\, dz = \int_{C_2} f(z)\, dz$$

However if a closed curve C contains non-interesting closed curves $C_1, C_2, \ldots, C_n$, then by introducing cross-cuts, it can be shown that

$$\oint_C f(z)dz = \oint_{C_1} f(z)dz + \oint_{C_2} f(z)dz + \cdots + \oint_{C_n} f(z)dz .$$

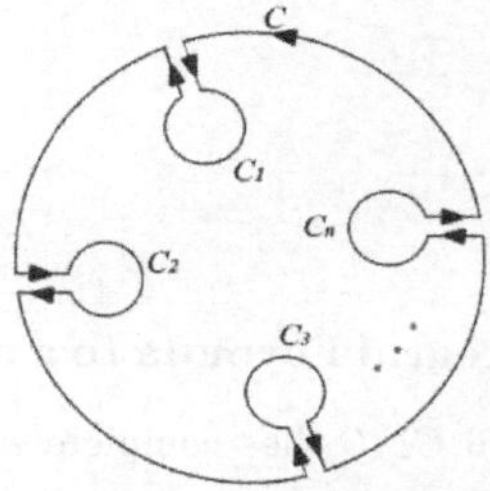

Fig. 5

4.3. Cauchy's Integral Formula

If $f(z)$ is analytic inside and on a simple closed curve C that encloses a simply connected region R and if 'a' is any point in R, then

$f(a) = \frac{1}{2\pi i} \oint_C \frac{f(z)}{(z-a)} dz$, where C is described in the anticlockwise sense.

Proof: Since $f(z)$ is analytic on and inside C, $\frac{f(z)}{(z-a)}$ is also analytic on and inside C, except at the point $z = a$. Now we draw a small circle C' with centre at $z = a$ and radius ρ, lying completely inside C. The function $\frac{f(z)}{(z-a)}$ is analytic in the doubly connected region bounded by C and C'.

∴ By Cauchy's Extended theorem, we have

$$\oint_C \frac{f(z)}{z-a} dz = \oint_{C'} \frac{f(z)}{z-a} dz \tag{1}$$

If z is any point on C', then $|z - a| = \rho$ and hence $z - a = \rho e^{i\theta}$ or $z = a + \rho e^{i\theta}$, $dz = i\rho e^{i\theta}\, d\theta$.

$$\therefore \oint_C \frac{f(z)}{z-a} dz = \int_0^{2\pi} \frac{f(a+\rho e^{i\theta})}{\rho e^{i\theta}}\, i\rho e^{i\theta}\, d\theta$$

(when z moves around C' once completely, θ varies from 0 to 2π)

$$= i \int_0^{2\pi} f(a + \rho e^{i\theta})\, d\theta \tag{2}$$

(2) is true, however small the circle C' is and hence true when $\rho \to 0$. Taking limits of (2) as $\rho \to 0$, we get

$$\oint_{C'} \frac{f(z)}{z-a} dz = i \int_0^{2\pi} f(a)\, d\theta = i\, f(a)\, 2\pi \qquad (3)$$

Using (3) in (1) we have

$$\oint_C \frac{f(z)}{z-a} dz = 2\pi i\, f(a)$$

$$\therefore \quad f(a) = \frac{1}{2\pi i} \oint_C \frac{f(z)}{z-a} dz.$$

Extension of Cauchy's Integral Formula to a doubly Connected Region

If $f(z)$ is analytic on C_1 and C_2 (C_2 lies completely within C_1) and in the annular region R between C_1 and C_2 and if 'a' is any point in R, then

$$f(a) = \frac{1}{2\pi i} \oint_{C_1} \frac{f(z)}{z-a} dz - \frac{1}{2\pi i} \oint_{C_2} \frac{f(z)}{z-a} dz.$$

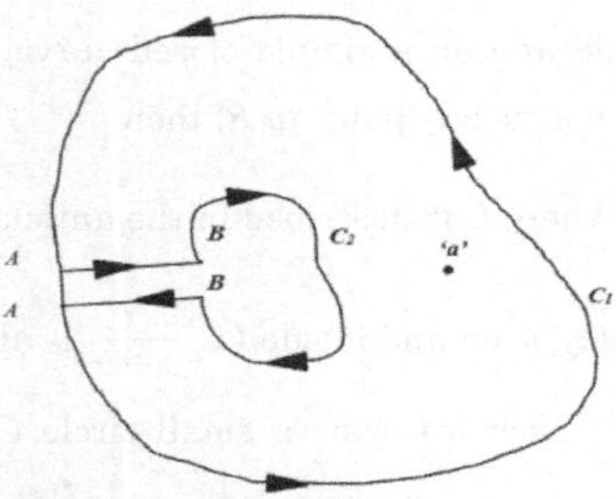

Fig. 6

Proof: We convert the doubly connected region to a simply connected region by introducing the cross-cut AB. Here 'a' lies in this region (Fig. 6).

By Cauchy's integral formula, $f(a) = \frac{1}{2\pi i} \oint_C \frac{f(z)}{z-a} dz$, where C includes C_1 described in the anticlockwise sense, C_2 in the clockwise sense, AB and BA.

$$f(a) = \frac{1}{2\pi i}\left[\oint_{C_1} \frac{f(z)}{z-a} dz + \oint_{C_2} \frac{f(z)}{z-a} dz + \int_{AB} \frac{f(z)}{z-a} dz + \int_{BA} \frac{f(z)}{z-a} dz\right]$$

$$= \frac{1}{2\pi i}\left[\oint_{C_1} \frac{f(z)}{z-a} dz + \oint_{C_2} \frac{f(z)}{z-a} dz\right] \text{ (since the last two integrals cancel each other)}$$

$$= \frac{1}{2\pi i} \oint_{C_1} \frac{f(z)}{z-a} dz - \frac{1}{2\pi i} \oint_{C_2} \frac{f(z)}{z-a} dz.$$

Cauchy's Integral Formulas for the Derivatives

By Cauchy's integral formula, we have $f(a) = \frac{1}{2\pi i}\oint_C \frac{f(z)}{z-a}dz$ (1)

Differentiating partially both sides of (1) with respect to 'a' performing the differentiation within the integration symbol in the R.H.S., we get

$$f'(a) = \frac{1!}{2\pi i}\oint_C \frac{f(z)}{z-a}dz \qquad (2)$$

Proceeding further, we get

$$f''(a) = \frac{2!}{2\pi i}\oint_C \frac{f(z)}{(z-a)^2}dz \quad \text{etc.}$$

$$f^{(n)}(a) = \frac{n\,!}{2\pi i}\oint_C \frac{f(z)}{(z-a)^{n+1}}dz.$$

Example 1: If C is the unit circle $|z| = 1$, show that $\int_C f(z)dz = 0$, where $f(z) = e^z \sin z$.

Solution : Since e^z and $\sin z$ are analytic, $e^z \sin z$ is analytic on and within C. Hence by Cauchy's theorem, $\int_C f(z)\, dz = 0$.

Example 2: Evaluate $\oint_C (x^2 - y^2 + 2ixy)dz$, where C is the contour $|z| = 1$.

Solution : $f(z) = x^2 - y^2 + 2ixy = (x + iy)^2 = z^2$ is analytic everywhere within and on $|z| = 1$.

Hence by Cauchy's integral theorem, $\oint_C f(z)\, dz = 0$.

$\therefore \oint_C (x^2 - y^2 + 2ixy)dz = 0$.

Example 3: Evaluate $\int_C \frac{z^2+1}{z^2-1}dz$, where C is $|z - 1| = 1$, using Cauchy's integral formula.

Solution : Since $z^2 - 1 = (z - 1)(z + 1)$, we see that $z = 1$ lies inside the given curve C and $z = -1$ lies outside C.

$$\therefore \int_C \frac{z^2+1}{z^2-1}dz = \int_C \frac{z^2+1}{(z-1)(z+1)}dz$$

$$= \int_C \frac{\left(\frac{z^2+1}{z+1}\right)}{(z-1)}dz \qquad (1)$$

Let $f(z) = \frac{z^2+1}{z+1}$. Clearly $f(z)$ is analytic inside C. Hence by Cauchy's integral formula (1) gives $\int_C \frac{f(z)}{(z-1)} dz = 2\pi i \,.\, f(1)$

$\therefore \quad \int_C \frac{z^2+1}{z^2-1} dz = 2\pi i \times \left[\frac{1+1}{1+1}\right] = 2\pi i.$

Example 4: Evaluate $\int_C \frac{7z-1}{z^2-3z-4} dz$, where C is the ellipse $x^2 + 4y^2 = 4$, using Cauchy's integral formula.

Solution : Let $\varphi(z) = \frac{7z-1}{z^2-3z-4} = \frac{7z-1}{(z-4)(z+1)}$. Since C is the ellipse $x^2 + 4y^2 = 4$, both the points $z = 4$ and $z = -1$ lie inside and on C.

$$\therefore \quad \int_C \frac{7z-1}{(z-4)(z+1)} dz = \int_{C_1} \frac{\left(\frac{7z-1}{z+1}\right)}{z-4} dz + \int_{C_2} \frac{\left(\frac{7z-1}{z-4}\right)}{z+1} dz.$$

Let $f(z) = \frac{7z-1}{z+1}$ and $g(z) = \frac{7z-1}{z-4}$.

Then by Cauchy's integral formula,

$$\int_C \varphi(z) dz = 2\pi i \,.\, f(4) + 2\pi i \,.\, g(-1)$$

$$= 2\pi i \,.\, \frac{27}{5} + 2\pi i \,.\, \frac{8}{5} = 14\pi i.$$

Example 5: Evaluate $\int_C \frac{\sin \pi z^2 + \cos \pi z^2}{(z-1)(z-2)} dz$, where C is the circle $|z| = 3$.

Solution : Let $\varphi(z) = \frac{\sin \pi z^2 + \cos \pi z^2}{(z-1)(z-2)}$. Since C is the circle $|z| = 3$, both the points $z = 1$ and $z = 2$ lie inside C.

$$\therefore \int_C \frac{\sin \pi z^2 + \cos \pi z^2}{(z-1)(z-2)} dz = \int_{C_1} \frac{\left(\frac{\sin \pi z^2 + \cos \pi z^2}{z-2}\right)}{z-1} dz + \int_{C_1} \frac{\left(\frac{\sin \pi z^2 + \cos \pi z^2}{z-1}\right)}{z-2} dz.$$

Let $f(z) = \frac{\sin \pi z^2 + \cos \pi z^2}{z-2}$ and $g(z) = \frac{\sin \pi z^2 + \cos \pi z^2}{z-1}$.

Then by Cauchy's integral formula,

$$\int_C \varphi(z) dz \quad = 2\pi i \,.\, f(1) + 2\pi i \,.\, g(2)$$

$$= -2\pi i\,(\sin \pi + \cos \pi) + 2\pi i(\sin 4\pi + \cos 4\pi) \;=\; -2\pi i(-1) + 2\pi i(1) = 4\pi i.$$

Example 6: Evaluate $\int_C \frac{4-3z}{z(z-1)(z-2)} dz$, where C is the circle $|z| = \frac{3}{2}$ using Cauchy's integral formula.

Solution : Let $\varphi(z) = \frac{4-3z}{z(z-1)(z-2)}$.

The points $z = 0$ and $z = 1$ lie inside C, but the point $z = 2$ lies outside C.

$$\therefore \quad \int_C \frac{4-3z}{z(z-1)(z-2)} dz = \int_{C_1} \frac{\left(\frac{4-3z}{(z-1)(z-2)}\right)}{z} dz + \int_{C_2} \frac{\left(\frac{4-3z}{z(z-2)}\right)}{z-1} dz.$$

Let $f(z) = \frac{4-3z}{(z-1)(z-2)}$ and $g(z) = \frac{4-3z}{z(z-2)}$.

Then by Cauchy's integral formula,

$$\int_C \varphi(z)dz = 2\pi i . f(0) + 2\pi i . g(1)$$

$$= 2\pi i .2 + 2\pi i .(-1) = 2\pi i.$$

Example 7: Evaluate $\int_C \frac{\cos \pi z}{z^2-1} dz$ where C is a rectangle with vertices at (a) $2 \pm i$, $-2 \pm i$ (b) $-i$, $2-i, 2+i, i$. **[AU, *Nov/Dec 2017*]**

Solution:

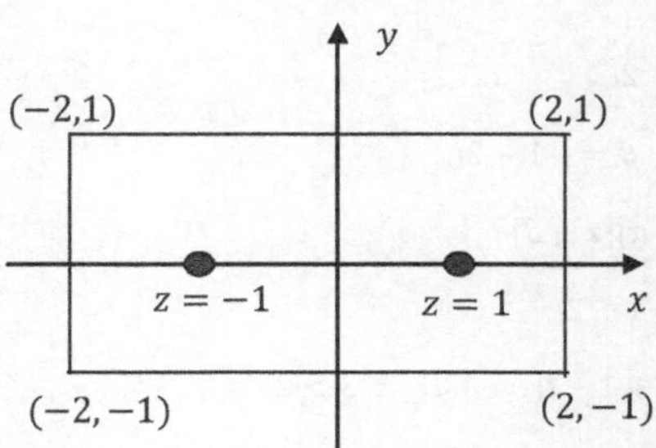

(a) C is the rectangle with vertices $(2,1)$, $(2,-1)$, $(-2,1)$ and $(-2,-1)$.

Now, $\int_C \frac{\cos \pi z}{z^2-1} dz = \int_C \frac{\cos \pi z}{(z-1)(z+1)} dz.$

Clearly, $z = 1$ and $z = -1$ lie inside the rectangle C.

$$\int_C \frac{\cos \pi z}{(z-1)(z+1)} dz = \int_{C_1} \frac{\left(\frac{\cos \pi z}{z+1}\right)}{z-1} dz + \int_{C_2} \frac{\left(\frac{\cos \pi z}{z-1}\right)}{z+1} dz$$

Let $f(z) = \frac{\cos \pi z}{z+1}$ and $g(z) = \frac{\cos \pi z}{z-1}$.

Then by Cauchy's integral formula,

$$\int_C \frac{\cos \pi z}{z^2-1} dz = 2\pi i \,.\, f(1) + 2\pi i \,.\, g(-1)$$

$$= \pi i \cos \pi - \pi i \cos(-\pi)$$

$$= -\pi i - \pi i\,(-1) = 0.$$

(b) C is the rectangle with vertices $(0,-1)$, $(2,-1)$, $(2,1)$ and $(0,1)$.

$z = 1$ lies inside C and $z = -1$ lies outside C.

We have, $\int_C \frac{\cos \pi z}{z^2-1} dz = \int_C \frac{\left(\frac{\cos \pi z}{z+1}\right)}{z-1} dz$

Then by Cauchy's integral formula, $\int_C \frac{\cos \pi z}{z^2-1} dz = 2\pi i\, f(1)$

$$= 2\pi i \times \frac{\cos \pi}{2} = -\pi i.$$

Example 8: Evaluate $\int_C \frac{z+4}{z^2+2z+5} dz$, where C is $|z+1+i| = 2$. ***[AU, Nov/Dec* 2014]**

Solution: Let $z^2 + 2z + 5 = 0$.

$$z = \frac{-2 \pm \sqrt{4-20}}{2} = \frac{-2 \pm 4i}{2} = -1 \pm 2i.$$

i.e., $z = -1+2i, -1-2i$.

Let $\alpha = -1 + 2i$ and $\beta = -1 - 2i$.

$\therefore\ z^2 + 2z + 5 = [z-\alpha][z-\beta]$.

Consider $z = \alpha = -1 + 2i$,

$|z+1+i| = |-1+2i+1+i| = |3i| = 3 > 2.$

$\therefore\ z = \alpha$ lies outside C.

Consider $z = \beta = -1 - 2i$,

$|z+1+i| = |-1-2i+1+i| = |-i| = 1 < 2.$

$\therefore\ z = \beta$ lies inside C.

$$\int_C \frac{z+4}{z^2+2z+5} dz = \int_C \frac{z+4}{[z-\alpha][z-\beta]} dz = \int_C \frac{\left(\frac{z+4}{z-\alpha}\right)}{z-\beta} dz.$$

Let $f(z) = \frac{z+4}{z-\alpha}$.

Then by Cauchy's integral formula,

$$\int_C \frac{z+4}{z^2+2z+5}\,dz = 2\,\pi i\,.f(\beta) \;=\; 2\,\pi i\,.\frac{\beta+4}{\beta-\alpha}$$

$$= 2\,\pi i\,.\frac{-1-2i+4}{-1-2i-[-1+2i]} \;=\; \frac{\pi}{2}(2i-3).$$

Example 9: Evaluate $\int_C \frac{z+1}{z^2+2z+4}\,dz$, where C is $|z+1+i| = 2$. ***[AU, Nov/Dec* 2017]**

Solution: Let $z^2 + 2z + 4 = 0$.

$$z \;=\; \frac{-2\pm\sqrt{4-16}}{2} = \frac{-2\pm 2\sqrt{3}i}{2} \;=\; -1 \pm \sqrt{3}i.$$

i.e., $z \;=\; -1+\sqrt{3}i, -1-\sqrt{3}i.$

Let $\alpha = -1+\sqrt{3}i$ and $\beta = -1-\sqrt{3}i.$

$\therefore\;\; z^2 + 2z + 4 \;=\; [z-\alpha][z-\beta].$

Consider $z \;=\; \alpha \;=\; -1+\sqrt{3}i,$

$|z+1+i| \;=\; \left|-1+\sqrt{3}i+1+i\right| \;=\; \left|(1+\sqrt{3})i\right| > 2.$

$\therefore\; z \;=\; \alpha$ lies outside C.

Consider $z \;=\; \beta \;=\; -1-\sqrt{3}i,$

$|z+1+i| \;=\; \left|-1-\sqrt{3}i+1+i\right| \;=\; \left|(1-\sqrt{3})i\right| < 2.$

$\therefore\; z \;=\; \beta$ lies inside C.

$$\int_C \frac{z+1}{z^2+2z+4}\,dz = \int_C \frac{z+1}{[z-\alpha][z-\beta]}\,dz = \int_C \frac{\left(\frac{z+1}{z-\alpha}\right)}{z-\beta}\,dz.$$

Let $f(z) = \frac{z+1}{z-\alpha}$.

Then by Cauchy's integral formula,

$$\int_C \frac{z+1}{z^2+2z+4}\,dz = 2\,\pi i\,.f(\beta) \;=\; 2\,\pi i\,.\frac{\beta+1}{\beta-\alpha}$$

$$= 2\,\pi i\,.\frac{-1-\sqrt{3}i+1}{-1-\sqrt{3}i-[-1+\sqrt{3}i]} \;=\; 2\,\pi i\,.\frac{-\sqrt{3}i}{-2\sqrt{3}i]} = \pi i.$$

Example 10: If C is the positively oriented circle $|z| = 2$, show that(i)$\int_C \frac{e^z}{(z-1)^2}\,dz = 2\pi i\, e$, (ii) $\int_C \frac{e^z}{(z-1)^3}\,dz = \pi i\, e$, (iii) $\int_C \frac{e^z}{(z-1)^4}\,dz = \frac{\pi i\, e}{3}$.

Solution : The point $z = 1$ lies inside $|z| = 2$ and also $f(z) = e^z$ is analytic.

By Cauchy's integral formula for derivatives,

$$\int_C \frac{e^z}{(z-1)^2}\, dz = 2\pi i\, f'(1) = 2\pi i\, e,$$

$$\int_C \frac{e^z}{(z-1)^3}\, dz = \frac{2\pi i}{2!} f''(1) = \pi i\, e,$$

and $$\int_C \frac{e^z}{(z-1)^4}\, dz = \frac{2\pi}{3!} i\, f'''(1) = \frac{\pi i\, e}{3}.$$

Example 11: Evaluate $\int_C \frac{z}{(z-1)^3} dz$, where C is the circle $|z| = 2$, using Cauchy's integral formula.

Solution : Since C is the circle $|z| = 2$, the point $z = 1$ lies inside C. Let $f(z) = z$.

By Cauchy's integral formula for derivatives,

$$\int_C \frac{z}{(z-1)^3} dz = \frac{2\,\pi i}{2!} f''(1)$$

$$= \pi i \times 0 \quad [\because f'(z) = 1 \text{ and } f''(z) = 0]$$

$$= 0.$$

Example 12: Evaluate $\int_C \frac{z+1}{(z-1)(z-2)^2} dz$, where C is the circle $|z-2| = \frac{1}{2}$, using Cauchy's integral formula. ***[AU, April/May 2017]***

Solution : Let $(z-1)(z-2)^2 = 0$.

i.e., $z = 1$ and $z = 2$.

The point $z = 1$ lies outside C and the point $z = 2$ lies inside C.

$$\therefore \int_C \frac{z+1}{(z-1)(z-2)^2} dz = \int_C \frac{\left(\frac{z+1}{z-1}\right)}{(z-2)^2} dz$$

Let $f(z) = \frac{z+1}{z-1}$.

Then by Cauchy's integral formula for derivatives,

$$\therefore \int_C \frac{z+1}{(z-1)(z-2)^2} dz = 2\,\pi i f'(2)$$

$$= 2\,\pi i \left\{\frac{d}{dz}\left(\frac{z+1}{z-1}\right)\right\}_{z=2}$$

$$= 2\,\pi i\left\{\frac{(z-1)-(z+1)}{(z-1)^2}\right\}_{z=2}$$

$$= 2\,\pi i\left\{\frac{-2}{(z-1)^2}\right\}_{z=2} \quad = -4\,\pi i.$$

Example 13: Evaluate $\int_C \frac{z-1}{(z+1)^2(z-2)}dz$, where C is $|z - i| = 2$, using Cauchy's integral formula. ***[AU, May/June 2014]***

Solution : Let $(z + 1)^2(z - 2) = 0$.

i.e., $z = -1$ and $z = 2$.

The point $z = -1$ lies inside C and the point $z = 2$ lies outside C.

$$\therefore \quad \int_C \frac{z-1}{(z+1)^2(z-2)}dz = \int_C \frac{\frac{z-1}{z-2}}{(z+1)^2}dz$$

Let $f(z) = \frac{z-1}{z-2}$.

Then by Cauchy's integral formula for derivatives,

$$\therefore \quad \int_C \frac{z-1}{(z+1)^2(z-2)}dz = 2\,\pi i f'(-1)$$

$$= 2\,\pi i\left\{\frac{d}{dz}\left(\frac{z-1}{z-2}\right)\right\}_{z=-1}$$

$$= 2\,\pi i\left\{\frac{-1}{(z-2)^2}\right\}_{z=-1} \quad = -\frac{2\,\pi i}{9}.$$

Example 14: Evaluate using Cauchy's integral formula: $\int_C \frac{z+1}{(z-3)(z-1)}dz$, where C is the circle $|z| = 2$. ***[AU, May/June 2016]***

Solution : We see that $z = 1$ lies inside the given curve C and $z = 3$ lies outside C.

$$\therefore \quad \int_C \frac{z+1}{(z-3)(z-1)}dz = \int_C \frac{\frac{z+1}{(z-3)}}{(z-1)}dz \qquad (1)$$

Let $f(z) = \frac{z+1}{(z-3)}$. Clearly $f(z)$ is analytic inside C.

Hence by Cauchy's integral formula, (1) gives $\int_C \frac{f(z)}{(z-1)}dz = 2\,\pi i\,.\,f(1)$

$$\therefore \quad \int_C \frac{z+1}{(z-3)(z-1)}dz = 2\,\pi i \times \left[\frac{1+1}{1-3}\right] = -2\,\pi i.$$

Example 15: Evaluate $\int_C \frac{\tan\left(\frac{z}{2}\right)}{(z-a)^2}\, dz$, where $-2 < a < 2$ and C is the boundary of the square whose sides lie along $x = \pm 2$ and $y = \pm 2$ described in the positive sense.

Solution: **[AU, Nov/Dec 2015]**

Let $f(z) = \tan\left(\frac{z}{2}\right)$. Then $f(z)$ is analytic on and inside C, since $f'(z) = \frac{1}{2}\sec^2\left(\frac{z}{2}\right)$ does not exist at $z = \pm\pi, \pm 3\pi, \cdots$, which lie outside C.

It is given that $-2 < a < 2$ and so the point $z = a$ lies inside C. Hence by Cauchy's integral formula,

$$\int_C \frac{\tan\left(\frac{z}{2}\right)}{(z-a)^2}\, dz = 2\pi i\, [f'(z)]_{z=a}$$

$$= 2\pi i \left[\frac{1}{2}\sec^2\left(\frac{z}{2}\right)\right]_{z=a}$$

$$= 2\pi i \times \frac{1}{2}\sec^2\left(\frac{a}{2}\right) = \pi i \sec^2\left(\frac{a}{2}\right).$$

Example 16: Evaluate $\int_C \frac{e^z}{(z^2+\pi^2)^2}\, dz$, where C is $|z| = 4$.

Solution : Let $(z^2 + \pi^2)^2 = 0$.

$\Rightarrow z = \pm\pi i$ (twice). Both the points lie inside C.

Now, $(z^2 + \pi^2)^2 = (z + \pi i)^2 (z - \pi i)^2$.

$$\therefore \int_C \frac{e^z}{(z+\pi i)^2(z-\pi i)^2}\, dz = \int_{C_1} \frac{\frac{e^z}{(z+\pi i)^2}}{(z-\pi i)^2}\, dz + \int_{C_2} \frac{\frac{e^z}{(z-\pi i)^2}}{(z+\pi i)^2}\, dz$$

Let $f(z) = \frac{e^z}{(z+\pi i)^2}$ and $g(z) = \frac{e^z}{(z-\pi i)^2}$

Then by Cauchy's integral formula for derivatives,

$$\int_C \frac{e^z}{(z^2+\pi^2)^2}\, dz = 2\pi i f'(\pi i) + 2\pi i g'(-\pi i)$$

$$= 2\pi i \left\{\frac{d}{dz}\left(\frac{e^z}{(z+\pi i)^2}\right)\right\}_{z=\pi i} + 2\pi i \left\{\frac{d}{dz}\left(\frac{e^z}{(z-\pi i)^2}\right)\right\}_{z=-\pi i}$$

$$= 2\pi i \left\{\frac{e^z(z+\pi i-2)}{(z+\pi i)^3}\right\}_{z=\pi i} + 2\pi i \left\{\frac{e^z(z-\pi i-2)}{(z-\pi i)^3}\right\}_{z=-\pi i}$$

$$= \left(\frac{\pi i-1}{2\pi^2}\right) + \left(\frac{\pi i+1}{2\pi^2}\right) = \frac{i}{\pi}.$$

EXERCISES

1. Evaluate $\int_C \frac{\sin 3z}{z+\frac{\pi}{2}}\,dz$ if C is the circle $|z| = 2$.
2. Find $\int_C \frac{dz}{z^3(z-1)}$ where C is $|z-1| = \frac{1}{2}$.
3. If C is the positively oriented circle $|z| = 1$ and k is a real constant, prove that $\int_C \frac{e^{kz}}{z}\,dz = 2\pi i$.
4. Evaluate $\int_C \frac{\sin \pi z^2 + \cos \pi z^2}{(z-2)(z-3)}\,dz$, where C is the circle $|z| = 4$.
5. Evaluate $\int_C \frac{e^z}{z^2+4}\,dz$, where C is the positively oriented circle $|z-i| = 2$.
6. Evaluate $\int_C \frac{\sin\left(\frac{\pi z}{4}\right)}{z^2-1}\,dz$, where C is the circle $x^2 + y^2 - 2x = 0$.
7. Evaluate $\int_C \frac{z+1}{z^3-2z^2}\,dz$, where C is the circle $|z-2-i| = 2$.
8. Evaluate $\int_C \frac{e^{iz}}{z^3}\,dz$, where C is the circle $|z| = 2$.
9. Evaluate the following :

   (i) $\int_C \frac{\sin z}{\left(z-\frac{\pi}{2}\right)^2}\,dz$, where C is the circle $|z| = 2$.

   (ii) $\int_C \frac{\tan z}{\left(z-\frac{\pi}{4}\right)^2}\,dz$, where C is the circle $|z| = 1$.
10. Evaluate $\int_C \frac{z^3+1}{(3z+1)^3}\,dz$, where C is $|z| = 1$.
11. Evaluate $\int_C \frac{e^z\,dz}{(z+2)(z+1)^2}$, where C is $|z| = 3$.
12. Evaluate $\int_C \frac{dz}{z^2(z-1)}$, where C is (i) $|z| = \frac{3}{4}$ (ii) $|z| = \frac{3}{2}$.
13. Evaluate $\int_C \frac{dz}{z^2(z^2+4)}$, where C is $|z+2i| = 3$.
14. If $f(a) = \int_C \frac{3z^2+7z+1}{z-a}\,dz$, where C is the positively oriented circle $|z| = 2$, find the values of $f'(1-i)$, $f''(1-i)$.
15. If $\int_C \frac{4z^2+z+5}{z-a}\,dz$, where C is $|z| = 2$, find the values $f(1)$, $f(i)$, $f'(-1)$ and $f''(-i)$.

ANSWERS

1. $2\pi i$ 2. $2\pi i$ 4. $-4\pi i$ 5. $\frac{\pi}{2}e^{2i}$

6. $\frac{\pi i}{\sqrt{2}}$ 7. $\frac{3\pi}{2}i$ 8. $-\pi i$ 9. (i) $2\pi i$ (ii) $4\pi i$ 10. $\frac{-4\pi i}{27}$

11. $\frac{2\pi i}{e^2}$ 12. (i) $-2\pi i$ (ii) 0 13. -2π

14. $\pi(12 + 26i)$; $12\pi i$ 15. $20\pi i$; $2\pi(i-1)$; $-14\pi i$; $16\pi i$.

Series Expansions of Functions of Complex Variable-Power Series

In this section we study a specific infinite series known as power series. These series arise in solutions of second order differential equations where coefficients are variables. Power series methods are considered efficient tools for solving various physical and engineering problems.

Definition: A series of the form $\sum_{n=0}^{\infty} c_n(z-a)^n = c_0 + c_1(z-a) + c_2(z-a)^2 + \cdots$ is called a *power series* in powers of $z-a$, where z is a complex variable , the constants $c_0, c_1, c_2, \ldots$ are called the *coefficients* and the constant a is called the *centre* of the series.

It can be proved that there exists a positive number R such that the power series given above converges for $|z-a| < R$ diverges for every $|z-a| > R$, while it may or may not converge $|z-a| = R$. This means that the power series converges at all points inside the circle $|z-a| = R$, diverges at all points outside the circle and may or may not converge on the circle. Due to this interpretation, R is called the *radius of convergence* of the above series and the circle $|z-a| = R$ is called the *circle of convergence*. In particular, the series $\sum_{n=0}^{\infty} c_n z^n$ is a power series about the origin.

Note: An important property of power series is that "every power series represents an analytic function inside its circle of convergence". Conversely, every analytic function has a power series representation called Taylor's series. Analytic functions can also be represented by another type of series, called Laurent series, which consists of positive and negative integral powers of the independent variable. These series are useful in evaluating complex integrals.

4.4. Taylor's and Laurent's Series

***Taylor's Series* (*Taylor's theorem*):** If $f(z)$ is analytic inside a circle C_0 with centre at $'a'$ and radius r_0 , then at each point z inside C_0,

$$f(z) = f(a) + \frac{(z-a)}{1!}f'(a) + \frac{(z-a)^2}{2!}f''(a) + \cdots + \frac{(z-a)^n}{n!}f^{(n)}(a) + \cdots + \infty.$$

Proof:

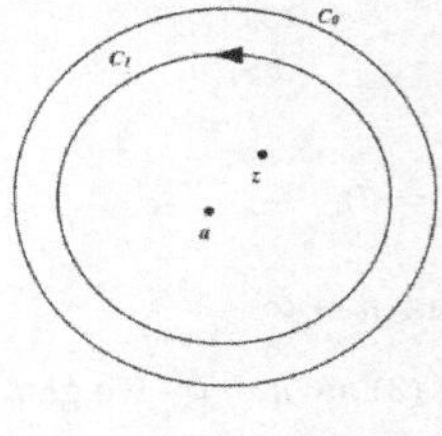

Fig. 7

Let C_1 be any circle with centre at $'a'$ and radius $r_1 < r_0$, containing the point z (Fig.7). Let w be any point on C_1. Then, by Cauchy's integral formula,

$$f(z) = \frac{1}{2\pi i}\int_{C_1} \frac{f(w)}{(w-z)}\, dw \tag{1}$$

$$\text{and } f^{(n)}(a) = \frac{n!}{2\pi i}\int_{C_1} \frac{f(w)}{(w-a)^{n+1}}\, dw \tag{2}$$

Now, $\frac{1}{w-z} = \frac{1}{(w-a)-(z-a)} = \frac{1}{w-a}\left\{1-\left(\frac{z-a}{w-a}\right)\right\}^{-1}$

$$= \frac{1}{w-a}\left[\left\{1+\left(\frac{z-a}{w-a}\right)+\left(\frac{z-a}{w-a}\right)^2+\cdots+\left(\frac{z-a}{w-a}\right)^{n-1}\right\}+\left(\frac{z-a}{w-a}\right)^n \frac{1}{1-\left(\frac{z-a}{w-a}\right)}\right]$$

$$since\ 1+\alpha+\alpha^2+\cdots+\alpha^{n-1} = \frac{1-\alpha^n}{1-\alpha}\ and\ so\ 1+\alpha+\alpha^2+\cdots+\alpha^{n-1}+\frac{\alpha^n}{1-\alpha} = \frac{1}{1-\alpha}$$

$$\frac{1}{w-z} = \frac{1}{(w-a)}+\frac{z-a}{(w-a)^2}+\frac{(z-a)^2}{(w-a)^3}+\cdots+\frac{(z-a)^{n-1}}{(w-a)^n}+\frac{(z-a)^n}{(w-a)^n}\frac{1}{w-z}$$

$$\frac{1}{2\pi i}\int_{C_1}\frac{f(w)}{(w-z)}\,dw = \frac{1}{2\pi i}\int_{C_1}\frac{f(w)}{(w-a)}\,dw + (z-a)\frac{1}{2\pi i}\int_{C_1}\frac{f(w)}{(w-a)^2}\,dw$$

$$+\cdots+(z-a)^{n-1}\frac{1}{2\pi i}\int_{C_1}\frac{f(w)}{(w-a)^n}\,dw + R_n,$$

where $R_n = \frac{1}{2\pi i}\int_{C_1}\left(\frac{z-a}{w-a}\right)^n \frac{f(w)}{(w-z)}\,dz$.

From (1) and (2), $f(z) = f(a) + \frac{(z-a)}{1!}f'(a) + \frac{(z-a)^2}{2!}f''(a) + \cdots + \frac{(z-a)^{n-1}}{(n-1)!}f^{(n-1)}(a) + R_n$ (3)

Let $|z-a| = r$ and $|w-a| = r_1$.

$\therefore\ |w-z| = |(w-a)-(z-a)| \geq r_1 - r$

Let M be the maximum value of $|f(w)|$ on C_1.

Then $|R_n| \leq \frac{r^n}{2\pi}\int_0^{2\pi}\frac{M}{r_1-r}\,\frac{r_1\,d\theta}{{r_1}^n}$ $\{w = r_1 e^{i\theta}\}$

$$= \frac{r_1 M}{r_1 - r}\left(\frac{r}{r_1}\right)^n$$

Since $\frac{r}{r_1} < 1$, $\lim_{n\to\infty}\left(\frac{r}{r_1}\right)^n = 0.$

$\therefore\ |R_n| \to \infty$ and so $R_n \to \infty$ as $n \to \infty$.

Taking limits on both sides of (3) as $n \to \infty$, we get

$$f(z) = f(a) + \frac{(z-a)}{1\,!}f'(a) + \frac{(z-a)^2}{2\,!}f''(a) + \cdots + \frac{(z-a)^n}{n\,!}f^{(n)}(a) + \cdots \infty.$$

Note: Putting $a = 0$ in Taylor's series, we get

$$f(z) = f(0) + \frac{z}{1\,!}f'(0) + \frac{z^2}{2\,!}f''(0) + \cdots + \frac{z^n}{n\,!}f^{(n)}(0) + \cdots \infty.$$

This series is called *Maclurin's series.*

For the Taylor's series expansion of a function $f(z)$ about a point, we require $f(z)$ to be analytic at that point. When $f(z)$ fails to be analytic at a point and if it is analytic in a neighbourhood of that point, then $f(z)$ can be expanded in the form of a series called *Laurent series*.

Laurent's Series (*Laurent's Theorem*): If C_1 and C_2 are two concentric circles with centre at 'a' and radii r_1 and r_2 $(r_1 > r_2)$ and if $f(z)$ is analytic on C_1 and C_2 and throughout the annular region R between them, then at each point z in R,

$$f(z) = \sum_{n=0}^{\infty} a_n (z-a)^n + \sum_{n=1}^{\infty} b_n (z-a)^{-n},$$

where $\quad a_n = \frac{1}{2\pi i}\oint_{C_1}\frac{f(w)}{(w-a)^{n+1}}\,dw,\ n = 0, 1, 2, \cdots$

and $\quad b_n = \frac{1}{2\pi i}\oint_{C_2}\frac{f(w)}{(w-a)^{-n+1}}\,dw,\ n = 1, 2, \cdots$

Proof:

By extension of Cauchy's integral formula to a doubly connected region,

$$f(z) = \frac{1}{2\pi i}\oint_{C_1}\frac{f(w)}{(w-z)}\,dw - \frac{1}{2\pi i}\oint_{C_2}\frac{f(w)}{(w-z)}\,dw = I_1 + I_2, \text{ say} \qquad (1)$$

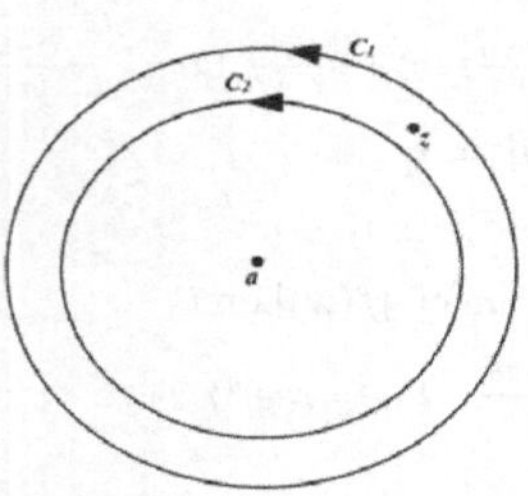

In I_1, $\frac{1}{w-z} = \frac{1}{(w-a)-(z-a)}$

$$= \frac{1}{(w-a)}\left\{1-\left(\frac{z-a}{w-a}\right)\right\}^{-1}$$

$$= \frac{1}{(w-a)} + \frac{z-a}{(w-a)^2} + \frac{(z-a)^2}{(w-a)^3} + \cdots + \frac{(z-a)^{n-1}}{(w-a)^n} + \frac{(z-a)^n}{(w-a)^n}\frac{1}{w-z} \quad (2)$$

Since $|z-a| < |w-a|$ when w is on C_1.

In I_2, $-\frac{1}{w-z} = \frac{1}{(z-a)-(w-a)}$

$$= \frac{1}{(z-a)}\left\{1-\left(\frac{w-a}{z-a}\right)\right\}^{-1}$$

$$= \frac{1}{(z-a)} + \frac{w-a}{(z-a)^2} + \frac{(w-a)^2}{(z-a)^3} + \cdots + \frac{(w-a)^{n-1}}{(z-a)^n} + \frac{(w-a)^n}{(z-a)^n}\frac{1}{z-w}, \quad (3)$$

Since $|w-a| < |z-a|$ when w is on C_2.

Using the expansion (2) in I_1, we have

$$I_1 = \frac{1}{2\pi i}\oint_{C_1}\frac{f(w)}{(w-a)}\,dw + \frac{z-a}{2\pi i}\oint_{C_1}\frac{f(w)}{(w-a)^2}\,dw + \cdots + \frac{(z-a)^{n-1}}{2\pi i}\oint_{C_1}\frac{f(w)}{(w-a)^n}\,dw + R_n$$

$$= a_0 + a_1(z-a) + \cdots + a_{n-1}(z-a)^{n-1} + R_n,$$

where $R_n = \frac{(z-a)^n}{2\pi i}\oint_{C_1}\frac{f(w)}{(w-a)^n(w-z)}\,dw \quad (4)$

Using the expansion (3) in I_2, we have

$$I_2 = \frac{1}{2\pi i}\oint_{C_2}\frac{f(w)}{(z-a)}\,dw + \frac{1}{2\pi i}\oint_{C_2}\frac{(w-a)f(w)}{(z-a)^2}\,dw + \cdots + \frac{1}{2\pi i}\oint_{C_2}\frac{(w-a)^{n-1}f(w)}{(z-a)^n}\,dw + S_n$$

$$= b_1(z-a)^{-1} + b_2(z-a)^{-2} + \cdots + b_n(z-a)^{-n} + S_n,$$

where $S_n = \frac{1}{2\pi i(z-a)^n}\oint_{C_2}\frac{(w-a)^n f(w)}{(z-w)}\,dw \quad (5)$

If $|z-a| = r$, then $r_2 < r < r_1$.

$R_n \to 0$ as $n \to \infty$, as in Taylor's theorem.

Let M be the maximum value of $|f(w)|$ on C_2 .

Then $|S_n| \leq \frac{1}{2\pi r^n}\int_0^{2\pi}\frac{r_2^n n}{r-r_2} r_2 d\theta$ (since on C_2, $w = r_2 e^{i\theta}$)

$$= \frac{Mr_2}{r-r_2}\left(\frac{r_2}{r}\right)^n$$

$\therefore |S_n| \to 0$ *and so* $S_n \to 0$ *as* $n \to \infty$, since $r_2 < r$.

Using (4) and (5) in (1) and taking limits as $n \to \infty$, we have

$f(z) = \sum_{n=0}^{\infty} a_n (z-a)^n + \sum_{n=1}^{\infty} b_n (z-a)^{-n}$.

Note

(i) If $f(z)$ is analytic inside C_2, then the Laurent's series reduces to the Taylor series of $f(z)$ with centre 'a', since in this case all the coefficients of negative powers in Laurent's series are zero.

(ii) The part $\sum_{n=0}^{\infty} a_n (z-a)^n$, consisting of positive integral powers of $(z-a)$ is called the *analytic part* of the Laurent's series, while $\sum_{n=1}^{\infty} b_n (z-a)^{-n}$, consisting of negative integral powers of $(z-a)$ is called the *principal part* of the Laurent's series.

Example 1: Find the Taylor's series expansion of $f(z) = \cos z$ about $z = \frac{\pi}{2}$ and give the region of convergence.

Solution:

The Taylor's series for $f(z)$ about $z = \frac{\pi}{2}$ is

$$f(z) = f\left(\frac{\pi}{2}\right) + \frac{z-\frac{\pi}{2}}{1\,!} f'\left(\frac{\pi}{2}\right) + \frac{\left(z-\frac{\pi}{2}\right)^2}{2\,!} f''\left(\frac{\pi}{2}\right) + \cdots$$

$$f(z) = \cos z, \quad f\left(\frac{\pi}{2}\right) = 0,$$

$$f'(z) = -\sin z, \quad f'\left(\frac{\pi}{2}\right) = -1,$$

$$f''(z) = -\cos z, \quad f''\left(\frac{\pi}{2}\right) = 0,$$

$$f'''(z) = \sin z, \quad f'''\left(\frac{\pi}{2}\right) = 1,$$

$$f^{(4)}(z) = \cos z, \quad f^{(4)}\left(\frac{\pi}{2}\right) = 0,$$

$$f^{(5)}(z) = -\sin z, \quad f^{(5)}\left(\frac{\pi}{2}\right) = -1, \text{ etc.}$$

Hence the Taylor's series of $f(z)$ at $z = \frac{\pi}{2}$ is given by

$$f(z) = \left(z - \frac{\pi}{2}\right)(-1) + \frac{\left(z - \frac{\pi}{2}\right)^2}{2\,!} \times 0 + \frac{\left(z - \frac{\pi}{2}\right)^3}{3\,!} \times 1 + \frac{\left(z - \frac{\pi}{2}\right)^5}{5\,!} \times (-1) + \cdots$$

$$\therefore \cos z = -\left(z - \frac{\pi}{2}\right) + \frac{\left(z - \frac{\pi}{2}\right)^3}{3\,!} - \frac{\left(z - \frac{\pi}{2}\right)^5}{5\,!} + \cdots$$

The region of convergence is the circle $\left|z - \frac{\pi}{2}\right| < \infty$.

Example 2: Find the Maclaurin's series for (i) e^z (ii) $\sin z$ (iii) $\log(1 + z)$.

Solution:

(i) $f(z) = e^z$.

$\therefore f^{(n)}(z) = e^z$. Hence $f^{(n)}(0) = e^0 = 1$ for all n.

$$\therefore e^z = f(0) + \frac{f'(0)\,z}{1\,!} + \frac{f''(0)\,z^2}{2\,!} + \cdots = 1 + \frac{z}{1\,!} + \frac{z^2}{2\,!} + \cdots + \frac{z^n}{n\,!} + \cdots$$

This expansion is valid for all z in the complex plane, that is $|z| < \infty$.

(ii) $f(z) = \sin z, \quad f(0) = 0,$

$f'(z) = \cos z, \quad f'(0) = 1,$

$f''(z) = -\sin z, \ f''(0) = 0,$

$f'''(z) = -\cos z, f'''(0) = -1,$

$f^{(4)}(z) = \sin z, \ f^{(4)}(0) = 0,$

$f^{(5)}(z) = \cos z, \ f^{(5)}(0) = 1,$

$\vdots$

$f^{(2n)}(0) = 0, \quad f^{(2n+1)}(0) = (-1)^n.$

$$f(z) = f(0) + \frac{f'(0)\,z}{1\,!} + \frac{f''(0)\,z^2}{2\,!} + \cdots$$

$$\therefore \sin z = z - \frac{z^3}{3\,!} + \frac{z^5}{5\,!} - \cdots$$

(iii) $f(z) = \log(1 + z), \quad f(0) = 0,$

$f'(z) = \frac{1}{1+z}, \quad f'(0) = 1,$

$f''(z) = -\frac{1}{(1+z)^2}, \quad f''(0) = -1,$

$f'''(z) = \frac{2}{(1+z)^3}, \qquad f'''(0) = 2,$

$f^{(4)}(z) = -\frac{6}{(1+z)^4}, \qquad f^{(4)}(0) = -6$, etc.

$$\therefore \log(1+z) = \frac{1}{1!}z - \frac{1}{2!}z^2 + \frac{2}{3!}z^3 - \frac{6}{4!}z^4 + \cdots$$

$$= z - \frac{z^2}{2} + \frac{z^3}{3} - \frac{z^4}{4} + \cdots$$

The expansion is valid if $|z| < 1$.

Example 3: Expand $f(z) = \frac{z+1}{(z-3)(z-4)}$ as a Taylor's series about the point $z = 2$.

Solution : Given $f(z) = \frac{z+1}{(z-3)(z-4)}$.

By partial fraction, $f(z) = \frac{A}{z-3} + \frac{B}{z-4} = \frac{A(z-4)+B(z-3)}{(z-3)(z-4)}$

$\therefore \quad z + 1 = A(z-4) + B(z-3)$ (1)

Putting $z = 4$ and $z = 3$ in (1), we get $B = 5$ and $A = -4$.

$$\therefore f(z) = \frac{-4}{z-3} + \frac{5}{z-4}$$

Let $u = z - 2$ i.e., $z = u + 2$

$$\therefore f(z) = \frac{-4}{u-1} + \frac{5}{u-2} = \frac{-4}{-(1-u)} + \frac{5}{-2\left(1-\frac{u}{2}\right)}$$

$$= 4(1-u)^{-1} - \frac{5}{2}\left(1 - \frac{u}{2}\right)^{-1}$$

$$= 4[1 + u + u^2 + u^3 + \cdots] - \frac{5}{2}\left[1 + \frac{u}{2} + \left(\frac{u}{2}\right)^2 + \left(\frac{u}{2}\right)^3 + \cdots\right]$$

$$\therefore f(z) = 4[1 + (z-2) + (z-2)^2 + \cdots] - \frac{5}{2}\left[1 + \frac{z-2}{2} + \left(\frac{z-2}{2}\right)^2 + \cdots\right].$$

Example 4: Find the Taylor's and Laurent's series which represent the function $\frac{z^2-1}{(z+2)(z+3)}$ when (i) $|z| < 2$ (ii) $2 < |z| < 3$ (iii) $|z| > 3$.

Solution: ***[AU, Nov/Dec 2017, April/May 2015, May/June 2014]***

Let $f(z) = \frac{z^2-1}{(z+2)(z+3)}$.

By partial fraction,

$f(z) = A + \frac{B}{z+2} + \frac{C}{z+3}$ [∵Nr& Dr of $f(z)$ has the same power.]

$= \frac{A(z+2)(z+3)+B(z+3)+C(z+2)}{(z+2)(z+3)}$

$\therefore z^2 - 1 = A(z+2)(z+3) + B(z+3) + C(z+2)$ (1)

Putting $z = -2$ and $z = -3$ in (1), we get $B = 3$ and $C = -8$.

Equating coefficient of z^2 in (1), we get $A = 1$.

$\therefore f(z) = 1 + \frac{3}{z+2} - \frac{8}{z+3}$.

(i) $|z| < 2$

$\Rightarrow \left|\frac{z}{2}\right| < 1$ and $\left|\frac{z}{3}\right| < \frac{2}{3} < 1$

$$\therefore f(z) = 1 + \frac{3}{2\left(1+\frac{z}{2}\right)} - \frac{8}{3\left(1+\frac{z}{3}\right)}$$

$$= 1 + \frac{3}{2}\left(1+\frac{z}{2}\right)^{-1} - \frac{8}{3}\left(1+\frac{z}{3}\right)^{-1}$$

$$= 1 + \frac{3}{2}\left[1 - \frac{z}{2} + \left(\frac{z}{2}\right)^2 - \left(\frac{z}{2}\right)^3 + \cdots\right] - \frac{8}{3}\left[1 - \frac{z}{3} + \left(\frac{z}{3}\right)^2 - \left(\frac{z}{3}\right)^3 + \cdots\right]$$

$$= 1 + \frac{3}{2}\left[1 - \frac{z}{2} + \frac{z^2}{4} - \cdots\right] - \frac{8}{3}\left[1 - \frac{z}{3} + \frac{z^2}{9} - \cdots\right]$$

$$= \left(1 + \frac{3}{2} - \frac{8}{3}\right) + \left(-\frac{3}{4} + \frac{8}{9}\right)z + \left(\frac{3}{8} - \frac{8}{27}\right)z^2 + \left(-\frac{3}{16} - \frac{8}{3\times 27}\right)z^3 + \cdots$$

$$= \frac{-1}{6} + \left(-\frac{3}{2^2} + \frac{8}{3^2}\right)z + \left(\frac{3}{2^3} - \frac{8}{3^3}\right)z^2 + \left(-\frac{3}{2^4} + \frac{8}{3^4}\right)z^3 + \cdots$$

$$\therefore f(z) = \frac{-1}{6} + \sum_{n=1}^{\infty}(-1)^{n+1}\left(\frac{8}{3^{n+1}} - \frac{3}{2^{n+1}}\right)z^n.$$

(ii) $2 < |z| < 3$

$\Rightarrow \left|\frac{2}{z}\right| < 1$ and $\left|\frac{z}{3}\right| < 1$

$$\therefore f(z) = 1 + \frac{3}{z\left(1+\frac{2}{z}\right)} - \frac{8}{3\left(1+\frac{z}{3}\right)}$$

$$= 1 + \frac{3}{z}\left(1+\frac{2}{z}\right)^{-1} - \frac{8}{3}\left(1+\frac{z}{3}\right)^{-1}$$

$$\therefore f(z) = 1 + \frac{3}{z}\left[1 - \frac{2}{z} + \left(\frac{2}{z}\right)^2 - \left(\frac{2}{z}\right)^3 + \cdots\right] - \frac{8}{3}\left[1 - \frac{z}{3} + \left(\frac{z}{3}\right)^2 - \left(\frac{z}{3}\right)^3 + \cdots\right]$$

(iii) $|z| > 3$

$$\Rightarrow \left|\frac{z}{3}\right| > 1 \text{ i.e., } \left|\frac{3}{z}\right| < 1 \text{ and } \left|\frac{2}{z}\right| < \frac{2}{3} < 1$$

$$\therefore f(z) = 1 + \frac{3}{z\left(1+\frac{2}{z}\right)} - \frac{8}{z\left(1+\frac{3}{z}\right)}$$

$$= 1 + \frac{3}{z}\left(1 + \frac{2}{z}\right)^{-1} - \frac{8}{z}\left(1 + \frac{3}{z}\right)^{-1}$$

$$\therefore f(z) = 1 + \frac{3}{z}\left[1 - \frac{2}{z} + \left(\frac{2}{z}\right)^2 - \left(\frac{2}{z}\right)^3 + \cdots\right] - \frac{8}{z}\left[1 - \frac{3}{z} + \left(\frac{3}{z}\right)^2 - \left(\frac{3}{z}\right)^3 + \cdots\right]$$

$$= 1 + \frac{3}{z} - \frac{6}{z^2} + \frac{12}{z^3} - \frac{24}{z^4} - \cdots - \frac{8}{z} + \frac{24}{z^2} - \frac{72}{z^3} + \frac{216}{z^4} - \cdots$$

$$\therefore f(z) = 1 - \frac{5}{z} + \frac{18}{z^2} - \frac{60}{z^3} + \frac{192}{z^4} - \cdots$$

Example 5: Expand $f(z) = \frac{1}{z^2-3z+2}$ in a Laurent's series valid for (i) $|z| > 2$ (ii) $|z| < 1$ (iii) $1 < |z| < 2$ (iv) $0 < |z-1| < 1$. ***[AU, Nov/Dec 2017, Nov/Dec 2014]***

Solution : Given $f(z) = \frac{1}{z^2-3z+2} = \frac{1}{(z-1)(z-2)}$

By partial fraction, $f(z) = \frac{A}{z-1} + \frac{B}{z-2} = \frac{A(z-2)+B(z-1)}{(z-1)(z-2)}$

$$\therefore 1 = A(z-2) + B(z-1) \qquad (1)$$

Putting $z = 1$ and $z = 2$ in (1), we get $A = -1$ and $B = 1$.

$$\therefore f(z) = \frac{-1}{z-1} + \frac{1}{z-2}$$

(i) $|z| > 2$

$$\Rightarrow \left|\frac{z}{2}\right| > 1 \text{ i.e., } \left|\frac{2}{z}\right| < 1 \text{ and } \left|\frac{1}{z}\right| < \frac{1}{2} < 1$$

$$\therefore f(z) = \frac{-1}{z\left(1-\frac{1}{z}\right)} + \frac{1}{z\left(1-\frac{2}{z}\right)}$$

$$= \frac{-1}{z}\left(1 - \frac{1}{z}\right)^{-1} + \frac{1}{z}\left(1 - \frac{2}{z}\right)^{-1}$$

$$= \frac{-1}{z}\left[1+\frac{1}{z}+\left(\frac{1}{z}\right)^2+\left(\frac{1}{z}\right)^3+\cdots\right]+\frac{1}{z}\left[1+\frac{2}{z}+\left(\frac{2}{z}\right)^2+\left(\frac{2}{z}\right)^3+\cdots\right].$$

$$= \frac{-1}{z}-\frac{1}{z^2}-\frac{1}{z^3}-\frac{1}{z^4}-\cdots+\frac{1}{z}+\frac{2}{z^2}+\frac{4}{z^3}+\frac{8}{z^4}+\cdots$$

$$\therefore f(z) = \frac{1}{z^2}+\frac{3}{z^3}+\frac{7}{z^4}+\cdots$$

(ii) $|z| < 1$

$\Rightarrow \left|\frac{z}{2}\right| < 1$

$$\therefore f(z) = \frac{1}{1-z}+\frac{1}{-2\left(1-\frac{z}{2}\right)}$$

$$= (1-z)^{-1}-\frac{1}{2}\left(1-\frac{z}{2}\right)^{-1}$$

$$\therefore f(z) = [1+z+z^2+z^3+\cdots]-\frac{1}{2}\left[1+\frac{z}{2}+\left(\frac{z}{2}\right)^2+\left(\frac{z}{2}\right)^3+\cdots\right]$$

$$= \frac{1}{2}+\frac{3}{4}z+\frac{7}{8}z^2+\frac{15}{16}z^3+\cdots$$

(iii) $1 < |z| < 2$

$\Rightarrow \left|\frac{1}{z}\right| < 1$ and $\left|\frac{z}{2}\right| < 1$

$$\therefore f(z) = \frac{-1}{z\left(1-\frac{1}{z}\right)}+\frac{1}{-2\left(1-\frac{z}{2}\right)}$$

$$= \frac{-1}{z}\left(1-\frac{1}{z}\right)^{-1}-\frac{1}{2}\left(1-\frac{z}{2}\right)^{-1}$$

$$\therefore f(z) = \frac{-1}{z}\left[1+\frac{1}{z}+\left(\frac{1}{z}\right)^2+\left(\frac{1}{z}\right)^3+\cdots\right]-\frac{1}{2}\left[1+\frac{z}{2}+\left(\frac{z}{2}\right)^2+\left(\frac{z}{2}\right)^3+\cdots\right]$$

(iv) $0 < |z-1| < 1$

Let $u = z-1$ i.e., $z = u+1$.

$\therefore 0 < |u| < 1$

We have,

$$f(z) = \frac{-1}{z-1}+\frac{1}{z-2}$$

$$= \frac{-1}{u}+\frac{1}{u-1} = \frac{-1}{u}+\frac{-1}{1-u} = \frac{-1}{u}-(1-u)^{-1}$$

$$= \frac{-1}{u}-[1+u+u^2+u^3+\cdots] = \frac{-1}{u}-1-u-u^2-u^3-\cdots$$

$$\therefore f(z) = \frac{-1}{z-1}-1-(z-1)-(z-1)^2-(z-1)^3-\cdots$$

Example 6: Find the Laurent's series expansion of $f(z) = \frac{1}{z^2+4z+3}$, valid in the regions $|z| < 1$ and $0 < |z+1| < 2$. **[*AU, April/May 2017*]**

Solution : Given $f(z) = \frac{1}{z^2+4z+3} = \frac{1}{(z+1)(z+3)}$

By partial fraction, $f(z) = \frac{A}{z+1} + \frac{B}{z+3} = \frac{A(z+3)+B(z+1)}{(z+1)(z+3)}$

$\therefore\ 1 = A(z+3) + B(z+1)$ (1)

Putting $z = -1$ and $z = -3$ in (1), we get $A = \frac{1}{2}$ and $B = \frac{-1}{2}$.

$$\therefore f(z) = \frac{\frac{1}{2}}{z+1} + \frac{\frac{-1}{2}}{z+3} = \frac{1}{2(z+1)} - \frac{1}{2(z+3)}$$

(i) $|z| < 1$

$\Rightarrow\ \left|\frac{z}{3}\right| < 1$

$$\therefore f(z) = \frac{1}{2(z+1)} - \frac{1}{6\left(1+\frac{z}{3}\right)}$$

$$= \frac{1}{2}(1+z)^{-1} - \frac{1}{6}\left(1+\frac{z}{3}\right)^{-1}$$

$$\therefore f(z) = \frac{1}{2}[1 - z + z^2 - z^3 + \cdots] - \frac{1}{6}\left[1 - \frac{z}{3} + \left(\frac{z}{3}\right)^2 - \left(\frac{z}{3}\right)^3 + \cdots\right]$$

$$= \frac{1}{3} - \frac{4}{9}z + \frac{13}{27}z^2 - \frac{40}{81}z^3 + \cdots$$

(ii) $0 < |z+1| < 2$

Let $u = z+1$ i.e., $z = u - 1$.

$\therefore\ 0 < |u| < 2 \Rightarrow\ \left|\frac{u}{2}\right| < 1$

We have,

$$f(z) = \frac{1}{2(z+1)} - \frac{1}{2(z+3)}$$

$$= \frac{1}{2u} - \frac{1}{2(u+2)} = \frac{1}{2u} - \frac{1}{4\left(1+\frac{u}{2}\right)} = \frac{1}{2u} - \frac{1}{4}\left(1+\frac{u}{2}\right)^{-1}$$

$$= \frac{1}{2u} - \frac{1}{4}\left[1 + \frac{u}{2} + \left(\frac{u}{2}\right)^2 + \left(\frac{u}{2}\right)^3 + \cdots\right]$$

$$\therefore f(z) = \frac{1}{2(z+1)} - \frac{1}{4}\left[1 + \frac{z+1}{2} + \left(\frac{z+1}{2}\right)^2 + \left(\frac{z+1}{2}\right)^3 + \cdots\right].$$

Example 7: Expand Laurent's series of $f(z) = \frac{z}{z^2-3z+2}$, in the region (i) $|z| < 1$ (ii) $1 < |z| < 2$ (iii) $|z| > 2$ (iv) $|z-1| < 1$. ***[AU, May/June 2016, Nov/Dec 2015]***

Solution:

Given $f(z) = \frac{z}{z^2-3z+2} = \frac{z}{(z-1)(z-2)}$

By partial fraction, $f(z) = \frac{A}{z-1} + \frac{B}{z-2} = \frac{A(z-2)+B(z-1)}{(z-1)(z-2)}$

$\therefore\ z = A(z-2) + B(z-1)$ (1)

Putting $z = 1$ and $z = 2$ in (1), we get $A = -1$ and $B = 2$.

$$\therefore f(z) = \frac{-1}{z-1} + \frac{2}{z-2}$$

(i) $|z| < 1$

$\Rightarrow\ \left|\frac{z}{2}\right| < 1$

$$\therefore f(z) = \frac{1}{(1-z)} + \frac{2}{-2\left(1-\frac{z}{2}\right)}$$

$$= (1-z)^{-1} - \left(1-\frac{z}{2}\right)^{-1}$$

$$\therefore f(z) = [1 + z + z^2 + z^3 + \cdots] - \left[1 + \frac{z}{2} + \left(\frac{z}{2}\right)^2 + \left(\frac{z}{2}\right)^3 + \cdots\right]$$

$$= \frac{z}{2} + \frac{3z^2}{4} + \frac{7z^3}{8} + \cdots$$

(ii) $1 < |z| < 2$

$\Rightarrow \left|\frac{1}{z}\right| < 1$ and $\left|\frac{z}{2}\right| < 1$

$$\therefore f(z) = \frac{-1}{z\left(1-\frac{1}{z}\right)} + \frac{2}{-2\left(1-\frac{z}{2}\right)}$$

$$= \frac{-1}{z}\left(1-\frac{1}{z}\right)^{-1} - \left(1-\frac{z}{2}\right)^{-1}$$

$$\therefore f(z) = \frac{-1}{z}\left[1 + \frac{1}{z} + \left(\frac{1}{z}\right)^2 + \left(\frac{1}{z}\right)^3 + \cdots\right] - \left[1 + \frac{z}{2} + \left(\frac{z}{2}\right)^2 + \left(\frac{z}{2}\right)^3 + \cdots\right].$$

(iii) $|z| > 2$

$\Rightarrow \left|\frac{z}{2}\right| > 1$ i.e., $\left|\frac{2}{z}\right| < 1$ and $\left|\frac{1}{z}\right| < \frac{1}{2} < 1$

$$\therefore f(z) = \frac{-1}{z\left(1-\frac{1}{z}\right)} + \frac{2}{z\left(1-\frac{2}{z}\right)} = \frac{-1}{z}\left(1-\frac{1}{z}\right)^{-1} + \frac{2}{z}\left(1-\frac{2}{z}\right)^{-1}$$

$$\therefore f(z) = \frac{-1}{z}\left[1+\frac{1}{z}+\left(\frac{1}{z}\right)^2+\left(\frac{1}{z}\right)^3+\cdots\right]+\frac{2}{z}\left[1+\frac{2}{z}+\left(\frac{2}{z}\right)^2+\left(\frac{2}{z}\right)^3+\cdots\right]$$

$$= \frac{1}{z}\left[1+\frac{3}{z}+\frac{7}{z^2}+\frac{15}{z^3}+\cdots\right]$$

(iv) $|z-1| < 1$

Let $u = z-1$ i.e., $z = u+1$.

$\therefore |u| < 1$

We have,

$$f(z) = \frac{-1}{z-1} + \frac{2}{z-2}$$

$$= \frac{-1}{u} + \frac{2}{u-1} = \frac{-1}{u} + \frac{-2}{1-u}$$

$$= \frac{-1}{u} - 2(1-u)^{-1} = \frac{-1}{u} - 2[1+u+u^2+u^3+\cdots]$$

$$\therefore f(z) = \frac{-1}{z-1} - 2[1+(z-1)+(z-1)^2+(z-1)^3+\cdots].$$

Example 8: Find the Laurent's series expansion of the function $f(z) = \frac{z-1}{(z+2)(z+3)}$, valid in the region $2 < |z| < 3$.

Solution: Resolving into partial fractions,

$$f(z) = \frac{z-1}{(z+2)(z+3)} = \frac{-3}{z+2} + \frac{4}{z+3}$$

Given region is $2 < |z| < 3$.

$$\Rightarrow \left|\frac{2}{z}\right| < 1 \text{ and } \left|\frac{z}{3}\right| < 1$$

$$\text{Hence } f(z) = \frac{-3}{z\left(1+\frac{2}{z}\right)} + \frac{4}{3\left(1+\frac{z}{3}\right)}$$

$$= \frac{-3}{z}\left(1+\frac{2}{z}\right)^{-1} + \frac{4}{3}\left(1+\frac{z}{3}\right)^{-1}$$

$$= \frac{-3}{z}\left[1-\left(\frac{2}{z}\right)+\left(\frac{2}{z}\right)^2-\cdots\right]+\frac{4}{3}\left[1-\left(\frac{z}{3}\right)+\left(\frac{z}{3}\right)^2-\cdots\right]$$

$$= \frac{-3}{z}\sum_{n=0}^{\infty}(-1)^n\left(\frac{2}{z}\right)^n + \frac{4}{3}\sum_{n=0}^{\infty}(-1)^n\left(\frac{z}{3}\right)^n.$$

Example 9: Find the Laurent's series expansion of the $f(z)=\frac{z}{(z+1)(z+2)}$ in (i) $|z|<1$, (ii) $1<|z|<2$, (iii) $|z|>2$ and (iv) $|z+1|<1$.

Solution:

Resolving into partial fractions,

$$f(z)=\frac{z}{(z+1)(z+2)}=\frac{-1}{z+1}+\frac{2}{z+2}.$$

(i) $|z|<1 \Rightarrow \left|\frac{z}{2}\right|<1.$

$$f(z)=\frac{-1}{1+z}+\frac{2}{2\left(1+\frac{z}{2}\right)}\left[\because \left|\frac{z}{2}\right|<\frac{1}{2}<1\right]$$

$$=-1(1+z)^{-1}+\left(1+\frac{z}{2}\right)^{-1}$$

$$=-1[1-z+z^2-z^3+\cdots]+\left[1-\left(\frac{z}{2}\right)+\left(\frac{z}{2}\right)^2-\left(\frac{z}{2}\right)^3+\cdots\right]$$

$$=\frac{z}{2}-\frac{3}{4}z^2+\frac{7}{8}z^3-\cdots$$

(ii) $1<|z|<2$

$$\Rightarrow \left|\frac{1}{z}\right|<1 \text{ and } \left|\frac{z}{2}\right|<1$$

$$\therefore\ f(z)=\frac{-1}{z\left(1+\frac{1}{z}\right)}+\frac{2}{2\left(1+\frac{z}{2}\right)}$$

$$=\frac{-1}{z}\left(1+\frac{1}{z}\right)^{-1}+\left(1+\frac{z}{2}\right)^{-1}$$

$$=\left[-\left(\frac{1}{z}\right)+\left(\frac{1}{z}\right)^2-\left(\frac{1}{z}\right)^3+\cdots\right]+\left[1-\left(\frac{z}{2}\right)+\left(\frac{z}{2}\right)^2-\left(\frac{z}{2}\right)^3+\cdots\right]$$

$$=\sum_{n=1}^{\infty}\frac{(-1)^n}{z^n}+\sum_{n=0}^{\infty}(-1)^n\left(\frac{z}{2}\right)^n.$$

(iii) $|z|>2$

i.e., $\left|\frac{2}{z}\right|<1$ and so $\left|\frac{1}{z}\right|<1.$

$$\therefore\ f(z)=\frac{-1}{z\left(1+\frac{1}{z}\right)}+\frac{2}{z\left(1+\frac{2}{z}\right)}$$

$$=\frac{-1}{z}\left(1+\frac{1}{z}\right)^{-1}+\frac{2}{z}\left(1+\frac{2}{z}\right)^{-1}$$

$$=\frac{-1}{z}\left[1-\left(\frac{1}{z}\right)+\left(\frac{1}{z}\right)^2-\left(\frac{1}{z}\right)^3+\cdots\right]+\frac{2}{z}\left[1-\left(\frac{2}{z}\right)+\left(\frac{2}{z}\right)^2-\left(\frac{2}{z}\right)^3\ \ldots\right]$$

$$= \sum_{n=1}^{\infty} \frac{(-1)^n}{z^n} + \sum_{n=0}^{\infty} (-1)^n \left(\frac{2}{z}\right)^{n+1}.$$

(iv) $|z+1| < 1$.

Let $u = z + 1$ i.e., $z = u - 1$.

$\therefore |u| < 1$

$$\therefore f(z) = \frac{-1}{u} + \frac{2}{u+1}$$

$$= \frac{-1}{u} + 2[1+u]^{-1} = \frac{-1}{u} + 2[1 - u + u^2 - u^3 + \cdots]$$

$$= \frac{-1}{z+1} + 2[1 - (z+1) + (z+1)^2 - (z+1)^3 + \cdots]$$

$$= \frac{-1}{z+1} + 2\sum_{n=0}^{\infty} (-1)^n (z+1)^n.$$

Example 10: Find the Laurent's series expansion of $f(z) = \frac{7z-2}{z\,(z-2)(z+1)}$ in $1 < |z+1| < 3$.

Solution : Resolving into partial fractions,

$$f(z) = \frac{7z-2}{z\,(z-2)(z+1)} = \frac{1}{z} + \frac{2}{z-2} - \frac{3}{z+1}.$$

The given region is $1 < |z+1| < 3$.

Let $u = z + 1$ i.e., $z = u - 1$.

$\therefore 1 < |u| < 3$

$$\Rightarrow \left|\frac{1}{u}\right| < 1 \text{ and } \left|\frac{u}{3}\right| < 1$$

$$\therefore f(z) = \frac{1}{u-1} + \frac{2}{u-3} - \frac{3}{u}$$

$$= \frac{1}{u\left[1-\frac{1}{u}\right]} + \frac{2}{-3\left[1-\frac{u}{3}\right]} - \frac{3}{u}$$

$$= \frac{1}{u}\left[1 - \frac{1}{u}\right]^{-1} - \frac{2}{3}\left[1 - \frac{u}{3}\right]^{-1} - \frac{3}{u}$$

$$= \frac{1}{u}\left[1 + \frac{1}{u} + \left(\frac{1}{u}\right)^2 + \cdots\right] - \frac{2}{3}\left[1 + \frac{u}{3} + \left(\frac{u}{3}\right)^2 + \cdots\right] - \frac{3}{u}$$

$$= \frac{1}{(z+1)}\left[1 + \frac{1}{z+1} + \left(\frac{1}{z+1}\right)^2 + \cdots\right] - \frac{2}{3}\left[1 + \frac{z+1}{3} + \left(\frac{z+1}{3}\right)^2 + \cdots\right] - \frac{3}{z+1}$$

$$= -\frac{2}{z+1} + \sum_{n=1}^{\infty} \frac{1}{(z+1)^{n+1}} - \frac{2}{3}\sum_{n=0}^{\infty} \left(\frac{z+1}{3}\right)^n.$$

Example 11: Expand in a Laurent's series for $f(z) = \frac{e^z}{(z-1)^2}$ about $z = 1$.

Solution : Given $f(z) = \frac{e^z}{(z-1)^2}$.

Let $u = z - 1$ i.e., $z = u + 1$.

$$\therefore f(z) = \frac{e^{u+1}}{u^2} = \frac{e^1}{u^2} e^u$$

$$= \frac{e^1}{u^2}\left[1 + \frac{u}{1!} + \frac{u^2}{2!} + \frac{u^3}{3!} + \frac{u^4}{4!} + \cdots\right]$$

$$= e\left[\frac{1}{u^2} + \frac{1}{u} + \frac{1}{2!} + \frac{u}{3!} + \frac{u^2}{4!} + \cdots\right] = e\left[\frac{1}{(z-1)^2} + \frac{1}{z-1} + \frac{1}{2!} + \frac{z-1}{3!} + \frac{(z-1)^2}{4!} + \cdots\right].$$

Example 12: Find the Laurent's series expansion of $f(z) = \frac{3z-2}{z\,(z^2-4)}$ valid in the region $2 < |z+2| < 4$.

Solution : Resolving into partial fractions,

$$f(z) = \frac{3z-2}{z\,(z^2-4)} = \frac{1}{2}\frac{1}{z} + \frac{1}{2}\frac{1}{z-2} - \frac{1}{z+2}\,.$$

The given region is $2 < |z+2| < 4$.

Let $u = z + 2$ i.e., $z = u - 2$.

$\therefore 2 < |u| < 4$

$\Rightarrow \left|\frac{2}{u}\right| < 1$ and $\left|\frac{u}{4}\right| < 1$.

$$f(z) = \frac{1}{2}\frac{1}{u-2} + \frac{1}{2}\frac{1}{u-4} - \frac{1}{u}\,.$$

$$= \frac{1}{2u\left[1-\frac{2}{u}\right]} + \frac{1}{-8\left[1-\frac{u}{4}\right]} - \frac{1}{u}$$

$$= \frac{1}{2u}\left[1-\frac{2}{u}\right]^{-1} - \frac{1}{8}\left[1-\frac{u}{4}\right]^{-1} - \frac{1}{u}$$

$$= \frac{1}{2u}\left[1+\frac{2}{u}+\left(\frac{2}{u}\right)^2+\cdots\right] - \frac{1}{8}\left[1+\frac{u}{4}+\left(\frac{u}{4}\right)^2+\cdots\right] - \frac{1}{u}$$

$$= \frac{1}{2(z+2)}\left[1+\frac{2}{z+2}+\left(\frac{2}{z+2}\right)^2+\cdots\right] - \frac{1}{8}\left[1+\frac{z+2}{4}+\left(\frac{z+2}{4}\right)^2+\cdots\right] - \frac{1}{z+2}.$$

EXERCISES

1. Expand $f(z) = z\,e^z$ as a Taylor's series about the point $z = 1$. Also determine the region of convergence.

2. Show that $f(z) = \frac{1+2z}{z^2+z^3} = \frac{1}{z^2} + \frac{1}{z} - \frac{1}{z} - 1 + z - z^2 + \cdots$

3. Show that the Laurent's expansion for $\frac{1}{2-z} + \frac{1}{z-1}$ valid in the ring $1 < |z| < 2$ is $\sum_{n=1}^{\infty} \frac{1}{2^{n+1}} z^n + \sum_{n=1}^{\infty} \frac{1}{z^n}$.

4. Expand $f(z) = \frac{z+3}{z(z^2-z-2)}$ in powers of z, where (i) $0 < |z| < 1$, (ii) $1 < |z| < 2$, (iii) $|z| > 2$.

5. Expand $\frac{1}{z(z^2-3z+2)}$ as a power series valid in $1 < |z| < 2$.

6. Expand $\frac{1}{(z-1)(z-2)}$ as a power series in z valid in (i) $|z| < 1$, (ii) $1 < |z| < 2$, (iii) $|z| > 2$.

7. Obtain Laurent's series expansion for $f(z) = \frac{1}{(z+1)(z+3)}$ valid in the following regions: (i) $|z| < 1$, (ii) $|z| > 3$, (iii) $1 < |z| < 3$, (iv) $0 < |z+1| < 2$.

8. Represent the function $f(z) = \frac{z}{(z-1)(z-3)}$ by a series of positive and negative powers of $(z-1)$ which converges to $f(z)$ when $0 < |z-1| < 2$.

9. Obtain the expansion of the function $f(z) = \frac{z-1}{z^2}$ in Taylor's series in powers of $z-1$ and give the region of validity and Laurent's series for the domain $|z-1| > 1$.

ANSWERS

1. $e\left[1 + \frac{2(z-1)}{1!} + \frac{3(z-1)^2}{2!} + \cdots\right]$

4. (i) $-\frac{3}{2z} + \frac{2}{3}\sum_{n=1}^{\infty}(-1)^n z^n - \frac{5}{12}\sum_{n=0}^{\infty}\left(\frac{z}{2}\right)^n$

(ii) $-\frac{3}{2z} + \frac{2}{3z}\sum_{n=1}^{\infty}(-1)^n \frac{1}{z^n} - \frac{5}{12}\sum_{n=0}^{\infty}\left(\frac{z}{2}\right)^n$

(iii) $-\frac{3}{2z} + \frac{2}{3}\sum_{n=1}^{\infty}(-1)^n \frac{1}{z^n} + \frac{5}{6z}\sum_{n=0}^{\infty}\left(\frac{2}{z}\right)^n$

5. $-\sum_{n=0}^{\infty}\left(\frac{1}{z^{n+2}}+\frac{z^{n-1}}{2^{n+1}}\right)$

6. (i) $\sum_{n=0}^{\infty}\left(1-\frac{1}{2^{n+1}}\right)z^n$

(ii) $-\sum_{n=0}^{\infty}\left(\frac{1}{z^{n+1}}+\frac{z^n}{2^{n+1}}\right)$

(iii) $\sum_{n=1}^{\infty}\frac{2^n-1}{z^{n+1}}$.

7. (i) $\frac{1}{3}-\frac{4}{9}z+\frac{13}{27}z^2-\frac{40}{81}z^3+\cdots$

(ii) $\frac{1}{z^2}-\frac{4}{z^3}+\frac{13}{z^4}-\frac{40}{z^5}+\cdots$

(iii) $\cdots-\frac{1}{2z^4}+\frac{1}{2z^3}-\frac{1}{2z^2}+\frac{1}{2z}-\frac{1}{6}+\frac{z}{18}-\frac{z^2}{54}+\frac{z^3}{162}-\frac{z^6}{1536}+\cdots$

(iv) $\frac{1}{2(z+1)}-\frac{1}{4}+\frac{z+1}{8}-\frac{(z+1)^2}{16}+\cdots$

8. $-\frac{1}{2(z-1)}-\frac{3}{4}\sum_{n=0}^{\infty}\left(\frac{z-1}{2}\right)^n$

9. $(z-1)-2(z-1)^2+3(z-1)^3-\cdots$ in $|z-1|<1$; $\sum_{n=1}^{\infty}(-1)^{n+1}\frac{n}{(z-1)^n}$.

4.5. Singularities

Zero of an analytic function: A *zero* of an analytic function $f(z)$ is a value of z such that $f(z) = 0$.

If $f(z)$ is analytic in the neighbourhood of $z = a$, then by Taylor's theorem $f(z) = \sum_{n=0}^{\infty} a_n (z-a)^n$ where $a_n = \frac{f^{(n)}(a)}{n!}$.

Hence $f(z) = a_0 + a_1(z-a) + a_2(z-a)^2 + \cdots + a_n(z-a)^n + \cdots\infty$. If $a_0 = a_1 = a_2 = \cdots = a_{n-1} = 0$ but $a_n \neq 0$, then $f(z)$ is said to have a *zero of order n* at $z = a$.

A zero of order one is called a *simple zero.*

Since $a_n = \frac{f^{(n)}(a)}{n!}$, for a *zero of order n* at $z = a$, $f(a) = f'(a) = \cdots = f^{(n-1)}(a) = 0$ but $f^{(n)}(a) \neq 0$. Thus in the neighbourhood of the zero at $z = a$ of order n,

$$f(z) = a_n(z-a)^n + a_{n+1}(z-a)^{n+1} + \cdots = (z-a)^n\left[a_n + a_{n+1}(z-a) + \cdots\right]$$

$= (z-a)^n \phi(z)$ where $\phi(z) = a_n + a_{n+1}(z-a) + \cdots$, $\phi(z)$ is analytic and non- zero in the neighbourhood of $z = a$.

Singularity: A *singularity* of a function $f(z)$ is a point at which the function ceases to be analytic.

i.e. A point at which $f(z)$ is not analytic is called a *singular point* or *singularity* of $f(z)$.

Isolated Singularity: Let $z = a$ be a singularity of $f(z)$. The point z = a is called an *isolated singularity* of $f(z)$, if there is no other singularity in its neighborhood. [If 'a' is not a singularity and we can find $\delta > 0$ such that $|z-a| = \delta$ encloses no singularity, then 'a' is called a regular point or ordinary point of $f(z)$.]

For example, the isolated singularities of $f(z) = \frac{z+1}{z^3(z^2+1)}$ are z = 0 and z = $\pm i$. As another example, the function $g(z) = \frac{1}{\sin\left(\frac{\pi}{2}\right)}$ is not analytic at the points where $z = \pm\frac{1}{n}$ $(n = 1, 2, 3, \cdots)$. Thus it has an infinite number of isolated singularities all lying on the real axis between $z = -1$ and $z = 1$. It is to be noted that the origin $z = 0$ is a singular point, but it is not isolated because every neighbourhood of $z = 0$ contains other singularities, $z = \pm\frac{1}{n}$ for large values of n, of the function.

Note: If a function has only a finite number of singularities in a region, then those singularities are necessarily isolated.

Types of Isolated Singularities

We can classify isolated singularities into (i) Removable singularity, (ii) Pole, (iii) Essential singularity.

(i) Removable singularity: Let $'a'$ be an isolated singularity of $f(z)$. Then $'a'$ is called a *removable singularity* of $f(z)$ if the principal part of $f(z)$ at $z = a$ in its Laurent's series expansion has no terms.

A singularity $z = a$ is a removable singularity of $f(z)$ if $\lim_{z \to a} f(z)$ exists.

For example, $f(z) = \frac{\sin z}{z}$ has removable singularity at $z = 0$ since $\lim_{z \to 0} \frac{\sin z}{z} = 1$.

The Laurent's series expansion of $f(z)$ is given by,

$$f(z) = \frac{\sin z}{z} = \frac{1}{z}\left[z - \frac{z^3}{3!} + \frac{z^5}{5!} - \cdots\right] = 1 - \frac{z^2}{3!} + \frac{z^4}{5!} - \cdots$$

It has no terms containing negative powers of z.

(ii) Pole: Let $'a'$ be an isolated singularity of $f(z)$. The point $z = a$ is called a *pole* of $f(z)$ if the principal part of $f(z)$ at $z = a$ in its Laurent's series expansion has a finite number of terms.

If $f(z) = \sum_{n=1}^{\infty} a_n (z-a)^n + \sum_{n=1}^{m} \frac{b_n}{(z-a)^n}$ then $z = a$ is a *pole of order m.*

A pole of order one is called a *simple pole.*

A singularity $z = a$ is a pole of $f(z)$ if $\lim_{z \to a} f(z) = \infty$.

For example, consider $f(z) = \frac{z^2+a^2}{z-a}$. Here $\lim_{z \to a} f(z) = \infty$. So $f(z)$ has a pole at $z = a$.

As another example, the function $f(z) = \frac{\sin(z-a)}{(z-a)^4}$ has a pole at $z = a$ since

$$\begin{aligned} f(z) &= \frac{\sin(z-a)}{(z-a)^4} \\ &= \frac{1}{(z-a)^4}\left[(z-a) - \frac{1}{3!}(z-a)^3 + \frac{1}{5!}(z-a)^5 + \cdots\right] \\ &= \frac{1}{(z-a)^3} - \frac{1}{3!\,(z-a)} + \frac{1}{5!}(z-a) - \cdots \end{aligned}$$

The principal part contains only two terms and hence $z = a$ is a pole of order 2.

(iii) Essential singularity: Let $'a'$ be an isolated singularity of $f(z)$. The point $z = a$ is called an *essential singularity* of $f(z)$ at $z = a$ if the principal part of $f(z)$ at $z = a$ has an infinite number of terms.

A singularity $z = a$ is an essential singularity of $f(z)$ if $\lim_{z \to a} f(z)$ does not exist.

For example, the function $f(z) = e^{1/(z-a)}$ has an essential singularity at $z = a$ since $\lim_{z \to a} f(z) = \lim_{z \to a} e^{1/(z-a)}$ does not exist.

Similarly, $f(z) = \sin\left(\frac{1}{z-a}\right) = \frac{1}{z-a} - \frac{1}{3!\,(z-a)^3} + \frac{1}{5!\,(z-a)^5} + \cdots$ has infinite number of terms in negative powers of $z - a$ and hence $z = a$ is an essential singularity.

Note 1: If an analytic function $f(z)$ has a pole of order m at $z = a$, then $\frac{1}{f(z)}$ has a zero of order m at $z = a$ and conversely.

Note 2: A function $f(z)$ is said to be a *meromorphic function* if it is analytic except at a finite number of poles.

For example, $f(z) = \frac{e^{2(z-1)}}{(z-1)^3}$ is analytic except at $z = 1$ and $z = 1$ is a pole of order 3, since $f(z) = \left[\frac{1}{(z-1)^3} + \frac{2}{(z-1)^2} + \frac{2}{(z-1)} + \frac{4}{3} + \frac{2}{3}(z-1) + \cdots\right]$. Hence $f(z)$ is a meromorphic function.

Note 3: A function $f(z)$ which is analytic everywhere in the finite z-plane is called an *entire function* or *integral function*. The functions e^z, $\sin z$, $\cosh z$ are examples of entire functions.

4.6. Residues

The coefficient of $\frac{1}{z-a}$ in the Laurent's series expansion of $f(z)$ around an isolated singularity is called *residue* of $f(z)$ at $z = a$, denoted by $[\text{Res}\, f(z)]_{z=a} = b_1 = \frac{1}{2\pi i} \oint_C f(z)\, dz$.

Cauchy's Residue Theorem

Let $f(z)$ be analytic within and on a simple closed contour C, except at a finite number of poles $z_1, z_2, \cdots, z_n$ within C and let $R_1, R_2, \cdots, R_n$ be respectively the residues of $f(z)$ at these poles, then

$$\oint_C f(z)\, dz = 2\pi i\,(R_1 + R_2 + \cdots + R_n).$$

i.e., $\oint_C f(z)\, dz = 2\pi i$ [Sum of the residues at the poles within C]

Proof:

Let $\gamma_1, \gamma_2, \cdots, \gamma_n$ be the circles with centres $z_1, z_2, \cdots, z_n$ respectively and radii so small that they all lie within C and do not overlap. Then $f(z)$ is analytic within the region enclosed by the curve C and these circles. Hence by Cauchy's theorem for multi-connected region, we have $\oint_C f(z)\, dz = \oint_{\gamma_1} f(z)\, dz + \oint_{\gamma_2} f(z)\, dz + \cdots + \oint_{\gamma_n} f(z)\, dz$.

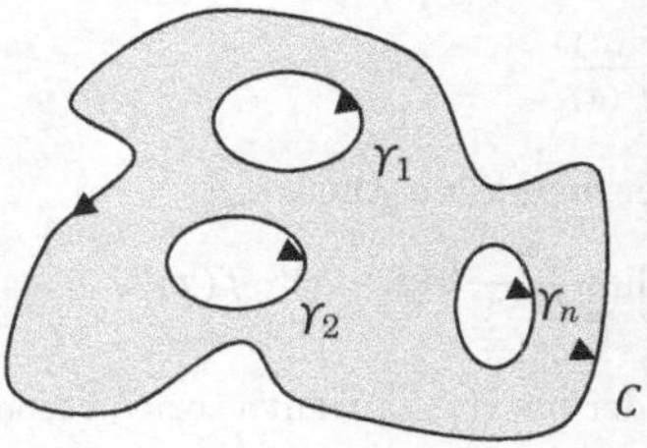

By definition of residues,

$$\frac{1}{2\pi i}\oint_{\gamma_j} f(z)\,dz = [Res\, f(z)]_{z=z_j} = R_j.$$

$$\therefore \quad \oint_{\gamma_j} f(z)\,dz = 2\pi i R_j$$

Hence $\oint_C f(z)\,dz = 2\pi i\,[R_1 + R_2 + \cdots + R_n]$.

Calculation of Residues

1. If $f(z)$ has a simple pole at $z = a$ (i.e., pole of order 1), then

$$[\text{Res}\, f(z)]_{z=a} = \lim_{z\to a}(z-a)\,f(z).$$

Since $z = a$ is a pole of order 1, the Laurent's series becomes

$$f(z) = a_0 + a_1(z-a) + a_2(z-a)^2 + \cdots + \frac{b_1}{z-a}.$$

Multiplying both sides by $(z-a)$, we get

$$(z-a)\,f(z) = a_0(z-a) + a_1(z-a)^2 + a_2(z-a)^3 + \cdots + b_1$$

$$\therefore \lim_{z\to a}(z-a)\,f(z) = b_1 = [\text{Res}\, f(z)]_{z=a}.$$

2. If $z = a$ *is a simple pole of* $f(z) = \dfrac{\varphi(z)}{\psi(z)}$, where $\psi(a) = 0$, but $\varphi(a) \neq 0$, then

$$[\text{Res}\, f(z)]_{z=a} = \frac{\varphi(a)}{\psi'(a)}.$$

Here $\lim_{z\to a}(z-a)f(z) = \lim_{z\to a}\dfrac{(z-a)\,\varphi(z)}{\psi(z)}$.

$$= \lim_{z\to a}\frac{(z-a)\,[\varphi(a)+(z-a)\varphi'(a)+\cdots]}{\psi(a)+(z-a)\psi'(a)+\cdots}$$

$$= \lim_{z\to a}\frac{(z-a)\,\varphi(a)+(z-a)^2\varphi'(a)+\cdots}{(z-a)\psi'(a)+\cdots}$$

$\therefore \quad [\text{Res}\, f(z)]_{z=a} \quad = \dfrac{\varphi(a)}{\psi'(a)}.$

3. If $f(z)$ has a pole of order m at $z = a$, then

$$[\text{Res}\, f(z)]_{z=a} = \frac{1}{(m-1)!} \lim_{z \to a} \frac{d^{m-1}}{dz^{m-1}} [(z-a)^m f(z)] .$$

Since $z = a$ is a pole of order m , the Laurent's series becomes

$$f(z) = \sum_{n=0}^{\infty} a_n (z-a)^n + \sum_{n=1}^{m} b_n (z-a)^{-n}$$

Multiplying both sides by $(z-a)^m$, we get

$$(z-a)^m f(z) = \sum_{n=0}^{\infty} a_n (z-a)^{m+n} + \sum_{n=1}^{m} b_n (z-a)^{m-n}$$

$$= \sum_{n=0}^{\infty} a_n (z-a)^{m+n} + b_1 (z-a)^{m-1} + b_2 (z-a)^{m-2} + \cdots + b_m.$$

Differentiating both sides $(m-1)$ times w.r.t. z and taking limit as $z \to a$ we get

$$\lim_{z \to a} \frac{d^{m-1}}{dz^{m-1}} [(z-a)^m f(z)] = (m-1)!\, b_1$$

$$\because\ b_1 = [\text{Res}\, f(z)]_{z=a} = \frac{1}{(m-1)!} \lim_{z \to a} \frac{d^{m-1}}{dz^{m-1}} [(z-a)^m f(z)].$$

Example 1: Find out the zeros and classify the nature of singularities of the following functions, if any

(a) $f(z) = \frac{z-2}{z^2} \sin\left(\frac{1}{z-1}\right)$ (b) $f(z) = z^2 \sin z.$

Solution:

(a) The zeros of $f(z)$ are obtained by equating the numerator of $f(z)$ to zero.

i.e., $(z-2) \sin\left(\frac{1}{z-1}\right) = 0$

Hence $z - 2 = 0$ or $\sin\left(\frac{1}{z-1}\right) = 0.$

i.e., $z = 2$ or $\frac{1}{z-1} = n\pi \ (n = 0, \pm 1, \pm 2, \cdots).$

Hence the zeros of $f(z)$ are $z = 2,\ z = 1 + \frac{1}{n\pi} \quad (n = 0, \pm 1, \pm 2, \cdots).$

Also, $\frac{z-2}{z^2} \sin\left(\frac{1}{z-1}\right) = \frac{z-2}{z^2}\left[\frac{1}{z-1} - \frac{1}{3!(z-1)^3} + \frac{1}{5!(z-1)^5} - \cdots\right]$

Since $\lim_{z \to 1} f(z)$ does not exist, $z = 1$ is an essential singularity.

The poles of $f(z)$ are obtained by equating the denominator of $f(z)$ to zero.

Hence $z = 0$ is a pole of order 2.

(b) $f(z) = z^2 \sin z = z^2\left[z - \frac{z^3}{3!} + \frac{z^5}{5!} - \frac{z^7}{7!} + \cdots\right]$

$$= z^3\left[1 - \frac{z^2}{3!} + \frac{z^4}{5!} - \frac{z^6}{7!} + \cdots\right] = z^3 g(z), \text{where } g(z) = 1 - \frac{z^2}{3!} + \frac{z^4}{5!} - \frac{z^6}{7!} + \cdots.$$

Now, $g(z)$ is analytic and $g(0) \neq 0$.

$\therefore z = 0$ is a zero of order 3.

Also, $\sin z = 0$ gives $z = n\pi$ $(n = 0, \pm 1, \pm 2, \cdots)$ is a simple zero.

Example 2: State the nature of singularity of $f(z) = (z-1)\sin\frac{1}{z}$ at $z = 0$.

Solution : Let $f(z) = (z-1)\sin\left(\frac{1}{z}\right)$

$$= (z-1)\left[\frac{1}{z} - \frac{1}{3!z^3} + \frac{1}{5!z^5} - \cdots\right]$$

$$= 1 - \frac{1}{z} - \frac{1}{3!z^2} + \frac{1}{3!z^3} + \frac{1}{5!z^4} - \frac{1}{5!z^5} + \cdots$$

The Laurent's series of $f(z)$ at $z = 0$ contains infinitely many terms of negative powers of z. Hence $z = 0$ is an essential singularity of $f(z)$.

Example 3: Find the singularities of the following functions and classify the singularities.

(i) $z\,e^{1/z}$ (ii) $\frac{z^2}{1+z}$ (iii) $\frac{z^2}{(4+z^2)^5}$ (iv) $\frac{z+1}{z^2-2z}$

Solution:

(i) Let $f(z) = z\,e^{1/z}$

Then $f(z) = z\left[1 + \frac{1}{1!z} + \frac{1}{2!z^2} + \frac{1}{3!z^3} + \cdots\right]$

$$= z + \frac{1}{1!} + \frac{1}{2!z} + \frac{1}{3!z^2} + \cdots$$

The Laurent's series of $f(z)$ at $z = 0$ contains infinitely many terms of negative powers of z. Hence $z = 0$ is an essential singularity of $f(z)$.

(ii) Let $f(z) = \frac{z^2}{1+z}$.

Here, $z = -1$ is a simple pole since $\lim_{z\to -1} (z+1)\,f(z) = 1 \neq 0$.

(iii) $f(z) = \frac{z^2}{(4+z^2)^5}$

$= \frac{z^2}{((z+2i)(z-2i))^5} = \frac{z^2}{(z+2i)^5\,(z-2i)^5}$.

$\therefore \lim_{z\to 2i} (z-2i)^5 f(z) = \frac{-4}{(4i)^5} \neq 0$.

Similarly, $\lim_{z\to -2i} (z+2i)^5 f(z) \neq 0$.

$\therefore z = 2i$ and $z = -2i$ are poles of order 5.

(iv) Let $f(z) = \frac{z+1}{z^2-2z}$.

Then $f(z) = \frac{z+1}{z(z-2)}$ and so $z = 0$ and $z = 2$ are singularities.

Since $\lim_{z\to 0} z\,f(z) = \lim_{z\to 0} \frac{z+1}{z-2} = -\frac{1}{2} \neq 0$, and $\lim_{z\to 2}(z-2)f(z) = \lim_{z\to 2} \frac{z+1}{z} = \frac{3}{2} \neq 0$,

we have $z = 0$ and $z = 2$ are simple poles.

Example 4: Calculate the residues of $f(z) = \frac{z+1}{z^2-2z}$ at its poles.

Solution : Given $f(z) = \frac{z+1}{z^2-2z}$.

Poles are given by equating the denominator to 0.

i.e., $z\,(z-2) = 0$.

$\Rightarrow z = 0$ and $z = 2$ are simple poles.

$\therefore [\text{Res}\, f(z)]_{z=0} = \lim_{z\to 0} z\,f(z)$

$= \lim_{z\to 0} z\frac{z+1}{z(z-2)}$

$= \lim_{z\to 0} \frac{z+1}{z-2} = \frac{-1}{2}$.

$[\text{Res}\, f(z)]_{z=2} = \lim_{z\to 2} (z-2)\,f(z)$

$= \lim_{z\to 2} \frac{z+1}{z} = \frac{3}{2}$.

Example 5: Calculate the residue of $f(z) = \frac{e^{2z}}{(z+1)^2}$ at its pole.

Solution: Given that $f(z) = \frac{e^{2z}}{(z+1)^2}$.

Then $z = -1$ is a pole of order 2.

$$\therefore \ [\text{Res}\, f(z)]_{z=-1} = \frac{1}{(2-1)!} \lim_{z\to-1} \frac{d^{2-1}}{dz^{2-1}} [(z+1)^2 f(z)]$$

$$= \frac{1}{1!} \lim_{z\to-1} \frac{d}{dz} \left[(z+1)^2 \frac{e^{2z}}{(z+1)^2}\right]$$

$$= \lim_{z\to-1} \frac{d}{dz} [e^{2z}] = \lim_{z\to-1} [2e^{2z}] = 2e^{-2}.$$

Example 6: Find residues of $f(z) = \frac{1}{(z^2+1)^2}$ about each singularity.

Solution: Given $f(z) = \frac{1}{(z^2+1)^2} = \frac{1}{(z-i)^2\,(z+i)^2}$.

The singularities of $f(z)$ are given by $(z-i)^2(z+i)^2 = 0$.

Hence $z = i$ and $z = -i$ are poles of order 2.

$$[\text{Res}\, f(z)]_{z=i} = \frac{1}{(2-1)!} \lim_{z\to i} \frac{d}{dz} [(z-i)^2 f(z)]$$

$$= \lim_{z\to i} \frac{d}{dz} \left[(z-i)^2 \frac{1}{(z-i)^2\,(z+i)^2}\right] = \lim_{z\to i} \frac{d}{dz} \left[\frac{1}{(z+i)^2}\right]$$

$$= \lim_{z\to i} \left[\frac{-2}{(z+i)^3}\right] = \frac{-2}{(2i)^3} = \frac{-i}{4}.$$

$$[\text{Res}\, f(z)]_{z=-i} = \frac{1}{(2-1)!} \lim_{z\to-i} \frac{d}{dz} [(z+i)^2 f(z)]$$

$$= \lim_{z\to-i} \frac{d}{dz} \left[(z+i)^2 \frac{1}{(z-i)^2\,(z+i)^2}\right] = \lim_{z\to-i} \frac{d}{dz} \left[\frac{1}{(z-i)^2}\right]$$

$$= \lim_{z\to-i} \left[\frac{-2}{(z-i)^3}\right] = \frac{i}{4}.$$

Example 7: Find the residues of the function $\frac{\cot \pi z}{(z-a)^2}$.

Solution: Let $f(z) = \frac{\cot \pi z}{(z-a)^2}$. Then $f(z) = \frac{\cos \pi z}{(z-a)^2 \sin \pi z}$.

$\therefore$ The poles of $f(z)$ are given by $(z-a)^2 \sin \pi z = 0$.

Hence $z = a$ is a double pole and $\sin \pi z = 0$ gives $\pi z = n\pi \ (n = 0, \pm 1, \pm 2, \cdots)$.

Thus $z = a$ is a double pole, and $z = n \ (n = 0, \pm 1, \pm 2, \cdots)$ are simple poles.

$$\therefore \ [\text{Res}\, f(z)]_{z=a} = \lim_{z\to a} \frac{d}{dz} \left((z-a)^2 \frac{\cot \pi z}{(z-a)^2}\right)$$

$$= \lim_{z \to a} \frac{d}{dz} [\cot \pi z]$$

$$= \lim_{z \to a} (-\pi \ \mathrm{cosec}^2 \pi z)$$

$$= -\pi \ \mathrm{cosec}^2 \pi a.$$

Also, $f(z) = \frac{\cos \pi z}{(z-a)^2 \sin \pi z} = \frac{\varphi(z)}{\psi(z)}$, where $\varphi(z) = \frac{\cos \pi z}{(z-a)^2}$ and $\psi(z) = \sin \pi z$.

Note that $\varphi(n) \neq 0$ and $\psi(n) = 0$ for $n = 0, \pm 1, \pm 2, \cdots$

Hence $[\mathrm{Res}\, f(z)]_{z=n} = \frac{\varphi(n)}{\psi'(n)} = \frac{\frac{\cos \pi n}{(n-a)^2}}{\pi \cos \pi n} = \frac{1}{\pi (n-a)^2}$.

Example 8: Calculate the residues of $\cot z$ at the poles.

Solution : Let $f(z) = \cot z$.

Then $f(z) = \frac{\cos z}{\sin z}$ and the poles are given by $\sin z = 0$.

Hence the poles of $f(z)$ are $z = \pm n\pi$, where $n = 0, 1, 2, \cdots$. Since each of the these is a simple pole, we have $[\mathrm{Res}\, f(z)]_{z = \pm n\pi} = \lim_{z \to \pm n\pi} \left[\frac{\cos z}{\cos z}\right] = 1$.

Example 9: Find the residues of $f(z) = \frac{z^2}{(z-1)^2 (z-2)}$ at its isolated singularities, using Laurent's series expansion.

Solution: The poles of $f(z)$ are given by $(z-1)^2 (z-2) = 0$.

$\therefore z = 1$ is a pole of order 2 and $z = 2$ is a simple pole of $f(z)$.

By partial fractions, we have

$$f(z) = \frac{z^2}{(z-1)^2 (z-2)} = \frac{A}{z-1} + \frac{B}{(z-1)^2} + \frac{C}{z-2} = \frac{A(z-1)(z-2) + B(z-2) + C(z-1)^2}{(z-1)^2 (z-2)}$$

$$\therefore \quad z^2 = A(z-1)(z-2) + B(z-2) + C(z-1)^2 \qquad (1)$$

Putting $z = 1$ and $z = 2$ in (1), we get $B = -1$ and $C = 4$.

Equating coefficient of z^2 in (1), we get $A + C = 1$. i.e., $A = -3$

$$\therefore \quad f(z) = \frac{-3}{z-1} - \frac{1}{(z-1)^2} + \frac{4}{z-2}$$

To find the residue of $f(z)$ at $z = 1$, we expand $f(z)$ in powers of $z - 1$.

Let $u = z - 1$ i.e., $z = u + 1$.

$$\therefore \quad f(z) = \frac{-3}{u} - \frac{1}{u^2} + \frac{4}{u-1}$$

$$= \frac{-3}{u} - \frac{1}{u^2} - 4[1-u]^{-1}$$

$$= \frac{-3}{u} - \frac{1}{u^2} - 4[1+u+u^2+u^3+\cdots]$$

$$= \frac{-3}{z-1} - \frac{1}{(z-1)^2} - 4[1+(z-1)+(z-1)^2+(z-1)^3+\cdots],$$

the expansion being valid in $0 < |z-1| < 1$.

$\therefore$ $[\text{Res } f(z)]_{z=1}$ = coefficient of $\frac{1}{z-1}$ in the expansion of $f(z) = -3$.

To find the residue of $f(z)$ at $z = 2$, we have to expand $f(z)$ about $z = 2$ in powers of $z - 2$.

Let $v = z - 2$ i.e., $z = v + 2$.

$$f(z) = \frac{-3}{v+1} - \frac{1}{[v+1]^2} + \frac{4}{v}$$

$$= -3[1+v]^{-1} - [1+v]^{-2} + \frac{4}{v}$$

$$= -3[1-v+v^2-v^3+\cdots] - [1-2v+3v^3-\cdots] + \frac{4}{v}$$

$$= -3[1-(z-2)+(z-2)^2-(z-2)^3+\cdots] - [1-2(z-2)+\ 3(z-2)^2-\cdots] + \frac{4}{z-2},$$

the expansion being valid in $0 < |z-2| < 1$.

$\therefore$ $[\text{Res } f(z)]_{z=2}$ = coefficient of $\frac{1}{z-2} = 4$.

Example 10: Obtain the Laurent's series expansion for $f(z) = \frac{9z^2-4z+1}{(z-1)(2z-1)(z+2)}$ in the region $1 < |z-1| < 3$. Hence, find the residue of the function at $z = 1$.

Solution: Let $f(z) = \frac{9z^2-4z+1}{(z-1)(2z-1)(z+2)} = \frac{A}{z-1} + \frac{B}{2z-1} + \frac{C}{z+2}$

$$= \frac{A\,(2z-1)\,(z+2) + B\,(z-1)\,(z+2) + C\,(z-1)\,(2z-1)}{(z-1)\,(2z-1)\,(z+2)}$$

Then, $\quad 9z^2 - 4z + 1 = A(2z-1)(z+2) + B(z-1)(z+2) + C(z-1)(2z-1)$

Putting $z = 1,\ z = \frac{1}{2}$ and $z = -2$, we get $A = 2, B = -1$ and $C = 3$.

$$\therefore \quad \frac{9z^2-4z+1}{(z-1)(2z-1)(z+2)} = \frac{2}{z-1} + \frac{-1}{2z-1} + \frac{3}{z+2}.$$

Given region is $1 < |z-1| < 3$.

Let $u = z - 1$ i.e., $z = u + 1$.

$\therefore 1 < |u| < 3$

$\Rightarrow \left|\frac{1}{u}\right| < 1$ and $\left|\frac{u}{3}\right| < 1$

$$\begin{aligned}
\therefore \quad f(z) &= \frac{2}{u} + \frac{-1}{2u+1} + \frac{3}{u+3} \\
&= \frac{2}{u} + \frac{-1}{2u\left(1+\frac{1}{2u}\right)} + \frac{3}{3\left(1+\frac{u}{3}\right)} \\
&= \frac{2}{u} + \frac{-1}{2u}\left(1+\frac{1}{2u}\right)^{-1} + \left(1+\frac{u}{3}\right)^{-1} \\
&= \frac{2}{u} + \frac{-1}{2u}\left[1 - \frac{1}{2u} + \left(\frac{1}{2u}\right)^2 + \cdots\right] + \left[1 - \frac{u}{3} + \left(\frac{u}{3}\right)^2 + \cdots\right] \\
&= \frac{2}{z-1} + \frac{-1}{2(z-1)}\left[1 - \frac{1}{2(z-1)} + \frac{1}{4(z-1)^2} + \cdots\right] + \left[1 - \frac{z-1}{3} + + \cdots\right]
\end{aligned}$$

$\therefore [\text{Res}\, f(z)]_{z=1}$ = coefficient of $\frac{1}{z-1}$ in the expansion of $f(z) = \frac{3}{2}$.

Example 11: Evaluate $\int_C \frac{z^2}{(z-1)^2\,(z+2)} dz$, where C is the circle $|z| = 3$ using Cauchy's residue theorem. **[*AU, April/May 2015*]**

Solution: Let $f(z) = \frac{z^2}{(z-1)^2\,(z+2)}$

Poles of $f(z)$ are given by, $(z-1)^2\,(z+2) = 0$.

$\Rightarrow z = 1$ (pole of order 2) and $z = -2$ (simple pole).

Both the poles lie inside C.

$$\therefore \quad [\text{Res}\, f(z)]_{z=1} = \frac{1}{(2-1)!} \lim_{z\to 1} \frac{d^{2-1}}{dz^{2-1}} \left[(z-1)^2\ f(z)\right]$$

$$= \frac{1}{1!} \lim_{z\to 1} \frac{d}{dz}\left[(z-1)^2\ \frac{z^2}{(z-1)^2\,(z+2)}\right]$$

$$= \lim_{z\to1} \frac{d}{dz}\left[\frac{z^2}{z+2}\right]$$

$$= \lim_{z\to1}\left[\frac{(z+2).2z - z^2.1}{(z+2)^2}\right] = \lim_{z\to1}\frac{z^2+4z}{(z+2)^2} = \frac{5}{9}.$$

$$[\text{Res } f(z)]_{z=-2} = \lim_{z\to-2}(z+2)\,f(z)$$

$$= \lim_{z\to-2}(z+2)\frac{z^2}{(z-1)^2\,(z+2)}$$

$$= \lim_{z\to-2}\frac{z^2}{(z-1)^2} = \frac{4}{9}.$$

∴ By Cauchy's residue theorem,

$\int_C f(z)\,dz = 2\pi i \times$ sum of residues.

$$= 2\pi i\left[\frac{5}{9}+\frac{4}{9}\right] = 2\pi i$$

Example 12: Evaluate $\int_C \frac{z-1}{(z+1)^2\,(z-2)}\,dz$, where C is the circle $|z-i| = 2$ using Cauchy's residue theorem.

Solution:

Let $f(z) = \frac{z-1}{(z+1)^2(z-2)}$

Poles of $f(z)$ are given by, $(z+1)^2\,(z-2) = 0$.

⇒ $z = -1$ (pole of order 2) and $z = 2$ (simple pole).

Given that C is the circle $|z-i| = 2$.

$z = -1 \Rightarrow |z-i| = |-1-i| = \sqrt{(-1)^2+(-1)^2} = \sqrt{2} < 2.$

∴ $z = -1$ lies inside C.

$z = 2 \Rightarrow |z-i| = |2-i| = \sqrt{2^2+(-1)^2} = \sqrt{5} > 2.$

∴ $z = 2$ lies outside C.

Hence, $[\text{Res } f(z)]_{z=-1} = \frac{1}{(2-1)!}\lim_{z\to-1}\frac{d^{2-1}}{dz^{2-1}}\,[(z+1)^2\,f(z)]$

$$= \frac{1}{1!}\lim_{z\to-1}\frac{d}{dz}\left[(z+1)^2\frac{z-1}{(z+1)^2(z-2)}\right]$$

$$= \lim_{z\to-1}\frac{d}{dz}\left[\frac{z-1}{z-2}\right]$$

$$= \lim_{z\to -1}\left[\frac{(z-2).1-(z-1).1}{(z-2)^2}\right] = \lim_{z\to -1}\frac{-1}{(z-2)^2} = \frac{-1}{9}.$$

∴ By Cauchy's residue theorem,

$$\int_C f(z)\,dz = 2\pi i \times \text{sum of residues} = \frac{-2\pi i}{9}.$$

Example 13: Evaluate $\int_C \frac{z^2-2z}{(z+1)^2(z^2+4)}dz$, where C is the circle $|z| = 10$ using Cauchy's residue theorem.

Solution : Let $f(z) = \frac{z^2-2z}{(z+1)^2(z^2+4)}$

Poles of $f(z)$ are given by, $(z+1)^2(z^2+4) = 0$

$\Rightarrow$ $z = -1$ (pole of order 2) and $z = \pm 2i$ (simple poles).

Now, $z^2 + 4 = (z+2i)(z-2i)$

All the poles lie inside the given circle $|z| = 10$.

$$\therefore \quad [\text{Res } f(z)]_{z=-1} = \frac{1}{(2-1)!}\lim_{z\to -1}\frac{d^{2-1}}{dz^{2-1}}[(z+1)^2 f(z)]$$

$$= \frac{1}{1!}\lim_{z\to -1}\frac{d}{dz}\left[(z+1)^2\frac{z^2-2z}{(z+1)^2(z^2+4)}\right]$$

$$= \lim_{z\to -1}\frac{d}{dz}\left[\frac{z^2-2z}{z^2+4}\right]$$

$$= \lim_{z\to -1}\left[\frac{(z^2+4).(2z-2)-(z^2-2z).2z}{(z^2+4)^2}\right]$$

$$= \lim_{z\to -1}\frac{2z^2+8z-8}{(z^2+4)^2}.$$

$$\therefore \quad R_1 = -\frac{14}{25}.$$

$$[\text{Res } f(z)]_{z=2i} = \lim_{z\to 2i}(z-2i)f(z)$$

$$= \lim_{z\to 2i}(z-2i)\frac{z^2-2z}{(z+1)^2(z+2i)(z-2i)}$$

$$= \lim_{z\to 2i}\frac{z^2-2z}{(z+1)^2(z+2i)} = \frac{1+i}{3i+4}$$

$$\therefore \quad R_2 = \frac{7+i}{25}.$$

Similarly, $[\text{Res } f(z)]_{z=-2i}$, $R_3 = \frac{7-i}{25}$.

∴ By Cauchy's residue theorem,

$\int_C f(z)\,dz \;=\; 2\pi i \times [R_1 + R_2 + R_3].$

$= 2\pi i \times \left[-\frac{14}{25} + \frac{7+i}{25} + \frac{7-i}{25}\right] = 0. \quad \frac{dz}{(z^2+4)^2}$

Example 14: Evaluate $\int_C \frac{dz}{(z^2+4)^2}$, where C is the circle $|z - i| = 2$.

Solution: Let $f(z) = \frac{1}{(z^2+4)^2}$.

Poles of $f(z)$ are given by $(z^2 + 4)^2 = 0$.

Then $z = 2i, -2i$ are the poles of $f(z)$ of order 2.

The given curve is $|z - i| = 2$.

Now, $|2i - i| = |i| = 1 < 2$ and $|-2i - i| = |-3i| = 3 > 2$.

∴ $z = 2i$ lies inside C and $z = -2i$ lies outside C.

$$\therefore\ [\text{Res}\, f(z)]_{z=2i} = \frac{1}{(2-1)!} \lim_{z\to 2i} \frac{d^{2-1}}{dz^{2-1}} [(z-2i)^2 f(z)]$$

$$= \lim_{z\to 2i} \frac{d}{dz}\left[(z-2i)^2 \frac{1}{(z-2i)^2 (z+2i)^2}\right]$$

$$= \lim_{z\to 2i} \frac{d}{dz}\left[\frac{1}{(z+2i)^2}\right]$$

$$= \lim_{z\to 2i} \left[\frac{-2}{(z+2i)^3}\right]$$

$$= \frac{-2}{(4i)^3} = \frac{1}{32i}.$$

∴ By Cauchy's residue theorem, $\int_C f(z)\,dz = 2\pi i \times \frac{1}{32i} = \frac{\pi}{16}$.

Example 15: Evaluate $\int_C \frac{z-3}{z^2+2z+5}\, dz$, where C is the circle (i) $|z| = 1$ (ii) $|z + 1 - i| = 2$, (iii) $|z + 1 + i| = 2$.

Solution : Let $f(z) = \frac{z-3}{z^2+2z+5}$.

Then the poles are given by $z^2 + 2z + 5 = 0$.

i.e., $z = \frac{-2 \pm \sqrt{4-20}}{2} = -1 \pm 2i$.

Note that the poles $-1 + 2i$, $-1 - 2i$ are simple.

$z^2 + 2z + 5 = [z - (-1 + 2i)]\,[z - (-1 - 2i)]$

(i) C is the circle $|z| = 1$.

Now $|-1+2i| = \sqrt{1+4} = \sqrt{5} > 1$,

And $|-1-2i| = \sqrt{1+4} = \sqrt{5} > 1$.

$\therefore$ The poles $z = -1 \pm 2i$ lie outside C.

Hence by Cauchy's integral theorem,

$$\int_C f(z)dz = \int_C \frac{z-3}{z^2+2z+5}\, dz = 0.$$

(ii) C is the circle $|z+1-i| = 2$.

Now $|-1+2i+1-i| = |i| = 1 < 2$.

$\therefore z = -1+2i$ lies inside the circle $|z+1-i| = 2$.

Also, $|-1-2i+1-i| = |-3i| = 3 > 2$.

$\therefore z = -1-2i$ lies outside C.

$$[\text{Res}\, f(z)]_{z=-1+2i} = \lim_{z \to (-1+2i)} (z+1-2i)\, f(z)$$

$$= \lim_{z \to (-1+2i)} (z+1-2i) \left[\frac{z-3}{z^2+2z+5}\right]$$

$$= \lim_{z \to (-1+2i)} \left[\frac{z-3}{z+1+2i}\right] = \frac{-1+2i-3}{-1+2i+1+2i}$$

$$= \frac{i-2}{2i}.$$

By Cauchy's residue theorem,

$$\int_C f(z)dz = 2\pi i\, [\text{sum of residues}]$$

$$= 2\pi i \left(\frac{i-2}{2i}\right) = \pi(i-2).$$

(iii) C is the circle $|z+1+i| = 2$.

Now, $|-1+2i+1+i| = |3i| = 3 > 2$

$\therefore -1+2i$ lies outside C.

Also, $|-1-2i+1+i| = |-i| = 1 < 2$.

$\therefore -1-2i$ lies inside C.

$\therefore$ $[\text{Res } f(z)]_{z=-1-2i} = \lim_{z \to (-1-2i)} (z+1+2i) f(z)$

$$= \lim_{z \to (-1-2i)} \left(\frac{z-3}{z+1-2i}\right)$$

$$= \frac{-1-2i-3}{-1-2i+1-2i} = \frac{i+2}{2i}.$$

$\therefore$ By Cauchy's residue theorem,

$$\int_C f(z)dz = 2\pi i \text{ [sum of residues]}$$

$$= 2\pi i \left(\frac{i+2}{2i}\right) = \pi(i+2).$$

Example 16 : Evaluate $\int_C z^2 e^{\frac{1}{z}} dz$, where C is $|z| = 1$.

Solution: Let $f(z) = z^2 e^{\frac{1}{z}}$.

Then $$f(z) = z^2 \left[1 + \frac{1}{1!z} + \frac{1}{2!\,z^2} + \frac{1}{3!z^3} + \cdots\right]$$

$$= z^2 + \frac{z}{1!} + \frac{1}{2!} + \frac{1}{3!z} + \cdots$$

$\therefore$ $[\text{Res } f(z)]_{z=0}$ = coefficient of $\frac{1}{z} = \frac{1}{3!} = \frac{1}{6}$.

$\therefore$ By Cauchy's residue theorem,

$$\int_C f(z)dz = 2\pi i \times \frac{1}{6} = \frac{\pi i}{3}.$$

EXERCISES

1. Find the singularities of the function $f(z) = \frac{4z}{(z-3)^4\,(z+2)^2}$.

2. Classify the singularities of $f(z) = \frac{\sin 2z}{(z^4+4)\,(z^2+1)}$.

3. Find the singularities of (i) $f(z) = \operatorname{cosec} z - \frac{1}{z}$ (ii) $e^{\frac{1}{z}}$.

4. Show that $z = 0$ is an essential singularity of

   (i) $z\,e^{\frac{1}{z}}$ (ii) $\cosh\left(\frac{1}{z}\right)$ (iii) $\sin\frac{1}{z}$.

5. What is the nature of the singularity $z = 0$ of the function $f(z) = \frac{\sin z - z}{z^3}$.

6. Classify the singularity of $(z) = \frac{e^{\frac{1}{z}}}{(z-a)^2}$. Find the residue of $\frac{1}{(z^2+a^2)^2}$ at $z = ia$.

7. Find the residue at $z = 0$ of (i) $\frac{1}{z(e^z-1)}$ (ii) $\frac{1}{z-\sin z}$ (iii) $\frac{1+z}{1-\cos z}$.

8. Test for singularity of $\frac{1}{z^2+1}$ and hence find the corresponding residues.

9. Evaluate $\int_C \frac{\sin^6 z}{\left(z-\frac{\pi}{6}\right)^3} dz$, where C is the circle $|z| = 1$.

10. Evaluate $\int_C \frac{z}{z^2+1} dz$, where C is the circle $|z + i| = 1$.

11. Evaluate $\int_C \frac{4-3z}{z\,(z-1)(z-2)} dz$, where C is the circle $|z| = \frac{3}{2}$.

12. Evaluate $\int_C \frac{3z^2+z}{z^2-1} dz$, where C is the circle $|z - 1| = 1$.

13. Evaluate $\int_C \frac{(3z^2+z+1)\,dz}{(z^2-1)(z+3)}$, where C is the circle $|z - i| = 2$.

14. Evaluate $\int_C \frac{(z-1)\,dz}{(z+1)^2(z-2)}$, where C is the circle $|z + i| = 2$.

15. Evaluate $\int_C \frac{z\,dz}{(z-1)\,(z-2)^2}$, where C is the circle $|z - 2| = \frac{1}{2}$.

16. Evaluate $\int_C \frac{dz}{(z^2+9)^3}$, where C is the circle $|z - i| = 2$.

17. If C is the positively oriented circle $|z| = 2$, show that

(i) $\int_C \frac{e^{2z}}{(1+z)^3} dz = \frac{4\pi i}{e^2}$ (ii) $\int_C \frac{z-2}{z(z-1)} dz = 2\pi i$.

18. If C is the positively oriented circle $|z| = 3$, show that

(i) $\int_C \frac{z+1}{z^2-2z} dz = 2\pi i$, (ii) $\int_C \frac{1}{z^3(z+4)} dz = \frac{\pi i}{32}$.

19. Evaluate $\int_C \frac{e^z\,dz}{(z^2+\pi^2)^2}$ where C is the circle $|z| = 4$ by using Cauchy's residue theorem.

20. If C is the boundary of the square whose sides along the lines $x = \pm 2$ and $y = \pm 2$ and described in the positive sense, find the value of $\int_C \frac{\tan \frac{z}{2}}{(z-1-i)^2} dz$.

ANSWERS

1. 3 is a pole of order 4 and -2 is a double pole. 2. $\pm i$, $\pm 2i$, simple poles

3. (i) $z = 0$ is a removable singularity

(ii) $z = 0$ is an essential singularity.

5. $z = 0$ is a removable singularity 6. $z = 0$ is a pole of order 2 7. $\frac{1}{4ia^3}$

8. (i) $-\frac{1}{2}$ (ii) $\frac{3}{10}$ (iii) 2 9. $\frac{i}{2}$.

10. $\frac{21\pi i}{16}$ 11. πi 12. $2\pi i$ 13. $4\pi i$ 14. $\frac{-\pi i}{4}$

15. $\frac{-2\pi i}{9}$ 16. $-2\pi i$ 17. $\frac{1}{648}$ 20. $\frac{i}{\pi}$ 21. $\pi i \sec^2\left(\frac{1+i}{2}\right)$.

4.7. Evaluation of Real Definite Integrals by Contour Integration

Many important definite integrals can be evaluated by applying the Cauchy's Residue theorem to properly chosen integrals. The contours chosen will consist of straight lines and circular arcs. The process of integrating along a contour is termed as *contour integration.*

Type I: Integration around the unit circle. Integrals of the type $\int_0^{2\pi} F(\cos\theta, \sin\theta)\, d\theta$, where $F(\cos\theta, \sin\theta)$ is a rational function of $\cos\theta$ and $\sin\theta$.

Such integrals can be reduced to complex line integrals by the substitution $z = e^{i\theta}$ such that $dz = ie^{i\theta}\, d\theta$ or $d\theta = \frac{dz}{iz}$.

Also, $\cos\theta = \frac{1}{2}\left[e^{i\theta} + e^{-i\theta}\right] = \frac{1}{2}\left(z + \frac{1}{z}\right)$ and

$$\sin\theta = \frac{1}{2i}\left[e^{i\theta} - e^{-i\theta}\right] = \frac{1}{2i}\left(z - \frac{1}{z}\right).$$

As θ varies from0 to 2π, z moves once round the unit circle in the anti-clockwise direction.

$$\therefore \int_0^{2\pi} F(\cos\theta, \sin\theta)\, d\theta = \oint_C F\left[\frac{1}{2}\left(z + \frac{1}{z}\right), \frac{1}{2i}\left(z - \frac{1}{z}\right)\right] \frac{dz}{iz} = \oint_C f(z)\, dz,$$

Where the contour C is the unit circle $|z| = 1$ and $f(z)$ is a rational function of z. Then $\oint_C f(z)\, dz$ is evaluated using Cauchy's residue theorem. Hence *the integral is equal to $2\pi i$ times the sum of the residues at the poles of $f(z)$ which are within C.*

Example 1: Evaluate $\int_0^{2\pi} \frac{1}{2+\cos\theta}\, d\theta$ using contour integration. **[AU, *Nov/Dec 2017*]**

Solution: Let $I = \int_0^{2\pi} \frac{d\theta}{2+\cos\theta}$.

Put $z = e^{i\theta}$. Then $dz = ie^{i\theta}\, d\theta$. $\therefore d\theta = \frac{dz}{iz}$.

Now, $\cos\theta = \frac{1}{2}\left(z + \frac{1}{z}\right) = \frac{z^2+1}{2z}$.

$$\therefore I = \int_C \frac{dz/iz}{2+\left(\frac{z^2+1}{2z}\right)} = \frac{1}{i}\int_C \frac{dz}{z\left[\frac{4z+z^2+1}{2z}\right]}$$

$$= \frac{2}{i}\int_C \frac{dz}{z^2+4z+1}, \text{ where } C \text{ is unit circle } |z| = 1.$$

Let $f(z) = \dfrac{1}{z^2+4z+1}$.

The poles of $f(z)$ are given by the zeros of $z^2 + 4z + 1 = 0$.

$$\therefore z = \frac{-4\pm\sqrt{16-4}}{2} = -2 \pm \sqrt{3}.$$

Now, $z = -2+\sqrt{3}$ is the only simple pole which lies inside C.

$$\therefore [\text{Res } f(z)]_{z=-2+\sqrt{3}} = \lim_{z\to(-2+\sqrt{3})} (z+2-\sqrt{3})\, f(z)$$

$$= \lim_{z\to(-2+\sqrt{3})} \left[(z+2-\sqrt{3})\frac{1}{(z+2-\sqrt{3})(z+2+\sqrt{3})}\right]$$

$$= \lim_{z\to(-2+\sqrt{3})} \left(\frac{1}{z+2+\sqrt{3}}\right) = \frac{1}{2\sqrt{3}}$$

$\therefore$ By Cauchy's residue theorem,

$$\int_C f(z)dz = 2\pi i \times \frac{1}{2\sqrt{3}} = \frac{\pi i}{\sqrt{3}}.$$

$$\therefore \quad \int_0^{2\pi} \frac{d\theta}{2+\cos\theta} = \frac{2}{i}\,\frac{\pi i}{\sqrt{3}} = \frac{2\pi}{\sqrt{3}}.$$

Example 2: Show that $\displaystyle\int_0^{2\pi} \frac{d\theta}{a+b\cos\theta} = \int_0^{2\pi} \frac{d\theta}{a+b\sin\theta} = \frac{2\pi}{\sqrt{a^2-b^2}},\ a > b > 0.$

Solution: Let $I_1 = \displaystyle\int_0^{2\pi} \frac{d\theta}{a+b\cos\theta}$.

Put $z = e^{i\theta}$. Then $dz = ie^{i\theta}\, d\theta$. $\therefore d\theta = \dfrac{dz}{iz}$.

Also, $\cos\theta = \dfrac{1}{2}\left(z+\dfrac{1}{z}\right) = \dfrac{z^2+1}{2z}$.

$$\therefore I_1 = \int_C \frac{dz/iz}{a+b\left(\frac{z^2+1}{2z}\right)} = \frac{2}{i}\int_C \frac{dz}{2az+b(z^2+1)}$$

$$= \frac{2}{i}\int_C \frac{dz}{bz^2+2az+b}, \text{ where } C \text{ is the unit circle } |z| = 1.$$

Let $f(z) = \dfrac{1}{bz^2+2az+b}$.

The poles of $f(z)$ are given by the zeros of $bz^2 + 2az + b = 0$.

$$\therefore \quad z = \frac{-2a \pm \sqrt{4a^2 - 4b^2}}{2b} = \frac{-a \pm \sqrt{a^2 - b^2}}{b}.$$

Let $\alpha = \frac{-a + \sqrt{a^2 - b^2}}{b}$ and $\beta = \frac{-a - \sqrt{a^2 - b^2}}{b}$.

$$\therefore f(z) = \frac{1}{b\,(z-\alpha)\,(z-\beta)}.$$

Since $a > b > 0$, we have $a^2 > b^2$. $\therefore\ a^2 - b^2 > 0$.

Hence $|\beta| = \left|\frac{-a - \sqrt{a^2 - b^2}}{b}\right| = \left|\frac{a + \sqrt{a^2 - b^2}}{b}\right| = \frac{a}{b} + \frac{\sqrt{a^2 - b^2}}{b} > 1.$

Since α and β are roots of $bz^2 + 2az + b = 0$, product of roots $= \alpha\beta = \frac{b}{b} = 1$.

$\therefore |\alpha\beta| = 1. \ \therefore\ |\alpha| = \frac{1}{|\beta|} < 1 \quad [\because\ |\beta| > 1].$

$\therefore\ \alpha = \frac{-a + \sqrt{a^2 - b^2}}{b}$ is a simple pole which lies inside C and $\beta = \frac{-a - \sqrt{a^2 - b^2}}{b}$ is a simple pole which lies outside C.

$$\therefore [\text{Res}\, f(z)]_{z=\alpha} = \lim_{z\to\alpha}(z-\alpha)f(z)$$

$$= \lim_{z\to\alpha}(z-\alpha)\frac{1}{b\,(z-\alpha)(z-\beta)}$$

$$= \lim_{z\to\alpha}\frac{1}{b\,(z-\beta)} = \frac{1}{b\,(\alpha-\beta)} = \frac{1}{2b\,\frac{\sqrt{a^2-b^2}}{b}} = \frac{1}{2\sqrt{a^2-b^2}}.$$

$\therefore$ By Cauchy's residue theorem,

$$\int_C f(z)dz = 2\pi i \times \frac{1}{2\sqrt{a^2-b^2}} = \frac{\pi i}{\sqrt{a^2-b^2}}.$$

$$\therefore \quad I_1 = \frac{2}{i}\,\frac{\pi i}{\sqrt{a^2-b^2}} = \frac{2\pi}{\sqrt{a^2-b^2}}.$$

Let $\quad I_2 = \int_0^{2\pi} \frac{d\theta}{a + b\sin\theta}$.

We have $\quad \sin\theta = \frac{1}{2i}\left(z - \frac{1}{z}\right) = \frac{z^2-1}{2iz}$.

$$\therefore \quad I_2 = \int_C \frac{dz/iz}{a + b\left(\frac{z^2-1}{2iz}\right)} = 2\int_C \frac{dz}{2iaz + b(z^2-1)}$$

$$= 2\int_C \frac{dz}{bz^2+2iaz-b}.$$

Let $\quad g(z) = \dfrac{1}{bz^2+2iaz-b}.$

The poles of $g(z)$ are given by the zeros of $bz^2 + 2iaz - b = 0$.

$\therefore \quad z = \dfrac{-2ia \pm \sqrt{-4a^2+4b^2}}{2b} = \dfrac{-ia \pm \sqrt{b^2-a^2}}{b}.$

Since $a > b > 0$, we have $b^2 < a^2$. Therefore, $b^2 - a^2 < 0$.

$\therefore \quad z = \dfrac{-ia \pm i\sqrt{a^2-b^2}}{b}$

Let $\quad \alpha = \dfrac{-ia + i\sqrt{a^2-b^2}}{b}$ and $\beta = \dfrac{-ia - i\sqrt{a^2-b^2}}{b}.$

$\therefore \; f(z) = \dfrac{1}{b\,(z-\alpha)(z-\beta)}.$

Now, $|\beta| = \dfrac{a+\sqrt{a^2-b^2}}{b} > 1$. $\therefore \beta$ lies outside C.

α and β are roots of $bz^2 + 2iaz - b = 0$ and so product of roots $= \alpha\beta = \frac{-b}{b} = -1$.

$\therefore |\alpha\beta| = 1. \;\; \therefore \; |\alpha| = \frac{1}{|\beta|} < 1.$

Hence $\alpha = \dfrac{-ia + i\sqrt{a^2-b^2}}{b}$ lies inside C.

$$\therefore [\text{Res}\, f(z)]_{z=\alpha} = \lim_{z\to\alpha} (z-\alpha)\, f(z)$$

$$= \lim_{z\to\alpha}(z-\alpha)\,\frac{1}{b\,(z-\alpha)\,(z-\beta)}$$

$$= \lim_{z\to\alpha} \frac{1}{b\,(z-\beta)} = \frac{1}{b\,(\alpha-\beta)}.$$

$$= \frac{1}{2i\,\sqrt{a^2-b^2}}.$$

$\therefore$ By Cauchy's residue theorem,

$$\int_C f(z)dz = 2\pi i \times \frac{1}{i2\sqrt{a^2-b^2}} = \frac{\pi}{\sqrt{a^2-b^2}}.$$

Hence $\quad I_2 = 2 \times \dfrac{\pi}{\sqrt{a^2-b^2}} = \dfrac{2\pi}{\sqrt{a^2-b^2}}.$

Example 3: Using contour integration evaluate $\int_0^{2\pi} \frac{d\theta}{13+5\sin\theta}$.

Solution: ***[AU, Nov/Dec 2017, May/June 2014, Nov/Dec 2014]***

Let $I = \int_0^{2\pi} \frac{d\theta}{13+5\sin\theta}$.

Put $z = e^{i\theta}$. Then $dz = i\, e^{i\theta}\, d\theta$. $\therefore d\theta = \frac{dz}{iz}$.

Now, $\cos\theta = \frac{1}{2}\left(z+\frac{1}{z}\right) = \frac{z^2+1}{2z}$ and $\sin\theta = \frac{1}{2i}\left(z-\frac{1}{z}\right) = \frac{z^2-1}{2iz}$.

$$\therefore \quad I = \int_C \frac{\frac{dz}{iz}}{13+5\left(\frac{z^2-1}{2iz}\right)}, \text{ where } C \text{ is the unit circle } |z| = 1.$$

$$= \int_C \frac{2}{5z^2+26iz-5}\, dz \qquad (1)$$

Let $f(z) = \frac{2}{5z^2+26iz-5} = \frac{2}{(5z+i)(z+5i)}$

$$= \frac{2}{5\left(z+\frac{i}{5}\right)(z+5i)}.$$

$\therefore$ The poles of $f(z)$ are given by, $5\left(z+\frac{i}{5}\right)(z+5i) = 0$

$\Rightarrow \quad z = -\frac{i}{5}$ and $z = -5i$.

Now, $z = -\frac{i}{5}$ is the only simple pole which lies inside C.

$$\therefore [\text{Res}\, f(z)]_{z=-\frac{i}{5}} = \lim_{z\to -\frac{i}{5}} \left(z+\frac{i}{5}\right) f(z)$$

$$= \lim_{z\to -\frac{i}{5}} \left(z+\frac{i}{5}\right) \frac{2}{5\left(z+\frac{i}{5}\right)(z+5i)}$$

$$= \lim_{z\to -\frac{i}{5}} \frac{2}{5\,(z+5i)}$$

$$= \frac{2}{5\left(-\frac{i}{5}+5i\right)} = \frac{2}{5\left(\frac{-i+25i}{5}\right)} = \frac{2}{24i} = \frac{1}{12i}.$$

$\therefore$ By Cauchy's residue theorem,

$$\int_C f(z)dz = 2\pi i \times \frac{1}{12i} = \frac{\pi}{6}.$$

$\therefore$ (1) gives $\int_0^{2\pi} \frac{d\theta}{13+5\sin\theta} = \frac{\pi}{6}$.

Example 4: Evaluate $\int_0^{2\pi} \frac{d\theta}{13+12\cos\theta}$ by using contour integration.

Solution: ***[AU, May/June 2016]***

Let $I = \int_0^{2\pi} \frac{d\theta}{13+12\cos\theta}$.

Put $z = e^{i\theta}$. Then $dz = ie^{i\theta}\,d\theta$. $\therefore d\theta = \frac{dz}{iz}$.

Also, $\cos\theta = \frac{1}{2}\left(z+\frac{1}{z}\right) = \frac{z^2+1}{2z}$.

$$\therefore I = \int_C \frac{dz/iz}{13+12\left(\frac{z^2+1}{2z}\right)} = \frac{2}{i}\int_C \frac{dz}{13z+6(z^2+1)}$$

$$= \frac{1}{i}\int_C \frac{dz}{6z^2+13z+6}, \text{ where } C \text{ is the unit circle } |z| = 1. \quad (1)$$

$$\text{Let } f(z) = \frac{1}{6z^2+13z+6} = \frac{1}{6z^2+9z+4z+6} = \frac{1}{(3z+2)(2z+3)}$$

$$= \frac{1}{6\left(z+\frac{2}{3}\right)\left(z+\frac{3}{2}\right)}.$$

The poles of $f(z)$ are given by, $z = -\frac{2}{3}$ and $z = -\frac{3}{2}$.

Now, $z = -\frac{2}{3}$ is the only simple pole which lies inside C.

$$\therefore [\text{Res } f(z)]_{z=-\frac{2}{3}} = \lim_{z\to-\frac{2}{3}} \left(z+\frac{2}{3}\right) f(z)$$

$$= \lim_{z\to-\frac{2}{3}} \left(z+\frac{2}{3}\right)\frac{1}{6\left(z+\frac{2}{3}\right)\left(z+\frac{3}{2}\right)}$$

$$= \lim_{z\to-\frac{2}{3}} \frac{1}{6\left(z+\frac{3}{2}\right)}$$

$$= \frac{1}{6\left(-\frac{2}{3}+\frac{3}{2}\right)} = \frac{1}{5}.$$

$\therefore$ By Cauchy's residue theorem,

$$\int_C f(z)dz = 2\pi i \times \frac{1}{5} = \frac{2\pi i}{5}.$$

$$\therefore (1) \text{ gives } \int_0^{2\pi} \frac{d\theta}{13+12\cos\theta} = \frac{1}{i}\times\frac{2\pi i}{5} = \frac{2\pi}{5}.$$

Example 5: Evaluate $\int_0^{2\pi} \frac{d\theta}{1-2x\cos\theta+x^2}$, $0<x<1$. ***[AU, April/May 2017]***

Solution: Let $I = \int_0^{2\pi} \frac{d\theta}{1-2x\cos\theta+x^2}$.

Put $z = e^{i\theta}$. $\therefore d\theta = \frac{dz}{iz}$ and $\cos\theta = \frac{z^2+1}{2z}$.

$$\therefore\ I = \int_C \frac{dz/iz}{1-2x\left(\frac{z^2+1}{2z}\right)+x^2} = \frac{2}{i}\int_C \frac{dz}{2z-2xz^2-2x+2x^2z}$$

$$= -\frac{2}{i}\int_C \frac{dz}{2xz^2-2z+2x-2x^2z}$$

$$= i\int_C \frac{dz}{xz^2-z(1+x^2)+x}\text{, where } C \text{ is the unit circle } |z|=1.$$

Let $f(z) = \frac{1}{xz^2-z(1+x^2)+x}$.

The poles of $f(z)$ are given by the zeros of $xz^2 - z(1+x^2) + x = 0$.

$$\therefore\quad z = \frac{(1+x^2)\pm\sqrt{(1+x^2)^2-4x^2}}{2x}$$

$$= \frac{(1+x^2)\pm\sqrt{1+x^4-2x^2}}{2x} = \frac{(1+x^2)\pm\sqrt{(1-x^2)^2}}{2x}$$

$$= \frac{(1+x^2)\pm(1-x^2)}{2x} = \frac{1}{x},\ x.$$

$$\therefore f(z) = \frac{1}{x\left(z-\frac{1}{x}\right)(z-x)}$$

Here both the poles are simple.

Now, since $0<x<1$, $z=x$ lies inside C. Also, since $\frac{1}{x}>1$, $z=\frac{1}{x}$ lies outside C.

$$\therefore [\text{Res}\, f(z)]_{z=x} = \lim_{z\to a}(z-x)\, f(z)$$

$$= \lim_{z\to a}(z-x)\frac{1}{x\left(z-\frac{1}{x}\right)(z-x)}$$

$$= \lim_{z\to a}\frac{1}{x\left(z-\frac{1}{x}\right)} = \frac{1}{x\left(x-\frac{1}{x}\right)} = \frac{1}{x^2-1}.$$

$\therefore$ By Cauchy's residue theorem,

$$\int_C f(z)dz = 2\pi i\times\frac{1}{x^2-1} = \frac{2\pi i}{x^2-1}.$$

$$\therefore\ I = i\times\frac{2\pi i}{x^2-1} = \frac{2\pi}{1-x^2}.$$

Example 6: Evaluate $\int_0^{2\pi} \frac{\cos 2\theta}{5-4\cos\theta}\, d\theta$.

Solution : Let $I = \int_0^{2\pi} \frac{\cos 2\theta}{5-4\cos\theta}\, d\theta$.

Then I = Real part of $\int_0^{2\pi} \frac{e^{i2\theta}}{5-4\cos\theta}\, d\theta$

= Real part of $\int_0^{2\pi} \frac{(e^{i\theta})^2}{5-4\cos\theta}\, d\theta$

Put $z = e^{i\theta}$. Then $d\theta = \frac{dz}{iz}$.

Now $\cos\theta = \frac{1}{2}\left(z + \frac{1}{z}\right) = \frac{z^2+1}{2z}$.

$\therefore \quad I =$ Real part of $\int_C \frac{z^2}{5-4\left(\frac{z^2+1}{2z}\right)} \frac{dz}{iz}$, where C is the circle $|z| = 1$.

= Real part of $\frac{1}{i} \int_C \frac{z^2}{5z-2z^2-2}\, dz$

= Real part of $-\frac{1}{i} \int_C \frac{z^2}{2z^2-5z+2}\, dz$ (1)

Let $f(z) = \frac{z^2}{2z^2-5z+2} = \frac{z^2}{(2z-1)(z-2)}$

$= \frac{z^2}{2\left(z-\frac{1}{2}\right)(z-2)}$.

$\therefore$ The poles of $f(z)$ are given by, $2\left(z-\frac{1}{2}\right)(z-2) = 0$

$\Rightarrow \quad z = \frac{1}{2}$ and $z = 2$.

Now, $z = \frac{1}{2}$ is the only simple pole which lies inside C.

$\therefore \quad [\text{Res } f(z)]_{z=\frac{1}{2}} = \lim_{z\to\frac{1}{2}} \left(z-\frac{1}{2}\right) f(z)$

$= \lim_{z\to\frac{1}{2}} \left(z-\frac{1}{2}\right) \frac{z^2}{2\left(z-\frac{1}{2}\right)(z-2)}$

$= \lim_{z\to\frac{1}{2}} \frac{z^2}{2(z-2)}$

$$= \frac{\frac{1}{4}}{2\left(\frac{1}{2}-2\right)} = -\frac{1}{12}.$$

Hence by Cauchy's residue theorem,

$$\int_C f(z)dz = 2\pi i \times \left(-\frac{1}{12}\right) = -\frac{\pi i}{6}.$$

$$\therefore \qquad I = \text{Real part of } \frac{-1}{i}\left(-\frac{\pi i}{6}\right)$$

$$= \text{Real part of } \left(\frac{\pi}{6}\right) = \frac{\pi}{6}.$$

Example 7: Evaluate $\int_0^{2\pi} \frac{\cos 2\theta \, d\theta}{5+4\cos\theta}$ and $\int_0^{2\pi} \frac{\sin 2\theta \, d\theta}{5+4\cos\theta}$.

Solution : Let $I = \int_0^{2\pi} \frac{e^{i2\theta}}{5+4\cos\theta} d\theta.$

Put $z = e^{i\theta}$. $\therefore d\theta = \frac{dz}{iz}.$

Now $\cos\theta = \frac{z^2+1}{2z}.$

$\therefore \ I = \frac{1}{i}\int_C \frac{z^2}{2z^2+5z+2} dz$, where C is unit circle $|z| = 1$.

Let $f(z) = \frac{z^2}{2z^2+5z+2}$.

The poles of $f(z)$ are given by the zeros of $2z^2 + 5z + 2 = 0$.

$\Rightarrow z = -\frac{1}{2}, -2.$

Now, $z = -\frac{1}{2}$ is the only simple pole which lies inside C.

Hence $\quad f(z) = \frac{z^2}{2\left(z+\frac{1}{2}\right)(z+2)}$ and

$$[\text{Res } f(z)]_{z=-\frac{1}{2}} = \lim_{z\to-\frac{1}{2}} \left(z+\frac{1}{2}\right) f(z)$$

$$= \frac{1}{2} \lim_{z\to-\frac{1}{2}} \frac{z^2}{z+2}$$

$$= \frac{1}{2}\left(\frac{\frac{1}{4}}{-\frac{1}{2}+2}\right) = \frac{1}{12}.$$

$\therefore$ By Cauchy's residue theorem,

$$\int_C \frac{z^2}{2z^2+5z+2}\, dz = \frac{2\pi i}{12} = \frac{\pi i}{6}.$$

$$\therefore \qquad I = \frac{1}{i} \times \frac{\pi i}{6} = \frac{\pi}{6}.$$

i.e., $\int_0^{2\pi} \frac{e^{i2\theta}}{5+4\cos\theta}\, d\theta = \frac{\pi}{6}.$

Equating the real and imaginary parts, we have

$$\int_0^{2\pi} \frac{\cos 2\theta\, d\theta}{5+4\cos\theta} = \frac{\pi}{6} \quad \text{and} \quad \int_0^{2\pi} \frac{\sin 2\theta\, d\theta}{5+4\cos\theta} = 0.$$

Example 8: Evaluate $\int_0^{2\pi} \frac{\cos 3\theta}{5-4\cos\theta} d\theta.$ ***[AU, Nov/Dec 2015]***

Solution : Let $I = \int_0^{2\pi} \frac{e^{i3\theta}}{5-4\cos\theta}\, d\theta.$

Then $\quad I = \text{Real part of} \int_0^{2\pi} \frac{e^{i3\theta}}{5-4\cos\theta}\, d\theta$

$$= \text{Real part of} \int_0^{2\pi} \frac{\left(e^{i\theta}\right)^3}{5-4\cos\theta}\, d\theta$$

Put $z = e^{i\theta}$. Then $d\theta = \frac{dz}{iz}$.

Now $\cos\theta = \frac{1}{2}\left(z + \frac{1}{z}\right) = \frac{z^2+1}{2z}$.

$$\therefore \qquad I = \text{Real part of} \int_C \frac{z^3}{5-4\left(\frac{z^2+1}{2z}\right)} \frac{dz}{iz}, \text{ where } C \text{ is the circle } |z| = 1.$$

$$= \text{Real part of } \frac{1}{i} \int_C \frac{z^3}{5z-2z^2-2}\, dz$$

$$= \text{Real part of } -\frac{1}{i} \int_C \frac{z^3}{2z^2-5z+2}\, dz \qquad (1)$$

Let $\quad f(z) = \frac{z^3}{2z^2-5z+2} = \frac{z^3}{(2z-1)(z-2)} = \frac{z^3}{2\left(z-\frac{1}{2}\right)(z-2)}.$

$\therefore$ The poles of $f(z)$ are given by, $\quad 2\left(z-\frac{1}{2}\right)(z-2) = 0$

$\Rightarrow \quad z = \frac{1}{2}$ and $z = 2$.

Now, $z = \frac{1}{2}$ is the only simple pole which lies inside C.

$\therefore \quad [\text{Res } f(z)]_{z=\frac{1}{2}} = \lim_{z \to \frac{1}{2}} \left(z - \frac{1}{2}\right) f(z)$

$$= \lim_{z \to \frac{1}{2}} \left(z - \frac{1}{2}\right) \frac{z^3}{2\left(z-\frac{1}{2}\right)(z-2)}$$

$$= \lim_{z \to \frac{1}{2}} \frac{z^3}{2(z-2)} = \frac{\frac{1}{8}}{2\left(\frac{1}{2}-2\right)} = -\frac{1}{24}.$$

Hence by Cauchy's residue theorem,

$$\int_C f(z)dz = 2\pi i \times \left(-\frac{1}{24}\right) = -\frac{\pi i}{12}.$$

$\therefore \quad I = \text{Real part of } \frac{-1}{i}\left(-\frac{\pi i}{12}\right)$

$$= \text{Real part of } \left(\frac{\pi}{12}\right) = \frac{\pi}{12}.$$

Example 9: Show that $\int_0^{2\pi} \frac{\sin^2\theta \, d\theta}{a+b\cos\theta} = \frac{2\pi}{b^2}\left(a - \sqrt{a^2-b^2}\right), (a > b > 0).$

Solution : Let $I = \int_0^{2\pi} \frac{\sin^2\theta \, d\theta}{a+b\cos\theta}.$

Put $z = e^{i\theta}$. $\therefore d\theta = \frac{dz}{iz}$. Also, $\sin\theta = \frac{1}{2i}\left(z - \frac{1}{z}\right)$ and $\cos\theta = \frac{1}{2}\left(z + \frac{1}{z}\right)$.

$\therefore I = \int_C \frac{\frac{-1}{4}\left(z-\frac{1}{z}\right)^2 dz/iz}{a+\frac{b}{2}\left(z+\frac{1}{z}\right)} = \frac{-1}{2i}\int_C \frac{\left(z^2-1\right)^2 dz}{z^2\,(bz^2+2az+b)}$, where C is the unit circle $|z| = 1$.

Let $f(z) = \frac{\left(z^2-1\right)^2}{z^2\,(bz^2+2az+b)}$.

The poles of $f(z)$ are given by the zeros of $z^2\,(bz^2 + 2az + b) = 0$.

i.e., $z^2 = 0$ and $bz^2 + 2az + b = 0$. Hence $z = 0$ is a pole of order 2 and

$z = \frac{-2a\pm\sqrt{4a^2-4b^2}}{2b} = \frac{-a\pm\sqrt{a^2-b^2}}{b}$ are simple poles.

Let $\alpha = \frac{-a+\sqrt{a^2-b^2}}{b}$ and $\beta = \frac{-a-\sqrt{a^2-b^2}}{b}$.

Then $f(z) = \frac{\left(z^2-1\right)^2}{z^2 b\,(z-\alpha)\,(z-\beta)}$.

Now, since $a > b$, $a^2 - b^2 > 0$ and $\frac{a}{b} > 1$.

$\therefore |\beta| = \frac{a+\sqrt{a^2-b^2}}{b} > 1$ and so $z = \beta$ lies outside the circle C. Since $\alpha\beta = 1$, we have $|\alpha\beta| = 1$.

$\therefore |\alpha| = \frac{1}{|\beta|} < 1$.

$\therefore z = \alpha$ lies inside C and it is a simple pole.

$$\therefore [\text{Res}\, f(z)]_{z=0} = \frac{1}{(2-1)!} \lim_{z\to 0} \frac{d}{dz}[(z-0)^2 f(z)]$$

$$= \lim_{z\to 0} \frac{d}{dz}\left[\frac{z^2(z^2-1)^2}{z^2\,(bz^2+2az+b)}\right]$$

$$= \lim_{z\to 0} \frac{d}{dz}\left(\frac{(z^2-1)^2}{bz^2+2az+b}\right)$$

$$= \lim_{z\to 0}\left[\frac{(bz^2+2az+b)\,4z\,(z^2-1)-(z^2-1)^2(2bz+2a)}{(bz^2+2az+b)^2}\right]$$

$$= -\frac{2a}{b^2}.$$

We have $f(z) = \frac{(z^2-1)^2}{b\,(z-\alpha)(z-\beta)z^2}$.

$$\therefore [\text{Res}\, f(z)]_{z=\alpha} = \lim_{z\to\alpha} (z-\alpha)\, f(z)$$

$$= \lim_{z\to\alpha} \frac{(z-\alpha)(z^2-1)^2}{b\,(z-\alpha)\,(z-\beta)\,z^2}$$

$$= \lim_{z\to\alpha} \frac{(z^2-1)^2}{b\,(z-\beta)\,z^2}$$

$$= \frac{(\alpha^2-1)^2}{b\,(\alpha-\beta)\,\alpha^2}$$

$$= \frac{(\alpha^2-\alpha\beta)^2}{b\,(\alpha-\beta)\,\alpha^2}, \text{ since } \alpha\beta = 1.$$

$$= \frac{\alpha^2(\alpha-\beta)^2}{b\,(\alpha-\beta)\,\alpha^2} = \frac{(\alpha-\beta)}{b} = \frac{2\sqrt{a^2-b^2}}{b^2}$$

$\therefore$ By Cauchy's residue theorem,

$$\int_C f(z)dz = 2\pi i\left[\frac{-2a}{b^2} + \frac{2\sqrt{a^2-b^2}}{b^2}\right]$$

$$= \frac{4\pi i}{b^2}\left[-a+\sqrt{a^2-b^2}\right].$$

$$\therefore \qquad I = -\frac{1}{2i}\,\frac{4\pi i}{b^2}\left[-a+\sqrt{a^2-b^2}\right]$$

$$= \frac{-2\pi}{b^2}\left[-a+\sqrt{a^2-b^2}\right]$$

$$= \frac{2\pi}{b^2}\left[a-\sqrt{a^2-b^2}\right].$$

Example 10: Evaluate the integral $\int_0^{2\pi} \frac{\sin^2\theta}{5+4\cos\theta}\, d\theta$.

Solution : Let $I = \int_0^{2\pi} \frac{\sin^2\theta}{5+4\cos\theta}\, d\theta$.

Let $z = e^{i\theta}$. $\therefore d\theta = \frac{dz}{iz}$.

Also, $\cos\theta = \frac{z^2+1}{2z}$ and $\sin^2\theta = \frac{1-\cos 2\theta}{2}$ = Real part of $\left(\frac{1-e^{i2\theta}}{2}\right)$

$$= \text{Real part of} \left(\frac{1-z^2}{2}\right)$$

$$\therefore \quad I = \text{Real part of} \int_C \frac{\frac{1-z^2}{2}}{5-4\left(\frac{z^2+1}{2z}\right)} \frac{dz}{iz}$$

$= \text{Real part of } \frac{1}{2i} \int_C \frac{1-z^2}{2z^2+5z+2}\, dz$, where C is the unit circle $|z| = 1$.

$$\text{Let} \quad f(z) = \frac{1-z^2}{2z^2-5z+2} = \frac{1-z^2}{(2z+1)(z+2)}$$

$$= \frac{1-z^2}{2\left(z+\frac{1}{2}\right)(z+2)}.$$

$\therefore$ The poles of $f(z)$ are $z = -\frac{1}{2}$ and $z = -2$.

Clearly, $z = -\frac{1}{2}$ is the only simple pole which lies inside C.

$$\therefore \quad [\text{Res } f(z)]_{z=-\frac{1}{2}} = \lim_{z\to -\frac{1}{2}} \left(z+\frac{1}{2}\right) f(z)$$

$$= \lim_{z\to -\frac{1}{2}} \left(z+\frac{1}{2}\right) \frac{1-z^2}{2\left(z+\frac{1}{2}\right)(z+2)}$$

$$= \lim_{z\to -\frac{1}{2}} \frac{1-z^2}{2(z+2)}$$

$$= \frac{1}{2}\left(\frac{1-\frac{1}{4}}{\frac{-1}{2}+2}\right) = \frac{1}{4}.$$

Hence by Cauchy's residue theorem,

$$\int_C f(z)dz = 2\pi i \times \left(\frac{1}{4}\right) = \frac{\pi i}{2}.$$

$$\therefore \quad I = \text{Real part of } \frac{1}{2i}\left(\frac{\pi i}{2}\right)$$

$$= \text{Real part of} \left(\frac{\pi}{4}\right) = \frac{\pi}{4}.$$

EXERCISES

1. Evaluate $\int_0^{2\pi} \frac{d\theta}{2+\cos\theta}$.
2. Using contour integration evaluate $\int_0^{2\pi} \frac{d\theta}{5-\sin\theta}$.
3. Evaluate $\int_0^{2\pi} \frac{d\theta}{5+4\cos\theta}$.
4. Evaluate $\int_0^{2\pi} \frac{d\theta}{5+3\cos\theta}$.
5. Prove that $\int_{-\pi}^{\pi} \frac{a\cos\theta}{a+\cos\theta}\, d\theta = 2\pi a\left(1-\frac{a}{\sqrt{a^2-1}}\right)$, where $a > 1$.
6. Evaluate $\int_0^{2\pi} \frac{1}{a^2-2a\cos\theta+1}\, d\theta$.
7. Prove that $\int_0^{2\pi} \frac{d\theta}{(5-4\cos\theta)^2} = \frac{40\pi}{27}$.
8. Prove that $\int_0^{2\pi} \frac{d\theta}{17-8\cos\theta} = \frac{2\pi}{15}$.
9. Prove the following:

   (i) $\int_0^{2\pi} \frac{d\theta}{(a+b\cos\theta)^2} = \frac{2\pi a}{(a^2-b^2)^{\frac{3}{2}}}$, where $a > |b| > 0$.

   (ii) $\int_0^{2\pi} \frac{d\theta}{(5-3\sin\theta)^2} = \frac{5\pi}{32}$.
10. Prove that $\int_0^{2\pi} \frac{\sin^2\theta\, d\theta}{a+b\cos\theta} = \frac{2\pi}{b^2}\left(a-\sqrt{a^2-b^2}\right)$, $a > b > 0$.
11. Evaluate $\int_0^{\pi} \frac{a\, d\theta}{a^2+\sin^2\theta}$ $(a > 0)$.
12. Evaluate $\int_0^{\pi} \frac{1}{a^2+\cos^2\theta}\, d\theta$.
13. Prove that $\int_0^{\pi} \frac{\cos 2\theta\, d\theta}{1-2a\cos\theta+a^2} = \frac{\pi a^2}{1-a^2}$ $(-1 < a < 1)$.
14. Prove that $\int_0^{\pi} \frac{d\theta}{(a+\cos\theta)^2} = \frac{\pi a}{(a^2-1)^{\frac{3}{2}}}$, $a > 0$.

ANSWERS

1. $\frac{2\pi}{\sqrt{3}}$ 2. $\frac{\pi}{\sqrt{6}}$ 3. $\frac{2\pi}{3}$ 4. $\frac{\pi}{2}$ 6. $\frac{2\pi}{1-a^2}$, $0 < a < 1$ 11. $\frac{\pi}{\sqrt{1-a^2}}$ 12. $\frac{\pi}{a\sqrt{1+a^2}}$.

Type II: Integrals of the form $\int_{-\infty}^{\infty} \frac{P(x)}{Q(x)} dx$, where $P(x)$ and $Q(x)$ are polynomials in x.

This integral exists, if

(*i*) the degree of $Q(x)$ is at least 2 greater than the degree of $P(x)$.

(*ii*) $Q(x) \neq 0$ for any real value of x.

To evaluate this integral, we evaluate $\int_C \frac{P(z)}{Q(z)} dz$, where C is the closed contour consisting of the real axis from $-R$ to $+R$ and the semicircle S above the real axis having this line as diameter, by using Cauchy's residue theorem and then letting $R \to \infty$. The contour C is shown in Fig.

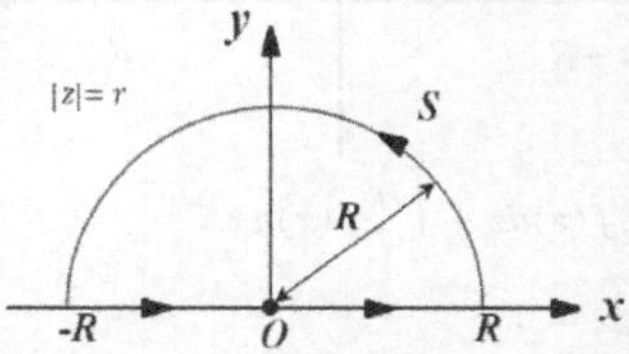

We state a lemma before proceeding to indicate the method of evaluating the integral.

Cauchy's Lemma

If $f(z)$ is a continuous function such that $|zf(z)| \to 0$ uniformly as $|z| \to \infty$, then $\int_S f(z)dz \to 0$, as $R \to \infty$ where S is the semi - circle $|z| = R$ above the real axis.

Proof:

$$\left|\int_S f(z)dz\right| = \left|\int_S z\, f(z)\, \frac{1}{z} dz\right|$$

$$\leq \int_S |z\, f(z)| \frac{1}{|z|}\, |dz| = \int_0^{\pi} |z\, f(z)| \frac{1}{R}\, d\theta$$

[When $|z| = R$, $z = Re^{i\theta}$ and $dz = Rie^{i\theta} d\theta$]

$\therefore \lim_{R\to\infty} \left|\int_S f(z)dz\right| \leq \lim_{R\to\infty} |z\, f(z)| \leq 0$, by the given condition.

$\therefore \left|\int_S f(z)dz\right| = 0$ and hence $\int_S f(z)dz \to 0$, as $R \to \infty$. Hence proved.

To evaluate the integrals of the form $\int_{-\infty}^{\infty} f(x)dx$, where $f(x) = \frac{P(x)}{Q(x)}$, we consider a closed curve C, consisting of

(*i*) the semi - circle Γ given by $|z| = R$, $Im(z) > 0$ and

(*ii*) the line segment of the real axis from $-R$ to R.

Since $f(z)$ has only a finite number of singular points, we can choose R so large that the singular points of $f(z)$ are all contained in the interior of C.

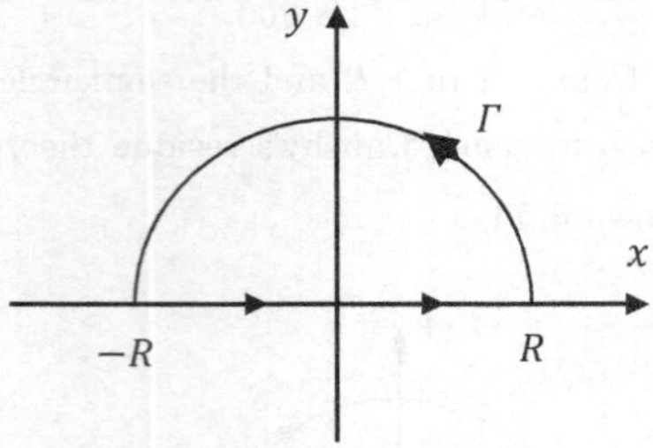

$\therefore$ We have $\int_C f(z)dz = \int_\Gamma f(z)dz + \int_{-R}^{+R} f(x)dx$.

By Lemma , $\lim_{R\to\infty} \int_\Gamma f(z)dz = 0$.

Also, $\lim_{R\to\infty} \int_{-R}^{+R} f(x)dx = \int_{-\infty}^{\infty} f(x)dx$.

$\therefore$ $\int_{-\infty}^{\infty} f(x)dx = \int_C f(z)dz$ which can be evaluated using Cauchy's residue theorem.

Note: We consider the case that $f(z)$ has no poles on the real axis.

Example 1: Calculate $\int_0^\infty \frac{x^2}{(x^2+1)^2} dx$.

Solution:

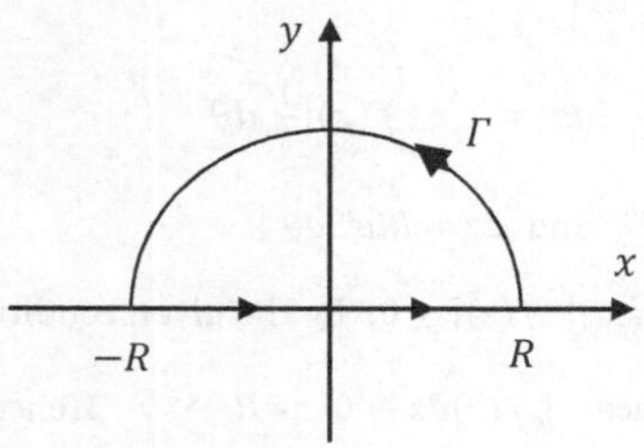

Let $f(x) = \frac{x^2}{(x^2+1)^2}$.

Consider the integral $\int_C f(z)dz$, where $f(z) = \frac{z^2}{(z^2+1)^2}$ and C is the contour consisting of a large upper semi-circle Γ of radius R and centre at origin along with the part of the real axis from $-R$ to $+R$.

$$\therefore \int_C f(z)\,dz = \int_\Gamma f(z)dz + \int_{-R}^{+R} f(x)\,dx \quad (1)$$

The poles of $f(z)$ are given by the zeros of $(z^2+1)^2 = 0$.

$\Rightarrow \quad z = \pm i$ (twice)

$\therefore \quad z = i$ is the only pole of order 2, which lies inside C.

Now, $\quad z^2+1 = (z+i)(z-i)$.

$$\therefore [\text{Res } f(z)]_{z=i} = \frac{1}{(2-1)!} \lim_{z\to i} \frac{d}{dz}[(z-i)^2 f(z)]$$

$$= \lim_{z\to i} \frac{d}{dz}\left[(z-i)^2 \frac{z^2}{(z-i)^2(z+i)^2}\right]$$

$$= \lim_{z\to i} \frac{d}{dz}\left[\frac{z^2}{(z+i)^2}\right]$$

$$= \lim_{z\to i}\left[\frac{(z+i)^2 2z - 2z^2(z+i)}{(z+i)^4}\right]$$

$$= \lim_{z\to i}\left[\frac{2zi}{(z+i)^3}\right] = \frac{-2}{(2i)^3} = -\frac{1}{(-4i)} = \frac{1}{4i}.$$

$\therefore$ By Cauchy's residue theorem,

$$\int_C f(z)dz = 2\pi i \times \frac{1}{4i} = \frac{\pi}{2}.$$

$\therefore$ (1) gives, $\int_\Gamma f(z)dz + \int_{-R}^{+R} f(x)dx = \frac{\pi}{2}$.

Taking limit as $R \to \infty$, $\int_\Gamma f(z)dz \to 0$ [By Cauchy's lemma].

$\therefore \quad \int_{-\infty}^{\infty} f(x)dx = \frac{\pi}{2}$.

i.e., $\int_{-\infty}^{\infty} \frac{x^2}{(x^2+1)^2} dx = \frac{\pi}{2}$.

Since $\frac{x^2}{(x^2+1)^2}$ is even, we get $\int_0^{\infty} \frac{x^2}{(x^2+1)^2} dx = \frac{1}{2}.\frac{\pi}{2} = \frac{\pi}{4}$.

Example 2: Evaluate by contour integration $\int_0^{\infty} \frac{dx}{(1+x^2)^2}$.

Solution: ***[AU, Nov/Dec 2017, May/June 2014]***

Let $(x) = \frac{1}{(x^2+1)^2}$.

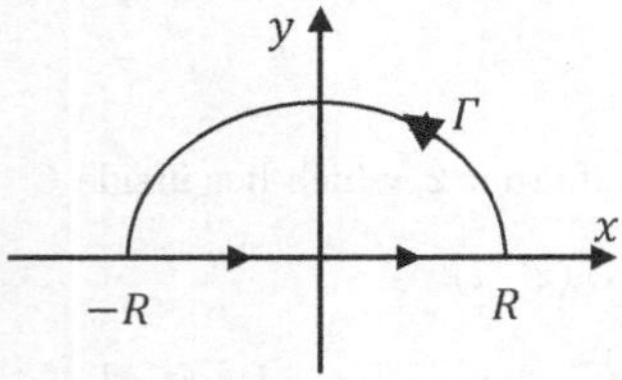

Consider the integral $\int_C f(z)dz$, where $f(z) = \frac{1}{(z^2+1)^2}$ and C is the contour consisting of a large upper semi-circle Γ of radius R and centre at origin along with the part of the real axis from $-R$ to $+R$.

$\therefore \int_C f(z)dz = \int_\Gamma f(z)dz + \int_{-R}^{+R} f(x)dx$ (1)

The poles of $f(z)$ are given by the zeros of $(z^2+1)^2 = 0$.

$\Rightarrow z = \pm i$ (twice)

$\therefore z = i$ is the only pole of order 2, which lies inside C.

Now, $z^2 + 1 = (z+i)(z-i)$.

$$\therefore [\text{Res}\, f(z)]_{z=i} = \frac{1}{(2-1)!} \lim_{z\to i} \frac{d}{dz} [(z-i)^2 f(z)]$$

$$= \lim_{z\to i} \frac{d}{dz} \left[(z-i)^2 \frac{1}{(z-i)^2(z+i)^2}\right]$$

$$= \lim_{z\to i} \frac{d}{dz} \left[\frac{1}{(z+i)^2}\right]$$

$$= \lim_{z\to i} \left[\frac{-2}{(z+i)^3}\right] = \frac{-2}{(2i)^3} = -\frac{1}{(-4i)} = \frac{1}{4i}.$$

$\therefore$ By Cauchy's residue theorem,

$$\int_C f(z)dz = 2\pi i \times \frac{1}{4i} = \frac{\pi}{2}.$$

$\therefore$ (1) gives, $\int_\Gamma f(z)dz + \int_{-R}^{+R} f(x)dx = \frac{\pi}{2}$.

Taking limit as $R \to \infty$, $\int_\Gamma f(z)dz \to 0$ [By Cauchy's lemma].

$$\therefore \int_{-\infty}^{\infty} f(x)dx = \frac{\pi}{2}$$

i.e., $\int_{-\infty}^{\infty} \frac{dx}{(x^2+1)^2} = \frac{\pi}{2}$

$$\therefore \int_0^{\infty} \frac{dx}{(x^2+1)^2} = \frac{\pi}{4}.$$

Example 3: Evaluate $\int_0^\infty \frac{dx}{(x^2+a^2)^2}$, $(a > 0)$ using contour integration.

Solution: ***[AU, April/May 2015]***

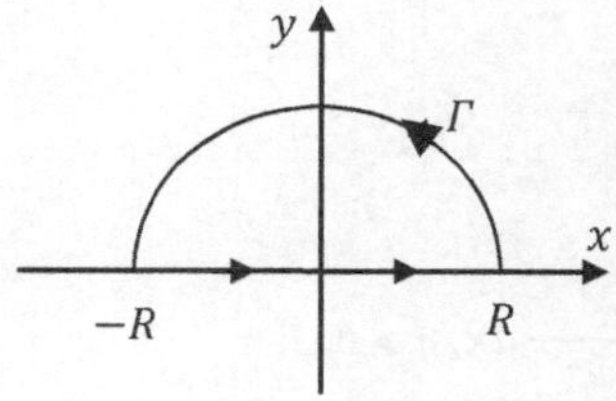

Let $f(x) = \frac{1}{(x^2+a^2)^2}$.

Consider the integral $\int_C f(z)dz$, where $f(z) = \frac{1}{(z^2+a^2)^2}$ and C is the contour consisting of a large upper semi-circle Γ of radius R and centre at origin along with the part of the real axis from $-R$ to $+R$.

$\therefore \int_C f(z)dz = \int_\Gamma f(z)dz + \int_{-R}^{+R} f(x)dx$ (1)

The poles of $f(z)$ are given by the zeros of $(z^2 + a^2)^2 = 0$.

$\Rightarrow z = \pm ai$ (twice)

$\therefore$ $z = ai$ is the only pole of order 2, which lies inside C.

Now, $z^2 + a^2 = (z + ai)(z - ai)$.

$$\therefore [\text{Res}\, f(z)]_{z=ai} = \frac{1}{(2-1)!} \lim_{z\to ai} \frac{d}{dz}[(z - ai)^2 f(z)]$$

$$= \lim_{z\to ai} \frac{d}{dz}\left[(z - ai)^2 \frac{1}{(z-ai)^2(z+ai)^2}\right]$$

$$= \lim_{z\to ai} \frac{d}{dz}\left[\frac{1}{(z+ai)^2}\right]$$

$$= \lim_{z\to ai}\left[\frac{-2}{(z+ai)^3}\right] = \frac{-2}{(2ai)^3} = -\frac{1}{(-4a^3i)} = \frac{1}{4a^3i}.$$

$\therefore$ By Cauchy's residue theorem,

$$\int_C f(z)dz = 2\pi i \times \frac{1}{4a^3i} = \frac{\pi}{2a^3}.$$

$\therefore$ (1) gives, $\int_\Gamma f(z)dz + \int_{-R}^{+R} f(x)dx = \frac{\pi}{2}$.

Taking limit as $R \to \infty$, $\int_{\Gamma} f(z)dz \to 0$ [By Cauchy's lemma].

$$\therefore \quad \int_{-\infty}^{\infty} f(x)dx = \frac{\pi}{2a^3}.$$

$$\text{i.e.,} \int_{-\infty}^{\infty} \frac{dx}{(x^2+a^2)^2} = \frac{\pi}{2a^3}.$$

$$\therefore \quad \int_{0}^{\infty} \frac{dx}{(x^2+a^2)^2} = \frac{\pi}{4a^3}.$$

Example 4: Evaluate $\int_0^{\infty} \frac{x^2}{(x^2+a^2)^3} dx,\ a > 0$.

Solution : Let $f(x) = \frac{x^2}{(x^2+a^2)^3}$.

Consider the integral $\int_C f(z)dz$, where $f(z) = \frac{z^2}{(z^2+a^2)^3}$ and C is the contour consisting of a sufficiently large upper semi-circle Γ with center origin and radius R and part of the real axis from $-R$ to $+R$.

$$\therefore \int_C f(z)dz = \int_{\Gamma} f(z)dz + \int_{-R}^{R} f(x)dx \qquad (1)$$

The poles of $f(z)$ are given by the zeroes of $(z^2 + a^2)^3 = 0$.

$\Rightarrow \quad z = \pm ai$ (three times)

$\therefore \quad z = ai$ is the only pole of order 3 which lies within C.

Now, $z^2 + a^2 = (z + ai)(z - ai)$.

$$\therefore [\text{Res } f(z)]_{z=ai} = \frac{1}{(3-1)!} \lim_{z \to ai} \frac{d^{3-1}}{dz^{3-1}} [(z - ai)^3 f(z)]$$

$$= \frac{1}{2!} \lim_{z \to ai} \frac{d^2}{dz^2} \left[(z - ai)^3 \frac{z^2}{(z-ai)^3 (z+ai)^3}\right]$$

$$= \frac{1}{2} \lim_{z \to ai} \frac{d^2}{dz^2} \left[\frac{z^2}{(z+ai)^3}\right] = \frac{1}{2} \lim_{z \to ai} \frac{d}{dz} \left[\frac{-z^2+2aiz}{(z+ai)^4}\right]$$

$$= \frac{1}{2} \lim_{z \to ai} \left[\frac{2z^2-2a^2-8aiz}{(z+ai)^5}\right] = \frac{1}{2} \left[\frac{-2a^2-2a^2+8a^2}{(2ai)^5}\right]$$

$$= \frac{1}{2} \frac{4a^2}{(2^5 a^5 i)} = \frac{1}{16a^3 i}.$$

$\therefore$ By Cauchy's residue theorem,

$$\int_C f(z)dz = 2\pi i \times \frac{1}{16a^3 i} = \frac{\pi}{8a^3}.$$

$\therefore$ (1) gives, $\int_\Gamma f(z)dz + \int_{-R}^{+R} f(x)dx = \frac{\pi}{8a^3}$.

Taking limit as $R \to \infty$, $\int_\Gamma f(z)dz \to 0$ [By Cauchy's lemma].

$\therefore \int_{-\infty}^{\infty} f(x)dx = \frac{\pi}{8a^3}$.

i.e., $\int_{-\infty}^{\infty} \frac{x^2}{(x^2+a^2)^3} dx = \frac{\pi}{8a^3}$.

Example 5: Use calculus of residues to find $\int_0^\infty \frac{1}{(x^2+a^2)(x^2+b^2)} dx$ where $a, b > 0$.

Solution: **[AU, *Nov/Dec 2017*]**

Let $f(x) = \frac{1}{(x^2+a^2)(x^2+b^2)}$.

Consider the integral $\int_C f(z)dz$, where $f(z) = \frac{1}{(z^2+a^2)(z^2+b^2)}$ and C is the contour consisting of a large upper semi-circle Γ with centre origin and radius R, and part of the real axis from $-R$ to R.

$\therefore \int_C f(z)dz = \int_\Gamma f(z)dz + \int_{-R}^{+R} f(x)dx \qquad (1)$

The poles of $f(z)$ are given by the zeros of $(z^2 + a^2)(z^2 + b^2) = 0$.

$\Rightarrow z = \pm ai$ and $z = \pm bi$.

$\therefore z = ai$ and $z = bi$ are simple poles which lie within C.

Now, $z^2 + a^2 = (z + ai)(z - ai)$ and $z^2 + b^2 = (z + bi)(z - bi)$.

$$[\text{Res } f(z)]_{z=ai} = \lim_{z\to ai} (z - ai) f(z) = \lim_{z\to ai} (z - ai)\frac{1}{(z-ai)(z+ai)(z^2+b^2)}$$

$$= \lim_{z\to ai} \frac{1}{(z+ai)(z^2+b^2)} = \frac{1}{2ai(-a^2+b^2)}$$

$$= \frac{1}{2ai(-a^2+b^2)} = \frac{-1}{2ai(a^2-b^2)}.$$

$$[\text{Res } f(z)]_{z=bi} = \lim_{z\to bi} (z - bi)f(z) = \lim_{z\to bi} (z - bi)\frac{1}{(z^2+a^2)(z-bi)(z+bi)}$$

$$= \lim_{z\to bi} \frac{1}{(z^2+a^2)(z+bi)} = \frac{1}{2bi(-b^2+a^2)}$$

$$= \frac{1}{2bi(a^2-b^2)}.$$

$\therefore$ By Cauchy's residue theorem,

$$\int_C f(z)dz = 2\pi i \times \left[\frac{-1}{2ai\,(a^2-b^2)} + \frac{1}{2bi\,(a^2-b^2)}\right] = \pi\left[\frac{-1}{a\,(a^2-b^2)} + \frac{1}{b\,(a^2-b^2)}\right]$$

$$= \pi\left[\frac{a-b}{ab\,(a^2-b^2)}\right] = \frac{\pi}{ab\,(a+b)}.$$

$\therefore$ (1) gives, $\int_\Gamma f(z)dz + \int_{-R}^{+R} f(x)dx \;=\; \frac{\pi}{ab\,(a+b)}.$

Taking limit as $R \to \infty$, $\int_\Gamma f(z)dz \to 0$ [By Cauchy's lemma].

$\therefore\; \int_{-\infty}^{\infty} f(x)dx \;=\; \frac{\pi}{ab\,(a+b)}.$

$$\therefore\; \int_{-\infty}^{\infty} \frac{1}{(x^2+a^2)(x^2+b^2)}\,dx \;=\; \frac{\pi}{ab\,(a+b)}.$$

$$\therefore\; \int_0^{\infty} \frac{1}{(x^2+a^2)(x^2+b^2)}\,dx \;=\; \frac{\pi}{2ab\,(a+b)}.$$

Example 6: Evaluate $\int_{-\infty}^{\infty} \frac{x^2\,dx}{(x^2+a^2)(x^2+b^2)}$, $a > 0, b > 0$.

Solution: Let $f(x) = \frac{x^2}{(x^2+a^2)(x^2+b^2)}$.

Consider the integral $\int_C f(z)dz$, where $f(z) = \frac{z^2}{(z^2+a^2)(z^2+b^2)}$ and C is the contour consisting of a large upper semi-circle Γ with centre origin and radius R, and part of the real axis from $-R$ to R.

$$\therefore\; \int_C f(z)dz \;=\; \int_\Gamma f(z)dz + \int_{-R}^{+R} f(x)dx \qquad (1)$$

The poles of $f(z)$ are given by the zeros of $(z^2 + a^2)(z^2 + b^2) = 0$.

$\Rightarrow\; z = \pm\, ai$ and $z = \pm\, bi$.

$\therefore z = ai$ and $z = bi$ are simple poles which lie within C.

Now, $z^2 + a^2 = (z + ai)(z - ai)$ and $z^2 + b^2 = (z + bi)(z - bi)$.

$$[\text{Res}\, f(z)]_{z=ai} \;=\; \lim_{z\to ai}\,(z - ai)\,f(z) = \lim_{z\to ai}(z - ai)\frac{z^2}{(z-ai)(z+ai)(z^2+b^2)}$$

$$= \lim_{z\to ai}\frac{z^2}{(z+ai)(z^2+b^2)} \;=\; \frac{-a^2}{2ai\,(-a^2+b^2)}$$

$$= \frac{-a}{2i(-a^2+b^2)} \;=\; \frac{a}{2i(a^2-b^2)}.$$

$$[\text{Res } f(z)]_{z=bi} = \lim_{z\to bi}(z-bi)f(z) = \lim_{z\to bi}(z-bi)\frac{z^2}{(z^2+a^2)(z-bi)(z+bi)}$$

$$= \lim_{z\to bi}\frac{z^2}{(z^2+a^2)(z+bi)} = \frac{-b^2}{2bi\,(-b^2+a^2)}$$

$$= \frac{-b}{2i(a^2-b^2)}.$$

$\therefore$ By Cauchy's residue theorem,

$$\int_C f(z)dz = 2\pi i \times \left[\frac{a}{2i(a^2-b^2)} + \frac{-b}{2i(a^2-b^2)}\right] = \pi\left[\frac{a}{(a^2-b^2)} - \frac{b}{(a^2-b^2)}\right]$$

$$= \pi\left[\frac{a-b}{(a^2-b^2)}\right] = \frac{\pi}{a+b}.$$

$\therefore$ (1) gives, $\int_\Gamma f(z)dz + \int_{-R}^{+R} f(x)dx = \frac{\pi}{a+b}.$

Taking limit as $R \to \infty$, $\int_\Gamma f(z)dz \to 0$ [By Cauchy's lemma].

$\therefore \int_{-\infty}^{\infty} f(x)dx = \frac{\pi}{a+b}.$

$\therefore \int_{-\infty}^{\infty} \frac{x^2}{(x^2+a^2)(x^2+b^2)}dx = \frac{\pi}{a+b}.$

Example 7: Evaluate $\int_0^{\infty} \frac{x^2 dx}{(x^2+4)(x^2+9)}$ using contour integration.

Solution : Let $f(x) = \frac{x^2}{(x^2+4)(x^2+9)}.$

Consider the integral $\int_C f(z)dz$, where $f(z) = \frac{z^2}{(z^2+4)(z^2+9)}$ and C is the contour consisting of a sufficiently large upper semi-circle Γ with centre origin and radius R and part of the real axis from $-R$ to R.

$\therefore \int_C f(z)dz = \int_\Gamma f(z)dz + \int_{-R}^{R} f(x)dx$ (1)

The poles of $f(z)$ are given by the zeros of $(z^2+4)(z^2+9) = 0.$

$\Rightarrow z = \pm 2i$ and $z = \pm 3i.$

$\therefore z = 2i$ and $z = 3i$ are the simple poles which lie within C.

Now, $z^2 + 4 = (z+2i)(z-2i)$ and $z^2 + 9 = (z+3i)(z-3i).$

$$[\text{Res } f(z)]_{z=2i} = \lim_{z\to 2i}(z-2i)f(z) = \lim_{z\to 2i}(z-2i)\frac{z^2}{(z-2i)(z+2i)(z^2+9)}$$

$$= \lim_{z \to 2i} \frac{z^2}{(z+2i)(z^2+9)} = \frac{-4}{4i(-4+9)} = \frac{i}{5}.$$

$$[\text{Res } f(z)]_{z=3i} = \lim_{z \to 3i}(z-3i)f(z) = \lim_{z \to 3i}(z-3i)\frac{z^2}{(z^2+4)(z-3i)(z+3i)}$$

$$= \lim_{z \to 3i} \frac{z^2}{(z^2+4)(z+3i)} = \frac{-9}{(-9+4)(6i)} = \frac{-3i}{10}.$$

$\therefore$ By Cauchy's residue theorem,

$$\int_C f(z)dz = 2\pi i \times \left(\frac{i}{5} - \frac{3i}{10}\right) = -2\pi\left[\frac{-1}{10}\right] = \frac{\pi}{5}.$$

Hence (1) gives, $\int_\Gamma f(z)dz + \int_{-R}^{+R} f(x)dx = \frac{\pi}{5}.$

Taking limit as $R \to \infty$, $\int_\Gamma f(z)dz \to 0$ [By Cauchy's lemma].

$\therefore \quad \int_{-\infty}^{\infty} f(x)dx = \frac{\pi}{5}.$

i.e., $\int_{-\infty}^{\infty} \frac{x^2 dx}{(x^2+4)(x^2+9)} = \frac{\pi}{5}.$

Since $\frac{x^2}{(x^2+4)(x^2+9)}$ is even, we get $\int_0^{\infty} \frac{x^2 dx}{(x^2+4)(x^2+9)} = \frac{1}{2}.\frac{\pi}{5} = \frac{\pi}{10}.$

Example 8: Evaluate $\int_0^{\infty} \frac{x^2 dx}{(x^2+9)(x^2+4)^2}$. ***[AU, Nov/Dec 2014]***

Solution : Let $f(x) = \frac{x^2}{(x^2+9)(x^2+4)^2}$.

Consider the integral $\int_C f(z)dz$, where $f(z) = \frac{z^2}{(z^2+9)(z^2+4)^2}$ and C is the contour consisting of a sufficiently large upper semi-circle Γ with centre origin and radius R and part of the real axis from $-R$ to R.

$\therefore \int_C f(z)dz = \int_\Gamma f(z)dz + \int_{-R}^{R} f(x)dx$ (1)

The poles of $f(z)$ are given by the zeros of $(z^2+9)(z^2+4)^2 = 0.$

$\Rightarrow z = \pm 3i$ and $z = \pm 2i$(twice)

$\therefore \quad z = 3i$ is a simple pole which lies within C and

$z = 2i$ is a double pole which lies within C.

Now, $z^2 + 9 = (z+3i)(z-3i)$ and $(z^2+4)^2 = (z+2i)^2(z-2i)^2$.

$$[\text{Res } f(z)]_{z=3i} = \lim_{z \to 3i}(z-3i)f(z) = \lim_{z \to 3i}(z-3i)\frac{z^2}{(z+3i)(z-3i)(z^2+4)^2}$$

$$= \lim_{z \to 3i} \frac{z^2}{(z+3i)(z^2+4)^2} = \frac{-9}{6i(-9+4)^2} = \frac{3i}{50}.$$

$$\therefore\ [\text{Res}\, f(z)]_{z=2i} = \frac{1}{(2-1)!}\lim_{z\to 2i}\frac{d}{dz}[(z-2i)^2 f(z)]$$

$$= \lim_{z\to 2i}\frac{d}{dz}\left[(z-2i)^2\frac{z^2}{(z^2+9)\,(z+2i)^2\,(z-2i)^2}\right]$$

$$= \lim_{z\to 2i}\frac{d}{dz}\left[\frac{z^2}{(z^2+9)\,(z+2i)^2}\right]$$

$$= \lim_{z\to 2i}\frac{(z^2+9)(z+2i)^2\,2z - z^2[2(z^2+9)(z+2i)+2z(z+2i)^2]}{(z^2+9)^2(z+2i)^4}$$

$$= \lim_{z\to 2i}\frac{(z^2+9)(z+2i)2z-2z^2[(z^2+9)+z(z+2i)]}{(z^2+9)^2(z+2i)^3}$$

$$= \frac{(-4+9)(4i)4i+8[(-4+9)+2i(4i)]}{(-4+9)^2(4i)^3} = \frac{-80-24}{25(-64)i}$$

$$= \frac{-104i}{1600} = \frac{-26i}{400}.$$

$\therefore$ By Cauchy's residue theorem,

$$\int_C f(z)dz = 2\pi i \times \left(\frac{3i}{50}-\frac{26i}{400}\right) = -2\pi\left[\frac{24-26}{400}\right] = \frac{\pi}{100}.$$

Hence (1) gives, $\int_\Gamma f(z)dz + \int_{-R}^{+R} f(x)dx = \frac{\pi}{100}.$

Taking limit as $R\to\infty$, $\int_\Gamma f(z)dz \to 0$ [By Cauchy's lemma].

$$\therefore\ \int_{-\infty}^{\infty} f(x)dx = \frac{\pi}{100}.$$

i.e., $\displaystyle\int_{-\infty}^{\infty}\frac{x^2\,dx}{(x^2+9)(x^2+4)^2} = \frac{\pi}{100}.$

Since $\frac{x^2}{(x^2+9)(x^2+4)^2}$ is even, we get $2\int_0^\infty \frac{x^2}{(x^2+9)(x^2+4)^2} = \frac{\pi}{100}.$

$$\int_0^\infty \frac{x^2}{(x^2+9)(x^2+4)^2} = \frac{1}{2}\cdot\frac{\pi}{100} = \frac{\pi}{200}.$$

Example 9: Prove that $\int_0^\infty \frac{dx}{x^4+a^4} = \frac{\pi\sqrt{2}}{4a^3}$, $(a>0)$.

Solution : Let $f(x) = \frac{1}{x^4+a^4}.$

Consider the integral $\int_C f(z)dz$, where $f(z) = \frac{1}{z^4+a^4}$ and C is the contour consisting of a sufficiently large semi - circle Γ with centre origin and radius R and part of the real axis $-R$ to R.

Then $\int_C f(z)dz = \int_\Gamma f(z)dz + \int_{-R}^{R} f(x)\,dx$ (1)

The poles of $f(z)$ are given by the zeros of $z^4+a^4 = 0$.

$z^4 = -a^4$

$= a^4 (\cos\pi + i\sin\pi).$

$\therefore z = a(\cos\pi + i\sin\pi)^{\frac{1}{4}}$

$= a(\cos[\pi + 2n\pi] + i\sin[\pi + 2n\pi])^{\frac{1}{4}}$

$= a\left(\cos\left(\frac{\pi+2n\pi}{4}\right) + i\sin\left(\frac{\pi+2n\pi}{4}\right)\right),\ n = 0, 1, 2, 3.$

i.e., $z = ae^{i\pi/4},\ ae^{i3\pi/4},\ ae^{i5\pi/4},\ ae^{i7\pi/4}.$

$z = ae^{i\pi/4}$ and $z = ae^{i3\pi/4}$ are the simple poles which lie within C.

$$\therefore [\text{Res}\, f(z)]_{z=ae^{i\pi/4}} = \lim_{z\to ae^{i\pi/4}} \left[\frac{1}{\frac{d}{dz}[z^4+a^4]}\right]$$

$$= \lim_{z\to ae^{i\pi/4}} \left[\frac{1}{4z^3}\right] = \frac{1}{4a^3\left(e^{\frac{i\pi}{4}}\right)^3} = \frac{1}{4a^3 e^{\frac{i3\pi}{4}}}$$

$$= \frac{1}{4a^3\left[\frac{-1}{\sqrt{2}} + i\frac{1}{\sqrt{2}}\right]} = \frac{(-1-i)\sqrt{2}}{8a^3}.$$

$$[\text{Res}\, f(z)]_{z=ae^{i3\pi/4}} = \lim_{z\to ae^{i3\pi/4}} \left[\frac{1}{4z^3}\right]$$

$$= \frac{1}{4a^3\left(e^{\frac{i3\pi}{4}}\right)^3} = \frac{1}{4a^3 e^{\frac{i9\pi}{4}}}$$

$$= \frac{1}{4a^3\left[\frac{1}{\sqrt{2}} + i\frac{1}{\sqrt{2}}\right]} = \frac{(1-i)\sqrt{2}}{8a^3}.$$

$\therefore$ By Cauchy's residue theorem,

$$\int_C f(z)dz = 2\pi i\left[\frac{(-1-i)\sqrt{2}}{8a^3} + \frac{(1-i)\sqrt{2}}{8a^3}\right]$$

$$= \frac{2\pi i}{4a^3}[-\sqrt{2}i] = \frac{\pi\sqrt{2}}{2a^3}.$$

$\therefore$ (1) gives, $\int_\Gamma f(z)dz + \int_{-R}^{+R} f(x)dx = \frac{\pi\sqrt{2}}{2a^3}.$

Taking limit as $R\to\infty$, $\int_\Gamma f(z)dz \to 0$ [By Cauchy's lemma].

$\therefore \int_{-\infty}^{\infty} f(x)dx = \frac{\pi\sqrt{2}}{2a^3}.$

i.e., $\int_{-\infty}^{\infty} \frac{dx}{x^4+a^4} = \frac{\pi\sqrt{2}}{2a^3}.$

Since $\frac{1}{x^4+a^4}$ is even, $\int_0^{\infty} \frac{dx}{x^4+a^4} = \frac{1}{2}\cdot\frac{\pi\sqrt{2}}{2a^3} = \frac{\pi\sqrt{2}}{4a^3}.$

EXERCISES

1. Evaluate $\int_0^\infty \frac{dx}{(1+x^2)^2}$.
2. Show that $\int_0^\infty \frac{dx}{(x^2+1)^3} = \frac{3\pi}{16}$.
3. Evaluate $\int_0^\infty \frac{x^2 dx}{(x^2+1)^2}$.
4. Show that $\int_0^\infty \frac{dx}{(x^2+a^2)^3} = \frac{3\pi}{16a^5}$, $a > 0$.
5. Evaluate $\int_{-\infty}^\infty \frac{x^2 dx}{(x^2+1)(x^2+4)}$ using contour integration.
6. Evaluate $\int_{-\infty}^\infty \frac{dx}{(x^2+1)(x^2+4)^2}$.
7. Evaluate $\int_0^\infty \frac{dx}{x^4+1}$.
8. Show that $\int_0^\infty \frac{x^2 dx}{x^4+a^4} = \frac{\pi}{2\sqrt{2}a}$, $a > 0$.
9. Evaluate $\int_0^\infty \frac{dx}{x^4+10x^2+9}$.

ANSWERS

1. $\frac{\pi}{4}$ 3. $\frac{\pi}{4}$ 5. $\frac{\pi}{3}$ 6. $\frac{5\pi}{144}$ 7. $\frac{\pi}{2\sqrt{2}}$ 9. $\frac{\pi}{24}$

Type III: Integrals of the form $\int_{-\infty}^\infty \frac{P(x)\sin mx}{Q(x)} dx$, $\int_{-\infty}^\infty \frac{P(x)\cos mx}{Q(x)} dx$, where $P(x)$ and $Q(x)$ are polynomials in x. This integral converges if

(i) $m > 0$

(ii) the degree of $Q(x)$ is greater than the degree of $P(x)$.

(iii) $Q(x)$ does not vanish for any any real value of x.

We state below a lemma, which will be used in the evaluation of such integrals.

Jordan's Lemma: If $f(z)$ is a continuous function such that $|zf(z)| \to 0$ uniformly as $|z| \to \infty$, then $\int_S e^{imz} f(z)dz \to 0$, as $R \to \infty$ where S is the semi - circle $|z| = R$ above the real axis and $m > 0$.

Proof : On the circle $|z| = R$, $z = Re^{i\theta}$

$$\left|\int_S e^{imz} f(z)dz\right| = \left|\int_S e^{imR(\cos\theta + i\sin\theta)} f(z)\, i\, Re^{i\theta}\, d\theta\right|$$

$$\leq \int_0^\pi \left|e^{imR(\cos\theta + i\sin\theta)}\right| |f(z)|\ R\, d\theta$$

$$\leq \int_0^{\pi} e^{-mRsin\theta} \ |f(z)|R\ d\theta$$

$$\leq 2\int_0^{\pi/2} e^{-mRsin\theta} \ |f(z)|R\ d\theta \qquad (1)$$

Now, $sin\theta \geq \frac{2\theta}{\pi}, for\ 0 \leq \theta \leq \pi/2$

$$\therefore \quad -mR\ sin\theta \leq -\frac{2mR\theta}{\pi},\ for\ 0 \leq \theta \leq \pi/2$$

$$\therefore \quad e^{-mR\ sin\theta} \leq e^{-\frac{2mR\theta}{\pi}},\ for\ 0 \leq \theta \leq \pi/2 \qquad (2)$$

Using (2) in (1), we have

$$\left|\int_S e^{imz} f(z)dz\right| \leq 2\int_0^{\pi/2} e^{-\frac{2mR\theta}{\pi}} \ |f(z)|R\ d\theta$$

$$\therefore \ \lim_{R\to\infty}\left|\int_S e^{imz} f(z)dz\right| \leq \int_0^{\pi/2} \lim_{R\to\infty}\left(2Re^{-\frac{2mR\theta}{\pi}}\right)\lim_{R\to\infty}[f(z)]\ d\theta$$

$$\leq \lim_{R\to\infty}[f(z)] \times \lim_{R\to\infty}\int_0^{\pi/2} 2Re^{-\frac{2mR\theta}{\pi}}\ d\theta$$

$$\leq \lim_{R\to\infty}[f(z)] \times \lim_{R\to\infty}\left[\frac{\pi}{m}(1-e^{-mR})\right]$$

$$\leq \frac{\pi}{m} \times 0, \text{ by the given condition.}$$

$\therefore \left|\int_S e^{imz} f(z)dz\right| = 0$ and hence $\int_S e^{imz} f(z)dz = 0$ as $R \to \infty$. Hence proved.

To evaluate this type of integrals, we consider $\int_C e^{imz} f(z)dz$ where $f(z) = \frac{P(z)}{Q(z)}$ round a contour C consisting of a semi-circle Γ with centre origin and radius R, large enough to include all the poles of the integrand in the upper half plane and also part of the real axis from $x = -R$ to $x = +R$.

By Cauchy's residue theorem,

$$\int_C e^{imz} f(z)dz = 2\pi i \times \text{sum of the residues}$$

$$= 2\pi i \sum R_j.$$

Now, $\int_C e^{imz} f(z)dz = \int_\Gamma e^{imz} f(z)dz + \int_{-R}^{+R} e^{imx} f(x)dx.$

Taking limit as $R \to \infty$, we have

$$\lim_{R\to\infty}\int_\Gamma e^{imz} f(z)dz + \int_{-\infty}^{\infty} e^{imx} f(x)dx = 2\pi i \sum R_j \cdot$$

By Jordan's lemma, $\lim_{R\to\infty} \int_\Gamma e^{imz} f(z)dz = 0.$

$\therefore \quad \int_{-\infty}^{\infty} e^{imx} f(x)dx = 2\pi i \sum R_j$

i.e., $\int_{-\infty}^{\infty} f(x)(\cos mx + i \sin mx)dx = 2\pi i \sum R_j$

i.e., $\int_{-\infty}^{\infty} f(x) \cos mx\, dx + i \int_{-\infty}^{\infty} f(x) \sin mx\, dx = 2\pi i \sum R_j$

Equating the real and imaginary parts, we get the values of the corresponding integrals.

Example 1: Evaluate $\int_0^{\infty} \frac{\cos mx}{a^2+x^2} dx$, $m > 0,\ a > 0.$

Solution : Consider the integral $\int_C \frac{e^{imz}}{a^2+z^2} dz$, where C is the contour consisting of a large enough upper semi – circle Γ with centre origin and radius R, and part of the real axis from $-R$ to R.

Let $f(z) = \frac{e^{imz}}{a^2+z^2}$.

Then $\int_C f(z)dz = \int_\Gamma f(z)dz + \int_{-R}^{R} f(x)dx.$ (1)

The poles of $f(z)$ are given by the zeros of $a^2 + z^2 = 0.$

$\Rightarrow \quad z = \pm ai.$

$\therefore \quad z = ai$ is a simple pole which lies within C.

$$[\text{Res } f(z)]_{z = ai} = \lim_{z\to ai} \left[\frac{e^{imz}}{2z}\right] = \frac{e^{-ma}}{2ai}.$$

$\therefore$ By Cauchy's residue theorem,

$$\int_C f(z)dz = \frac{2\pi i\ e^{-ma}}{2ai} = \frac{\pi e^{-ma}}{a}.$$

$\therefore$ (1) gives, $\int_\Gamma f(z)dz + \int_{-R}^{R} f(x)dx = \frac{\pi e^{-ma}}{a}.$

Taking limit as $R \to \infty$, $\int_\Gamma f(z)dz \to 0$ [By Jordan's lemma].

$$\therefore \quad \int_{-\infty}^{\infty} f(x)dx = \frac{\pi e^{-ma}}{a}.$$

Hence $\int_{-\infty}^{\infty} \frac{e^{imx}}{a^2+x^2} dx = \frac{\pi e^{-ma}}{a}.$

$$\int_{-\infty}^{\infty} \frac{\cos mx + i \sin mx}{a^2+x^2} dx = \frac{\pi e^{-ma}}{a}.$$

$$\therefore \int_{-\infty}^{\infty} \frac{\cos mx}{a^2+x^2} dx + i \int_{-\infty}^{\infty} \frac{\sin mx}{a^2+x^2} dx = \frac{\pi e^{-ma}}{a}.$$

Equating the real part, we get

$$\int_{-\infty}^{\infty} \frac{\cos mx}{a^2+x^2} dx = \frac{\pi e^{-ma}}{a}.$$

Since $\frac{\cos mx}{a^2+x^2}$ is even, $\int_{-\infty}^{\infty} \frac{\cos mx}{a^2+x^2} dx = 2 \int_{0}^{\infty} \frac{\cos mx}{a^2+x^2} dx.$

$$\therefore \qquad \int_{0}^{\infty} \frac{\cos mx}{a^2+x^2} dx = \frac{\pi e^{-ma}}{2a}.$$

Example 2: Show that $\int_{0}^{\infty} \frac{\cos mx}{(a^2+x^2)^2} dx = \frac{\pi}{4a^3}(1+ma)e^{-ma}, \quad (m > 0,\ a > 0).$

Solution: Let $f(z) = \frac{e^{imz}}{(a^2+z^2)^2}.$

Consider the integral $\int_C f(z)dz$ where C is the contour consisting of a large enough upper semi – circle Γ with centre origin and radius R, and part of the real axis from $-R$ to R.

$$\therefore \int_C f(z)dz = \int_\Gamma f(z)dz + \int_{-R}^{R} f(x)dx. \qquad (1)$$

The poles of $f(z)$ are given by the zeros of $(a^2+z^2)^2 = 0.$

$\Rightarrow \quad z = \pm\, ai$ (twice)

$\therefore \quad z = ai$ is a pole of order 2 which lies within C.

Now, $\quad z^2 + a^2 = (z+ai)(z-ai).$

$$[\text{Res } f(z)]_{z=ai} = \frac{1}{(2-1)!} \lim_{z\to ai} \frac{d}{dz}[(z-ai)^2 f(z)]$$

$$= \lim_{z\to ai} \frac{d}{dz}\left[(z-ai)^2 \frac{e^{imz}}{(z-ai)^2(z+ai)^2}\right]$$

$$= \lim_{z\to ai} \frac{d}{dz}\left[\frac{e^{imz}}{(z+ai)^2}\right]$$

$$= \lim_{z\to ai} . \frac{(z+ai)^2 im\, e^{imz} - e^{imz}\, 2(z+ai)}{(z+ai)^4}$$

$$= \lim_{z\to ai} . \frac{[(z+ai)im-2]e^{imz}}{(z+ai)^3} = \frac{[i(2ai)m-2]e^{im(ai)}}{(2ai)^3}$$

$$= \frac{2[-am-1]e^{-am}}{-8a^3 i} = \frac{-i[1+am]e^{-am}}{4a^3}.$$

$\therefore$ By Cauchy's residue theorem,

$$\int_C f(z)dz = 2\pi i \times (-i)(1+am)\,\frac{e^{-ma}}{4a^3}$$

$$= \frac{\pi}{2a^3}\,(1+am)\,e^{-am}.$$

$\therefore$ (1) gives, $\int_\Gamma f(z)dz + \int_{-R}^{R} f(x)dx = \frac{\pi}{2a^3}(1+am)e^{-am}.$

Taking limit as $R \to \infty$, $\int_\Gamma f(z)dz \to 0$ [By Jordan's lemma].

$$\therefore \int_{-\infty}^{\infty} e^{imx}\,\frac{1}{(a^2+x^2)^2}\,dx = \frac{\pi}{2a^3}(1+am)e^{-am}.$$

Equating the real parts on both sides,

$$\int_{-\infty}^{\infty} \frac{\cos mx}{(a^2+x^2)^2}\,dx = \frac{\pi}{2a^3}(1+am)e^{-am}.$$

Since $\frac{\cos mx}{(a^2+x^2)^2}$ is even, we have $\int_0^{\infty} \frac{\cos mx}{(a^2+x^2)^2}\,dx = \frac{1}{2}\frac{\pi}{2a^3}(1+am)e^{-am}.$

$$\therefore \quad \int_0^{\infty} \frac{\cos mx}{(a^2+x^2)^2}\,dx = \frac{\pi}{4a^3}(1+am)e^{-am}.$$

Example 3: Evaluate $\int_{-\infty}^{\infty} \frac{\cos x\,dx}{(x^2+a^2)(x^2+b^2)}$, $a > b > 0$ using contour integration.

Solution : ***[AU, Nov/Dec 2015]***

Let $f(z) = \frac{e^{iz}}{(z^2+a^2)(z^2+b^2)}$.

Consider $\int_C f(z)dz$, where C is the contour consisting of a large enough upper semi-circle Γ with centre origin and radius R and part of the real axis from $-R$ to R.

$$\therefore \int_C f(z)dz = \int_\Gamma f(z)dz + \int_{-R}^{R} f(x)dx \qquad (1)$$

The poles of $f(z)$ are given by the zeros of $(z^2+a^2)(z^2+b^2) = 0$.

$\Rightarrow z = \pm\, ai$ and $z = \pm\, bi$.

$\therefore z = ai$ and $z = bi$ are simple poles which lie within C.

Now, $z^2 + a^2 = (z+ai)(z-ai)$ and $z^2 + b^2 = (z+bi)(z-bi)$.

$$\therefore [\text{Res}\, f(z)]_{z=ai} = \lim_{z\to ai}(z-ai)\,\frac{e^{iz}}{(z-ai)(z+ai)(z^2+b^2)}$$

$$= \lim_{z\to ai} \frac{e^{iz}}{(z+ai)(z^2+b^2)}$$

$$= \frac{e^{i(ai)}}{2ai\,(-a^2+b^2)} = \frac{e^{-a}}{2ai\,(b^2-a^2)}.$$

$$[\text{Res}\, f(z)]_{z=bi} = \lim_{z\to bi} (z-bi)\,\frac{e^{iz}}{(z-bi)\,(z+bi)\,(z^2+a^2)}$$

$$= \lim_{z\to bi} \frac{e^{iz}}{(z+bi)\,(z^2+a^2)} = \frac{e^{i(bi)}}{2bi\,(-b^2+a^2)} = \frac{-\,e^{-b}}{2bi\,(b^2-a^2)}.$$

$\therefore$ By Cauchy's residue theorem,

$$\int_C f(z)dz = 2\pi i\left[\frac{e^{-a}}{2ai\,(b^2-a^2)} - \frac{e^{-b}}{2bi\,(b^2-a^2)}\right]$$

$$= \frac{\pi}{b^2-a^2}\left[\frac{e^{-a}}{a} - \frac{e^{-b}}{b}\right]$$

i.e.,
$$\int_C f(z)\,dz = \frac{\pi}{b^2-a^2}\left[\frac{e^{-a}}{a} - \frac{e^{-b}}{b}\right].$$

$\therefore$ (1) gives $\int_\Gamma f(z)\,dz + \int_{-R}^{R} f(x)\,dx = \frac{\pi}{b^2-a^2}\left[\frac{e^{-a}}{a} - \frac{e^{-b}}{b}\right].$

Taking limit as $R \to \infty$, $\int_\Gamma f(z)dz \to 0$ [By Jordan's lemma].

$$\int_{-\infty}^{\infty} f(x)\,dx = \frac{\pi}{b^2-a^2}\left(\frac{e^{-a}}{a} - \frac{e^{-b}}{b}\right)$$

$$\therefore \quad \int_{-\infty}^{\infty} \frac{e^{imx}\,dx}{(x^2+a^2)(x^2+b^2)} = \frac{\pi}{b^2-a^2}\left(\frac{e^{-a}}{a} - \frac{e^{-b}}{b}\right).$$

Hence $\int_{-\infty}^{\infty} \frac{(\cos x + i \sin x)\,dx}{(x^2+a^2)(x^2+b^2)} = \frac{\pi}{b^2-a^2}\left(\frac{e^{-a}}{a} - \frac{e^{-b}}{b}\right).$

Equating the real parts on both sides,

$$\int_{-\infty}^{\infty} \frac{\cos x\,dx}{(x^2+a^2)(x^2+b^2)} = \frac{\pi}{b^2-a^2}\left(\frac{e^{-a}}{a} - \frac{e^{-b}}{b}\right)$$

$$= \frac{\pi}{a^2-b^2}\left(\frac{e^{-b}}{b} - \frac{e^{-a}}{a}\right).$$

Example 4: Evaluate $\int_0^{\infty} \frac{x \sin mx}{x^2+a^2}\,dx$ where $a > 0, m > 0$. ***[AU, May/June 2016]***

Solution:

Let $(z) = \frac{z e^{imz}}{(z^2+a^2)}$.

Consider $\int_C f(z)dz$, where C is the contour consisting of a sufficiently large upper semi-circle Γ with centre origin and radius R, and part of the real axis from $-R$ to R.

$$\therefore \quad \int_C f(z)dz = \int_\Gamma f(z)dz + \int_{-R}^{R} f(x)dx \qquad (1)$$

The poles of $f(z)$ are given by the zeros of $(z^2 + a^2) = 0$.

$\Rightarrow \quad z = \pm ai$

$\therefore \quad z = ai$ is a simple pole which lies within C.

$z^2 + a^2 = (z + ai)(z - ai)$.

$$\therefore [\text{Res } f(z)]_{z=ai} = \lim_{z \to ai} \left[\frac{z\, e^{imz}}{2z}\right] = \lim_{z \to ai} \frac{e^{imz}}{2}$$

$$= \frac{e^{im\,(ai)}}{2} = \frac{e^{-ma}}{2}.$$

$\therefore$ By Cauchy's residue theorem,

$$\int_C f(z) dz = \frac{2\pi i\, e^{-ma}}{2} = \pi i\, e^{-ma}.$$

$\therefore$ (1) gives $\int_\Gamma f(z)\, dz + \int_{-R}^{R} f(x)\, dx = \pi i\, e^{-ma}$.

Taking limit as $R \to \infty$, $\int_\Gamma f(z) dz \to 0$ [By Jordan's lemma].

$$\int_{-\infty}^{\infty} f(x)\, dx = \pi i\, e^{-ma}.$$

$$\therefore \quad \int_{-\infty}^{\infty} \frac{x\, e^{imx}\, dx}{x^2+a^2} = \pi i\, e^{-ma}.$$

i.e., $\int_{-\infty}^{\infty} x \left(\frac{\cos mx + i \sin mx}{x^2+a^2}\right) dx = \pi i\, e^{-ma}$.

Equating the imaginary parts on both sides, we have

$$\int_{-\infty}^{\infty} \frac{x \sin mx\, dx}{x^2+a^2} = \pi\, e^{-ma}.$$

Since $\frac{x \sin mx}{x^2+a^2}$ is even, we have

$$\int_0^{\infty} \frac{x \sin mx\, dx}{x^2+a^2} = \frac{1}{2} \int_{-\infty}^{\infty} \frac{x \sin mx\, dx}{x^2+a^2} = \frac{\pi}{2} e^{-ma}.$$

Example 5: Evaluate $\int_0^{\infty} \frac{x \cos mx}{x^2+a^2}\, dx$.

Solution:

Proceeding as in Problem 4, we get $\int_{-\infty}^{\infty} \frac{x\,(\cos mx + i \sin mx)\, dx}{x^2+a^2} = \pi i e^{-ma}$.

Equating the real parts on both sides,

$$\int_0^{\infty} \frac{x \cos mx}{x^2+a^2}\, dx = 0.$$

Example 6: Evaluate $\int_{-\infty}^{\infty} \frac{x \sin x \, dx}{(x^2+1)(x^2+4)}$.

Solution:

Let $f(z) = \frac{ze^{iz}}{(z^2+1)(z^2+4)}$.

Consider the integral $\int_C f(z)dz$, where C is the contour consisting of a sufficiently large semi-circle Γ with centre origin and radius R, and part of the real axis from $-R$ to R.

$$\therefore \int_C f(z)dz = \int_\Gamma f(z)dz + \int_{-R}^{R} f(x)dx \qquad (1)$$

The poles of $f(z)$ are given by the zeros o f $(z^2+1)(z^2+4) = 0$.

$\Rightarrow \quad z = \pm\, i$ and $z = \pm\, 2i$.

$\therefore \quad z = i$ and $z = 2i$ are simple poles which lie within C.

$z^2+1 = (z+i)(z-i)$ and $z^2+4 = (z+2i)(z-2i)$.

$$\therefore \quad [\text{Res}\, f(z)]_{z=i} = \lim_{z\to i}(z-i)\, f(z)$$

$$= \lim_{z\to i}(z-i)\, \frac{z\, e^{iz}}{(z+i)(z-i)(z^2+4)}$$

$$= \lim_{z\to i} \frac{z\, e^{iz}}{(z+i)(z^2+4)}$$

$$= \frac{i\, e^{-1}}{2i\,(-1+4)} = \frac{1}{6e}.$$

$$\therefore \quad [\text{Res}\, f(z)]_{z=2i} = \lim_{z\to 2i}(z-2i)\, f(z)$$

$$= \lim_{z\to 2i}(z-2i)\, \frac{z\, e^{iz}}{(z^2+1)(z-2i)(z+2i)}$$

$$= \lim_{z\to 2i} \frac{z\, e^{iz}}{(z^2+1)(z+2i)}$$

$$= \frac{2i\, e^{-2}}{(-3)(4i)} = \frac{-1}{6}e^{-2} = \frac{-1}{6e^2}.$$

$\therefore$ By Cauchy's residue theorem,

$$\int_C f(z)dz = 2\pi i\left(\frac{1}{6e} - \frac{1}{6e^2}\right) = \frac{\pi i}{3}\left(\frac{1}{e} - \frac{1}{e^2}\right).$$

$\therefore$ (1) gives $\int_{\Gamma} f(z)\,dz + \int_{-R}^{R} f(x)\,dx = \frac{\pi i}{3}\left(\frac{1}{e} - \frac{1}{e^2}\right)$.

Taking limit as $R \to \infty$, $\int_{\Gamma} f(z)dz \to 0$ [By Jordan's lemma].

$\therefore$ $$\int_{-\infty}^{\infty} f(x)\,dx = \frac{\pi i}{3}\left(\frac{1}{e} - \frac{1}{e^2}\right).$$

$$\int_{-\infty}^{\infty} \frac{x e^{ix}\,dx}{(x^2+1)(x^2+4)} = \frac{\pi i}{3}\left(\frac{1}{e} - \frac{1}{e^2}\right).$$

i.e., $$\int_{-\infty}^{\infty} \frac{x(\cos x + i\sin x)\,dx}{(x^2+1)(x^2+4)} = \frac{\pi i}{3}\left(\frac{1}{e} - \frac{1}{e^2}\right).$$

Equating the imaginary parts on both sides, we have

$$\int_{-\infty}^{\infty} \frac{x\sin x\,dx}{(x^2+1)(x^2+4)} = \frac{\pi}{3}\left(\frac{1}{e} - \frac{1}{e^2}\right).$$

Example 7: Evaluate $\int_{-\infty}^{\infty} \frac{x\sin x\,dx}{(x^2+1)^2}$.

Solution: Let $f(z) = \frac{z e^{iz}}{(z^2+1)^2}$.

Consider the integral $\int_C f(z)dz$, where C is the contour consisting of a sufficiently large semi-circle Γ with centre origin and radius R, and part of the real axis from $-R$ to R.

$$\therefore \int_C f(z)dz = \int_{\Gamma} f(z)dz + \int_{-R}^{R} f(x)dx \qquad (1)$$

The poles of $f(z)$ are given by the zeros of $(z^2+1)^2 = 0$.

$\Rightarrow z = \pm i$ (twice)

$\therefore$ $z = i$ is a simple pole which lies within C.

$z^2 + 1 = (z+i)(z-i)$

$$\therefore \quad [\text{Res } f(z)]_{z=i} = \frac{1}{1!} \lim_{z\to i} \frac{d}{dz}[(z-i)^2 f(z)]$$

$$= \lim_{z\to i} \frac{d}{dz}\left[(z-i)^2 \frac{z e^{iz}}{(z-i)^2(z+i)^2}\right]$$

$$= \lim_{z\to i} \frac{d}{dz}\left[\frac{z e^{iz}}{(z+i)^2}\right]$$

$$= \lim_{z\to i}\left[\frac{(z+i)^2[e^{iz} + iz e^{iz}] - 2z e^{iz}(z+i)}{(z+i)^4}\right]$$

$$= \lim_{z\to i}\left[\frac{(z+i)e^{iz}(1+iz) - 2z e^{iz}}{(z+i)^3}\right] = \frac{e^{-1}}{4} = \frac{1}{4e}.$$

∴ By Cauchy's residue theorem,

$$\int_C f(z)dz = \frac{2\pi i\, e^{-1}}{4} = \frac{\pi i}{2e}.$$

∴ (1) gives $\int_\Gamma f(z)\, dz + \int_{-R}^{R} f(x)\, dx = \frac{\pi i}{2e}$.

Taking limit as $R \to \infty$, $\int_\Gamma f(z)dz \to 0$ [By Jordan's lemma].

$$\int_{-\infty}^{\infty} f(x)\, dx = \frac{\pi i}{2e}.$$

∴
$$\int_{-\infty}^{\infty} \frac{x e^{ix}}{(x^2+1)^2}\, dx = \frac{\pi i}{2e}.$$

i.e.,
$$\int_{-\infty}^{\infty} \frac{x(\cos x + i \sin x)\, dx}{(x^2+1)^2} = \frac{\pi i}{2e}.$$

Equating the imaginary parts on both sides, we have

$$\int_{-\infty}^{\infty} \frac{x \sin x\, dx}{(x^2+1)^2} = \frac{\pi}{2e}.$$

EXERCISES

1. Evaluate $\int_{-\infty}^{\infty} \frac{\cos ax}{x^2+b^2}\, dx \quad (a > 0; b > 0)$.
2. Evaluate $\int_0^{\infty} \frac{\cos ax}{x^2+1}\, dx \quad (a > 0)$.
3. Evaluate $\int_{-\infty}^{\infty} \frac{x \sin x}{(x^2+1)(x^2+4)}\, dx$.
4. Evaluate $\int_0^{\infty} \frac{x^3 \sin x}{(x^2+a^2)^2}\, dx,\ a > 2$.
5. Evaluate $\int_{-\infty}^{\infty} \frac{x^2 \cos mx}{(x^2+a^2)(x^2+b^2)}\, dx\ (a > 0; b > 0; m > 0)$.
6. Evaluate $\int_{-\infty}^{\infty} \frac{\sin x}{x^2+4x+5}\, dx$.
7. Evaluate $\int_0^{\infty} \frac{x \sin ax}{x^2+k^2}\, dx,\ a > 0$.

ANSWERS

1. $\frac{\pi}{b}\, e^{-ab}$ 2. $\frac{\pi}{2}\, e^{-a}$

3. $\frac{\pi}{3e^2}(e-1)$ 4. $-\frac{\pi}{4} e^{-a}(a-2)$.

5. $\frac{\pi}{2(a+b)}\, e^{-m}$ 6. $-\frac{\pi}{e} \sin 2$

7. $\frac{\pi}{2}\, e^{-ak}$.

4.8. Short Answer Questions

1. State Cauchy's integral theorem or Cauchy's fundamental theorem.

[*AU, April/May 2015, Nov/Dec 2015*]

If $f(z)$ is an analytic function and $f'(z)$ is continuous at each point within and on a simple closed curve C, then $\oint_C f(z)dz = 0$.

2. State Cauchy's integral formula

If $f(z)$ is analytic inside and on a simple closed curve C that encloses a simply connected region R and if 'a' is any point in R, then

$f(a) = \frac{1}{2\pi i} \oint_C \frac{f(z)}{(z-a)} dz$, where C is described in the anticlockwise sense.

3. Evaluate $\int_C \frac{dz}{z-2}$, C is the circle $|z| = 1$.

Solution: Clearly $z = 2$ lies outside the circle $|z| = 1$.

Let $f(z) = \frac{1}{z-2}$. Clearly $f(z)$ is analytic inside C.

$\therefore$ By Cauchy's fundamental theorem, $\int_C f(z)dz = 0$.

$\therefore \quad \int_C \frac{dz}{z-2} = 0.$

4. Evaluate $\int_C \frac{e^z}{z-2} dz$, where C is the unit circle with centre as origin.

Solution: **[AU, *April/May 2017*]**

Clearly $z = 2$ lies outside the circle $|z| = 1$.

Let $f(z) = \frac{e^z}{z-2}$. Clearly $f(z)$ is analytic inside C.

$\therefore$ By Cauchy's fundamental theorem, $\int_C f(z)dz = 0$.

$\therefore \quad \int_C \frac{e^z}{z-2} = 0.$

5. Evaluate $\int_C \frac{z}{z-2} dz$, where C is (a) $|z| = 1$, (b) $|z| = 3$. **[*AU, May/June 2014*]**

Solution:

(a) Clearly $z = 2$ lies outside the circle $|z| = 1$.

Let $f(z) = \frac{2}{z-2}$. Clearly $f(z)$ is analytic inside C.

$\therefore$ By Cauchy's fundamental theorem, $\int_C f(z)dz = 0$.

$\therefore \int_C \frac{z}{z-2}dz = 0.$

(b) Given that C is $|z| = 3$.

$\therefore$ The point $z = 2$ lies inside C.

Let $f(z) = z$. Clearly $f(z)$ is analytic inside C.

Hence by Cauchy's integral formula,

$$\int_C \frac{f(z)}{(z-2)}dz = 2\pi i . f(2) = 2\pi i . 2 = 4\pi i.$$

$$\therefore \quad \int_C \frac{z}{z-2}dz = 0.$$

6. Evaluate $\int_C \frac{z+1}{z^3-4z}dz$, where C is $|z| = 3/2$.

Solution: We have $z^3 - 4z = z(z^2-4) = z(z+2)(z-2)$.

Only the point $z = 0$ lies inside the curve $|z| = 3/2$.

$$\therefore \quad \int_C \frac{z+1}{z^3-4z}dz = \int_C \frac{\frac{z+1}{(z+2)(z-2)}}{z}dz$$

Let $f(z) = \frac{z+1}{(z+2)(z-2)} = \frac{z+1}{z^2-4}$. Clearly $f(z)$ is analytic inside C.

By Cauchy's integral formula,

$$\int_C \frac{f(z)}{z-0}dz = 2\pi i \times f(0)$$

$$= 2\pi i \times \left(\frac{z+1}{z^2-4}\right)_{z=0} = \frac{2\pi i}{-4} = \frac{-\pi i}{2}.$$

$$\therefore \quad \int_C \frac{z+1}{z^3-4z}dz = \frac{-\pi i}{2}.$$

7. Evaluate $\int_C \frac{e^z}{z-1}dz$ if 1 lies inside C.

Solution:

Given $z = 1$ lies inside C.

Let $f(z) = e^z$. Clearly $f(z)$ is analytic inside C.

Then by Cauchy's integral formula,

$$\int_C \frac{e^z}{z-1}dz = 2\pi i f(1) = 2\pi i e^1 = 2\pi e i.$$

8. Evaluate $\int_C \frac{z}{z-1} dz$, C is the circle $|z-1| = 1$. ***[AU, Nov/Dec 2017]***

Solution: Clearly $z = 1$ lies intside the circle $|z-1| = 1$.

Let $f(z) = z$. Clearly $f(z)$ is analytic inside C.

Then by Cauchy's integral formula,

$\int_C \frac{z}{z-1} dz = 2\pi i f(1) = 2\pi i.$

9. Evaluate $\int_C \frac{dz}{z^2 e^z}$, where C is the circle $|z| = 1$.

Solution: $\int_C \frac{dz}{z^2 e^z} = \int_C \frac{e^{-z} dz}{z^2}.$

Clearly the point $z = 0$ lies inside C.

Let $f(z) = e^{-z}$. Clearly $f(z)$ is analytic inside C.

∴ By Cauchy's formula for higher derivatives,

$$\int_C \frac{f(z)dz}{z^2} = 2\pi i \,.\, f'(0)$$

$$\int_C \frac{e^{-z}dz}{z^2} = -\,2\pi i\, e^0$$

i.e.,

$$\int_C \frac{dz}{z^2 e^z} = -2\pi i.$$

10. Evaluate $\int_C \frac{4z^2+4z+5}{z-3} dz$, where C is the ellipse $9x^2 + 4y^2 = 36$.

Solution:

Clearly $z = 3$ lies outside the ellipse $\frac{x^2}{2^2} + \frac{y^2}{3^2} = 1$. $[z = 3 \Rightarrow x = 3, y = 0]$

$f(z) = \frac{4z^2+4z+5}{z-3}$ is analytic in C.

∴ By Cauchy's integral theorem, $\int_C f(z)dz = 0$.

i.e., $\int_C \frac{4z^2+4z+5}{z-3} dz = 0.$

11. Evaluate $\int_C \frac{dz}{z^2+4}$, C is the ellipse $4x^2 + (y-2)^2 = 4$.

Solution: $z^2 + 4 = 0 \Rightarrow z = 2i, -2i.$

Clearly $z = 2i$ lies inside C and $z = -2i$ lies out side C.

$\int_C \frac{dz}{z^2+4} = \int_C \frac{\left(\frac{1}{z+2i}\right)}{z-2i} dz.$

Let $f(z) = \frac{1}{z+2i}$ be analytic in C.

$\therefore$ By Cauchy's integral formula,

$$\int_C \frac{f(z)}{z-2i} dz = 2\pi i . f(2i)$$

$$= 2\pi i . \left(\frac{1}{z+2i}\right)_{z=2i} = 2\pi i \left(\frac{1}{4i}\right) = \frac{\pi}{2}.$$

12. Write the Taylor's series expansion of $\cos z$ about $z = 0$.

Solution:

$$\begin{aligned}
\text{Let } (z) &= \cos z\,; & f(0) &= 1 \\
f'(z) &= -\sin z\,; & f'(0) &= 0 \\
f''(z) &= -\cos z\,; & f''(0) &= -1 \\
f'''(z) &= \sin z\,; & f'''(0) &= 0 \\
f^{v}(z) &= \cos z\,; & f^{v}(0) &= 1
\end{aligned}$$

$\vdots$

$$\therefore f(z) = \cos z = f(0) + (z-0)f'(0) + \frac{1}{2!}(z-0)^2 f''(0) + \cdots$$

$$= 1 + 0 + \frac{1}{2!}z^2(-1) + 0 + \frac{1}{4!}z^4(1) + \cdots$$

$$\cos z = 1 - \frac{z^2}{2!} + \frac{z^4}{4!} - \cdots$$

13. Find the Taylor's series expansion of $\sin z$ about $z = \frac{\pi}{4}$.

Solution:

$$\begin{aligned}
\text{Let } f(z) &= \sin z\,; & f\left(\tfrac{\pi}{4}\right) &= \tfrac{1}{\sqrt{2}} \\
f'(z) &= \cos z\,; & f'\left(\tfrac{\pi}{4}\right) &= \tfrac{1}{\sqrt{2}} \\
f''(z) &= -\sin z\,; & f''\left(\tfrac{\pi}{4}\right) &= -\tfrac{1}{\sqrt{2}}
\end{aligned}$$

$\vdots$

$$\therefore f(z) = \sin z = f\left(\frac{\pi}{4}\right) + \left(z - \frac{\pi}{4}\right) f'\left(\frac{\pi}{4}\right) + \frac{1}{2!}\left(z - \frac{\pi}{4}\right)^2 f''\left(\frac{\pi}{4}\right) + \cdots$$

$$= \frac{1}{\sqrt{2}} + \left(z - \frac{\pi}{4}\right)\frac{1}{\sqrt{2}} + \frac{1}{2!}\left(z - \frac{\pi}{4}\right)^2\left(-\frac{1}{\sqrt{2}}\right) + \cdots$$

$$= \frac{1}{\sqrt{2}}\left[1 + \left(z - \frac{\pi}{4}\right) - \frac{1}{2!}\left(z - \frac{\pi}{4}\right)^2 + \cdots\right].$$

14. Expand $f(z) = \frac{1}{z^2}$ as a Taylor's series about the point $z = 2$. **[*AU, May/June 2016*]**

Solution:

Let $u = z - 2$. i.e., $z = u + 2$.

$$f(z) = \frac{1}{z^2} = \frac{1}{(u+2)^2} = \frac{1}{4\left(1+\frac{u}{2}\right)^2}$$

$$= \frac{1}{4}\left(1+\frac{u}{2}\right)^{-2} = \frac{1}{4}\left[1 - 2\frac{u}{2} + 3\left(\frac{u}{2}\right)^2 - 4\left(\frac{u}{2}\right)^3 + \cdots\right]$$

$$= \frac{1}{4}\left[1 - (z-2) + 3\left(\frac{z-2}{2}\right)^2 - 4\left(\frac{z-2}{2}\right)^3 + \cdots\right].$$

15. Expand $\frac{z-1}{z+1}$ about $z = 1$. **[AU, *Nov/Dec 2017*]**

Solution: let $f(z) = \frac{z-1}{z+1}$.

Let $u = z - 1$ i.e., $z = u + 1$.

$$\therefore f(z) = \frac{u}{u+2} = \frac{u}{u\left(1+\frac{2}{u}\right)}$$

$$= \left(1+\frac{2}{u}\right)^{-1}$$

$$= 1 - \frac{2}{u} + \left(\frac{2}{u}\right)^2 - \left(\frac{2}{u}\right)^3 + \cdots$$

$$= 1 - \frac{2}{z-1} + \left(\frac{2}{z-1}\right)^2 - \left(\frac{2}{z-1}\right)^3 + \cdots$$

16. Obtain the Laurent's expansion for $f(z) = \frac{1}{z^2(1-z)}$ in $0 < |z| < 1$.

Solution: In $0 < |z| < 1$,

$$f(z) = \frac{1}{z^2(1-z)} = \frac{1}{z^2}(1-z)^{-1}$$

$$= \frac{1}{z^2}(1 + z + z^2 + z^3 + \cdots) = \left(\frac{1}{z^2} + \frac{1}{z} + 1 + z + \cdots\right).$$

17. Classify the singularities of $f(z) = \frac{\tan z}{z}$.

Solution: $f(z) = \frac{\tan z}{z} = \frac{\sin z}{z \cos z}$.

The singularities of $f(z)$ are given by $z \cos z = 0$.

i.e., $z = 0$ or $\cos z = 0$

$$z = 0 \text{ or } z = (2n+1)\frac{\pi}{2},\ n = 0, 1, 2, \ldots$$

Here $z = 0$ is a removable singularity and $z = (2n+1)\frac{\pi}{2}$ is a simple pole.

18. Identify the type of singularity of function $sin\left(\frac{1}{1-z}\right)$. **[*AU, April/May 2015*]**

Solution: Let $f(z) = sin\left(\frac{1}{1-z}\right)$.

The zeros are given by $\sin\left(\frac{1}{z-1}\right) = 0$.

i.e., $$\frac{1}{z-1} = n\pi \ (n = 0, \pm 1, \pm 2, \cdots).$$

Hence the zeros of $f(z)$ are $z = 1 + \frac{1}{n\pi}$ $(n = 0, \pm 1, \pm 2, \cdots)$.

Since $\lim_{z \to 1} f(z)$ does not exist, $z = 1$ is an essential singularity.

19. Classify the singular point of $f(z) = \frac{z-\sin z}{z^3}$.

Solution: Clearly $z = 0$ is a singular point of $f(z)$.

Now, $$\frac{z-\sin z}{z^3} = \frac{1}{z^3}\left[z - \left(z - \frac{z^3}{3!} + \frac{z^5}{5!} - \cdots\right)\right]$$
$$= \frac{1}{3!} - \frac{z^2}{5!} + \cdots$$

Since the Laurent's series expansion of $f(z)$ has no principal part, $z = 0$ is a removable singularity of $f(z)$.

20. Find the singular points of $f(z) = \frac{\sin z}{z}$. **[AU, *Nov/Dec 2017*]**

Solution: Clearly $z = 0$ is a singular point of $f(z)$.

Now, $$\frac{\sin z}{z} = \frac{1}{z}\left[z - \frac{z^3}{3!} + \frac{z^5}{5!} - \cdots\right]$$
$$= 1 - \frac{z^2}{3!} + \frac{z^4}{5!} - \cdots$$

Since the Laurent's series expansion of $f(z)$ has no principal part, $z = 0$ is a removable singularity of $f(z)$.

21. Find the poles of $\frac{z^3-1}{z^3+1}$.

Solution: The poles of $f(z) = \frac{z^3-1}{z^3+1}$ are the solutions of $z^3 + 1 = 0$.

i.e., $$z = -1, \frac{1 \pm i\sqrt{3}}{2}.$$

$\therefore$ The poles are $-1, \frac{1 \pm i\sqrt{3}}{2}$.

22. What are the poles of cosec z?

Solution:

The poles of $f(z) = \operatorname{cosec} z = \dfrac{1}{\sin z}$ are given by,

i.e., $\sin z = 0$

i.e., $z = n\pi;\ n$ is an integer.

23. Find the singularity and its type for $f(z) = ze^{1/z}$.

Solution: $f(z) = ze^{1/z} = z\left[1 + \frac{1}{z} + \frac{1}{2!}\left(\frac{1}{z}\right)^2 + \cdots\right]$

$$= z + 1 + \frac{1}{2!}\left(\frac{1}{z}\right) + \cdots$$

The Laurent's series expansion of $f(z)$ about $z = 0$ has an infinite number of negative powers of z.

$\therefore z = 0$ is an essential singular point of $f(z)$.

24. Find the residue at $z = 0$ for $f(z) = \dfrac{1-z}{z^2}$ from its Laurent's series.

Solution: The Laurent's series of $f(z)$ is $f(z) = \dfrac{1}{z^2} - \dfrac{1}{z}$.

$\therefore [\text{Res } f(z)]_{z=0} =$ coefficient of $\frac{1}{z}$ in the Laurent's series $= -1$.

25. Evaluate the residue of $f(z) = \tan z$ at its singularities. ***[AU, May/June 2016]***

Solution:

$$f(z) = \tan z = \frac{\sin z}{\cos z} = \frac{P(z)}{Q(z)} \text{ has simple poles at } z = \pm\frac{\pi}{2}, \pm 3\frac{\pi}{2}, \ldots$$

Now, $\dfrac{P(z)}{Q'(z)} = \dfrac{\sin z}{-\sin z} = -1$.

$$[\text{Res } f(z)]_{z = poles} = \lim_{z \to poles}\left(\frac{P(z)}{Q'(z)}\right) = -1.$$

26. Find the residue of $z\,e^{\frac{-2}{z}}$ at $z = 0$. ***[AU, Nov/Dec 2015]***

Solution: Clearly $z = 0$ is a singular point.

The Laurent's series expansion of $f(z) = z\,e^{\frac{-2}{z}}$ is

$$f(z) = z\,e^{\frac{-2}{z}} = z\left[1 - \frac{2}{z} + \frac{1}{2!}\left(\frac{-2}{z}\right)^2 + \cdots\right]$$

$$= z - 2 + 2\,\frac{1}{z} + \cdots$$

$\therefore [\text{Res } f(z)]_{z=0} =$ coefficient of $\frac{1}{z}$ in the Laurent's series expansion $= 2$.

27. Find the residue of $f(z) = \dfrac{\sin z}{z \cos z}$ at $z = 0$.

Solution: Given $f(z) = \dfrac{\sin z}{z \cos z} = \dfrac{P(z)}{Q(z)}$

$$[\text{Res}\, f(z)]_{z=0} = \lim_{z \to 0} \frac{P(z)}{Q'(z)} = \lim_{z \to 0} \frac{\sin z}{(-z\, \sin z + \cos z)} = \frac{0}{1} = 0.$$

28. Find the residue of $\dfrac{z\, e^z}{(z-1)^2}$ at $z = 1$.

Solution: Let $f(z) = \dfrac{z\, e^z}{(z-1)^2}$.

$$[\text{Res}\, f(z)]_{z=1} = \lim_{z \to 1} \frac{d}{dz}\left[(z-1)^2 \frac{z\, e^z}{(z-1)^2}\right] = \lim_{z \to 1} \frac{d}{dz}(z\, e^z)$$

$$= \lim_{z \to 1}(z\, e^z + e^z) = e + e = 2e.$$

29. Determine the residue of $f(z) = \dfrac{z+1}{(z-1)(z+2)}$ at $z = 1$. **[AU, *April/May 2017*]**

Solution: Let $f(z) = \dfrac{z+1}{(z-1)(z+2)}$.

$$[\text{Res}\, f(z)]_{z=1} = \lim_{z \to 1} [(z-1)\, f(z)] = \lim_{z \to 1} \frac{z+1}{z+2} = \frac{2}{3}.$$

30. Find the residue of $f(z) = \dfrac{z^2}{(z+1)^2 (z-2)}$ at $z = 2$. ***[AU, Nov/Dec 2014]***

Solution: Let $f(z) = \dfrac{z^2}{(z+1)^2 (z-2)}$.

$$[\text{Res}\, f(z)]_{z=2} = \lim_{z \to 2} [(z-2)\, f(z)] = \lim_{z \to 2} \frac{z^2}{(z+1)^2} = \frac{4}{9}.$$

31. State Cauchy's residue theorem. ***[AU, Nov/Dec 2017,May/June 2014]***

Let $f(z)$ be analytic within and on a simple closed contour C, except at a finite number of poles $z_1, z_2, \cdots, z_n$ within C and let $R_1, R_2, \cdots, R_n$ be respectively the residues of $f(z)$ at these poles, then

$$\oint_C f(z)\, dz = 2\pi i\, (R_1 + R_2 + \cdots + R_n).$$

i.e., $\oint_C f(z)\, dz = 2\pi i$ [Sum of the residues at the poles within C].

32. Evaluate $\int_C \tan z\, dz$ where C is $|z| = 2$.

Solution: Let $f(z) = \tan z = \dfrac{\sin z}{\cos z}$, which has simple poles at $z = \pm \frac{\pi}{2}, \pm 3\, \frac{\pi}{2}, \ldots$.

But $z = \pm \frac{\pi}{2}$ lie inside the given curve $|z| = 2$.

$$[\text{Res}\, f(z)]_{z=\frac{\pi}{2}} = \lim_{z \to \frac{\pi}{2}} \left(\frac{\sin z}{-\sin z}\right) = -1.$$

$$[\text{Res } f(z)]_{z=-\frac{\pi}{2}} = \lim_{z \to -\frac{\pi}{2}} \left(\frac{\sin z}{-\sin z}\right) = -1.$$

$\therefore$ By Residue theorem,

$\int_C \tan z \, dz = 2\pi i\,(-1-1) = -4\pi i.$

33. Evaluate $\int_C \frac{z+4}{z^2+2z+5}\, dz$ where C is the circle $|z| = 1$.

Solution: The singular points of $f(z) = \frac{z+4}{z^2+2z+5}$ are given by $z^2 + 2z + 5 = 0$.

$\therefore \; z = -1 \pm 2i.$

Clearly $z = -1 + 2i$ and $z = -1 - 2i$ lie outside C.

$\therefore$ By Cauchy's integral theorem, $\int_C \frac{z+4}{z^2+2z+5}\, dz = 0.$

CHAPTER 5

LAPLACE TRANSFORMS

5.1. Introduction

Laplace transform is a very powerful mathematical tool applied in various areas of engineering and science. It is an integral transform method which is particularly useful in solving linear ordinary differential equations. It finds very wide applications in various areas of physics, electrical engineering, control engineering, optics, mathematics and signal processing. The Laplace transform can be interpreted as a transformation from the time domain where inputs and outputs are functions of time to the frequency domain where inputs and outputs are functions of complex angular frequency. Laplace transform is a very effective mathematical tool to simplify very complex problems in the area of stability and control.

5.2. Laplace Transforms

Definition: Let $f(t)$be a function defined of the variable $t \geq 0$. Then the integral

$L[f(t)] = \int_0^\infty e^{-st} f(t)dt$ exists and is equal to $F(s)$ ie., $L[f(t)] = F(s)$ is called the

Laplace transform of $f(t)$, s may be a real or a complex number.

Note 1: Here the operator L is called the *Laplace transform operator* which transforms the function $f(t)$ into $F(s)$

Note 2 : Not all functions $f(t)$ are Laplace transformable. For a function $f(t)$ to be Laplace transformable it must satisfy the following existence conditions.

5.3. Existence Conditions

Piece-wise continuous function: A function $f(t)$ is said to be *piece-wise continuous* in any interval $[a, b]$, if it is defined on that interval and is such that the interval can be divided into a finite number of sub-intervals in each of which $f(t)$ is continuous.

Exponential order: A function$f(t)$is said to be of the *exponential order* if $\lim_{t\to\infty} \left(e^{-\alpha t} f(t)\right)$ is a finite quantity for some constant α.

Thus sufficient conditions for the existence of the Laplace transforms are

(*i*) $f(t)$ should be continuous or piece-wise continuous in the given closed interval $[a, b]$ where $a > 0$.

(*ii*) $f(t)$ should be of exponential order.

5.4. Transforms of Elementary Functions

1. $L[k] = \frac{k}{s},\ s > 0,$ where k is a constant.

Proof: $L[k] = \int_0^\infty e^{-st}\, k\, dt$

$$= k\left[\frac{e^{-st}}{-s}\right]_0^\infty$$

$$= k\left[0 - \frac{1}{-s}\right] = \frac{k}{s}. \qquad \left[\because e^{-st} \to 0 \text{ as } t \to \infty \text{ only when } s > 0\right]$$

Result: $L[0] = 0$ and $L[1] = \frac{1}{s}$.

2. $L[e^{-at}] = \frac{1}{s+a}$ if $s + a > 0$.

Proof: $L[e^{-at}] = \int_0^\infty e^{-st}.e^{-at}\, dt$

$$= \int_0^\infty e^{-(s+a)t}\, dt$$

$$= \left[\frac{e^{-(s+a)t}}{-(s+a)}\right]_0^\infty$$

$$= \left[0 - \frac{1}{-(s+a)}\right] = \frac{1}{s+a}.$$

3. $L[e^{at}] = \frac{1}{s-a}$ if $s - a > 0$.

Proof: $L[e^{at}] = \int_0^\infty e^{-st}.e^{at}\, dt$

$$= \int_0^\infty e^{-(s-a)t}\, dt$$

$$= \left[\frac{e^{-(s-a)t}}{-(s-a)}\right]_0^\infty$$

$$= \left[0 - \frac{1}{-(s-a)}\right]$$

$$= \frac{1}{s-a}.$$

4. $L[\sin at] = \frac{a}{s^2+a^2}$ if $s > 0$.

Proof : $L[\sin at] = \int_0^\infty e^{-st} \sin at \, dt$

$$= \left[\frac{e^{-st}}{s^2+a^2}(-s\sin at - a\cos at)\right]_0^\infty$$

$$= \left[0 - \frac{1}{s^2+a^2}(0-a)\right] = \frac{a}{s^2+a^2}.$$

5. $L[\cos at] = \frac{s}{s^2+a^2}$ if $s > 0$.

Proof : $L[\cos at] = \int_0^\infty e^{-st} \cos at \, dt$

$$= \left[\frac{e^{-st}}{s^2+a^2}(-s\cos at + a\sin at)\right]_0^\infty$$

$$= \left[0 - \frac{1}{s^2+a^2}(-s+0)\right] = \frac{s}{s^2+a^2}.$$

6. $L[\sinh at] = \frac{a}{s^2-a^2}$ if $s > |a|$.

Proof : $L[\sinh at] = L\left[\frac{1}{2}(e^{at} - e^{-at})\right]$

$$= \frac{1}{2}\left(L[e^{at}] - L[e^{-at}]\right), \quad \text{by linear property}$$

$$= \frac{1}{2}\left(\frac{1}{s-a} - \frac{1}{s+a}\right) \text{ if } s > a \text{ and } s > -a = \frac{a}{s^2-a^2} \text{ if } s > |a|.$$

7. $L[\cosh at] = \frac{s}{s^2-a^2}$ if $s > |a|$.

Proof : $L[\cosh at] = L\left[\frac{1}{2}(e^{at} + e^{-at})\right]$

$$= \frac{1}{2}\left(L[e^{at}] + L[e^{-at}]\right), \quad \text{by linear property}$$

$$= \frac{1}{2}\left(\frac{1}{s-a} + \frac{1}{s+a}\right) \text{ if } s > a \text{ and } s > -a = \frac{s}{s^2-a^2} \text{ if } s > |a|.$$

8. $L[t^n] = \frac{\Gamma(n+1)}{s^{n+1}}$ if $s > 0$ and $n > -1$.

Proof : $L[t^n] = \int_0^\infty e^{-st} t^n \, dt$

Put $st = x$. Then $dt = \frac{dx}{s}$.

$$\therefore L[t^n] = \int_0^\infty e^{-x} \left(\frac{x}{s}\right)^n \frac{dx}{s}$$

$$= \frac{1}{s^{n+1}} \int_0^\infty e^{-x}\, x^n\, dx$$

$$= \frac{\Gamma(n+1)}{s^{n+1}} \text{ if } s > 0 \text{ and } n+1 > 0. \qquad \left[\because \Gamma(n) = \int_0^\infty e^{-x}\, x^{n-1} dx\right]$$

Result : $L[t^n] = \frac{n!}{s^{n+1}}$ if n is a positive integer.

Proof : We know that, $L[t^n] = \frac{\Gamma(n+1)}{s^{n+1}}$

$$= \frac{n!}{s^{n+1}}. \qquad [\because n \text{ is a positive integer}]$$

Example 1 : Find the Laplace transform of $f(t) = \begin{cases} 1, & 0 \le t < 1 \\ -1, & t \ge 1 \end{cases}$.

Solution : $L[f(t)] = \int_0^\infty e^{-st} f(t) dt$

$$= \int_0^1 e^{-st} dt + \int_1^\infty e^{-st}(-1) dt$$

$$= \left[\frac{e^{-st}}{-s}\right]_0^1 - \left[\frac{e^{-st}}{-s}\right]_1^\infty = \left[\frac{e^{-s}}{-s} - \frac{1}{-s}\right] - \left[0 - \frac{e^{-s}}{-s}\right] = \frac{1-2e^{-s}}{s}.$$

Example 2 : Find $L[f(t)]$, if $f(t) = \begin{cases} 0, & \text{for} \quad 0 < t < 1 \\ (t-1)^2, & \text{for} \quad t \ge 1 \end{cases}$.

Solution : $L[f(t)] = \int_0^\infty e^{-st}\, f(t)\, dt$

$$= \int_0^1 e^{-st}\, 0\, dt + \int_1^\infty e^{-st}(t-1)^2\, dt$$

$$= \int_1^\infty (t^2 - 2t + 1)\, e^{-st}\, dt$$

$$= \left[(t^2 - 2t + 1)\left(\frac{e^{-st}}{-s}\right) - (2t-2)\left(\frac{e^{-st}}{s^2}\right) + (2)\left(\frac{e^{-st}}{-s^3}\right)\right]_1^\infty$$

$$= \left[(0) - \left(0 - 0 - 2\,\frac{e^{-s}}{s^3}\right)\right] \quad \text{by } L' \text{ Hospital's rule}$$

$$= 2\frac{e^{-s}}{s^3}.$$

Example 3 : Find $L[f(t)]$, if $f(t) = \begin{cases} \sin at, 0 < t < \frac{\pi}{a} \\ 0, \quad t > \frac{\pi}{a} \end{cases}$.

Solution : $[f(t)] = \int_0^{\pi/a} e^{-st} \sin at\, dt + \int_{\pi/a}^\infty e^{-st}.0\, dt$

$$= \left[\frac{e^{-st}}{s^2+a^2}(-s \sin at - a \cos at)\right]_0^{\frac{\pi}{a}}$$

$$= \left[\frac{e^{\frac{-\pi s}{a}}}{s^2+a^2} (0 - a(-1)) - \frac{1}{s^2+a^2} (0 - a) \right]$$

$$= \frac{a}{s^2+a^2} \left(e^{\frac{-\pi s}{a}} + 1 \right).$$

5.5. Basic Properties

1. Linear Property

$L[af(t) + bg(t)] = aL[f(t)] + bL[g(t)]$, where a and b are constants.

Proof : $L[af(t) + bg(t)] = \int_0^{\infty} e^{-st} [af(t) + bg(t)]\, dt$

$$= a \int_0^{\infty} e^{-st} f(t)dt + b \int_0^{\infty} e^{-st} g(t)dt = aL[f(t)] + bL[g(t)].$$

Results

1. $L[af(t) - bg(t)] = aL[f(t)] - bL[g(t)]$
2. $L[f(t) + g(t)] = L[f(t)] + L[g(t)]$.
3. $L[af(t)] = aL[f(t)]$.

2. Change of Scale Property

If $L[f(t)] = F(s)$, then $L[\, f(at)\,] = \frac{1}{a} F\left(\frac{s}{a}\right)$.

Proof : By definition,

$L[f(at)] = \int_0^{\infty} e^{-st} f(at)dt$ (1)

Put $at = x$. Then $dt = \frac{dx}{a}$.

When $t = 0$, $x = 0$. When $t = \infty$, $x = \infty$.

$\therefore$ (1) gives $L[f(at)] = \int_0^{\infty} e^{-s\left(\frac{x}{a}\right)} f(x) \frac{dx}{a}$

$$= \frac{1}{a} \int_0^{\infty} e^{-\left(\frac{s}{a}\right)x} f(x)dx = \frac{1}{a} F\left(\frac{s}{a}\right).$$

Result : If $L[f(t)] = F(s)$, then $L\left[f\left(\frac{t}{a}\right) \right] = a\, F(as)$.

Example 1 : Find the Laplace transform of $(2 + 5t)^2$.

Solution : $L[(2 + 5t)^2] = L[4 + 20t + 25t^2]$

$$= 4L[1] + 20L[t] + 25L[t^2] = \frac{4}{s} + \frac{20}{s^2} + \frac{50}{s^3}.$$

Example 2 : Find the Laplace transforms of $(e^t - 3)^2$.

Solution : $L[(e^t-3)^2] = L[e^{2t} - 6e^t + 9]$

$$= L[e^{2t}] - 6L[e^t] + 9L[1]$$

$$= \frac{1}{s-2} - \frac{6}{s-1} + \frac{9}{s} = \frac{2(2s^2-8s+9)}{s(s-1)(s-2)}.$$

Example 3: Find the Laplace transform of $\sin 2t - 3\cosh 3t + e^{-7t}$.

Solution : $L[\sin 2t - 3\cosh 3t + e^{-7t}]$

$$= L[\sin 2t] - 3L[\cosh 3t] + L[e^{-7t}]$$

$$= \frac{2}{s^2+4} - 3\frac{s}{s^2-9} + \frac{1}{s+7} = \frac{2}{s^2+4} - \frac{3s}{s^2-9} + \frac{1}{s+7}.$$

Example 4 : Find the Laplace transforms of the following:

(i) $\sin 6t \cos 3t$ (ii) $\cos^2 t$

Solution:

(i) $L[\sin 6t \cos 3t] = L\left[\frac{1}{2}(\sin 9t + \sin 3t)\right]$ $\left[\because \sin A \cos B = \frac{1}{2}[\sin(A+B) + \sin(A-B)]\right]$

$$= \frac{1}{2}\{L[\sin 9t] + L[\sin 3t]\}$$

$$= \frac{1}{2}\left\{\frac{9}{s^2+81} + \frac{3}{s^2+9}\right\} = \frac{6(s^2+27)}{(s^2+81)(s^2+9)}.$$

(ii) $L[\cos^2 t] = L\left[\frac{1}{2}(1+\cos 2t)\right]$ $\left[\because \cos^2 x = \frac{1}{2}(1+\cos 2x)\right]$

$$= \frac{1}{2}(L[1] + L[\cos 2t])$$

$$= \frac{1}{2}\left(\frac{1}{s} + \frac{s}{s^2+4}\right) = \frac{s^2+2}{s(s^2+4)}.$$

Example 5 : Find $L[a^t]$.

Solution : $L[a^t] = L\left[e^{\log a^t}\right] = L\left[e^{t\log a}\right] = L\left[e^{(\log a)t}\right] = \frac{1}{s-\log a}.$

EXERCISES

1. Find the Laplace transform of the following functions:

(i) $(t+2t^2)^2$ (ii) $(a+bt)^2$ (iii) $t^3 + 6t - 7 + \sqrt{t}$

(iv) $\sqrt[3]{t}$ (v) $\frac{(\sqrt{t}+1)(2-\sqrt{t})}{\sqrt{t}}$

2. Find the Laplace transform of each of the following:

(i) $e^{at} + b$ (ii) $e^{2t-3} + \sqrt{\pi}$ (iii) $2\cosh 3t - \sin 2t$

(iv) $e^{-2t} + 3t^2 - 2\sin 3t$ (v) $\sin\left(\frac{2n\pi t}{T}\right)$ (vi) $\cos(\omega t - \theta)$.

3. Find the Laplace transform of the following:

(i) $\sin^2 3t$ (ii) $(\sin t - \cos t)^2$ (iii) $\cos^3 2t$

(iv) $\sin 5t \sin 3t$ (v) $\sin 5t \cos 2t$ (vi) $\sin^2 t \cos^3 t$.

ANSWERS

1. (i) $\frac{50}{s(s^2+100)}$ (ii) $\frac{a^2}{s} + \frac{2b^2}{s^3} + \frac{2ab}{s^2}$ (iii) $\frac{6}{s^4} + \frac{6}{s^2} - \frac{7}{s} + \frac{1}{2}\frac{\sqrt{\pi}}{s^{3/2}}$

(iv) $\frac{1}{3}\frac{\Gamma\left(\frac{1}{3}\right)}{s^{4/3}}$ (v) $\frac{1}{s} - \frac{1}{2}\frac{\sqrt{\pi}}{s^{3/2}} + \frac{2\sqrt{\pi}}{\sqrt{s}}$

2. (i) $e^b \frac{1}{s-a}$ (ii) $e^{-3}\frac{1}{s-2} + \frac{\sqrt{\pi}}{s}$ (iii) $\frac{2s}{s^2-9} - \frac{2}{s^2+4}$

(iv) $\frac{1}{s+2} + \frac{6}{s^3} - \frac{6}{s^2+9}$ (v) $\frac{2n\pi}{T}\frac{1}{s^2+\frac{4n^2\pi^2}{T^2}}$ (vi) $\cos\theta\frac{s}{s^2+\omega^2} + \sin\theta\frac{\omega}{s^2+\omega^2}$

3. (i) $\frac{1}{2}\left[\frac{1}{s} - \frac{s}{s^2+36}\right]$ (ii) $\frac{1}{s} - \frac{2}{s^2+4}$ (iii) $\frac{s(s^2+28)}{(s^2+4)(s^2+36)}$

(iv) $\frac{30s}{(s^2+4)(s^2+64)}$ (v) $\frac{1}{2}\left[\frac{7}{s^2+49} + \frac{3}{s^2+9}\right]$ (vi) $\frac{1}{16}\left[\frac{2s}{s^2+1} - \frac{s}{s^2+9} - \frac{s}{s^2+29}\right]$.

3. Multiplication by t^n

If $L[f(t)] = F(s)$, then $L[t^n f(t)] = (-1)^n \frac{d^n}{ds^n}[F(s)] = (-1)^n \frac{d^n}{ds^n}\{L[f(t)]\}$.

Result: If $L[f(t)] = F(s)$, then $L[t\,f(t)] = \frac{-d}{ds}F(s) = -\frac{d}{ds}\{L[f(t)]\}$.

Proof: By definition, $F(s) = L[f(t)] = \int_0^\infty e^{-st} f(t)dt$ (1)

Differentiating (1) with respect to s, we get

$$\frac{d}{ds}F(s) = \frac{d}{ds}\left(\int_0^\infty e^{-st} f(t)dt\right)$$

$$= \int_0^\infty \frac{\partial}{\partial s}[e^{-st} f(t)]dt,$$

Since there are two independent variables s and t,

$$= \int_0^\infty f(t)e^{-st}[-t]dt$$

$$= -\int_0^\infty e^{-st}[tf(t)]dt = -L[tf(t)]$$

$$\therefore L[tf(t)] = -\frac{d}{ds}F(s).$$

Example 1: Find the Laplace transforms of the following functions:

(i) $t\sin at$ (ii) $t\cos at$ (iii) $t^2\cos\omega t$ (iv) $t^2\sinh at$.

Solution: (i) $L[\,t\sin at\,] = -\frac{d}{ds}L[\,\sin at\,] = -\frac{d}{ds}\left[\frac{a}{s^2+a^2}\right]$

$$= -a\left[\frac{-1}{(s^2+a^2)^2}\times 2s\right] = \frac{2as}{(s^2+a^2)^2}.$$

(ii) $L[\,t\cos at\,] = -\frac{d}{ds}L[\,\cos at\,] = -\frac{d}{ds}\left[\frac{s}{s^2+a^2}\right]$

$$= \left[\frac{(s^2+a^2)-s.2s}{(s^2+a^2)^2}\right] = \frac{s^2-a^2}{(s^2+a^2)^2}.$$

(iii) $L[\,t^2\cos\omega t\,] = (-1)^2\frac{d^2}{ds^2}L\,[\cos\omega t]$

$$= \frac{d^2}{ds^2}\left[\frac{s}{s^2+\omega^2}\right] \qquad (1)$$

Now $\frac{d}{ds}\left[\frac{s}{s^2+\omega^2}\right] = \frac{(s^2+\omega^2).1-s.2s}{(s^2+\omega^2)^2} = \frac{\omega^2-s^2}{(s^2+\omega^2)^2}$

$$\frac{d^2}{ds^2}\left[\frac{s}{s^2+\omega^2}\right] = \frac{d}{ds}\left[\frac{\omega^2-s^2}{(s^2+\omega^2)^2}\right]$$

$$= \frac{(s^2+\omega^2)^2(-2s)-(\omega^2-s^2)2(s^2+\omega^2).2s}{(s^2+\omega^2)^4}$$

$$= \frac{-(s^2+\omega^2)[2s(s^2+\omega^2)+4s(\omega^2-s^2)]}{(s^2+\omega^2)^4}$$

$$= \frac{2s^3-6\omega^2 s}{(s^2+\omega^2)^3}.$$

(iv) $L[\,t^2\sinh at] = (-1)^2\frac{d^2}{ds^2}L[\,\sinh at]$

$$= \frac{d^2}{ds^2}\left[\frac{a}{s^2-a^2}\right]$$

$$= \frac{d}{ds}\left[\frac{-2as}{(s^2-a^2)^2}\right]$$

$$= -2a\left[\frac{(s^2-a^2)^2 1 - s2\,(s^2-a^2)2s}{(s^2-a^2)^4}\right]$$

$$= -2a\,(s^2-a^2)\left[\frac{(s^2-a^2)-4s^2}{(s^2-a^2)^4}\right]$$

$$= \frac{2a(3s^2+a^2)}{(s^2-a^2)^3}.$$

Example 2 : Evaluate: (i) $L[t\ \cos^2 3t]$ (ii) $L[t \sin 2t \sin t]$ (iii) $L[t\ \sin(2t+3)]$.

Solution: (i) $L\,[\,t\cos^2 3t] = -\frac{d}{ds}\,L\,[\cos^2 3t]$ (1)

Now $L[t\cos^2 3t] = L\left[\frac{1+\cos 6t}{2}\right] = \frac{1}{2}\left[\frac{1}{s}+\frac{s}{s^2+36}\right].$

$\therefore$ (1) gives $L\,[t\ \cos^2 3t] = \frac{-d}{ds}\left[\frac{1}{2}\left(\frac{1}{s}+\frac{s}{s^2+36}\right)\right]$

$$= \frac{-1}{2}\left[\frac{-1}{s^2}+\frac{(s^2+36)\,1-s.2s}{(s^2+36)^2}\right]$$

$$= \frac{1}{2}\left[\frac{1}{s^2}+\frac{s^2-36}{(s^2+36)^2}\right].$$

(ii) $L[\,t\ \sin 2t\ \sin t] = -\frac{d}{ds}\,L\,[\,\sin 2t\ \sin t\,]$ (1)

$$L\,[\sin 2t\ \sin t] = L\left[\frac{1}{2}(\cos t - \cos 3t)\right] = \frac{1}{2}\left[\frac{s}{s^2+1}-\frac{s}{s^2+9}\right].$$

$\therefore$ (1) Gives $L[t\ \sin 2t\ \ \sin t] = -\frac{d}{ds}\left[\frac{1}{2}\left(\frac{s}{s^2+1}-\frac{s}{s^2+9}\right)\right]$

$$= -\frac{1}{2}\left[\frac{s^2+1-2s^2}{(s^2+1)^2}-\frac{s^2+9-2s^2}{(s^2+9)^2}\right]$$

$$= \frac{1}{2}\left[\frac{s^2-1}{(s^2+1)^2}-\frac{s^2-9}{(s^2+9)^2}\right].$$

(iii) $L\,[\,t\ \sin(2t+3)] = -\frac{d}{ds}\,L\,[\sin(2t+3)]$

$$= -\frac{d}{ds}\,L\,[\,\sin 2t\cos 3 + \cos 2t \sin 3\,]$$

$$= -\frac{d}{ds}\,[\cos 3\,L\,[\,\sin 2t] + \sin 3\,L\,[\cos 2t]]$$

$$= -\frac{d}{ds}\left[\cos 3\,\frac{2}{s^2+4} + \sin 3\,\frac{s}{s^2+4}\right]$$

$$= -2\,\cos 3\,\frac{d}{ds}\left[\frac{1}{s^2+4}\right] - \sin 3\,\frac{d}{ds}\left[\frac{s}{s^2+4}\right]$$

$$= -2\cos 3\left[\frac{-2s}{(s^2+4)^2}\right] - \sin 3\left[\frac{s^2+4-2s^2}{(s^2+4)^2}\right]$$

$$= \frac{4s\cos 3}{(s^2+4)^2} + \frac{(s^2-4)\sin 3}{(s^2+4)^2}.$$

EXERCISES

1. Find the Laplace transforms of the following functions.

   (i) $t\cos 2t$ (ii) $2t\cos 3t$ (iii) $t^2\sin at$ (iv) $t^2\cos 3t$

   (v) $\sin kt - kt\cos kt$ (vi) $t\sinh 3t$ (vii) $t^2\cosh at$.

2. Find the Laplace transforms of the following functions

   (i) $t(3\sin 2t - 2\cos 2t)$ (ii) $t\sin^2 at$ (iii) $t\sin 3t\cos 2t$

   (iv) $t\cos(at+b)$ (v) $t\cos 2t\cos t$ (vi) $(t\sin at)^2$

   (vii) $t\cosh t\cos t$.

ANSWERS

1. (i) $\frac{s^2-4}{(s^2+4)^2}$ (ii) $\frac{2(s^2-9)}{(s^2+9)^2}$ (iii) $\frac{6as^2-2a^3}{(s^2+a^2)^3}$ (iv) $\frac{2s^3-54s}{(s^2+9)^3}$

   (v) $\frac{2k^3}{(s^2+k^2)^2}$ (vi) $\frac{6s}{(s^2-9)^2}$ (vii) $\frac{2s^3+6a^2s}{(s^2-a^2)^3}$.

2. (i) $\frac{8+12s-2s^2}{(s^2+4)^2}$ (ii) $\frac{1}{2}\left[\frac{1}{s^2} - \frac{s^2-4a^2}{(s^2+4a^2)^2}\right]$ (iii) $\frac{5s}{(s^2+25)^2} + \frac{s}{(s^2+1)^2}$

   (iv) $\cos b\left(\frac{s^2-a^2}{(s^2+a^2)^2}\right) - \sin b\left(\frac{2as}{(s^2+a^2)^2}\right)$ (v) $\frac{1}{2}\left[\frac{s^2-9}{(s^2+9)^2} + \frac{s^2-1}{(s^2+1)^2}\right]$

   (vi) $\frac{1}{s^3} + \frac{s(12a^2-s^2)}{(s^2+4a^2)^3}$ (vii) $\frac{1}{2}\left[\frac{s^2-2s}{(s^2-2s+2)^2} + \frac{s^2+2s}{(s^2+2s+2)^2}\right]$.

4. Division by t

If $L[f(t)] = F(s)$, then $L\left[\frac{f(t)}{t}\right] = \int_s^\infty F(s)\,ds = \int_s^\infty L[f(t)]\,ds$, provided $\lim_{t\to 0}\frac{f(t)}{t}$ exists.

Proof: By definition, $F(s) = \int_0^\infty e^{-st}f(t)dt$.

Integrating both sides with respect to s from s to ∞.

$$\int_s^\infty F(s)\,ds = \int_s^\infty \int_0^\infty e^{-st}f(t)dt\,ds$$

By changing the order of integration, we have

$$\int_s^\infty F(s)\,ds = \int_0^\infty \int_s^\infty e^{-st}f(t)\,ds\,dt$$

$$= \int_0^\infty \left[\frac{e^{-st}}{-t}\right]_s^\infty f(t)\ dt$$

$$= \int_0^\infty \left[0 - \left(\frac{e^{-st}}{-t}\right)\right] f(t)\ dt$$

$$= \int_0^\infty e^{-st} \left[\frac{f(t)}{t}\right] dt = L\left[\frac{f(t)}{t}\right].$$

Corollary : $L\left[\frac{f(t)}{t^2}\right] = \int_s^\infty \int_s^\infty F(s)\ ds\ ds = \int_s^\infty \int_s^\infty L[f(t)]\ ds\ ds$

Example 1 : Find $L\left[\frac{1-e^{-t}}{t}\right]$.

Solution : $L\left[\frac{1-e^{-t}}{t}\right] = \int_s^\infty L\left[1 - e^{-t}\right] ds$

$$= \int_s^\infty \left[\frac{1}{s} - \frac{1}{s+1}\right] ds$$

$$= \left[\log s - \log(s+1)\right]_s^\infty$$

$$= \left[\log \frac{s}{s+1}\right]_s^\infty = \left[\log \frac{s}{s\,(1+{}^1/_s)}\right]_s^\infty$$

$$= \left[\log \frac{1}{1+{}^1/_s}\right]_s^\infty = \log 1 - \log \frac{1}{1+{}^1/_s}$$

$$= 0 - \log \frac{s}{s+1} = \log \frac{s+1}{s}.$$

Example 2 : Find the Laplace transform for the following $L\left[\frac{1-\cos t}{t}\right]$.

Solution: ***[AU, Nov/Dec 2014]***

$$L\left[\frac{1-\cos t}{t}\right] = \int_s^\infty L\left[1 - \cos t\right] ds$$

$$= \int_s^\infty \left[\frac{1}{s} - \frac{s}{s^2+1}\right] ds$$

$$= \left[\log s - \frac{1}{2}\log(s^2+1)\right]_s^\infty = \left[\log s - \log\sqrt{s^2+1}\right]_s^\infty$$

$$= \left[\log \frac{s}{\sqrt{s^2+1}}\right]_s^\infty = \left[\log \frac{s}{s\sqrt{1+1/s^2}}\right]_s^\infty$$

$$= \left[\log \frac{1}{\sqrt{1+1/s^2}}\right]_s^\infty$$

$$= \log 1 - \log \frac{1}{\sqrt{1+1/s^2}} = 0 - \log \frac{S}{\sqrt{S^2+1}} = \log \frac{\sqrt{S^2+1}}{S}.$$

Example 3 : Find $L\left[\frac{\cos at}{t}\right]$.

Solution : $\lim_{t\to 0} \frac{\cos at}{t} = \frac{1}{0} = \infty.$

$\therefore L\left[\frac{\cos at}{t}\right]$ does not exist.

Example 4 : Evaluate $L\left[\frac{e^{-at}-e^{-bt}}{t}\right]$.

Solution:
$$L\left[\frac{e^{-at}-e^{-bt}}{t}\right] = \int_s^\infty L\left[e^{-at} - e^{-bt}\right] ds$$

$$= \int_s^\infty \left[\frac{1}{s+a} - \frac{1}{s+b}\right] ds$$

$$= \left[\log(s+a) - \log(s+b)\right]_s^\infty = \left[\log\frac{s+a}{s+b}\right]_s^\infty$$

$$= \left[\log\frac{s\left(1+\frac{a}{s}\right)}{s\left(1+\frac{b}{s}\right)}\right]_s^\infty = \left[\log\frac{\left(1+\frac{a}{s}\right)}{\left(1+\frac{b}{s}\right)}\right]_s^\infty = 0 - \log\frac{1+\frac{a}{s}}{1+\frac{b}{s}}$$

$$= -\log\frac{s+a}{s+b} = \log\frac{s+b}{s+a}.$$

Example 5 : Find $L\left[\frac{\cos at - \cos bt}{t}\right]$. ***[AU, April/May 2015]***

Solution :
$$L\left[\frac{\cos at - \cos bt}{t}\right] = \int_s^\infty L\left[\cos at - \cos bt\right] ds$$

$$= \int_s^\infty \left[\frac{s}{s^2+a^2} - \frac{s}{s^2+b^2}\right] ds$$

$$= \left[\frac{1}{2}\log(s^2+a^2) - \frac{1}{2}\log(s^2+b^2)\right]_s^\infty$$

$$= \left[\frac{1}{2}\log\frac{S^2+a^2}{S^2+b^2}\right]_s^\infty = \frac{1}{2}\left[\log\frac{S^2\left(1+\frac{a^2}{s^2}\right)}{S^2\left(1+\frac{b^2}{s^2}\right)}\right]_s^\infty$$

$$= \frac{1}{2}\left[\log 1 - \log\frac{1+\frac{a^2}{s^2}}{1+\frac{b^2}{s^2}}\right]$$

$$= \frac{1}{2}\left[0 - \log\left(\frac{s^2+a^2}{s^2+b^2}\right)\right] = \frac{1}{2}\log\left(\frac{s^2+b^2}{s^2+a^2}\right)$$

$$= \log\left(\sqrt{\frac{s^2+b^2}{s^2+a^2}}\right).$$

Example 6: Evaluate $L\left[\frac{1-\cos 2t}{t^2}\right]$.

Solution : $L\left[\frac{1-\cos 2t}{t^2}\right] = \int_s^\infty \int_s^\infty L[1-\cos 2t]\, ds\, ds$

$$= \int_s^\infty \int_s^\infty \left[\frac{1}{s} - \frac{s}{s^2+4}\right] ds\ ds$$

$$= \int_s^\infty \left[\log s - \frac{1}{2}\log(s^2+4)\right]_s^\infty ds$$

$$= \int_s^\infty \left[\log \frac{s}{\sqrt{s^2+4}}\right]_s^\infty ds$$

$$= \int_s^\infty \left[0 - \log\frac{s}{\sqrt{s^2+4}}\right] ds = \int_s^\infty \log\left(\frac{\sqrt{s^2+4}}{s}\right) ds$$

$$= \frac{1}{2}\int_s^\infty \log\left(\frac{s^2+4}{s^2}\right) ds = \frac{1}{2}\int_s^\infty \log\left(1+\frac{4}{s^2}\right) ds$$

$$= \frac{1}{2}\left\{\left[s.\log\left(1+\frac{4}{s^2}\right)\right]_s^\infty - \int_s^\infty s.\frac{1}{1+\frac{4}{s^2}}\left(\frac{-8}{s^3}\right) ds\right\}, \quad \text{integration by parts}$$

$$= \frac{1}{2}\left\{0 - s\log\left(1+\frac{4}{s^2}\right) + 8\int_s^\infty \frac{1}{s^2+4} ds\right\}$$

$$= \frac{1}{2}\left\{-s\log\left(\frac{s^2+4}{s^2}\right) + 8\left[\frac{1}{2}\tan^{-1}\frac{s}{2}\right]_s^\infty\right\}$$

$$= -\frac{s}{2}\log\left(\frac{s^2+4}{s^2}\right) + 2\left[\frac{\pi}{2} - \tan^{-1}\frac{s}{2}\right] = -\frac{s}{2}\log\left(\frac{s^2+4}{s^2}\right) + 2\tan^{-1}\frac{2}{s}.$$

EXERCISES

1. Find $L\left[\frac{1-e^t}{t}\right]$.
2. Find $L\left[\frac{1-\cos at}{t}\right]$.
3. Find $L\left[\frac{\sinh 3t}{t}\right]$
4. Find $L\left[\frac{\sin^2 7t}{t}\right]$.
5. Find $L\left[\frac{e^{-t}-e^{-3t}}{2}\right]$.
6. Evaluate $L\left[\frac{e^{2t}-\cos 3t}{t}\right]$.
7. Evaluate $L\left[\frac{\sin 3t\ \cos t}{t}\right]$.
8. Evaluate $L\left[\frac{\cos 2t-\cos 3t}{t}\right]$.
9. Evaluate $L\left[\frac{e^{-4t}\sin 6t}{t}\right]$.
10. Evaluate $L\left[\frac{e^{-t}\sin^2 t}{t}\right]$.
11. Evaluate $L\left[\frac{1-\cos t}{te^{2t}}\right]$.
12. Evaluate $L\left[e^{-t}\left(\frac{1-\cos 2t}{t^2}\right)\right]$.
13. Evaluate $L\left[\frac{\sin 2t\ \sin t}{t}\right]$.
14. Evaluate $L\left[\left(\frac{\sin t}{t}\right)^2\right]$.

ANSWERS

1. $\log \frac{s-1}{s}$ 2. $\log \frac{\sqrt{s^2+a^2}}{s}$ 3. $\log \sqrt{\frac{s+3}{s-3}}$

4. $\frac{1}{2}\log \frac{\sqrt{s^2+196}}{s}$ 5. $\log \frac{s+3}{s+1}$ 6. $\log \frac{\sqrt{s^2+9}}{s-2}$

7. $\frac{1}{2}\left[\pi - \tan^{-1}\frac{s}{4} - \tan^{-1}\frac{s}{2}\right]$ 8. $\frac{1}{2}\log \frac{s^2+9}{s^2+4}$ 9. $\cot^{-1}\frac{s+4}{6}$

10. $\frac{1}{2}\log \frac{\sqrt{(s+1)^2+4}}{s}$ 11. $\log \frac{\sqrt{(s-2)^2+1}}{s-2}$ 12. $\frac{-(s+1)}{2}\log\left[\frac{(s+1)^2+4}{(s+1)^2}\right] + 2\tan^{-1}\frac{2}{s+1}$

13. $\frac{1}{14}\log \frac{s^2+9}{s^2+1}$ 14. $\frac{s}{4}\log\left(\frac{s^2}{s^2+4}\right) + \cot^{-1}\frac{s}{2}$.

5.6. Shifting Theorems

1. First Shifting Property

If $L[f(t)] = F(s)$, then $L[e^{at} f(t)] = F(s-a)$.

Proof: By definition,

$$L[e^{at} f(t)] = \int_0^\infty e^{-st} e^{at} f(t)\, dt$$

$$= \int_0^\infty e^{-(s-a)t} f(t)\, dt$$

$$= F(s-a).$$

Similarly, we can prove the following result.

Result: If $L[f(t)] = F(s)$, then $L[e^{-at} f(t)] = F(s+a)$.

Note: The working rule for the above properties are

1. $L[e^{at} f(t)] = L[f(t)]_{s \to s-a}$
2. $L[e^{-at} f(t)] = L[f(t)]_{s \to s+a}$

2. Second Shifting Property

If $L[f(t)] = F(s)$, then $L[f(t-a)\, u(t-a)] = e^{-as} F(s)$.

Proof: By definition,

$$L[f(t-a)\, u(t-a)] = \int_0^\infty e^{-st} f(t-a)\, u(t-a)\, dt$$

$$= \int_0^a e^{-st} f(t-a)\, u(t-a)\, dt + \int_a^\infty e^{-st} f(t-a)\, u(t-a)\, dt$$

$$= 0 + \int_a^\infty e^{-st} f(t-a)\, dt, \quad \text{by the definition } u(t-a) \qquad (1)$$

Put $t - a = x$. Then $dt = dx$.

When $t = a$, $x = 0$. When $t = \infty$, $x = \infty$.

$\therefore$ (1) gives, $L[f(t-a)\,u(t-a)] = \int_0^{\infty} e^{-s(x+a)} f(x)\,dx$

$$= e^{-as} \int_0^{\infty} e^{-sx} f(x)\,dx = e^{-as} F(s).$$

Note : The working rule for the above property is

$$L[f(t-a)\,u(t-a)] = e^{-as} L[f(t)].$$

Remark : If $f(t) = \begin{cases} g(t-a), & t \geq a \\ 0, & t < a \end{cases}$, then $L[f(t)] = e^{-as} L[g(t)]$.

Example 1: Find the Laplace transforms of the following functions:

(i) $\frac{e^{3t}}{\sqrt{t}}$ (ii) $e^{at} \sin bt$

Solution: (i) $L\left[\frac{e^{3t}}{\sqrt{t}}\right] = L\left[e^{3t} t^{-1/2}\right]$

$$= L\left[t^{-1/2}\right]_{s \to s-3}$$

$$= \left[\frac{\Gamma(1/2)}{s^{1/2}}\right]_{s \to s-3} = \frac{\sqrt{\pi}}{\sqrt{s-3}}.$$

(ii) $L[e^{at} \sin bt] = L[\sin bt]_{s \to s-a}$

$$= \left[\frac{b}{s^2 + b^2}\right]_{s \to s-a}$$

$$= \frac{b}{(s-a)^2 + b^2}.$$

Example 2 : Find the Laplace transforms of the following functions:

(i) $(1+te^{-t})^3$ (ii) $e^{-t} \sin^2 t$.

Solution:

(i) $L[(1+te^{-t})^3] = L[1 + 3te^{-t} + 3t^2 e^{-2t} + t^3 e^{-3t}]$

$$= L[1] + 3L[e^{-t} t] + 3L[e^{-2t} t^2] + L[e^{-3t} t^3]$$

$$= L[1] + 3L[t]_{s \to s+1} + 3L[t^2]_{s \to s+2} + L[t^3]_{s \to s+3}$$

$$= \frac{1}{s} + 3\left[\frac{1}{s^2}\right]_{s \to s+1} + 3\left[\frac{2!}{s^3}\right]_{s \to s+2} + L\left[\frac{3!}{s^4}\right]_{s \to s+3}$$

$$= \frac{1}{s} + \frac{3}{(s+1)^2} + \frac{6}{(s+2)^3} + \frac{6}{(s+3)^4}.$$

(ii) $L[e^{-t}\sin^2 t] = L[\sin^2 t]_{s \to s+1}$

$$= L\left[\frac{1-\cos 2t}{2}\right]_{s \to s+1}$$

$$= \frac{1}{2}\left[\frac{1}{s} - \frac{s}{s^2+4}\right]_{s \to s+1} = \frac{1}{2}\left[\frac{1}{s+1} - \frac{s+1}{(s+1)^2+4}\right].$$

Example 3 : Evaluate $L[e^{2t}\sin 2t \sin t]$.

Solution: $L[e^{2t}\sin 2t \sin t] = L[\sin 2t \sin t]_{s \to s-2}$

$$= L\left[\frac{1}{2}(\cos t - \cos 3t)\right]_{s \to s-2}$$

$$= \frac{1}{2}\left[\frac{s}{s^2+1} - \frac{s}{s^2+9}\right]_{s \to s-2}$$

$$= \frac{1}{2}\left[\frac{s-2}{(s-2)^2+1} - \frac{s-2}{(s-2)^2+9}\right].$$

Example 4: Find the Laplace transform for the following $t\,e^{-t}\cos t$.

Solution: $L[t\,e^{-t}\cos t] = L[e^{-t}\,t\cos t]$ ***[AU, Nov/Dec 2014]***

$$= L[t\cos t]_{s \to s+1} \qquad (1)$$

Now, $L[t\cos t] = -\frac{d}{ds}L[\cos t]$

$$= -\frac{d}{ds}\left[\frac{s}{s^2+1}\right] = \frac{s^2-1}{(s^2+1)^2}.$$

$\therefore$ (1) gives $L[t\,e^{-t}\cos t] = \left[\frac{s^2-1}{(s^2+1)^2}\right]_{s \to s+1}$

$$= \frac{(s+1)^2-1}{[(s+1)^2+1]^2}$$

$$= \frac{s^2+2s}{[s^2+2s+2]^2}.$$

Example 5: Find the Laplace transform of $f(t) = t\,e^{-3t}\cos 2t$.

Solution: **[AU, *April/May2017, May/June 2014*]**

$L[t\,e^{-3t}\cos 2t] = L[e^{-3t}\,t\cos 2t]$

$$= L[t\cos 2t]_{s \to s+3} \qquad (1)$$

Now, $L[t\cos 2t] = -\frac{d}{ds}L[\cos 2t]$

$$= -\frac{d}{ds}\left[\frac{s}{s^2+4}\right] = \frac{s^2-4}{(s^2+4)^2}.$$

$\therefore$ (1) gives, $L\,[\,e^{-3t}\ t\ \cos 2t] = \left[\frac{s^2-4}{\left(s^2+4\right)^2}\right]_{s\to s+3}$

$$= \frac{(s+3)^2-4}{[(s+3)^2+4\,]^2} = \frac{s^2+6s+5}{[s^2+6s+13\,]^2}.$$

Example 6: Find Laplace transform of $f(t) = t\,e^{-3t}\ \sin 3t$. ***[AU, Nov/Dec 2017]***

Solution: $L\,[t\,e^{-3t}\ \sin 3t] = L\,[e^{-3t}\ t\ \sin 3t]$

$$= L\,[t\ \sin 3t]_{s\to s+3} \qquad (1)$$

Now, $L[\,t\ \sin 3t] = -\frac{d}{ds}L\,[\sin 3t]$

$$= -\frac{d}{ds}\left[\frac{3}{s^2+9}\right] = \frac{6s}{(s^2+9)^2}.$$

$\therefore$ (1) gives, $L\,[\,e^{-3t}\ t\ \cos 2t] = \left[\frac{6s}{\left(s^2+9\right)^2}\right]_{s\to s+3}$

$$= \frac{6(s+3)}{[(s+3)^2+9\,]^2} = \frac{6(s+3)}{[s^2+6s+18\,]^2}.$$

Example 7 : Evaluate $L\,[\,t^2\ e^{-t}\ \cos t\,]$. ***[AU, May/June 2016]***

Solution: $L\,[t^2 e^{-t}\ \cos t\,] = L\,[\,e^{-t}\ t^2 \cos t\,]$

$$= L\,[\,t^2 \cos t\,]_{s\to s+1} \qquad (1)$$

Now, $L[\,t^2 \cos t] = (-1)^2\frac{d^2}{ds^2}L\,[\cos t\,]$

$$= \frac{d^2}{ds^2}\left[\frac{s}{s^2+1}\right]$$

$$= \frac{d}{ds}\left[\frac{1-s^2}{(s^2+1\,)^2}\right] = \frac{2s\ (s^2-3)}{(s^2+1)^3}.$$

$\therefore$ (1) gives, $L\,[t^2\ e^{-t}\ \cos t] = \left[\frac{2\,s\,(s^2-3)}{\left(s^2+1\right)^3}\right]_{s\to s+1}$

$$= \frac{2(s+1)[(s+1)^2-3]}{[(s+1)^2+1]^3}$$

$$= \frac{2\,[s^3+3s^2-2\,]}{[\,s^2+2s+2\,]^3}.$$

Example 8: Find the Laplace transform of $t^2 e^{-3t}\ \cos t$. ***[AU, Nov/Dec 2015]***

Solution: $L\,[t^2 e^{-3t}\ \cos t\,] = L\,[\,e^{-3t}\ t^2 \cos t]$

$$= L\,[t^2 \cos t]_{s \to s+3} \qquad (1)$$

Now, $L[t^2 \cos t\,] \;=\; \frac{2s(s^2-3)}{(s^2+1)^3}$.(By example 7)

$\therefore$ (1) gives $L\,[t^2 e^{-3t}\,\cos t] = \left[\frac{2\,s\,(s^2-3)}{(s^2+1)^3}\right]_{s \to s+3}$

$$= \frac{2\,(s+3)[\,(s+3)^2-3]}{[(s+3)^2+1]^3} = \frac{2\,[s^3+9s^2+24s+18]}{[s^2+6s+10]^3}.$$

Example 9: Find the Laplace transforms of the following functions

(i) $f(t) = \begin{cases} (t-1)^2, & \text{for } t > 1 \\ 0, & for\;\; 0 < t < 1 \end{cases}$

(ii) $f(t) = \begin{cases} \cos\left(t - \frac{2\pi}{3}\right), & t > \frac{2\pi}{3} \\ 0 & , \; t < \frac{2\pi}{3}. \end{cases}$

Solution :

(i) Let $a = 1$ and $g(t-1) = (t-1)^2$. Then $g(t) = t^2$.

By second shifting theorem, $L\,[\,f(t)\,] = e^{-as}\,L\,[\,g(\,t\,)]$

$$= e^{-s}L[t^2] = \frac{2e^{-s}}{s^3}$$

(ii) Let $a = \frac{2\pi}{3}$ and $g\left(t - \frac{2\pi}{3}\right) = \cos\left(t - \frac{2\pi}{3}\right)$. Then $g(t) = \cos t$.

By second shifting theorem, $L\,[f(t)] \;=\; e^{-as}\,L\,[g(t)] \;=\; e^{\frac{-2\pi s}{3}}\,L\,[\cos t]$

$$= e^{\frac{-2\pi s}{3}}\,\frac{s}{s^2+1} = \frac{s\,e^{\frac{-2\pi s}{3}}}{s^2+1}.$$

Example 10 : Evaluate $L\,[e^{-3t}\,u\,(t-2)]$.

Solution: $L[e^{-3t}u(t-2)] \;=\; L[u(t-2)]_{s \to s+3}$

$$= \left[\frac{e^{-2s}}{s}\right]_{s \to s+3}$$

$$= \frac{e^{-2s\,(s+3)}}{s+3}.$$

Example 11 : Find $L\left[\frac{e^{-3t}\sin 2t}{t}\right]$.

Solution : $\mathrm{L}\left[\frac{e^{-3t}\sin 2t}{t}\right] = L\left[\frac{\sin 2t}{t}\right]_{s \to s+3} \qquad (1)$

Now, $L\left[\frac{\sin 2t}{t}\right] = \int_s^\infty L[\sin 2t]\, ds$

$$= \int_s^\infty \frac{2}{s^2+4}\, ds = \left[\tan^{-1}\frac{s}{2}\right]_s^\infty$$

$$= \cot^{-1}\frac{s}{2} = \tan^{-1}\frac{2}{s}.$$

$\therefore$ (1) gives $L\left[\frac{e^{-3t}\sin 2t}{t}\right] = \left[\tan^{-1}\frac{2}{s}\right]_{s\to s+3} = \tan^{-1}\frac{2}{s+3}.$

EXERCISES

1. Find the Laplace transforms of the following functions

   (i) $e^{-at}t^n$ (ii) $e^{at}t^n$ (iii) $e^{-3t}[e^{4t}+t^4+7]$

2. Find the Laplace transforms of the following functions

   (i) $e^{-2t}\cosh^3 2t$ (ii) $\sinh t \sin\frac{\sqrt{3}}{2}t$ (iii) $e^{-t}\sin t\cos t$

   (iv) $e^{-2t}\cos^2 t$ (v) $e^{at}\cos(2t+3)$

3. Find the Laplace transforms of the following functions

   (i) $e^{-2t}\sin 5t\cos 2t$ (ii) $\frac{\cos 3t\cos 2t}{e^{3t}}$ (iii) $e^t\left(\cosh 2t + \frac{1}{2}\sin 2t\right)$

4. Find the Laplace transforms of the following functions

   (i) $f(t) = \begin{cases}(t-a)^2, & t \geq a\\ 0, & t < a\end{cases}$ (ii) $f(t) = \begin{cases}\sin(t-\pi/6), & t > \pi/6\\ 0, & otherwise\end{cases}$

   (iii) $(t-1)^2 u_1(t)$ (iv) $\sin t\, u_\pi(t)$.

5. Find the Laplace transforms of the following functions

   (i) $t\,e^{-t}\sin t$ (ii) $\frac{t^2\cos t}{e^t}$

   (iii) $e^{-2t}(6-3t)\sin t$ (iv) $e^{3t}t\cosh 2t$.

ANSWERS

1. (i) $\frac{\Gamma(n+1)}{(s+a)^{n+1}}$ (ii) $\frac{\Gamma(n+1)}{(s-a)^{n+1}}$ (iii) $\frac{1}{s-1}+\frac{24}{(s+3)^5}+\frac{7}{s+3}$

2. (i) $\frac{1}{8}\left(\frac{1}{s-4}+\frac{3}{s}+\frac{3}{s+4}+\frac{1}{s+8}\right)$ (ii) $\frac{1}{2}\left[\frac{\sqrt{3}/2}{(s-1)^2+\frac{3}{4}}-\frac{\sqrt{3}/2}{(s+1)^2+\frac{3}{4}}\right]$ (iii) $\frac{1}{(s+1)^2+4}$

   (iv) $\frac{1}{2}\left[\frac{1}{s+2}+\frac{s+2}{(s+2)^2+4}\right]$ (v) $\frac{(s-a)\cos 3-2\sin 3}{(s-a)^2+4}$

3. (i) $\frac{1}{2}\left[\frac{7}{(s+2)^2+49}+\frac{3}{(s+2)^2+9}\right]$ (ii) $\frac{1}{2}\left[\frac{s+3}{(s+3)^2+25}+\frac{s+3}{(s+3)^2+1}\right]$

   (iii) $\frac{s-1}{(s-1)^2-4}+\frac{1}{(s-1)^2+4}$

4. (i) $\frac{2e^{-as}}{s^3}$ (ii) $\frac{e^{-\frac{\pi s}{6}}}{s^2+1}$ (iii) $\frac{2}{s^3}e^{-s}$ (iv) $\frac{-e^{-\pi s}}{s^2+1}$.

5. (i) $\frac{2(s+1)}{(s+1)^2+1}$ (ii) $\frac{2(s+1)(s^2+2s-2)}{(s^2+2s+2)^3}$ (iii) $\frac{6}{(s^2+4s+5)^2}-\frac{6(s+2)}{(s^2+4s+5)^2}$

   (iv) $\frac{(s-3)^2+4}{((s-3)^2-4)^2}$

5.7. Laplace Transforms of Derivatives and Integrals

Laplace Transforms of Derivatives

If $L[f(t)] = F(s)$, then $L[f'(t)] = sF(s) - f(0)$ provided $\lim_{t\to\infty} e^{-st} f(t) = 0$.

Proof: By definition, $L[f'(t)] = \int_0^\infty e^{-st} f'(t)dt$

$$= \int_0^\infty e^{-st} d(f(t))$$

$$= [e^{-st} f(t)]_0^\infty - \int_0^\infty f(t)e^{-st}(-s)dt, \text{ integration by parts}$$

$$= 0 - f(0) + s\int_0^\infty e^{-st} f(t)dt$$

$$= -f(0) + s\,F(s) = s\,F(s) - f(0).$$

In general, we can prove that

$$L\left[f^{(n)}(t)\right] = s^n F(s) - s^{n-1}f(0) - s^{n-2}f'(0) - \cdots - f^{(n-1)}(0).$$

Laplace Transforms of Integrals

If $L[f(t)] = F(s)$, then $L\left[\int_0^t f(t)dt\right] = \frac{1}{s}F(s) = \frac{L[f(t)]}{s}$.

Proof: Let $g(t) = \int_0^t f(t)\,dt$.

Then $g(0) = 0$ and $g'(t) = f(t)$

By the previous theorem, $L[g'(t)] = s\,L[g(t)] - g(0)$

i.e., $L[f(t)] = s\,L\left[\int_0^t f(t)dt\right] - 0$

$$L\left[\int_0^t f(t)dt\right] = \frac{1}{s}F(s).$$

In general, we can prove that $L\left[\int_0^t\int_0^t \cdots \int_0^t f(t)(dt)^n\right] = \frac{1}{s^n}F(s)$.

Example 1: If $[\sin\sqrt{t}] = \frac{\sqrt{\pi}}{2\, s^{\frac{3}{2}}} e^{\frac{-1}{4s}}$, prove that $\left[\frac{\cos\sqrt{t}}{\sqrt{t}}\right] = \frac{\sqrt{\pi}}{\sqrt{s}} e^{\frac{-1}{4s}}$.

Solution: Let $f(t) = \sin\sqrt{t}$.

Then $f'(t) = \cos\sqrt{t}\frac{1}{2\sqrt{t}} = \frac{1}{2}\frac{\cos\sqrt{t}}{\sqrt{t}}$ and $f(0) = 0$.

We know that $L[f'(t)] = s\, L[f(t)] - f(0)$.

$$\therefore \quad L\left[\frac{1}{2}\frac{\cos\sqrt{t}}{\sqrt{t}}\right] = s\, L[\sin\sqrt{t}] - 0$$

$$\frac{1}{2}L\left[\frac{\cos\sqrt{t}}{\sqrt{t}}\right] = s\,.\frac{\sqrt{\pi}}{2\, s^{\frac{3}{2}}}\, e^{\frac{-1}{4s}}$$

$$L\left[\frac{\cos\sqrt{t}}{\sqrt{t}}\right] = \frac{\sqrt{\pi}}{\sqrt{s}}\, e^{\frac{-1}{4s}}.$$

Example 2: Find $L\left[\int_0^t \cos 3t\, dt\right]$.

Solution: $L\left[\int_0^t \cos 3t\, dt\right] = \frac{1}{s}\, L[\cos 3t]$, by Laplace transform of integrals

$$= \frac{1}{s}\,.\frac{s}{s^2+9} = \frac{1}{s^2+9}.$$

Example 3: Evaluate $L\left[\int_0^t \frac{\sin t}{t} dt\right]$. **[AU, Nov/Dec 2015]**

Solution: $L\left[\int_0^t \frac{\sin t}{t} dt\right] = \frac{1}{s}\, L\left[\frac{\sin t}{t}\right]$. (1)

Now, $L\left[\frac{\sin t}{t}\right] = \int_s^\infty L[\sin t]ds = \int_s^\infty \frac{1}{s^2+1}ds$

$$= [\tan^{-1} s]_s^\infty = \frac{\pi}{2} - \tan^{-1} s = \cot^{-1} s.$$

$\therefore$(1) gives $L\left[\int_0^t \frac{\sin t}{t} dt\right] = \frac{1}{s}.\cot^{-1} s.$

Example 4: Find $L\left[\int_0^t e^{-2t}\cos t\, dt\right]$ and $L\left[e^{-2t}\int_0^t \cos t\, dt\right]$.

Solution: $L\left[\int_0^t e^{-2t}\cos t\, dt\right] = \frac{1}{s}\, L[e^{-2t}\cos t]$

$$= \frac{1}{s}\{L[\cos t]\}_{s\to s+2}$$

$$= \frac{1}{s}\left\{\frac{s}{s^2+1}\right\}_{s\to s+2} = \frac{1}{s}\cdot\frac{s+2}{(s+2)^2+1}.$$

and $L\left[e^{-2t}\int_0^t \cos t\, dt\right] = \left\{L\left[\int_0^t \cos t\, dt\right]\right\}_{s\to s+2}$

$$= \left\{\frac{1}{s}.L[\cos t]\right\}_{s\to s+2}$$

$$= \left\{\frac{1}{s}\cdot\frac{s}{s^2+1}\right\}_{s\to s+2} = \frac{1}{(s+2)^2+1}.$$

Example 5: Find$L\left[\int_0^t \frac{\cos at-\cos bt}{t}\, dt\right]$.

Solution: $L\left[\int_0^t \frac{\cos at-\cos bt}{t}\, dt\right] = \frac{1}{s}\, L\left[\frac{\cos at-\cos bt}{t}\right]$, by L.T. of integrals

$$= \frac{1}{s}.\log\left(\sqrt{\frac{s^2+b^2}{s^2+a^2}}\right).$$

EXERCISES

1. Find the Laplace transforms of the following functions using the theorem on transform of derivatives.

   (i) e^{-at} (ii) $\sin at$ (iii) $\cos^2 t$

   (iv) $\sin^2 t$ (v) t^n (vi) $t\ \cosh at$.

2. Find the Laplace transforms of $\sqrt{\frac{t}{\pi}}$ and hence find $L\left[\frac{1}{\sqrt{\pi t}}\right]$.

3. Find $L[t\cos at]$ using the Laplace transforms of the derivatives. Hence find $L[\sin at - at\cos at]$and$L[\cos at - at\ \sin at]$.

4. Find $L\left[\int_0^t (\sin t + e^{-t})dt\right]$.

5. Find $L\left[e^{-t} + \int_0^t \sin t\, dt\right]$.

6. Evaluate$\int_0^t \frac{1-e^{-u}}{u}\, du$.

7. Find $L\left[\int_0^t \frac{e^{-t}\sin t}{t}\, dt\right]$.

8. Evaluate$\mathrm{L}\left[e^{-t}\int_0^t \frac{\sin t}{t}dt\right]$.

9. Evaluate$\mathrm{L}\left[\int_0^t te^{-t}\sin 4t\, dt\right]$.

10. Evaluate$\mathrm{L}\left[\frac{1}{t}\int_0^t e^{-t}\sin t\, dt\right]$.

11. Evaluate$\mathrm{L}\left[\cosh t\int_0^t e^{t}\cosh t\, dt\right]$.

ANSWERS

1. (i) $\frac{1}{s+1}$ (ii) $\frac{a}{s^2+a^2}$ (iii) $\frac{s^2+2}{s(s^2+4)}$

   (iv) $\frac{2}{s(s^2+4)}$ (v) $\frac{n!}{s^{n+1}}$ (vi) $\frac{s^2+a^2}{(s^2-a^2)^2}$.

2. $\frac{1}{2s^{3/2}}$; $\frac{1}{\sqrt{s}}$ 3. $\frac{s^2-a^2}{(s^2+a^2)^2}$; $\frac{2a^3}{(s^2+a^2)^2}$; $\frac{s(s^2-a^2)}{(s^2+a^2)^2}$ 4. $\frac{1}{s}\left[\frac{1}{s^2+1}+\frac{1}{s+1}\right]$

5. $\frac{1}{s+1}+\frac{1}{s}.\frac{1}{s^2+1}$ 6. $\frac{1}{s}\log\left(\frac{s+1}{s}\right)$ 7. $\frac{1}{s}\tan^{-1}\frac{1}{s-1}$

8. $\frac{1}{s+1}\cot^{-1}(s+1)$ 9. $\frac{8\,(s+1)}{s(s^2+2s+17)^2}$ 10. $\frac{1}{4}\log\frac{(s^2+2s+2)}{s^2}-\frac{1}{2}\cot^{-1}(s+1)$

11. $\frac{1}{2}\left[\frac{s-2}{(s-1)^2(s-3)}+\frac{s}{(s+1)^2(s-1)}\right]$

Evaluation of Integrals Using Laplace Transforms

Example 1: Evaluate $\int_0^{\infty} t\cos 2t\,dt$.

Solution: $\int_0^{\infty} t\cos 2t\,dt = \left[\int_0^{\infty} e^{-st}\, t\cos 2t\,dt\right]_{s=0}$

$$= L[t\cos 2t]_{s=0} = \left[\frac{s^2-4}{\left(s^2+4\right)^2}\right]_{s=0} = \frac{-1}{4}.$$

Example 2: Evaluate $\int_0^{\infty} te^{-2t}\sin 3t\,dt$ using Laplace transform **[AU, *Nov/Dec 2017*]**

Solution: Let $I = \int_0^{\infty} e^{-2t}\, t\sin 3t\,dt$

$$= L[t\sin 3t]_{s=2} \quad \ldots (1)$$

Now, $L[t\sin 3t] = -\frac{d}{ds}L[\sin 3t]$

$$= -\frac{d}{ds}\left[\frac{3}{s^2+9}\right] = \frac{6s}{(s^2+9)^2}.$$

$\therefore$ (1) gives, $I = \left[\frac{6s}{\left(s^2+9\right)^2}\right]_{s=2}$

$$= \frac{12}{(13)^2} = \frac{12}{169}.$$

Example 3: Prove that $\int_0^{\infty}\frac{e^{-t}-e^{-3t}}{t}\,dt = \log 3$.

Solution: $\int_0^{\infty}\frac{e^{-t}-e^{-3t}}{t}\,dt = \left[\int_0^{\infty} e^{-st}\left(\frac{e^{-t}-e^{-3t}}{t}\right)dt\right]_{s=0}$

$$= L\left[\frac{e^{-t}-e^{-3t}}{t}\right]_{s=0} \quad (1)$$

Now, $L\left[\frac{e^{-t}-e^{-3t}}{t}\right] = \int_s^{\infty} L[e^{-t}-e^{-3t}]\,ds$

$$= \int_s^{\infty}\left[\frac{1}{s+1}-\frac{1}{s+3}\right]ds$$

$$= [\log(s+1)-\log(s+3)]_s^{\infty}$$

$$= \left[\log\left(\frac{s+1}{s+3}\right)\right]_s^\infty = \left[\log\left(\frac{1+\frac{1}{s}}{1+\frac{3}{s}}\right)\right]_s^\infty$$

$$= \log 1 - \log\left(\frac{1+\frac{1}{s}}{1+\frac{3}{s}}\right)$$

$$= 0 - \log\left(\frac{s+1}{s+3}\right) = \log\left(\frac{s+3}{s+1}\right).$$

$\therefore$ (1) gives $\int_0^\infty \frac{e^{-t}-e^{-3t}}{t}\, dt = \left[\log\left(\frac{s+3}{s+1}\right)\right]_{s=0} = \log 3.$

Example 4: Use Laplace transform to evaluate $\int_0^\infty \left(\frac{\cos 2t - \cos 5t}{t}\right) dt.$

Solution: $\int_0^\infty \left(\frac{\cos 2t-\cos 5t}{t}\right) dt = \left[\int_0^\infty e^{-st}\left(\frac{\cos 2t-\cos 5t}{t}\right) dt\right]_{s=0}$

$$= L\left[\frac{\cos 2t - \cos 5t}{t}\right]_{s=0} \quad (1)$$

Now, $L\left[\frac{\cos 2t-\cos 5t}{t}\right] = \int_s^\infty L[\cos 2t - \cos 5t]\, ds$

$$= \int_s^\infty \left[\frac{s}{s^2+4} - \frac{s}{s^2+25}\right] ds$$

$$= \left[\frac{1}{2}\log(s^2+4) - \frac{1}{2}\log(s^2+25)\right]_s^\infty$$

$$= \left[\frac{1}{2}\log\left(\frac{s^2+4}{s^2+25}\right)\right]_s^\infty = \frac{1}{2}\left[\log\left(\frac{1+\frac{4}{s^2}}{1+\frac{25}{s^2}}\right)\right]_s^\infty$$

$$= \frac{1}{2}\left[\log 1 - \log\left(\frac{1+\frac{4}{s^2}}{1+\frac{25}{s^2}}\right)\right]$$

$$= \frac{1}{2}\left[0 - \log\left(\frac{s^2+4}{s^2+25}\right)\right] = \log\left(\sqrt{\frac{s^2+25}{s^2+4}}\right).$$

$\therefore$ (1) gives $\int_0^\infty \left(\frac{\cos 2t-\cos 5t}{t}\right) dt = \left[\log\left(\sqrt{\frac{s^2+25}{s^2+4}}\right)\right]_{s=0}$

$$= \log\left(\frac{5}{2}\right).$$

Example 5: Evaluate $\int_0^\infty \frac{e^t(\cos 2t - \cos 5t)}{t}\, dt.$ **[AU, April/May 2017]**

Solution: $\int_0^\infty \frac{e^t(\cos 2t-\cos 5t)}{t}\, dt = \left[\int_0^\infty e^t\left(\frac{\cos 2t-\cos 5t}{t}\right) dt\right]_{s=-1}$

$$= L\left[\frac{\cos 2t - \cos 5t}{t}\right]_{s=-1} \qquad (1)$$

Now, $L\left[\frac{\cos 2t - \cos 5t}{t}\right] = \int_s^\infty L[\cos 2t - \cos 5t]\, ds$

$$= \int_s^\infty \left[\frac{s}{s^2+4} - \frac{s}{s^2+25}\right] ds$$

$$= \left[\frac{1}{2}\log(s^2+4) - \frac{1}{2}\log(s^2+25)\right]_s^\infty$$

$$= \left[\frac{1}{2}\log\left(\frac{s^2+4}{s^2+25}\right)\right]_s^\infty = \frac{1}{2}\left[\log\left(\frac{1+\frac{4}{s^2}}{1+\frac{25}{s^2}}\right)\right]_s^\infty$$

$$= \frac{1}{2}\left[\log 1 - \log\left(\frac{1+\frac{4}{s^2}}{1+\frac{25}{s^2}}\right)\right]$$

$$= \frac{1}{2}\left[0 - \log\left(\frac{s^2+4}{s^2+25}\right)\right] = \log\left(\sqrt{\frac{s^2+25}{s^2+4}}\right).$$

$\therefore$ (1) gives $\int_0^\infty \frac{e^t(\cos 2t - \cos 5t)}{t}\, dt = \left[\log\left(\sqrt{\frac{s^2+25}{s^2+4}}\right)\right]_{s=-1} = \log\left(\frac{26}{5}\right)$

EXERCISES

1. Show that $\int_0^\infty e^{-t}\, t\cos t\, dt = 0$.
2. Evaluate $\int_0^\infty e^{-2t}\, t\sin 3t\, dt$.
3. Evaluate $\int_0^\infty \frac{t^3 \sin t}{e^t}\, dt$.
4. Evaluate $\int_0^\infty \frac{e^{-t}\sin^2 t}{t}\, dt$.
5. Prove that $\int_0^\infty \frac{e^{-2t} - e^{-4t}}{t}\, dt = \log 2$.
6. Prove that $\int_0^\infty e^{-t}\frac{\sin t}{t}\, dt = \frac{\pi}{4}$.
7. Prove that $\int_0^\infty \left(\frac{\cos at - \cos bt}{t}\right) dt = \log\left(\frac{b}{a}\right)$.
8. Use Laplace transform to evaluate $\int_0^\infty \frac{e^t(\cos 2t - \cos 5t)}{t}\, dt$.
9. Evaluate $\int_0^\infty \frac{1-\cos t}{t\, e^t}\, dt$.
10. Evaluate $\int_0^\infty \frac{e^{-\sqrt{2}t}\sin t \sinh t}{t}\, dt$.

ANSWERS

2. $\frac{12}{169}$ 3. 0 4. $\frac{1}{4}\log 5$ 8. $\log\sqrt{\frac{26}{5}}$

9. $\frac{1}{2}\log 2$ 10. $\frac{\pi}{8}$.

5.8. Laplace Transforms of Unit Step and Impulse Functions

Unit step function: The *unit step function* or *Heaviside's unit function* is defined as $u(t-a)=\begin{cases}0 & \text{for} \quad t<a\\ 1 & \text{for} \quad t\geq a\end{cases}$, where a is always positive. It is also denoted as $H(t-a)$.

Result: $L[u(t-a)] = \frac{e^{-as}}{s}$ if $s>0$.

Proof: $L[u(t-a)] = \int_0^\infty e^{-st}u_a(t)\,dt$

$$= \int_0^a e^{-st}u_a(t)dt + \int_a^\infty e^{-st}\,u_a\,(t)\,dt$$

$$= 0 + \int_a^\infty e^{-st}\,dt$$

$$= \left[\frac{e^{-st}}{-s}\right]_a^\infty$$

$$= \left[0-\frac{e^{-as}}{-s}\right] \text{ if } s>0$$

$$= \frac{e^{-as}}{s}.$$

In particular, $L[u(t)] = \frac{1}{s}$ if $s>0$.

Unit impulse function: The *unit impulse function* or *Dirac Delta function* $\delta(t-a)$ *is* defined as $\delta(t-a)=\begin{cases}\infty & \text{for} \quad t=a\\ 0 & \text{for} \quad t\neq a\end{cases}$ such that $\int_0^\infty \delta(t-a)\,dt = 1\,,(a\geq 0)$.

Result: $L[\,\delta\,(t-a)] = e^{-as}$.

Proof: $L\,[\,\delta\,(t-a)] \;=\; L[\lim_{h\to 0}\{f(t)\}] = \lim_{h\to 0}\{\,L\,[f(t)]\}$

$$= \lim_{h\to 0}\left\{\int_0^\infty e^{-st}f(t)\,dt\right\}$$

$$= \lim_{h\to 0}\left\{\int_0^a e^{-st}f(t)\,dt + \int_a^{a+h} e^{-st}f(t)\,dt + \int_{a+h}^\infty e^{-st}f(t)\,dt\right\}$$

$$= \lim_{h\to 0}\left\{0 + \int_a^{a+h} e^{-st}\,\frac{1}{h}\,dt + 0\right\}$$

$$= \lim_{h\to 0}\frac{1}{h}\left[\frac{e^{-st}}{-s}\right]_a^{a+h} = \lim_{h\to 0}\frac{1}{h}\left[\frac{e^{-s(a+h)}}{-s}-\frac{e^{-sa}}{-s}\right]$$

$$= \lim_{h\to 0}\frac{1}{h}\left[\frac{e^{-sa}\left(1-e^{-sh}\right)}{s}\right] = \frac{e^{-as}}{s}\lim_{h\to 0}\frac{1-e^{-sh}}{h}$$

$$= \frac{e^{-as}}{s}\lim_{h\to 0} s\,e^{-sh}, \text{ by L'Hospital's rule}$$

$$= \frac{e^{-as}}{s}s = e^{-as}.$$

5.9. Laplace Transform of Periodic Functions

Definition: A function $f(t)$is said to be a *periodic function,* if there exists a positive constant p such that $f(t+p) = f(t)$, for all values of t. The smallest positive value of p satisfying the condition $f(t+p) = f(t)$ is called the period of the function.

Laplace Transform of periodic function

If $f(t)$ is a periodic function with period p, then$L[f(t)] = \frac{1}{1-e^{-ps}}\int_0^p e^{-st}f(t)dt$.

Proof: By definition, $L[f(t)] = \int_0^\infty e^{-st}f(t)\,dt$

$$= \int_0^p e^{-st}f(t)dt + \int_p^\infty e^{-st}f(t)\,dt \qquad (1)$$

In the second integral, put $t = x+p$.

Then $dt = dx$ and x varies from 0 to ∞.

$\therefore$(1) gives $L[f(t)] = \int_0^p e^{-st}f(t)dt + \int_0^\infty e^{-s(x+p)}f(x+p)dx$

$$= \int_0^p e^{-st}f(t)dt + e^{-sp}\int_0^\infty e^{-sx}f(x)dx \,[\because f(x+p) = f(x)]$$

$$= \int_0^p e^{-st}f(t)dt + e^{-ps}L[f(t)],$$

On changing the dummy variable x to t.

$L[f(t)]\{1-e^{-ps}\} = \int_0^p e^{-st}f(t)dt$

$\therefore \quad L[f(t)] = \frac{1}{1-e^{-ps}}\int_0^p e^{-st}f(t)dt$.

Example 1: Find the Laplace transform of $f(t) = \begin{cases} t^2 \text{ if } 0 < t < 2 \\ 0 \text{ if } 2 < t < 4 \end{cases}$ and $f(t+4) = f(t)$.

Solution: The given function$f(t)$ is a periodic function with period $p = 4$.

The Laplace transform of a periodic function with period p is

$$L[f(t)] = \frac{1}{1-e^{-ps}}\int_0^p e^{-st}f(t)dt$$

$$= \frac{1}{1-e^{-4s}}\int_0^4 e^{-st}f(t)dt$$

$$= \frac{1}{1-e^{-4s}}\left[\int_0^2 e^{-st}t^2dt + \int_2^4 e^{-st}.0\,dt\right]$$

$$= \frac{1}{1-e^{-4s}}\int_0^2 t^2e^{-st}dt$$

$$= \frac{1}{1-e^{-4s}}\left[(t^2)\left(\frac{e^{-st}}{-s}\right) - (2t)\left(\frac{e^{-st}}{(-s)^2}\right) + (2)\left(\frac{e^{-st}}{(-s)^3}\right)\right]_0^2$$

$$= \frac{1}{1-e^{-4s}}\left[\left(\frac{-4e^{-2s}}{s} - \frac{4e^{-2s}}{s^2} - \frac{2e^{-2s}}{s^3}\right) - \left(0 - 0 - \frac{2}{s^3}\right)\right]$$

$$= \frac{1}{1-e^{-4s}}\left[\frac{2-2e^{-2s}-4s\,e^{-2s}-4s^2e^{-2s}}{s^3}\right].$$

Example 2: Find the Laplace transform of $f(t) = t$ in $0 < t < 2$ and $f(t+2) = f(t)$.

Solution: The given function $f(t)$ is periodic with period $p = 2$.

$$\therefore \quad L[f(t)] = \frac{1}{1-e^{-ps}}\int_0^p e^{-st}f(t)dt$$

$$= \frac{1}{1-e^{-2s}}\int_0^2 e^{-st}.t\,dt$$

$$= \frac{1}{1-e^{-2s}}\left[(t)\left(\frac{e^{-st}}{-s}\right) - (1)\left(\frac{e^{-st}}{(-s)^2}\right)\right]_0^2$$

$$= \frac{1}{1-e^{-2s}}\left[\left(\frac{-2e^{-2s}}{s} - \frac{e^{-2s}}{s^2}\right) - \left(0 - \frac{1}{s^2}\right)\right]$$

$$= \frac{1}{1-e^{-2s}}\left[\frac{1-e^{-2s}-2s\,e^{-2s}}{s^2}\right].$$

Example 3: Obtain the Laplace transform of the periodic saw tooth wave function given by $f(t) = \frac{\alpha t}{\omega}$ for $0 < t < \omega$ and $f(t+\omega) = f(t)$.

Solution: The given function is periodic with period $p = \omega$.

$$\therefore \quad L[f(t)] = \frac{1}{1-e^{-ps}}\int_0^p e^{-st}f(t)\,dt$$

$$= \frac{1}{1-e^{-\omega s}}\int_0^{\omega} e^{-st}\frac{\alpha t}{\omega}dt = \frac{1}{1-e^{-\omega s}}.\frac{\alpha}{\omega}\int_0^{\omega} te^{-st}dt$$

$$= \frac{\alpha}{\omega} \cdot \frac{1}{1-e^{-\omega s}} \left[(t)\left(\frac{e^{-st}}{-s}\right) - (1)\left(\frac{e^{-st}}{(-s)^2}\right) \right]_0^{\omega}$$

$$= \frac{\alpha}{\omega} \cdot \frac{1}{1-e^{-\omega s}} \left[\left(\frac{-\omega\, e^{-s\omega}}{s} - \frac{e^{-s\omega}}{s^2} \right) - \left(0 - \frac{1}{s^2} \right) \right]$$

$$= \frac{\alpha}{\omega} \cdot \frac{1}{1-e^{-\omega s}} \left[\frac{1 - s\,\omega\, e^{-s\omega} - e^{-s\omega}}{s^2} \right].$$

Example 4: Find the Laplace transform of $f(t)$ is defined by

$f(t) = \begin{cases} E & if\ 0 < t < \frac{a}{2} \\ -E & if\ \frac{a}{2} < t < a \end{cases}$ where $f(t + a) = f(t)$. **[*AU, Nov/Dec 2017, May/June 2016*]**

Solution: The given function $f(t)$ is periodic with period $p = a$.

For a periodic function,

$$L[f(t)] = \frac{1}{1-e^{-ps}} \int_0^p e^{-st} f(t) dt$$

$$= \frac{1}{1-e^{-as}} \int_0^a e^{-st} f(t) dt$$

$$= \frac{1}{1-e^{-as}} \left\{ \int_0^{\frac{a}{2}} e^{-st} E\, dt + \int_{\frac{a}{2}}^{a} e^{-st} (-E)\, dt \right\}$$

$$= \frac{1}{1-e^{-as}} \left\{ E \int_0^{\frac{a}{2}} e^{-st} dt - E \int_{\frac{a}{2}}^{a} e^{-st} dt \right\}$$

$$= \frac{E}{1-e^{-as}} \left\{ \left[\frac{e^{-st}}{-s} \right]_0^{\frac{a}{2}} - \left[\frac{e^{-st}}{-s} \right]_{\frac{a}{2}}^{a} \right\}$$

$$= \frac{E}{1-e^{-as}} \left\{ \left[-\frac{e^{-s\frac{a}{2}}}{s} + \frac{1}{s} \right] - \left[-\frac{e^{-as}}{s} + \frac{e^{-s\frac{a}{2}}}{s} \right] \right\}$$

$$= \frac{E}{s\,(1-e^{-as})} \left[1 - 2e^{-\frac{as}{2}} + e^{-as} \right]$$

$$= \frac{E}{s\left(1-e^{-\frac{as}{2}}\right)\left(1+e^{-\frac{as}{2}}\right)} \left[1 - e^{-\frac{as}{2}} \right]^2$$

$$= \frac{E}{s} \frac{1-e^{-\frac{as}{2}}}{1+e^{-\frac{as}{2}}} = \frac{E}{s} \frac{e^{\frac{as}{4}} - e^{-\frac{as}{4}}}{e^{\frac{as}{4}} + e^{-\frac{as}{4}}}$$

$$= \frac{E}{s} \tanh\left(\frac{as}{4}\right).$$

Example 5: Find the Laplace transform of "triangular wave" function $f(t)$defined by $f(t) = \begin{cases} t & ,if \ \ 0 \le t \le a \\ 2a - t, if & a \le t \le 2a \end{cases}$ and $f(t + 2a) = f(t)$ for all t. **[AU, Nov/Dec 2014]**

Solution: The given function is a periodic function with period $p = 2a$.

$$\therefore \ L\,[f(t)] = \frac{1}{1-e^{-ps}} \int_0^p e^{-st}\, f(t)\, dt$$

$$= \frac{1}{1-e^{-2as}} \int_0^{2a} e^{-st}\, f(t)\, dt$$

$$= \frac{1}{1-e^{-2as}} \left\{ \int_0^a t\, e^{-st}\, dt + \int_a^{2a} (2a-t) e^{-st}\, dt \right\}$$

$$= \frac{1}{1-e^{-2as}} \left\{ \left[(t)\left(\frac{e^{-st}}{-s}\right) - (1)\left(\frac{e^{-st}}{s^2}\right) \right]_0^a + \left[(2a-t)\left(\frac{e^{-st}}{-s}\right) - (-1)\left(\frac{e^{-st}}{s^2}\right) \right]_a^{2a} \right\}$$

$$= \frac{1}{1-e^{-2as}} \left\{ \left[\left(\frac{-ae^{-sa}}{s} - \frac{e^{-sa}}{s^2}\right) - \left(0 - \frac{1}{s^2}\right) \right] + \left[\left(0 + \frac{e^{-2as}}{s^2}\right) - \left(\frac{-ae^{-sa}}{s} + \frac{e^{-sa}}{s^2}\right) \right] \right\}$$

$$= \frac{1}{1-e^{-2as}} \left\{ \frac{1 - 2e^{-as} + e^{-2as}}{s^2} \right\}$$

$$= \frac{1}{(1-e^{-as})(1+e^{-as})} \frac{(1-e^{-as})^2}{s^2} = \frac{1}{s^2} \frac{1-e^{-as}}{1+e^{-as}}$$

$$= \frac{1}{s^2} \frac{e^{\frac{as}{2}} - e^{-\frac{as}{2}}}{e^{\frac{as}{2}} + e^{-\frac{as}{2}}} = \frac{1}{s^2} . \tanh\left(\frac{as}{2}\right).$$

Example 6: Find the Laplace transform of "triangular wave" function $f(t)$defined by $f(t) = \begin{cases} t & in \ \ 0 < t \le c \\ 2c - t & in \ \ c < t \le 2c \end{cases}$ and $f(t + 2c) = f(t)$ for all t. **[AU, April/May 2015]**

Solution: The given function is a periodic function with period $p = 2c$.

$$\therefore \ L[f(t)] = \frac{1}{1-e^{-ps}} \int_0^p e^{-st} f(t) dt$$

$$= \frac{1}{1-e^{-2cs}} \int_0^{2c} e^{-st} f(t) dt$$

$$= \frac{1}{1-e^{-2cs}} \left\{ \int_0^c t\, e^{-st}\, dt + \int_c^{2c} (2c-t) e^{-st}\, dt \right\}$$

$$= \frac{1}{1-e^{-2cs}} \left\{ \left[(t)\left(\frac{e^{-st}}{-s}\right) - (1)\left(\frac{e^{-st}}{s^2}\right) \right]_0^c + \left[(2c-t)\left(\frac{e^{-st}}{-s}\right) - (-1)\left(\frac{e^{-st}}{s^2}\right) \right]_c^{2c} \right\}$$

$$= \frac{1}{1-e^{-2cs}}\left\{\left[\left(\frac{-ce^{-sc}}{s}-\frac{e^{-sc}}{s^2}\right)-\left(0-\frac{1}{s^2}\right)\right]+\left[\left(0+\frac{e^{-2cs}}{s^2}\right)-\left(\frac{-ce^{-sc}}{s}+\frac{e^{-sc}}{s^2}\right)\right]\right\}$$

$$= \frac{1}{1-e^{-2cs}}\left\{\frac{1-2e^{-cs}+e^{-2cs}}{s^2}\right\}$$

$$= \frac{1}{(1-e^{-cs})(1+e^{-cs})}\cdot\frac{(1-e^{-cs})^2}{s^2}$$

$$= \frac{1}{s^2}\frac{1-e^{-cs}}{1+e^{-cs}} = \frac{1}{s^2}\frac{e^{\frac{cs}{2}}-e^{-\frac{cs}{2}}}{e^{\frac{cs}{2}}+e^{-\frac{cs}{2}}} = \frac{1}{s^2}.\tanh\left(\frac{cs}{2}\right).$$

Example7: Find the Laplace transform of $f(t)$, where $f(t) = \begin{cases} \frac{4E}{T}t - E\ , & 0 < t < \frac{T}{2} \\ 3E - \frac{4E}{T}t\ , & \frac{T}{2} < t < T \end{cases}$

and $f(t+T) = f(t)$ and E is a constant. ***[AU, Nov/Dec 2015]***

Solution: The given function is a periodic function with period $p = T$.

$$\therefore\ L[f(t)] = \frac{1}{1-e^{-ps}}\int_0^p e^{-st}f(t)dt$$

$$= \frac{1}{1-e^{-Ts}}\int_0^T e^{-st}f(t)dt$$

$$= \frac{1}{1-e^{-Ts}}\left\{\int_0^{\frac{T}{2}}\left(\frac{4E}{T}t-E\right)e^{-st}\,dt + \int_{\frac{T}{2}}^{T}\left(3E-\frac{4E}{T}t\right)e^{-st}\,dt\right\}$$

$$= \frac{1}{1-e^{-Ts}}\left\{\left[\left(\frac{4E}{T}t-E\right)\left(\frac{e^{-st}}{-s}\right)-\frac{4E}{T}\left(\frac{e^{-st}}{s^2}\right)\right]_0^{\frac{T}{2}}+\left[\left(3E-\frac{4E}{T}t\right)\left(\frac{e^{-st}}{-s}\right)-\left(-\frac{4E}{T}\right)\left(\frac{e^{-st}}{s^2}\right)\right]_{\frac{T}{2}}^{T}\right\}$$

$$= \frac{1}{1-e^{-Ts}}\left\{\left[\left(\frac{4E}{T}\frac{T}{2}-E\right)\left(\frac{e^{\frac{-sT}{2}}}{-s}\right)-\frac{4E}{T}\left(\frac{e^{\frac{-sT}{2}}}{s^2}\right)\right]-\left[\frac{E}{S}-\frac{4E}{Ts^2}\right]+\left(3E-\frac{4E}{T}T\right)\left(\frac{e^{-sT}}{-s}\right)+\left(\frac{4E}{T}\right)\left(\frac{e^{-sT}}{s^2}\right)\right.$$
$$\left.-\left[\left(3E-\frac{4E}{T}\frac{T}{2}\right)\left(\frac{e^{\frac{-sT}{2}}}{-s}\right)-\left(-\frac{4E}{T}\right)\left(\frac{e^{\frac{-sT}{2}}}{s^2}\right)\right]\right\}$$

$$= \frac{1}{1-e^{-Ts}}\left[\frac{-E}{s}e^{\frac{-sT}{2}}-\frac{4E}{Ts^2}e^{\frac{-sT}{2}}-\frac{E}{S}+\frac{4E}{Ts^2}+\frac{E}{s}e^{-sT}+\frac{4E}{Ts^2}e^{-sT}+\frac{E}{s}e^{\frac{-sT}{2}}+\frac{4E}{Ts^2}e^{\frac{-sT}{2}}\right]$$

$$= \frac{1}{1-e^{-Ts}}\left[-\frac{4E}{Ts^2}e^{\frac{-sT}{2}}-\frac{E}{S}+\frac{4E}{Ts^2}+\frac{E}{s}e^{-sT}+\frac{4E}{Ts^2}e^{-sT}-\frac{4E}{Ts^2}e^{\frac{-sT}{2}}\right]$$

$$= \frac{1}{1-e^{-Ts}}\left[-\frac{E}{S}(1-e^{-sT})+\frac{4E}{Ts^2}\left(1-2e^{\frac{-sT}{2}}+e^{-sT}\right)\right]$$

$$= \frac{1}{1-e^{-Ts}}\left[-\frac{E}{S}(1-e^{-sT})+\frac{4E}{Ts^2}\left(1-e^{\frac{-sT}{2}}\right)^2\right]$$

$$= -\frac{E}{S}+\frac{4E}{Ts^2}\frac{\left(1-e^{\frac{-sT}{2}}\right)^2}{1-e^{-Ts}}$$

$$= -\frac{E}{S}+\frac{4E}{Ts^2}\frac{\left(1-e^{\frac{-sT}{2}}\right)^2}{1^2-\left(e^{\frac{-sT}{2}}\right)^2}$$

$$= -\frac{E}{S}+\frac{4E}{Ts^2}\frac{\left(1-e^{\frac{-sT}{2}}\right)^2}{\left(1-e^{\frac{-sT}{2}}\right)\left(1+e^{\frac{-sT}{2}}\right)}$$

$$= -\frac{E}{S}+\frac{4E}{Ts^2}\frac{\left(1-e^{\frac{-sT}{2}}\right)}{\left(1+e^{\frac{-sT}{2}}\right)}$$

$$= -\frac{E}{S}+\frac{4E}{Ts^2}\frac{\left(e^{\frac{sT}{4}}-e^{\frac{-sT}{4}}\right)}{\left(e^{\frac{sT}{4}}+e^{\frac{-sT}{4}}\right)} \qquad = -\frac{E}{S}+\frac{4E}{Ts^2}\tanh\frac{sT}{4}.$$

Example 8: Find the Laplace transform of $f(t)=\begin{cases}\sin t, & for\ 0\le t\le\pi\\ 0 & , for\ \pi\ \le t<2\pi\end{cases}.$

Solution: The given function is periodic with period $p=2\pi$.

$$\therefore \quad L[f(t)] = \frac{1}{1-e^{-ps}}\int_0^p e^{-st}f(t)dt$$

$$= \frac{1}{1-e^{-2\pi s}}\int_0^{2\pi} e^{-st}f(t)dt$$

$$= \frac{1}{1-e^{-2\pi s}}\left\{\int_0^{\pi} e^{-st}\sin t\ dt+\int_{\pi}^{2\pi} e^{-st}.0.dt\right\}$$

$$= \frac{1}{1-e^{-2\pi s}}\left[\frac{e^{-st}}{s^2+1}(-s\ \sin t-\cos t)\right]_0^{\pi}$$

$$= \frac{1}{1-e^{-2\pi s}}\left[\frac{e^{-s\pi}}{s^2+1}(0-(-1))-\frac{1}{s^2+1}(0-1)\right]$$

$$= \frac{1}{1-e^{-2\pi s}}\cdot\frac{(1+e^{-\pi s})}{s^2+1}$$

$$= \frac{1}{(1-e^{-\pi s})(1+e^{-\pi s})}\cdot\frac{(1+e^{-\pi s})}{s^2+1} \quad = \frac{1}{(s^2+1)(1-e^{-\pi s})}.$$

Example9: Find the Laplace transform of $f(t)$, where $f(t) = \begin{cases} \sin \omega t, & \text{for } 0 < t < \frac{\pi}{\omega} \\ 0 \ , & \text{for } \frac{\pi}{\omega} < t < \frac{2\pi}{\omega} \end{cases}$

and $f(t+2\pi) = f(t)$. **[AU, May/June 2014]**

Solution: The given function is periodic with period $p = \frac{2\pi}{\omega}$.

$$\therefore L[f(t)] = \frac{1}{1-e^{-ps}} \int_0^p e^{-st} f(t) dt$$

$$= \frac{1}{1-e^{-\frac{2\pi s}{\omega}}} \int_0^{\frac{2\pi}{\omega}} e^{-st} \sin \omega t \ dt$$

$$= \frac{1}{1-e^{-\frac{2\pi s}{\omega}}} \left\{ \int_0^{\frac{\pi}{\omega}} e^{-st} \sin \omega t \ dt + \int_{\frac{\pi}{\omega}}^{\frac{2\pi}{\omega}} e^{-st} . 0 . dt \right\}$$

$$= \frac{1}{1-e^{-\frac{2\pi s}{\omega}}} \left[\frac{e^{-st}}{s^2+\omega^2} (-s \ \sin \omega t - \omega \ \cos \omega t) \right]_0^{\frac{\pi}{\omega}}$$

$$= \frac{1}{1-e^{-\frac{2\pi s}{\omega}}} \left[\frac{e^{-\frac{\pi s}{\omega}}}{s^2+\omega^2} (0 - \omega(-1)) - \frac{1}{s^2+\omega^2} (0-\omega) \right]$$

$$= \frac{1}{1-e^{-\frac{2\pi s}{\omega}}} \frac{\omega\left(1+e^{-\frac{\pi s}{\omega}}\right)}{s^2+\omega^2}$$

$$= \frac{\omega}{s^2+\omega^2} \frac{1+e^{-\frac{\pi s}{\omega}}}{\left(1-e^{-\frac{\pi s}{\omega}}\right)\left(1+e^{-\frac{\pi s}{\omega}}\right)} = \frac{\omega}{(s^2+\omega^2)\left(1-e^{-\frac{\pi s}{\omega}}\right)}.$$

Example 10: Find the Laplace transform of $f(t) = |\sin \omega t|, t \geq 0$.

Solution: Note that $f\left(t+\frac{\pi}{\omega}\right) = |\sin \omega(t+\pi/\omega)|$.

$$= |\sin \omega t| = f(t).$$

$\therefore$ $f(t) = |\sin \omega t|$ is a periodic function with period $p = \frac{\pi}{\omega}$.

Now $L[f(t)] = \frac{1}{1-e^{-ps}} \int_0^p e^{-st} f(t) dt$

$$= \frac{1}{1-e^{-\frac{\pi s}{\omega}}} \int_0^{\frac{\pi}{\omega}} e^{-st} |\sin \omega t| \ dt$$

$$= \frac{1}{1-e^{-\frac{\pi s}{\omega}}}\int_0^{\frac{\pi}{\omega}} e^{-st}\sin\omega t\, dt \left[\because \sin\omega t > 0 \text{ in } 0 \le t \le \frac{\pi}{\omega}\right]$$

$$= \frac{1}{1-e^{-\frac{\pi s}{\omega}}}\left[\frac{e^{-st}}{s^2+\omega^2}(-s\sin\omega t - \omega\cos\omega t)\right]_0^{\frac{\pi}{\omega}}$$

$$= \frac{1}{1-e^{-\frac{\pi s}{\omega}}}\left[\frac{e^{-\frac{\pi s}{\omega}}}{s^2+\omega^2}(0-\omega(-1)) - \frac{1}{s^2+\omega^2}(0-\omega)\right]$$

$$= \frac{1}{1-e^{-\frac{\pi s}{\omega}}}\frac{\omega\left(1+e^{-\frac{\pi s}{\omega}}\right)}{s^2+\omega^2} = \frac{\omega}{s^2+\omega^2}\frac{e^{\frac{\pi s}{2\omega}}+e^{-\frac{\pi s}{2\omega}}}{e^{\frac{\pi s}{2\omega}}-e^{-\frac{\pi s}{2\omega}}} = \frac{\omega}{s^2+\omega^2}\coth\left(\frac{\pi s}{2\omega}\right).$$

EXERCISES

1. Find the Laplace transform of $f(t) = kt$ in $0 < t < 1$.
2. Find the Laplace transform of the periodic function $f(t) = 1$ when $0 < t < \frac{a}{2}$ and $f(t) = -1$ when $\frac{a}{2} < t < a$ and $f(t+a) = f(t)$ for $t > 0$.
3. Find the Laplace transform of a "square wave" given by $f(t) = \begin{cases} E, & \text{for } 0 < t < \frac{\tau}{2} \\ -E, & \text{for } \frac{\tau}{2} < t < \tau \end{cases}$ and $f(t+\tau) = f(t)$.
4. Find the Laplace transform of $f(t) = \begin{cases} t, & 0 < t < 1 \\ 0, & 1 < t < 2 \end{cases}$ and $f(t+2) = f(t)$.
5. Find the Laplace transform of $f(t) = e^{-t}$ for $0 \le t < 2$ and $f(t+2) = f(t)$
6. Find the Laplace transform of $f(t) = k\sin\frac{t}{2}$, for $0 < t < 2\pi$.
7. Find the Laplace transform of $f(t) = \begin{cases} a\sin\omega t, & \text{in } 0 \le t \le \frac{\pi}{\omega} \\ 0, & \text{in } \frac{\pi}{\omega} \le t \le \frac{2\pi}{\omega} \end{cases}$.
8. Find the Laplace transform of $f(t) = \begin{cases} 0, & \text{for } 0 < t < \frac{\pi}{\omega} \\ -\sin\omega t, & \text{for } \frac{\pi}{\omega} < t < \frac{2\pi}{\omega} \end{cases}$.
9. Find the Laplace transform of $f(t) = \begin{cases} t, & \text{in } 0 < t < \pi \\ \pi - t, & \text{in } \pi < t < 2\pi \end{cases}$.
10. Find the Laplace transform of $f(t) = \begin{cases} \cos t, & \text{for } 0 < t < \pi \\ 0, & \text{for } \pi < t < 2\pi \end{cases}$.
11. Find the Laplace transform of $f(t)$, where $f(t) = \begin{cases} t & \text{in } 0 < t \le a \\ 2a - t & \text{in } a < t \le 2a \end{cases}$ and $f(t+2a) = f(t)$. ***[AU, Nov/Dec 2017]***

ANSWERS

1. $\frac{k}{s^2} - \frac{k\,e^{-s}}{s(1-e^{-s})}$ 2. $\frac{1}{s}\tanh\left(\frac{sa}{4}\right)$ 3. $\frac{E}{s}\tanh\left(\frac{s\tau}{4}\right)$

4. $\frac{1}{1-e^{-2s}}\left[\frac{1-e^{-s}-se^{-s}}{s^2}\right]$ 6. $\frac{2k}{4s^2+1}\coth(\pi s)$ 7. $\frac{\omega a}{(s^2+\omega^2)\left(1-e^{-\frac{\pi s}{\omega}}\right)}$

8. $\frac{\omega}{(s^2+\omega^2)\left(e^{\frac{\pi s}{\omega}}-1\right)}$ 9. $\frac{\left[\frac{1}{s^2}(e^{-\pi s}-1)^2+\frac{\pi}{s}e^{-\pi s}(e^{-\pi s}-1)\right]}{(1-e^{-2\pi s})}$

5.10. Initial and Final Value Theorems

Initial Value Theorem: If the Laplace transforms of $f(t)$ and $f'(t)$exist and $L[f(t)] = F(s)$, then $\lim_{t\to 0} f(t) = \lim_{s\to\infty} s\,F(s)$.

Proof: We know that $L\,[f'(t)] = s\,F(s) - f(0)$.

i.e., $\int_0^\infty e^{-st} f'(t)dt = s\,F(s) - f(0)$.

$\therefore \lim_{s\to\infty}\int_0^\infty e^{-st} f'(t)dt = \lim_{s\to\infty}[s\,F(s) - f(0)]$

i.e., $\int_0^\infty \lim_{s\to\infty}[e^{-st} f'(t)]dt = \lim_{s\to\infty} s.F(s) - f(0)$

$0 = \lim_{s\to\infty} s\,F(s) - f(0)\left[\because \lim_{s\to\infty} e^{-st} = 0\right]$

i.e., $f(0) = \lim_{s\to\infty} s\,F(s)$

$\therefore \lim_{t\to 0} f(t) = \lim_{s\to\infty} s\,F(s)$.

Final Value Theorem: If the Laplace transforms of $f(t)$and $f'^{(t)}$ exist and $L[f(t)] = F(s)$,then $\lim_{t\to\infty} f(t) = \lim_{s\to 0} s\,F(s)$.

Proof: We know that $L[f'(t)] = s\,F(s) - f(0)$.

i.e., $\int_0^\infty e^{-st} f'(t)dt = s\,F(s) - f(0)$

$\therefore \lim_{s\to 0}\int_0^\infty e^{-st} f'(t)dt = \lim_{s\to 0}[s\,F(s) - f(0)]$.

$\int_0^\infty \lim_{s\to 0}[e^{-st} f'(t)]dt = \lim_{s\to 0} s.F(s) - f(0)$

$\int_0^\infty f'^{(t)}dt = \lim_{s\to 0} s\,F(s) - f(0)$

$[f(t)]_0^\infty = \lim_{s\to 0} s\,F(s) - f(0)$

$\lim_{t\to\infty} f(t) - f(0) = \lim_{s\to 0} s\,F(s) - f(0)$

$\lim_{t\to\infty} f(t) = \lim_{s\to 0} s\,F(s)$.

Example 1: Verify the initial and final value theorem for the function $f(t) = e^{-2t}t^3$.

Solution : Let $f(t) = e^{-2t}\, t^3$.

Then $F(s) = L\,[e^{-2t}\, t^3] = L[t^3]_{s \to s+2}$

$$= \left[\frac{3!}{s^4}\right]_{s \to s+2} = \frac{6}{(s+2)^4}.$$

Now, $\lim\limits_{t\to 0} f(t) = \lim\limits_{t\to 0} e^{-2t}t^3 = 0$ (1)

$\lim\limits_{t\to \infty} f(t) = \lim\limits_{t\to \infty} e^{-2t}t^3 = 0,$ (2)

(by repeated applications of L' Hospitals rule)

$\lim\limits_{s\to 0} s\,F(s) = \lim\limits_{s\to 0} \frac{6s}{(s+2)^4} = 0$ (3)

and $\lim\limits_{s\to \infty} s\,F(s) = \lim\limits_{s\to \infty} \frac{6s}{(s+2)^4} = \lim\limits_{s\to \infty} \frac{6s}{s^4(1+2/s)^4}$

$= \lim\limits_{s\to \infty} \frac{6}{s^3(1+2/s)^4} = 0.$ (4)

From (1) and (4), $\lim\limits_{t\to 0} f(t) = 0 = \lim\limits_{s\to \infty} s\,F(s).$

Hence initial value theorem is verified.

From (2) and (3), $\lim\limits_{t\to \infty} f(t) = 0 = \lim\limits_{s\to 0} sF(s).$

Hence final value theorem is verified.

Example 2: Verify the initial and final value theorem for the function

$$f(t) = 1 + e^{-t}(\sin t + \cos t).$$

Solution: Let $f(t) = 1 + e^{-t}(\sin t + \cos t).$

Then $F(s) = L[1 + e^{-t}(\sin t + \cos t)]$

$= L[1] + L[e^{-t}(\sin t + \cos t)]$

$= \frac{1}{s} + L[\sin t + \cos t]_{s \to s+1}$

$= \frac{1}{s} + \left[\frac{1}{s^2+1} + \frac{s}{s^2+1}\right]_{s \to s+1}$

$= \frac{1}{s} + \left[\frac{s+1}{s^2+1}\right]_{s \to s+1} = \frac{1}{s} + \frac{s+2}{(s+1)^2+1}.$

Now, $\lim\limits_{t\to 0} f(t) = \lim\limits_{t\to 0} [1 + e^{-t}(\sin t + \cos t)] = 1 + 1 = 2$ (1)

$$\lim_{t\to\infty} f(t) = \lim_{t\to\infty} [1 + e^{-t}(\sin t + \cos t)] = 1 + 0 = 1 \quad (2)$$

$$\lim_{s\to 0} s\,F(s) = \lim_{s\to 0} s\left[\frac{1}{s} + \frac{s+2}{(s+1)^2+1}\right]$$

$$= \lim_{s\to 0}\left[1 + \frac{s(s+2)}{(s+1)^2+1}\right] = 1 + 0 = 1 \quad (3)$$

and $$\lim_{s\to\infty} sF(s) = \lim_{s\to\infty} s\left[\frac{1}{s} + \frac{(s+2)}{(s+1)^2+1}\right]$$

$$= \lim_{s\to\infty}\left[1 + \frac{s(s+2)}{(s+1)^2+1}\right]$$

$$= \lim_{s\to\infty}\left[1 + \frac{(1+2/s)}{(1+1/s)^2+\frac{1}{s^2}}\right]$$

$$= 1 + 1 = 2. \quad (4)$$

From (1) and (4), $\lim_{t\to 0} f(t) = 2 = \lim_{s\to\infty} s\,F(s)$.

Hence initial value theorem is verified.

From (2) and (3), $\lim_{t\to\infty} f(t) = 1 = \lim_{s\to 0} s\,F(s)$.

Hence final value theorem is verified.

EXERCISES

1. Verify the initial and final value theorems for the following functions
   (i) $f(t) = 1 - e^{-at}$
   (ii) $f(t) = e^{-t}(t+2)^2$
   (iii) $f(t) = t^2 e^{-4t}$
   (iv) $f(t) = 2e^{-3t}$.
2. If $L[f(t)] = \frac{1}{s(s+1)(s+2)}$, find $\lim_{t\to 0} f(t)$ and $\lim_{t\to\infty} f(t)$.
3. Find $f(0)$, if $F(s) = \frac{5s^2+6s+7}{s^2(s^2+8s+9)}$.

ANSWERS

2. 0, $\frac{1}{2}$ 3. 5

5.11. Inverse Laplace Transforms

If $L[f(t)] = F(s)$, then $f(t)$ is known as the *inverse Laplace transform* of $F(s)$ and is denoted by $L^{-1}[F(s)]$.

Sl.No.	$L[f(t)] = F(s)$	$L^{-1}[F(s)] = f(t)$
1.	$L[0] = 0$	$L^{-1}[0] = 0$
2.	$L[1] = \frac{1}{s}$	$L^{-1}\left[\frac{1}{s}\right] = 1$
3.	$L[t] = \frac{1}{s^2}$	$L^{-1}\left[\frac{1}{s^2}\right] = t$
4.	$L[t^n] = \frac{n!}{s^{n+1}}$ (*n*-positive integer)	$L^{-1}\left[\frac{n!}{s^{n+1}}\right] = t^n$ (or) $L^{-1}\left[\frac{1}{s^n}\right] = \frac{t^{n-1}}{(n-1)!}$
5.	$L[e^{at}] = \frac{1}{s-a}$	$L^{-1}\left[\frac{1}{s-a}\right] = e^{at}$
6.	$L[e^{-at}] = \frac{1}{s+a}$	$L^{-1}\left[\frac{1}{s+a}\right] = e^{-at}$
7.	$L[\sin at] = \frac{a}{s^2+a^2}$	$L^{-1}\left[\frac{a}{s^2+a^2}\right] = \sin at$ (or) $L^{-1}\left[\frac{1}{s^2+a^2}\right] = \frac{1}{a}\sin at$
8.	$L[\cos at] = \frac{s}{s^2+a^2}$	$L^{-1}\left[\frac{s}{s^2+a^2}\right] = \cos at$
9.	$L[\sinh at] = \frac{a}{s^2-a^2}$	$L^{-1}\left[\frac{a}{s^2-a^2}\right] = \sinh at$ (or) $L^{-1}\left[\frac{1}{s^2-a^2}\right] = \frac{1}{a}\sinh at$
10.	$L[\cosh at] = \frac{s}{s^2-a^2}$	$L^{-1}\left[\frac{s}{s^2-a^2}\right] = \cosh at$
11.	$L[t\sin at] = \frac{2as}{(s^2+a^2)^2}$	$L^{-1}\left[\frac{2as}{(s^2+a^2)^2}\right] = t\sin at$ (or) $L^{-1}\left[\frac{s}{(s^2+a^2)^2}\right] = \frac{1}{2a}t\sin at$
12.	$L[t\cos at] = \frac{s^2-a^2}{(s^2+a^2)^2}$	$L^{-1}\left[\frac{s^2-a^2}{(s^2+a^2)^2}\right] = t\cos at$
13.	$L[u_a(t)] = \frac{e^{-as}}{s}$	$L^{-1}\left[\frac{e^{-as}}{s}\right] = u_a(t)$
14.	$L[\delta(t-a)] = e^{-as}$	$L^{-1}[e^{-as}] = \delta(t-a)$
15.	$L[\delta(t)] = 1$	$L^{-1}(1) = \delta(t)$
16.	$L[e^{-at}f(t)] = F(s+a)$	$L^{-1}[F(s+a)] = e^{-at}L^{-1}[F(s)]$
17.	$L[e^{at}f(t)] = F(s-a)$	$L^{-1}[F(s-a)] = e^{at}L^{-1}[F(s)]$

Note :

1. Here L^{-1} is known as the inverse Laplace transformation operator.
2. Inverse Laplace transform is the reverse process of Laplace transform.
3. Inverse Laplace transform of $F(s)$need not exit for all $F(s)$.

The Laplace transform of a given function is found by the methods of ordinary integration and so it is rather easy to get. But finding the inverse Laplace transform of a given function is more difficult. Since $F(s)$ is a rational algebraic function, to find inverse Laplace transforms, we first express the given function of s into partial fractions which will, then, be recognizable as one of the following standard forms.

Properties of Inverse Laplace Transforms

1. Linear Property

$L^{-1}[a\,F(s) + b\,G(s)] = a\,L^{-1}[F(s)] + b\,L^{-1}[G(s)]$, where a and b are any two constants.

Proof: Let $L[f(t)] = F(s)$and $L[g(t)] = G(s)$.

From the linear property for Laplace transform, we have

$$L\,[a\,f(t) + b\,g(t)] = a\,L[f(t)] + b\,L[g(t)] = a\,F(s) + b\,G(s)$$

$$\therefore a\,f(t) + b\,g(t) = L^{-1}[a\,F(s) + b\,G(s)]$$

i.e., $a\,L^{-1}[F(s)] + b\,L^{-1}[G(s)] = L^{-1}[a\,F(s) + b\,G(s)]$.

2. First Shifting Property

(i) $L^{-1}[F(s+a)] = e^{-at}\,L^{-1}[F(s)]$

(ii) $L^{-1}[F(s-a)] = e^{at}\,L^{-1}[F(s)]$.

Proof: By the first shifting property of L. T., $L[e^{-at}f(t)] = F(s+a)$

$$\Rightarrow \quad e^{-at}f(t) = L^{-1}[F(s+a)]$$

$$\Rightarrow e^{-at}\,L^{-1}[F(s)] = L^{-1}[F(s+a)]$$

$$\therefore L^{-1}[F(s+a)] = e^{-at}\,L^{-1}[F(s)].$$

Similarly, $L^{-1}[F(s-a)] = e^{at}\,L^{-1}[F(s)]$.

3. Change of Scale Property

If $f(t) = L^{-1}[F(s)]$, then $L^{-1}[F(ks)] = \frac{1}{k}\,f\left(\frac{t}{k}\right)$.

Proof: The change of scale property for Laplace transform is, $L[f(at)] = \frac{1}{a} F\left(\frac{s}{a}\right)$.

Let $a = \frac{1}{k}$. Then $L\left[f\left(\frac{t}{k}\right)\right] = k\,F(ks)$

$\Rightarrow \quad f\left(\frac{t}{k}\right) = L^{-1}[k\,F(ks)] = kL^{-1}[\,F(ks)]$

i.e., $L^{-1}[\,F(ks)] = \frac{1}{k} f\left(\frac{t}{k}\right)$.

Example 1: Find the inverse Laplace transforms of $\frac{3}{s-5} + \frac{8s}{s^2+1}$.

Solution: $L^{-1}\left[\frac{3}{s-5} + \frac{8s}{s^2+1}\right] = 3L^{-1}\left[\frac{1}{s-5}\right] + 8L^{-1}\left[\frac{s}{s^2+1}\right]$

$$= 3e^{5t} + 8\cos t.$$

Example 2: Find the inverse Laplace transforms of $\frac{4\,(s^2-3)^2}{3s^4}$.

Solution: $L^{-1}\left[\frac{4(s^2-3)^2}{3s^4}\right] = L^{-1}\left[\frac{4}{3}\left(\frac{s^4-6s^2+9}{s^4}\right)\right]$

$$= \frac{4}{3}L^{-1}[1] - 8L^{-1}\left[\frac{1}{s^2}\right] + 12\,L^{-1}\left[\frac{1}{s^4}\right]$$

$$= \frac{4}{3}\,\delta(t) - 8t + 12\frac{t^3}{3!} = \frac{4}{3}\,\delta(t) - 8t + 2t^3.$$

Example 3: Find the inverse Laplace transforms of $\frac{4s+5}{(s-1)^2(s+2)}$.

Solution: By partial fractions,

$$\frac{4s+5}{(s-1)^2(s+2)} = \frac{A}{s-1} + \frac{B}{(s-1)^2} + \frac{C}{s+2} \tag{1}$$

$$= \frac{A(s-1)(s+2)+B(s+2)+C(s-1)^2}{(s-1)^2(s+2)}$$

$$\Rightarrow \quad 4s+5 = A(s-1)(s+2) + B(s+2) + C(s-1)^2 \tag{2}$$

Put $s = 1$ in (2), we get

$9 = 3B \Rightarrow B = 3.$

Put $s = -2$ in (2), we get

$-3 = 9C \Rightarrow C = -\frac{1}{3}$.

Equating the coefficients of s^2 on both sides of (2),

$0 = A + C \Rightarrow A = -C = \frac{1}{3}$.

$\therefore$ (1) gives $\frac{4s+5}{(s-1)^2(s+2)} = \frac{1/3}{s-1} + \frac{3}{(s-1)^2} + \frac{-1/3}{s+2}$

$$\therefore L^{-1}\left[\frac{4s+5}{(s-1)^2(s+2)}\right] = \frac{1}{3}L^{-1}\left[\frac{1}{s-1}\right] + 3L^{-1}\left[\frac{1}{(s-1)^2}\right] - \frac{1}{3}L^{-1}\left[\frac{1}{s+2}\right]$$

$$= \frac{1}{3}e^t + 3e^t t - \frac{1}{3}e^{-2t}.$$

Example4 : Find $L^{-1}\left[\frac{s}{(s^2+a^2)(s^2+b^2)}\right]$.

Solution: By partial fractions,

$$\frac{1}{(s^2+a^2)(s^2+b^2)} = \frac{1}{(u+a^2)(u+b^2)} \quad \text{where } u = s^2 \qquad (1)$$

$$\frac{1}{(u+a^2)(u+b^2)} = \frac{A}{u+a^2} + \frac{B}{u+b^2}$$

$$= \frac{A(u+b^2)+B(u+a^2)}{(u+a^2)(u+b^2)}$$

$$\Rightarrow 1 = A(u+b^2) + B(u+a^2) \qquad (2)$$

Put $u = -b^2$ in (2), $B = \frac{1}{a^2-b^2}$

Put $u = -a^2$ in (2), $A = \frac{1}{b^2-a^2} = \frac{-1}{a^2-b^2}$

$\therefore$ (1) gives $\frac{1}{(s^2+a^2)(s^2+b^2)} = \frac{\left(-\frac{1}{a^2-b^2}\right)}{(s^2+a^2)} + \frac{\left(\frac{1}{a^2-b^2}\right)}{(s^2+b^2)}$

$\therefore$ (1) gives $\frac{s}{(s^2+a^2)(s^2+b^2)} = \frac{\left(-\frac{1}{a^2-b^2}\right)s}{(s^2+a^2)} + \frac{\left(\frac{1}{a^2-b^2}\right)s}{(s^2+b^2)}$

$$= \frac{1}{a^2-b^2}\left[\frac{s}{s^2+b^2} - \frac{s}{s^2+a^2}\right]$$

$$\therefore L^{-1}\left[\frac{s}{(s^2+a^2)(s^2+b^2)}\right] = \frac{1}{a^2-b^2}\left\{L^{-1}\left[\frac{s}{s^2+b^2}\right] - L^{-1}\left[\frac{s}{s^2+a^2}\right]\right\}$$

$$= \frac{1}{a^2-b^2}(\cos bt - \cos at).$$

Example 5 : Find $L^{-1}\left[\frac{s}{(s^2+1)(s^2+4)}\right]$. **[*AU, April/May 2015*]**

Solution : By partial fractions,

$$\frac{1}{(s^2+1)(s^2+4)} = \frac{1}{(u+1)(u+4)} \quad \text{where } u = s^2 \qquad (1)$$

$$\frac{1}{(u+1)(u+4)} = \frac{A}{u+1} + \frac{B}{u+4}$$

$$= \frac{A\,(u+4)+B\,(u+1)}{(u+1)(u+4)}$$

$$\Rightarrow 1 = A(u+4) + B(u+1) \qquad (2)$$

Put $u = -4$ in (2), $B = \frac{-1}{3}$

Put $u = -1$ in (2), $A = \frac{1}{3}$

$\therefore$ (1) gives $\dfrac{1}{(s^2+1)(s^2+4)} = \dfrac{\left(\frac{1}{3}\right)}{(s^2+1)} + \dfrac{\left(\frac{-1}{3}\right)}{(s^2+4)}$

$$\therefore \quad \frac{s}{(s^2+1)(s^2+4)} = \frac{s}{3(s^2+1)} - \frac{s}{3(s^2+4)}$$

$$\therefore L^{-1}\left[\frac{s}{(s^2+1)(s^2+4)}\right] = \frac{1}{3}\left\{L^{-1}\left[\frac{s}{s^2+1}\right] - L^{-1}\left[\frac{s}{s^2+4}\right]\right\}$$

$$= \frac{1}{3}(\cos t - \cos 2t).$$

Example 6 : Find the inverse Laplace transform of the following :

(i) $\dfrac{s^2}{(s^2+a^2)^2}$ (ii) $\dfrac{1}{(s^2+a^2)^2}$ (iii) $\dfrac{s^3}{(s^2+a^2)^2}$.

Solution : (i) $L^{-1}\left[\dfrac{s^2}{(s^2+a^2)^2}\right] = \dfrac{1}{2}L^{-1}\left[\dfrac{(s^2-a^2)+(s^2+a^2)}{(s^2+a^2)^2}\right]$

$$= \frac{1}{2}L^{-1}\left[\frac{s^2-a^2}{(s^2+a^2)^2} + \frac{1}{s^2+a^2}\right]$$

$$= \frac{1}{2}\left\{L^{-1}\left[\frac{s^2-a^2}{(s^2+a^2)^2}\right] + L^{-1}\left[\frac{1}{s^2+a^2}\right]\right\}$$

$$= \frac{1}{2}\left[t\cos at + \frac{1}{a}\sin at\right]$$

$$= \frac{1}{2a}\,[\sin at + at\cos at]\,.$$

(ii) $L^{-1}\left[\dfrac{1}{(s^2+a^2)^2}\right] = L^{-1}\left[\dfrac{1}{2a^2}\left(\dfrac{(s^2+a^2)-(s^2-a^2)}{(s^2+a^2)^2}\right)\right]$

$$= \frac{1}{2a^2}\left\{L^{-1}\left[\frac{1}{s^2+a^2}\right] - L^{-1}\left[\frac{s^2-a^2}{(s^2+a^2)^2}\right]\right\}$$

$$= \frac{1}{2a^2}\left[\frac{1}{a}\sin at - t\cos at\right]$$

$$= \frac{1}{2a^3}\,[\sin at - at\cos at]\,.$$

(iii) $L^{-1}\left[\frac{s^3}{(s^2+a^2)^2}\right] = L^{-1}\left[\frac{s(s^2+a^2-a^2)}{(s^2+a^2)^2}\right]$

$$= L^{-1}\left[\frac{s}{s^2+a^2} - a^2\frac{s}{(s^2+a^2)^2}\right]$$

$$= \cos at - a^2\frac{1}{2a}\,t\sin at \;= \cos at - \frac{a}{2}\,t\sin at.$$

Example 7: Find $L^{-1}\left[\frac{s}{(s+1)^2(s^2+1)}\right]$.

Solution: By partial fractions,

$$\frac{s}{(s+1)^2(s^2+1)} = \frac{A}{s+1} + \frac{B}{(s+1)^2} + \frac{Cs+D}{s^2+1} \quad (1)$$

$$= \frac{A(s+1)(s^2+1)+B(s^2+1)+(Cs+D)(s+1)^2}{(s+1)^2(s^2+1)}$$

$$\Rightarrow\; s = A(s+1)(s^2+1) + B(s^2+1) + (Cs+D)(s+1)^2 \quad (2)$$

Put $s = -1$ in (2), we get

$-1 = 2B \;\Rightarrow B = -\frac{1}{2}$.

Equating the coefficients of s^3, s^2 and constant on both sides,

$0 = A + C$ (i)

$0 = A + B + D + 2C$ (ii)

$0 = A + B + D$ (iii)

(ii) – (iii) gives $C = 0$

(i) gives $A = 0$

(iii) gives $D = -B = \frac{1}{2}$.

$\therefore$ (1) gives $\frac{s}{(s+1)^2(s^2+1)} = \frac{-1/2}{(s+1)^2} + \frac{1/2}{s^2+1}$

Then $L^{-1}\left[\frac{s}{(s+1)^2(s^2+1)}\right] = -\frac{1}{2}L^{-1}\left[\frac{1}{(s+1)^2}\right] + \frac{1}{2}L^{-1}\left[\frac{1}{s^2+1}\right]$

$$= -\frac{1}{2}te^{-t} + \frac{1}{2}\sin t.$$

Example 8: Find the inverse Laplace transforms of the following functions:

(i) $\frac{2s}{(s+3)^2-2}$ (ii) $\frac{2s+7}{s^2-4s+13}$

Solution: (i) $L^{-1}\left[\frac{2s}{(s+3)^2-2}\right] = L^{-1}\left[\frac{2(s+3)-6}{(s+3)^2-2}\right]$

$$= e^{-3t}L^{-1}\left[\frac{2s-6}{s^2-2}\right]$$

$$= e^{-3t}\left\{2L^{-1}\left[\frac{s}{s^2-(\sqrt{2})^2}\right] - 6L^{-1}\left[\frac{1}{s^2-(\sqrt{2})^2}\right]\right\}$$

$$= e^{-3t}\left\{2\cosh\sqrt{2}t - \frac{6}{\sqrt{2}}\sinh\sqrt{2}t\right\}.$$

(ii) $L^{-1}\left[\frac{2s+7}{s^2-4s+13}\right] = L^{-1}\left[\frac{2s+7}{(s^2-2.2s+4)+9}\right]$

$$= L^{-1}\left[\frac{2s+7}{(s-2)^2+9}\right] = L^{-1}\left[\frac{2(s-2)+11}{(s-2)^2+9}\right]$$

$$= e^{2t}L^{-1}\left[\frac{2s+11}{s^2+3^2}\right]$$

$$= e^{2t}\left\{2L^{-1}\left[\frac{s}{s^2+3^2}\right] + 11L^{-1}\left[\frac{1}{s^2+3^2}\right]\right\}$$

$$= e^{2t}\left\{2\cos 3t + \frac{11}{3}\sin 3t\right\}.$$

Example 9: Evaluate $L^{-1}\left[\frac{14s+10}{49s^2+28s+13}\right]$.

Solution: $L^{-1}\left[\frac{14s+10}{49s^2+28s+13}\right] = \frac{1}{49}L^{-1}\left[\frac{14s+10}{s^2+\frac{4}{7}s+\frac{13}{49}}\right]$

$$= \frac{14}{49}L^{-1}\left[\frac{s+\frac{5}{7}}{\left(s^2+2.s.\frac{2}{7}+\frac{4}{49}\right)-\frac{4}{49}+\frac{13}{49}}\right]$$

$$= \frac{2}{7}L^{-1}\left[\frac{s+\frac{5}{7}}{\left(s+\frac{2}{7}\right)^2+\frac{9}{49}}\right] = \frac{2}{7}L^{-1}\left[\frac{\left(s+\frac{2}{7}\right)+\frac{3}{7}}{\left(s+\frac{2}{7}\right)^2+\left(\frac{3}{7}\right)^2}\right]$$

$$= \frac{2}{7}e^{-\frac{2}{7}t}L^{-1}\left[\frac{s+\frac{3}{7}}{s^2+\left(\frac{3}{7}\right)^2}\right]$$

$$= \frac{2}{7}e^{-\frac{2}{7}t}\left\{L^{-1}\left[\frac{s}{s^2+\left(\frac{3}{7}\right)^2}\right] + L^{-1}\left[\frac{\frac{3}{7}}{s^2+\left(\frac{3}{7}\right)^2}\right]\right\}$$

$$= \frac{2}{7}e^{-\frac{2}{7}t}\left\{\cos\frac{3}{7}t + \sin\frac{3}{7}t\right\}.$$

Example 10: Evaluate $L^{-1}\left[\frac{s}{s^4+s^2+1}\right]$.

Solution: Consider $\frac{s}{s^4+s^2+1} = \frac{s}{s^4+2s^2+1-s^2} = \frac{s}{(s^2+1)^2-s^2}$

$$= \frac{s}{(s^2+s+1)(s^2-s+1)}$$

$$= \frac{As+B}{s^2+s+1} + \frac{Cs+D}{s^2-s+1}, \quad \text{by partial fraction.}$$

$\Rightarrow \quad s = (As+B)(s^2-s+1)+(Cs+D)(s^2+s+1).$

Equating coefficients on both sides, we get

$A=0,\ B=-\frac{1}{2},\ C=0$, and $D=\frac{1}{2}$.

$$\therefore \frac{s}{s^4+s^2+1} = \frac{-\frac{1}{2}}{s^2+s+1} + \frac{\frac{1}{2}}{s^2-s+1}$$

$$\therefore L^{-1}\left[\frac{s}{s^4+s^2+1}\right] = -\frac{1}{2}L^{-1}\left[\frac{1}{s^2+s+1}\right] + \frac{1}{2}L^{-1}\left[\frac{1}{s^2-s+1}\right]$$

$$= -\frac{1}{2}L^{-1}\left[\frac{1}{\left(s+\frac{1}{2}\right)^2+\frac{3}{4}}\right] + \frac{1}{2}L^{-1}\left[\frac{1}{\left(s-\frac{1}{2}\right)^2+\frac{3}{4}}\right]$$

$$= -\frac{1}{2}e^{\frac{-t}{2}}L^{-1}\left[\frac{1}{s^2+\left(\frac{\sqrt{3}}{2}\right)^2}\right] + \frac{1}{2}e^{\frac{t}{2}}L^{-1}\left[\frac{1}{s^2+\left(\frac{\sqrt{3}}{2}\right)^2}\right]$$

$$= -\frac{1}{2}e^{\frac{-t}{2}}\frac{1}{\frac{\sqrt{3}}{2}}\sin\frac{\sqrt{3}}{2}t + \frac{1}{2}e^{\frac{t}{2}}\frac{1}{\frac{\sqrt{3}}{2}}\sin\frac{\sqrt{3}}{2}t$$

$$= -\frac{1}{\sqrt{3}}e^{\frac{-t}{2}}\sin\frac{\sqrt{3}}{2}t + \frac{1}{\sqrt{3}}e^{\frac{t}{2}}\sin\frac{\sqrt{3}}{2}t$$

$$= \frac{1}{\sqrt{3}}\sin\frac{\sqrt{3}}{2}t\left[e^{\frac{t}{2}} - e^{\frac{-t}{2}}\right] \quad = \frac{2}{\sqrt{3}}\sin\frac{\sqrt{3}}{2}t\sinh\frac{t}{2}.$$

Example 11: Evaluate $L^{-1}\left[\frac{1}{\sqrt{3s+7}}\right]$.

Solution: $L^{-1}\left[\frac{1}{\sqrt{3s+7}}\right] = \frac{1}{\sqrt{3}}L^{-1}\left[\frac{1}{\left(s+\frac{7}{3}\right)^{\frac{1}{2}}}\right]$

$$= \frac{1}{\sqrt{3}}e^{-\frac{7}{3}t}L^{-1}\left[\frac{1}{s^{\frac{1}{2}}}\right] = \frac{1}{\sqrt{3}}e^{-\frac{7}{3}t}\frac{t^{\frac{-1}{2}}}{\Gamma\left(\frac{1}{2}\right)}$$

$$= \frac{1}{\sqrt{3}}\frac{e^{-\frac{7}{3}t}t^{\frac{-1}{2}}}{\sqrt{\pi}} = \frac{1}{\sqrt{3\pi}}\frac{e^{-\frac{7}{3}t}}{\sqrt{t}}.$$

Example 12: Find $L^{-1}\left[\frac{s}{s^4+4a^4}\right]$.

Solution: $\frac{s}{s^4+4a^4} = \frac{1}{(s^2+2a^2)^2-(2as)^2} = \frac{1}{(s^2+2as+2a^2)(s^2-2as+2a^2)}$

$$= \frac{As+B}{s^2+2as+2a^2} + \frac{Cs+D}{s^2-2as+2a^2}, \text{ by partial fractions} \qquad (1)$$

$\Rightarrow s = (As+B)(s^2-2as+2a^2) + (Cs+D)(s^2+2as+2a^2)$

Equating coefficients on both sides, we get

$0 = A + C \Rightarrow C = -A$ (i)

$0 = -2aA + B + 2aC + D$ (ii)

$1 = 2a^2A - 2aB + 2a^2C + 2aD$ (iii)

$0 = 2a^2B + 2a^2D \Rightarrow B + D = 0 \Rightarrow D = -B$ (iv)

Using (i) and (iv) in (ii), we get $B = \frac{-1}{4a}$, $D = \frac{1}{4a}$.

(ii) gives $A = 0$ and so $C = 0$.

$\therefore$ (1) gives $\frac{s}{s^4+4a^4} = \frac{-\frac{1}{4a}}{s^2+2as+2a^2} + \frac{\frac{1}{4a}}{s^2-2as+2a^2}$

$$\therefore L^{-1}\left[\frac{s}{s^4+4a^4}\right] = \frac{1}{4a}L^{-1}\left[\frac{1}{(s-a)^2+a^2}\right] - \frac{1}{4a}L^{-1}\left[\frac{1}{(s+a)^2+a^2}\right]$$

$$= \frac{1}{4a}\left\{e^{at}\,L^{-1}\left[\frac{1}{s^2+a^2}\right] - e^{-at}\,L^{-1}\left[\frac{1}{s^2+a^2}\right]\right\}$$

$$= \frac{1}{4a}\left\{e^{at}\frac{1}{a}\sin at - e^{-at}\frac{1}{a}\sin at\right\}$$

$$= \frac{1}{4a^2}\sin at\,\{e^{at} - e^{-at}\}$$

$$= \frac{1}{4a^2}\sin at\; 2\sinh at$$

$$= \frac{1}{2a^2}\sin at \sinh at.$$

Example 13: If $L^{-1}\left[\frac{s}{(s^2+1)^2}\right] = \frac{t\sin t}{2}$, show that $L^{-1}\left[\frac{8s}{(4s^2+1)^2}\right] = \frac{t}{2}\sin\frac{t}{2}$.

Solution: Let $f(t) = L^{-1}\left[\frac{s}{(s^2+1)^2}\right] = \frac{t\sin t}{2}$.

$$\therefore \quad L^{-1}\left[\frac{8s}{(4s^2+1)^2}\right] = L^{-1}\left[\frac{4(2s)}{((2s)^2+1)^2}\right] = 4L^{-1}\left[\frac{2s}{((2s)^2+1)^2}\right]$$

$$= 4L^{-1}[\,F(2s)]\ \text{ where } F(s) = \frac{s}{(s^2+1)^2}$$

$$= 4\frac{1}{2}\,f\left(\frac{t}{2}\right) \text{ by change of scale property.}$$

$$= 2\,\frac{\left(\frac{t}{2}\right)}{2}\sin\frac{t}{2} = \frac{t}{2}\sin\frac{t}{2}\,.$$

Example 14: Given that $L^{-1}\left[\frac{s^2+4}{(s^2-4)^2}\right] = t\cosh 2t$, find $L^{-1}\left[\frac{s^2+1}{(s^2-1)^2}\right]$.

Solution: Let $L^{-1}\left[\frac{s^2+4}{(s^2-4)^2}\right] = t\cosh 2t.$

By change of scale property, $L^{-1}[\,F(2s)] = \frac{1}{2}f\left(\frac{t}{2}\right)$

i.e.,
$$L^{-1}\left[\frac{(2s)^2+4}{((2s)^2-4)^2}\right] = \frac{1}{2}\frac{t}{2}\cosh t$$

$$\Rightarrow \quad \frac{1}{4}L^{-1}\left[\frac{s^2+1}{(s^2-1)^2}\right] = \frac{t}{4}\cosh t$$

$$\Rightarrow \quad L^{-1}\left[\frac{s^2+1}{(s^2-1)^2}\right] = t\cosh t.$$

EXERCISES

1. Find the inverse Laplace transforms of the following:

   (i) $\frac{3}{s+4}$ (ii) $\frac{3s-2}{s^2+16}$ (iii) $\frac{1}{2s-5}$ (iv) $s^{\frac{-7}{2}}$ (v) $\frac{2s-5}{9s^2-25}$

2. Find the inverse Laplace transforms of each of the following functions:

   (i) $\frac{2s^2-4s+5}{s^3}$ (ii) $\frac{s^3-3s^2+4}{s^4}$ (iii) $\frac{3(s^2-2)^2}{2s^5}$

3. Find the inverse Laplace transforms of

   (i) $\frac{1}{(s-1)(s+2)(s+3)}$ (ii) $\frac{3s+5}{s^2-2s-3}$ (iii) $\frac{s+1}{(s-1)^2(s+2)}$

   (iv) $\frac{1}{s^2(s^2+1)}$ (v) $\frac{1}{s^2(s^2+1)(s^2+9)}$.

4. Find the inverse Laplace transforms of the following functions:

   (i) $\frac{3s+1}{(s+1)^4}$ (ii) $\frac{5}{(s+2)^5}$ (iii) $\frac{a^2}{s(s+a)^3}$ (iv) $\frac{2s^2+5s+2}{(s-2)^4}$ (v) $\frac{3s}{2s+9}$

   (vi) $\frac{s}{s^2+4s+8}$ (vii) $\frac{5s+7}{2s^2-4s+5}$ (viii) $\frac{1}{s^2+2s+5}$ (ix) $\frac{1}{\sqrt{2s+5}}$ (x) $\frac{1}{\sqrt[4]{16s-81}}$

   (xi) $\frac{s^2}{s^4+4a^4}$ (xii) $\frac{3s^2+16s+26}{s(s^2+4s+13)}$ (xiii) $\frac{s+1}{(s^2+2s+2)^2}$ (xiv) $\frac{1}{s^4+4}$ (xv) $\frac{1}{s^3-a^3}$

5. If $L^{-1}\left[\frac{s^2-1}{(s^2+1)^2}\right] = t\cos t$, show that $L^{-1}\left[\frac{9s^2-1}{(9s^2+1)^2}\right] = \frac{t}{9}\cos\frac{t}{3}$.

6. If $L^{-1}\left[\frac{s}{(s^2+1)^2}\right] = \frac{t}{2}\sin t$, find $L^{-1}\left[\frac{s}{(s^2+a^2)^2}\right]$.

7. If $L^{-1}\left[\frac{1}{(s^2+4)^2}\right] = \frac{1}{16}(\sin 2t - 2t\cos 2t)$, find $L^{-1}\left[\frac{1}{(s^2+9)^2}\right]$.

ANSWERS

1. (i) $3e^{-4t}$ (ii) $3\cos 4t - \frac{1}{2}\sin 4t$ (iii) $\frac{1}{2}e^{\frac{-5}{2}t}$

   (iv) $\frac{8t^{\frac{5}{2}}}{15\sqrt{\pi}}$ (v) $\frac{2}{9}\cos h\frac{5}{3}t - \frac{1}{3}\sin h\frac{5}{3}t$

2. (i) $2 - 4t + \frac{5}{2}t^2$ (ii) $1 - 3t + \frac{2}{3}t^3$ (iii) $\frac{3}{2} - 3t^2 + \frac{t^4}{4}$

3. (i) $\frac{1}{12}e^t - \frac{1}{3}e^{-2t} + \frac{1}{4}e^{-3t}$ (ii) $\frac{7}{2}e^{3t} - \frac{1}{2}e^{-t}$ (iii) $\frac{1}{9}e^t + \frac{2}{3}e^t t - \frac{1}{9}e^{-2t}$

   (iv) $t - \sin t$ (v) $\frac{1}{9}t - \frac{1}{8}\sin t + \frac{1}{216}\sin 3t$.

4. (i) $e^{-t}\left[\frac{3}{2}t^2 - \frac{1}{3}t^3\right]$ (ii) $\frac{5}{24}t^4 e^{-2t}$

   (iii) $\frac{1}{a} - \frac{e^{-at}}{a} - te^{-at} - \frac{a}{2}t^2e^{-at}$ (iv) $\frac{1}{6}e^{2t}(12t + 39t^2 + 20t^3)$

   (v) $\frac{3}{2} - \frac{27}{4}e^{\frac{-9}{2}t}$ (vi) $e^{-2t}[\cos 2t - \sin 2t]$

   (vii) $\frac{5}{2}e^t\cos\sqrt{\frac{3}{2}}t + \frac{6\sqrt{2}}{\sqrt{3}}e^t\sin\frac{\sqrt{3}}{\sqrt{2}}t$ (viii) $\frac{1}{2}e^{-t}\sin 2t$

   (ix) $\frac{1}{\sqrt{2\pi}}e^{\frac{-5}{2}t}t^{-\frac{1}{2}}$ (x) $\frac{1}{2\Gamma(1/4)}e^{\frac{81}{16}t}t^{\frac{-3}{4}}$

   (xi) $\frac{1}{2a}[\cos at\sinh at + \sin at\cos hat]$ (xii) $2 + e^{-2t}(\cos 3t + 2\sin 3t)$

   (xiii) $\frac{1}{2}e^{-t}t\sin t$ (xiv) $\frac{1}{4}(\sin t\cosh t - \cos t\ \sinh t)$

   (xv) $\frac{1}{3a^2}\left[e^{at} - e^{\frac{-at}{2}}\left\{\cos\frac{\sqrt{3}}{2}at + \sqrt{3}\sin\frac{\sqrt{3}}{2}at\right\}\right]$

6. $\frac{t}{2a}\sin at$

7. $\frac{1}{54}(\sin 3t - 3t\cos 3t)$.

4. Second Shifting Theorem

If $L^{-1}[F(s)] = f(t)$, then $L^{-1}[e^{-as}F(s)] = f(t-a)\,u(t-a)$.

Proof: By the second shifting property of L. T,

$$L[f(t-a)u(t-a)] = e^{-as}F(s)$$

$$\Rightarrow \quad L^{-1}[e^{-as}F(s)] = f(t-a)u(t-a).$$

Result 1: Another form of the second shifting property is

$$L^{-1}[e^{-as}F(s)] = \begin{cases} f(t-a), & t > a \\ 0 & , \quad t < a \end{cases}.$$

Result 2: To find the inverse Laplace transform of a function which contains the exponential term, apply the second shifting theorem.

Example 1: Find the inverse Laplace transforms of the following functions: (i) $\frac{e^{-3s}}{s^2+25}$ (ii) $\frac{s\,e^{-\frac{s}{2}}+\pi\,e^{-s}}{s^2+\pi^2}$.

Solution: (i) We have, $L^{-1}[e^{-as}\,F(s)] = f(t-a)u(t-a)$

$$\therefore L^{-1}\left[\frac{e^{-3s}}{s^2+25}\right] = f(t-3)u(t-3) \qquad (1)$$

Now, $\quad f(t) = L^{-1}[F(s)] = L^{-1}\left[\frac{1}{s^2+5^2}\right] = \frac{1}{5}\sin 5t.$

Equation (1) implies $L^{-1}\left[\frac{e^{-3s}}{s^2+25}\right] = \frac{1}{5}\sin 5(t-3)\; u(t-3)$

(or)

$$L^{-1}\left[\frac{e^{-3s}}{s^2+25}\right] = \begin{cases} \frac{1}{5}\sin 5(t-3)\; ; & t > 3 \\ 0 & ; \; t < 3 \end{cases}.$$

(ii) $L^{-1}\left[\frac{s\,e^{-\frac{s}{2}}+\pi\,e^{-s}}{s^2+\pi^2}\right] = L^{-1}\left[\frac{s\,e^{-\frac{s}{2}}}{s^2+\pi^2}\right] + \pi L^{-1}\left[\frac{e^{-s}}{s^2+\pi^2}\right] \qquad (2)$

We have, $L^{-1}[e^{-as}\,F(s)] = f(t-a)u(t-a)$

$$\therefore L^{-1}\left[\frac{s\,e^{-\frac{s}{2}}}{s^2+\pi^2}\right] = f\left(t-\frac{1}{2}\right)u\left(t-\frac{1}{2}\right) \qquad (3)$$

Now $f(t) = L^{-1}[F(s)] = L^{-1}\left[\frac{s}{s^2+\pi^2}\right] = \cos\pi t.$

Equation (3) implies $L^{-1}\left[\frac{s\,e^{-\frac{s}{2}}}{s^2+\pi^2}\right] = \cos\pi\left(t-\frac{1}{2}\right)\, u\left(t-\frac{1}{2}\right).$

Also, $L^{-1}\left[\frac{e^{-s}}{s^2+\pi^2}\right] = f(t-1)u(t-1) \qquad (4)$

Now $f(t) = L^{-1}[F(s)] = L^{-1}\left[\frac{1}{s^2+\pi^2}\right] = \frac{1}{\pi}\sin\pi t.$

Equation (4) implies $L^{-1}\left[\frac{e^{-s}}{s^2+\pi^2}\right] = \frac{1}{\pi}\sin\pi(t-1)\ u(t-1)$.

$\therefore$ (2) gives $L^{-1}\left[\frac{s\,e^{-\frac{s}{2}}+\pi\ e^{-s}}{s^2+\pi^2}\right] = \cos\pi\left(t-\frac{1}{2}\right)\ u(t-\frac{1}{2})+\sin\pi(t-1)\,u(t-1)$.

Example 2: Find $L^{-1}\left[\frac{e^{-s}}{(s-2)(s+3)}\right]$.

Solution: We have, $L^{-1}[e^{-as}\ F(s)] = f(t-a)u(t-a)$

$\therefore L^{-1}\left[\frac{(s+1)e^{-\pi s}}{s^2+2s+5}\right] = f(t-\pi)u(t-\pi)$ (1)

Now, $f(t) = L^{-1}[F(s)] = L^{-1}\left[\frac{1}{(s-2)(s+3)}\right]$ (2)

Consider $\frac{1}{(s-2)(s+3)} = \frac{A}{s-2}+\frac{B}{s+3}$

$= \frac{A(s+3)+B(s-2)}{(s-2)(s+3)}$

$\Rightarrow 1 = A(s+3)+B(s-2)$

Put $s = 2$, $1 = 5A \Rightarrow A = \frac{1}{5}$.

Put $s = -3$, $1 = -5B \Rightarrow B = -\frac{1}{5}$.

$\therefore \frac{1}{(s-2)(s+3)} = \frac{1}{5}\frac{1}{s-2} - \frac{1}{5}\frac{1}{s+3}$

(2) gives $f(t) = L^{-1}\left[\frac{1}{5}\frac{1}{s-2} - \frac{1}{5}\frac{1}{s+3}\right]$

$= \frac{1}{5}(e^{2t} - e^{-3t})$.

$\therefore$ (1) gives $L^{-1}\left[\frac{e^{-s}}{(s-2)(s+3)}\right] = \frac{1}{5}\left(e^{2(t-1)} - e^{-3(t-1)}\right)u(t-1)$.

Example3: Find $L^{-1}\left[\frac{(s+1)e^{-\pi s}}{s^2+2s+5}\right]$.

Solution: We have, $L^{-1}[e^{-as}\ F(s)] = f(t-a)u(t-a)$

$\therefore L^{-1}\left[\frac{(s+1)e^{-\pi s}}{s^2+2s+5}\right] = f(t-\pi)u(t-\pi)$ (1)

Now $f(t) = L^{-1}[F(s)] = L^{-1}\left[\frac{s+1}{s^2+2s+5}\right]$

$= L^{-1}\left[\frac{s+1}{(s+1)^2+4}\right]$

$$= e^{-t} L^{-1}\left[\frac{s}{s^2+4}\right]$$

$$= e^{-t}\cos 2t.$$

$\therefore$ (1) gives $L^{-1}\left[\frac{(s+1)e^{-\pi s}}{s^2+2s+5}\right] = e^{-(t-\pi)}\cos 2(t-\pi)\,u(t-\pi).$

Example 4: Evaluate $L^{-1}\left[\frac{e^{-2s}}{(s^2+2s+1)^2}\right]$. ***[AU, May/June 2016]***

Solution: We have, $L^{-1}[e^{-as}F(s)] = f(t-a)u(t-a)$

$$\therefore \quad L^{-1}\left[\frac{e^{-2s}}{(s^2+2s+1)^2}\right] = f(t-2)u(t-2) \quad (1)$$

Now $f(t) = L^{-1}[F(s)] = L^{-1}\left[\frac{1}{(s^2+2s+1)^2}\right]$

$$= L^{-1}\left[\frac{1}{(s+1)^4}\right] = e^{-t}L^{-1}\left[\frac{1}{s^4}\right] = e^{-t}\frac{t^3}{6}.$$

$\therefore$ (1) gives $L^{-1}\left[\frac{e^{-2s}}{(s^2+2s+1)^2}\right] = e^{-(t-2)}\frac{(t-2)^3}{6}u(t-2).$

Example 5: Find $L^{-1}\left[\frac{5-3e^{-3s}-2e^{-7s}}{s}\right]$.

Solution: $L^{-1}\left[\frac{5-3e^{-3s}-2e^{-7s}}{s}\right] = 5\,L^{-1}\left[\frac{1}{s}\right] - 3L^{-1}\left[\frac{e^{-3s}}{s}\right] - 2L^{-1}\left[\frac{e^{-7s}}{s}\right]$

$$= 5 - 3f(t-3)u(t-3) - f(t-7)2\,u(t-7),$$

By the second shifting theorem, where $f(t) = L^{-1}\left[\frac{1}{s}\right] = 1$.

$$\therefore L^{-1}\left[\frac{5-3e^{-3s}-2e^{-7s}}{s}\right] = 5 - 3u(t-3) - 2\,u(t-7) \quad (1)$$

Also, $u(t-3) = \begin{cases} 0, & 0 < t < 3 \\ 1, & t > 3 \end{cases}$

And $u(t-7) = \begin{cases} 0, & 0 < t < 7 \\ 1, & t > 7 \end{cases}$

$\therefore$ (1) gives $L^{-1}\left[\frac{5-3e^{-3s}-2e^{-7s}}{s}\right] = \begin{cases} 5-3.0-2.0, & 0<t<3 \\ 5-3.1-2.0, & 3<t<7 \\ 5-3.1-2.1, & t>7 \end{cases}$

$$= \begin{cases} 5, & 0<t<3 \\ 2, & 3<t<7 \\ 0, & t>7. \end{cases}$$

EXERCISES

Find the inverse Laplace transforms of the following functions.

(i) $\frac{e^{-\pi s}}{s+3}$ (ii) $\frac{1}{(s+2)^{5/2}}e^{-s}$ (iii) $\frac{s+4}{(s^2-4)e^{4s}}$

(iv) $\frac{1}{e^s\sqrt{s+1}}$ (v) $\frac{(2+5s)e^{-4s}}{s^2}$ (vi) $\frac{e^{-3s}}{s^2+4s+13}$

(vii) $\frac{s\,e^{-2s}}{s^2-5s+6}$ (viii) $\frac{e^{-4s}-e^{-7s}}{s^2}$

ANSWERS

(i) $e^{-3(t-\pi)}u_\pi(t)$ (ii) $\frac{4}{3\sqrt{\pi}}(t-1)^{\frac{3}{2}}e^{-2(t-1)}u_1(t)$

(iii) $[\cosh 2(t-4) + 2\sinh 2(t-4)].u_4(t)$ (iv) $\frac{1}{\sqrt{\pi}}e^{-(t-1)}(t-1)^{\frac{-1}{2}}u_1(t)$

(v) $(2t-3)u_4(t)$ (vi) $\frac{1}{3}e^{-2(t-3)}\sin 3(t-3)u_3(t)$

(vii) $\left[-2e^{2(t-2)} + 3e^{3(t-2)}\right]u_2(t)$ (viii) $\begin{cases} 0; & 0 < t < 4 \\ t-4; & 4 < t < 7 \\ 3; & t > 7. \end{cases}$

Inverse Laplace Transform of Derivatives

$L^{-1}[F'(s)] = -t\,L^{-1}[F(s)].$

Proof: We know that ,

$$L[tf(t)] = -\frac{d}{ds}F(s) = -F'(s)$$

$$\therefore \quad t\,f(t) = -L^{-1}[F'(s)]$$

i.e., $L^{-1}[F'(s)] = -t\,L^{-1}[F(s)]$.

Result: $L^{-1}[F(s)] = -\frac{1}{t}L^{-1}[F'(s)]$. This result is used to find inverse Laplace transforms of logarithmic functions, $\tan^{-1}$ and $\cot^{-1}$ functions.

Corollary: $L^{-1}\left[F^{(n)}(s)\right] = (-1)^n t^n L^{-1}[F(s)]$.

Example 1: Find $L^{-1}\left[\log\left(1-\frac{a}{s}\right)\right]$.

Solution: Let $F(s) = \log\left(1-\frac{a}{s}\right) = \log\left(\frac{s-a}{s}\right) = \log(s-a) - \log s$.

Then $\quad F'(s) = \frac{1}{s-a} - \frac{1}{s}$.

We know that $L^{-1}[F(s)] = \frac{-1}{t} L^{-1}[F'(s)]$

i.e., $\quad L^{-1}\left[\log\left(1-\frac{a}{s}\right)\right] = \frac{-1}{t} L^{-1}\left[\frac{1}{s-a} - \frac{1}{s}\right]$

$$= \frac{-1}{t}(e^{at} - 1) = \frac{1}{t}(1 - e^{at}).$$

Example 2: Find $L^{-1}\left[\log\left(\frac{s+1}{s-1}\right)\right]$.

Solution: Let $F(s) = \log\left(\frac{s+1}{s-1}\right) = \log(s+1) - \log(s-1)$.

Then $\quad F'(s) = \frac{1}{s+1} - \frac{1}{s-1}$.

We know that $L^{-1}[F(s)] = \frac{-1}{t} L^{-1}[F'(s)]$

i.e., $\quad L^{-1}\left[\log\left(\frac{s+1}{s-1}\right)\right] = \frac{-1}{t} L^{-1}\left[\frac{1}{s+1} - \frac{1}{s-1}\right]$

$$= \frac{-1}{t}[e^{-t} - e^{t}] = \frac{1}{t}[e^{t} - e^{-t}] = \frac{2}{t}\sinh t.$$

Example 3: Find $L^{-1}\left[\log\left(1-\frac{w^2}{s^2}\right)\right]$.

Solution: Let $F(s) = \log\left(1-\frac{w^2}{s^2}\right) = \log\left(\frac{s^2-w^2}{s^2}\right) = \log(s^2-w^2) - \log s^2$.

Then $F'(s) = \frac{1}{s^2-w^2}\, 2s - \frac{1}{s^2}\, 2s = \frac{2s}{s^2-w^2} - \frac{2}{s}$.

We know that $L^{-1}[F(s)] = \frac{-1}{t} L^{-1}[F'(s)]$.

$\therefore L^{-1}\left[\log\left(1-\frac{w^2}{s^2}\right)\right] = \frac{-1}{t} L^{-1}\left[\frac{2s}{s^2-w^2} - \frac{2}{s}\right]$

$$= \frac{-2}{t}(\cosh wt - 1) = \frac{2}{t}(1 - \cosh wt).$$

Example 4: Find $L^{-1}\left[\log\left(\frac{s^2+a^2}{s^2+b^2}\right)\right]$.

Solution: Let $F(s) = \log\left(\frac{s^2+a^2}{s^2+b^2}\right) = \log(s^2+a^2) - \log(s^2+b^2)$.

Then $\quad F'(s) = \frac{2s}{s^2+a^2} - \frac{2s}{s^2+b^2}$.

We know that $L^{-1}[F(s)] = \frac{-1}{t} L^{-1}[F'(s)]$.

$$\therefore \qquad L^{-1}\left[\log\left(\frac{s^2+a^2}{s^2+b^2}\right)\right] = -\frac{1}{t}L^{-1}\left[\frac{2s}{s^2+a^2} - \frac{2s}{s^2+b^2}\right]$$

$$= \frac{-2}{t}(\cos at - \cos bt) = \frac{2}{t}(\cos bt - \cos at).$$

Example 5: Find $L^{-1}\left[\log\frac{s^2+4}{(s-2)^2}\right]$ **[AU, *April/May2017*]**

Solution: Let $F(s) = \log\left(\frac{s^2+4}{(s-2)^2}\right) = \log(s^2+4) - \log((s-2)^2)$.

$$= \log(s^2+4) - 2\log(s-2).$$

Then $\quad F'(s) = \frac{2s}{s^2+4} - \frac{2}{(s-2)}$.

We know that $L^{-1}[F(s)] = \frac{-1}{t}L^{-1}[F'(s)]$.

$$\therefore \qquad L^{-1}\left[\log\left(\frac{s^2+4}{(s-2)^2}\right)\right] = -\frac{1}{t}L^{-1}\left[\frac{2s}{s^2+4} - \frac{2}{(s-2)}\right]$$

$$= \frac{-2}{t}(\cos 2t - e^{2t}) = \frac{2}{t}(e^{2t} - \cos 2t).$$

Example 6: Find $L^{-1}\left[\log\left(\frac{s(s+2)}{s^2+5}\right)\right]$.

Solution: Let $F(s) = \log\left(\frac{s(s+2)}{s^2+5}\right) = \log s + \log(s+2) - \log(s^2+5)$.

Then $\quad F'(s) = \frac{1}{s} + \frac{1}{s+2} - \frac{2s}{s^2+5}$.

We know that $L^{-1}[F(s)] = \frac{-1}{t}L^{-1}[F'(s)]$

i.e., $\quad L^{-1}\left[\log\left(\frac{s(s+2)}{s^2+5}\right)\right] = \frac{-1}{t}L^{-1}\left[\frac{1}{s} + \frac{1}{s+2} - \frac{2s}{s^2+5}\right]$

$$= \frac{-1}{t}\left(1 + e^{-2t} - 2\cos\sqrt{5}\,t\right).$$

Example 7: Find $L^{-1}\left[s\log\left(\frac{s-a}{s+a}\right) + b\right]$.

Solution: $L^{-1}\left[s\log\left(\frac{s-a}{s+a}\right) + b\right] = L^{-1}\left[s\log\left(\frac{s-a}{s+a}\right)\right] + bL^{-1}[1]$

$$= L^{-1}\left[s\log\left(\frac{s-a}{s+a}\right)\right] + b\delta(t) \qquad (1)$$

Let $\quad F(s) = s\log\left(\frac{s-a}{s+a}\right) = s[\log(s-a) - \log(s+a)]$.

Then $F'(s) = s\left[\frac{1}{s-a} - \frac{1}{s+a}\right] + [\log(s-a) - \log(s+a)]$

$$= s\frac{2a}{s^2-a^2} + \log\left(\frac{s-a}{s+a}\right) = \frac{2as}{s^2-a^2} + \log\left(\frac{s-a}{s+a}\right).$$

We know that $L^{-1}[F(s)] = \frac{-1}{t}L^{-1}[F'(s)]$ (2)

i.e., $L^{-1}\left[s\log\left(\frac{s-a}{s+a}\right)\right] = \frac{-1}{t}L^{-1}\left[\frac{2as}{s^2-a^2} + \log\left(\frac{s-a}{s+a}\right)\right]$

$$= \frac{-1}{t}\left(2a\cosh at + L^{-1}\left[\log\left(\frac{s-a}{s+a}\right)\right]\right). \quad (3)$$

Now, let $F(s) = \log\left(\frac{s-a}{s+a}\right) = \log(s-a) - \log(s+a)$.

Then $\quad F'(s) = \frac{1}{s-a} - \frac{1}{s+a}$.

Apply $F(s)$ in (2), we get

$$L^{-1}\left[\log\left(\frac{s-a}{s+a}\right)\right] = \frac{-1}{t}L^{-1}\left[\frac{1}{s-a} - \frac{1}{s+a}\right]$$

$$= \frac{-1}{t}[e^{at} - e^{-at}] = \frac{-2}{t}\sinh at.$$

$\therefore$ (3) gives $L^{-1}\left[s\log\left(\frac{s-a}{s+a}\right)\right] = \frac{-1}{t}\left\{2a\cosh at - \frac{2}{t}\sinh at\right\}$

$$= \frac{2}{t^2}\sinh at - \frac{2a}{t}\cosh at.$$

$\therefore$ (1) gives $L^{-1}\left[s\log\left(\frac{s-a}{s+a}\right) + b\right] = \frac{2}{t^2}\sinh at - \frac{2a}{t}\cosh at + b\,\delta(t).$

Example 8: Find $L^{-1}\left[\tan^{-1}\left(\frac{2}{s+1}\right)\right]$.

Solution: Let $F(s) = \tan^{-1}\left(\frac{2}{s+1}\right)$.

Then $F'(s) = \frac{1}{1+\left(\frac{2}{s+1}\right)^2} \times \left(\frac{-2}{(s+1)^2}\right)$

$$= \frac{(s+1)^2}{(s+1)^2+4} \times \left(\frac{-2}{(s+1)^2}\right) = \frac{-2}{(s+1)^2+4}.$$

Apply $L^{-1}[F(s)] = \frac{-1}{t}L^{-1}[F'(s)]$.

$$L^{-1}\left[\tan^{-1}\left(\frac{2}{s+1}\right)\right] = \frac{-1}{t}L^{-1}\left[\frac{-2}{(s+1)^2+4}\right]$$

$$= \frac{1}{t}e^{-t}L^{-1}\left(\frac{2}{s^2+4}\right) = \frac{1}{t}e^{-t}\sin 2t.$$

Example 9: Find $L^{-1}\left[\tan^{-1}\frac{2}{s^2}\right]$.

Solution: Let $F(s) = \tan^{-1}\frac{2}{s^2}$.

Then $\quad F'(s) = \frac{1}{1+\left(\frac{2}{s^2}\right)^2} \times 2\left(\frac{-2}{s^3}\right) = \frac{s^4}{s^4+4}\left(\frac{-4}{s^3}\right) = \frac{-4s}{s^4+4}.$

Apply $L^{-1}[F(s)] = \frac{-1}{t}L^{-1}[F'(s)]$.

$$L^{-1}\left[\tan^{-1}\left(\frac{2}{s^2}\right)\right] = \frac{-1}{t}L^{-1}\left[\frac{-4s}{s^4+4}\right] = \frac{4}{t}L^{-1}\left[\frac{s}{s^4+4}\right] \qquad (1)$$

Consider $\frac{s}{s^4+4} = \frac{s}{(s^2+2)^2-(2s)^2} = \frac{s}{(s^2+2s+2)(s^2-2s+2)}$

$$= \frac{1}{4}\left[\frac{1}{s^2-2s+2} - \frac{1}{s^2+2s+2}\right], \text{ by partial fractions}$$

$$= \frac{1}{4}\left[\frac{1}{(s-1)^2+1} - \frac{1}{(s+1)^2+1}\right].$$

$$\therefore \quad L^{-1}\left[\frac{s}{s^4+4}\right] = \frac{1}{4}\left\{L^{-1}\left[\frac{1}{(s-1)^2+1}\right] - L^{-1}\left[\frac{1}{(s+1)^2+1}\right]\right\}$$

$$= \frac{1}{4}\left\{e^t L^{-1}\left[\frac{1}{s^2+1}\right] - e^{-t}L^{-1}\left[\frac{1}{s^2+1}\right]\right\}$$

$$= \frac{1}{4}\{e^t \sin t - e^{-t}\sin t\}$$

$$= \frac{1}{4}\sin t\,(e^t - e^{-t}) = \frac{1}{2}\sin t\ \sinh t. \qquad (2)$$

Using (2) in (1), we have

$$L^{-1}\left[\tan^{-1}\left(\frac{2}{s^2}\right)\right] = \frac{4}{t}\frac{1}{2}\sin t \sinh t = \frac{2}{t}\sin t \sinh t.$$

Example 10: Find (i) $L^{-1}\left[\cot^{-1}\left(\frac{s}{w}\right)\right]$ (ii) $L^{-1}[\cot^{-1}(ws)]$.

Solution: (i) Let $F(s) = \cot^{-1}\frac{s}{w}$.

Then $\quad F'(s) = \frac{-1}{1+\left(\frac{s}{w}\right)^2} \times \left(\frac{1}{w}\right) = \frac{-w^2}{s^2+w^2}.\frac{1}{w} = \frac{-w}{s^2+w^2}$

Apply $\quad L^{-1}[F(s)] = \frac{-1}{t}L^{-1}[F'(s)]$.

$$L^{-1}\left[\cot^{-1}\left(\frac{s}{w}\right)\right] = \frac{-1}{t}L^{-1}\left[\frac{-w}{s^2+w^2}\right] = \frac{1}{t}\sin wt.$$

(ii) Let $F(s) = \cot^{-1}(ws)$.

Then $\quad F'(s) = \frac{-1}{1+(ws)^2}w = \frac{-w}{1+w^2s^2}.$

Apply $L^{-1}[F(s)] = \frac{-1}{t}L^{-1}[F'(s)]$

$$= L^{-1}[\cot^{-1}(ws)] = \frac{-1}{t}L^{-1}\left[\frac{-w}{1+w^2s^2}\right]$$

$$= \frac{w}{t}L^{-1}\left[\frac{1}{w^2s^2+1}\right]$$

$$= \frac{1}{wt}L^{-1}\left[\frac{1}{s^2+\left(\frac{1}{w}\right)^2}\right]$$

$$= \frac{1}{wt}.\frac{1}{1/w}\sin\left(\frac{1}{w}\,t\right) = \frac{1}{t}\,\sin\left(\frac{t}{w}\right).$$

Example 11: Find $L^{-1}\left[\tan^{-1}\left(\frac{s+2}{4}\right) + \cot^{-1}\left(\frac{s-2}{4}\right)\right]$.

Solution: Let $F(s) = \tan^{-1}\left(\frac{s+2}{4}\right) + \cot^{-1}\left(\frac{s-2}{4}\right)$.

Then $\quad F'(s) = \frac{1}{1+\left(\frac{s+2}{4}\right)^2}\frac{1}{4} + \frac{-1}{1+\left(\frac{s-2}{4}\right)^2}\frac{1}{4} = \frac{4}{(s+2)^2+16} - \frac{4}{(s-2)^2+16}$.

Apply $L^{-1}[F(s)] = -\frac{1}{t}L^{-1}[F'(s)]$.

$$L^{-1}\left[\tan^{-1}\left(\frac{s+2}{4}\right) + \cot^{-1}\left(\frac{s-2}{4}\right)\right] = -\frac{1}{t}\left\{L^{-1}\left[\frac{4}{(s+2)^2+16}\right] - L^{-1}\left[\frac{4}{(s-2)^2+16}\right]\right\}$$

$$= -\frac{1}{t}\left\{e^{-2t}L^{-1}\left[\frac{4}{s^2+16}\right] - e^{2t}L^{-1}\left[\frac{4}{s^2+16}\right]\right\}$$

$$= -\frac{1}{t}\{e^{-2t}\sin 4t - e^{2t}\sin 4t\}$$

$$= \frac{\sin 4t}{t}\{e^{2t} - e^{-2t}\} = \frac{2}{t}\sin 4t\sinh 2t\,.$$

Example 12: Evaluate $L^{-1}\left[\cot^{-1}\left(\frac{2}{s+1}\right)\right]$ **[AU, *Nov/Dec 2017*]**

Solution: Let $F(s) = \cot^{-1}\left(\frac{2}{s+1}\right)$.

Then $\quad F'(s) = \frac{-1}{1+\left(\frac{2}{s+1}\right)^2} \times \left(\frac{-2}{(s+1)^2}\right)$

$$= \frac{(s+1)^2}{(s+1)^2+4} \times \left(\frac{2}{(s+1)^2}\right) = \frac{2}{(s+1)^2+4}.$$

Apply $L^{-1}[F(s)] = \frac{-1}{t}L^{-1}[F'(s)]$.

$$L^{-1}\left[cot^{-1}\left(\frac{2}{s+1}\right)\right] = \frac{-1}{t}L^{-1}\left[\frac{2}{(s+1)^2+4}\right]$$

$$= \frac{-1}{t}e^{-t}L^{-1}\left(\frac{2}{s^2+4}\right) = \frac{-1}{t}e^{-t}\sin 2t.$$

EXERCISES

1. Find the inverse Laplace transforms of the following functions:

(i) $\log\left(1+\frac{a}{s}\right)$ (ii) $\log\left(\frac{s+b}{s+a}\right)$ (iii) $\log\left(1+\frac{a^2}{s^2}\right)$

(iv) $\log\left(\frac{s^2-a^2}{s^2-b^2}\right)$ (v) $\log\frac{s^2+1}{(s-1)^2}$ (vi) $s\log\frac{s-1}{s+1}-3$

(vii) $\tan^{-1}\left(\frac{s+a}{b}\right)$ (viii) $\tan^{-1}\left(\frac{2}{s}\right)$ (ix) $\cot^{-1}\left(\frac{2}{s+1}\right)$

(x) $\log\left(1-\frac{s}{a}\right)+\tan^{-1}\left(1+\frac{s}{a}\right)$.

ANSWERS

(i) $\frac{1}{t}(1-e^{-at})$ (ii) $\frac{1}{t}(e^{-at}-e^{-bt})$

(iii) $\frac{2}{t}(1-\cos at)$ (iv) $\frac{2}{t}[\cosh bt-\cosh at]$ (v) $\frac{2}{t}(e^t-\cos t)$

(vi) $\frac{2}{t^2}[\sinh t-t\ \tanh ht]-3\delta(t)$ (vii) $\frac{-1}{t}e^{-at}\sin bt$ (viii) $\frac{\sin 2t}{t}$

(ix) $\frac{-1}{t}e^{-t}\sin 2t$ (x) $\frac{-1}{t}[1-e^{at}-e^{-at}\sin at]$.

Inverse Laplace Transforms of Multiplication by s^n

$$L^{-1}[s^n F(s)]=\frac{d^n}{dt^n}L^{-1}[F(s)]$$

Result: $L^{-1}[s\,F(s)]=\frac{d}{dt}L^{-1}[F(s)]$ if $L^{-1}[F(s)]\to 0$ as $t\to 0$.

Proof: We have $L[f'(t)]=sF(s)-f(0)$.

If $L^{-1}[F(s)]\to 0$ as $t\to 0$ (i.e., $f(0)=0$), then

$$L[f'(t)]=sF(s)$$

i.e., $$f'(t)=L^{-1}[s\,F(s)]$$

$$L^{-1}[s\,F(s)]=\frac{d}{dt}L^{-1}[F(s)].$$

Inverse Laplace Transforms of Division by s

$$L^{-1}\left[\frac{F(s)}{s}\right]=\int_0^t L^{-1}[F(s)]\,dt.$$

Proof: We have $L\left[\int_0^t f(t)\,dt\right]=\frac{F(s)}{s}$

$\therefore L^{-1}\left[\frac{F(s)}{s}\right] = \int_0^t f(t)dt = \int_0^t L^{-1}[F(s)]dt.$

Inverse Laplace Transforms of Integrals

If $L^{-1}[F(s)] = f(t)$, then $L^{-1}\left[\int_s^\infty F(s)ds\right] = \frac{f(t)}{t} = \frac{1}{t}L^{-1}[F(s)]$.

Proof: By Property 4 of Laplace transform,

$$L\left[\frac{f(t)}{t}\right] = \int_s^\infty F(s)ds$$

Then $L^{-1}\left[\int_s^\infty F(s)ds\right] = \frac{f(t)}{t}$.

Example 1: Find $L^{-1}\left[\frac{s}{(s+1)(s-3)}\right]$.

Solution: $L^{-1}\left[\frac{s}{(s+1)(s-3)}\right] = L^{-1}\left[s.\frac{1}{(s+1)(s-3)}\right]$

$$= \frac{d}{dt}L^{-1}\left[\frac{1}{(s+1)(s-3)}\right] \qquad (1)$$

Consider $\frac{1}{(s+1)(s-3)} = \frac{A}{s+1} + \frac{B}{s-3}$

$$= \frac{A(s-3)+B(s+1)}{(s+1)(s-3)}$$

$\Rightarrow \quad 1 = A(s-3) + B(s+1)$

Put $s = -1$, we get $A = \frac{-1}{4}$.

Put $s = 3$, we get $B = \frac{1}{4}$.

$\therefore L^{-1}\left[\frac{1}{(s+1)(s-3)}\right] = L^{-1}\left[\frac{\frac{-1}{4}}{s+1} + \frac{\frac{1}{4}}{s-3}\right]$

$$= \frac{1}{4}(e^{3t} - e^{-t}).$$

Now $\lim_{t\to 0}\frac{1}{4}(e^{3t} - e^{-t}) = 0.$

Hence (1) gives $L^{-1}\left[\frac{s}{(s+1)(s-3)}\right] = \frac{d}{dt}\left[\frac{1}{4}(e^{3t} - e^{-t})\right]$

$$= \frac{1}{4}(3e^{3t} + e^{-t}).$$

Example 2: Find $L^{-1}\left[\frac{s^2}{(s^2+a^2)^2}\right]$.

Solution: $L^{-1}\left[\frac{s^2}{(s^2+a^2)^2}\right] = L^{-1}\left[s \cdot \frac{s}{(s^2+a^2)^2}\right]$

$$= \frac{d}{dt}\left\{L^{-1}\left[\frac{s}{(s^2+a^2)^2}\right]\right\}, \qquad (1)$$

Consider $L^{-1}\left[\frac{s}{(s^2+a^2)^2}\right] = \frac{1}{2a} t \sin at$

$$\lim_{t\to 0} L^{-1}\left[\frac{s}{(s^2+a^2)^2}\right] = \lim_{t\to 0} \frac{1}{2a} t \sin at = 0.$$

$\therefore$ (1) gives $L^{-1}\left[\frac{s^2}{(s^2+a^2)^2}\right] = \frac{d}{dt}\left[\frac{1}{2a} t \sin at\right]$

$$= \frac{1}{2a}[t \cos at\; a + \sin at]$$

$$= \frac{1}{2a}[\sin at + at \cos at].$$

Example 3: Find $L^{-1}\left[\frac{1}{s(s-2)}\right]$ and $L^{-1}\left[\frac{1}{s(s^2+9)}\right]$.

Solution: (i) $L^{-1}\left[\frac{1}{s(s-2)}\right] = \int_0^t L^{-1}\left[\frac{1}{s-2}\right] dt$

$$= \int_0^t e^{2t}\, dt = \left[\frac{e^{2t}}{2}\right]_0^t = \frac{1}{2}(e^{2t} - 1).$$

(ii) $L^{-1}\left[\frac{1}{s(s^2+9)}\right] = \int_0^t L^{-1}\left[\frac{1}{(s^2+9)}\right] dt$

$$= \int_0^t \frac{1}{3} \sin 3t\, dt = \frac{1}{3}\left[\frac{-\cos 3t}{3}\right]_0^t$$

$$= \frac{1}{9}[-\cos 3t + 1] = \frac{1}{9}[1-\cos 3t].$$

Example 4: Find $L^{-1}\left[\frac{s+2}{s^2(s+3)}\right]$.

Solution: $L^{-1}\left[\frac{s+2}{s^2(s+3)}\right] = L^{-1}\left[\frac{1}{s(s+3)} + \frac{2}{s^2(s+3)}\right]$

$$= \int_0^t L^{-1}\left[\frac{1}{s+3}\right] dt + 2\int_0^t \int_0^t L^{-1}\left[\frac{1}{s+3}\right] dtdt$$

$$= \int_0^t e^{-3t}\, dt + 2\int_0^t \int_0^t e^{-3t}\, dt\; dt$$

$$= \left[\frac{e^{-3t}}{-3}\right]_0^t + 2\int_0^t \left[\frac{e^{-3t}}{-3}\right]_0^t dt$$

$$= \frac{1}{3}(1 - e^{-3t}) + \frac{2}{3}\int_0^t (1 - e^{-3t})dt$$

$$= \frac{1}{3}(1 - e^{-3t}) + \frac{2}{3}\left[t - \frac{e^{-3t}}{-3}\right]_0^t$$

$$= \frac{1}{3}(1 - e^{-3t}) + \frac{2}{3}\left[\left(t + \frac{e^{-3t}}{3}\right) - \left(0 + \frac{1}{3}\right)\right]$$

$$= \frac{1}{9} - \frac{1}{9}e^{-3t} + \frac{2}{3}t.$$

Example 5: Find$L^{-1}\left[\frac{1}{(s^2+a^2)^2}\right]$.

Solution: $L^{-1}\left[\frac{1}{(s^2+a^2)^2}\right] = L^{-1}\left[\frac{1}{s} \cdot \frac{s}{(s^2+a^2)^2}\right]$

$$= \int_0^t L^{-1}\left[\frac{s}{(s^2+a^2)^2}\right] dt = \int_0^t \frac{1}{2a}\, t \sin at\, dt$$

$$= \frac{1}{2a}\left[t\left(\frac{-\cos at}{a}\right) - \left(\frac{-\sin at}{a^2}\right)\right] = \frac{1}{2a^3}[\sin at - at \cos at].$$

Example 6: Find $L^{-1}\left[\int_s^\infty \left(\frac{1}{s} - \frac{1}{s+1}\right) ds\right]$.

Solution: $L^{-1}\left[\int_s^\infty \left(\frac{1}{s} - \frac{1}{s+1}\right) ds\right] = \frac{1}{t}L^{-1}\left[\frac{1}{s} - \frac{1}{s+1}\right]$

$$= \frac{1}{t}(1 - e^{-t}).$$

Example 7: Find $L^{-1}\left[\int_s^\infty \left(\frac{u}{u^2+4} - \frac{u}{u^2+9}\right) du\right]$.

Solution: $L^{-1}\left[\int_s^\infty \left(\frac{u}{u^2+4} - \frac{u}{u^2+9}\right) du\right] = L^{-1}\left[\int_s^\infty \left(\frac{s}{s^2+4} - \frac{s}{s^2+9}\right) ds\right]$

$$= \frac{1}{t} L^{-1}\left[\frac{s}{s^2+4} - \frac{s}{s^2+9}\right]$$

$$= \frac{1}{t}[\cos 2t - \cos 3t].$$

Example 8: Find $L^{-1}\left[\int_s^\infty \log\left(\frac{s+2}{s+1}\right) ds\right]$.

Solution: $L^{-1}\left[\int_s^\infty \log\left(\frac{s+2}{s+1}\right) ds\right] = \frac{1}{t}L^{-1}\left[\log\left(\frac{s+2}{s+1}\right)\right]$ (1)

To find $L^{-1}\left[\log\left(\frac{s+2}{s+1}\right)\right]$, we can use the formula

$L^{-1}[F(s)] = \frac{-1}{t}L^{-1}[F'(s)].$ (2)

Put $F(s) = \log\left(\frac{s+2}{s+1}\right)$ in (2). Then $F(s) = \log(s+2) - \log(s+1)$

$$F'(s) = \frac{1}{s+2} - \frac{1}{s+1}.$$

$\therefore$ (2) gives $L^{-1}\left[\log\left(\frac{s+2}{s+1}\right)\right] = \frac{-1}{t}L^{-1}\left[\frac{1}{s+2} - \frac{1}{s+1}\right]$

$$= \frac{-1}{t}[e^{-2t} - e^{-t}] = \frac{1}{t}[e^{-t} - e^{-2t}]$$

Now (1) gives $L^{-1}\left[\int_s^\infty \log\left(\frac{s+2}{s+1}\right) ds\right] = \frac{1}{t}.\frac{1}{t}[e^{-t} - e^{-2t}] = \frac{1}{t^2}[e^{-t} - e^{-2t}]$.

EXERCISES

1. Find $L^{-1}\left[\frac{s}{(s+1)(s+2)}\right]$

2. Find $L^{-1}\left[\frac{s^2}{(s^2-w^2)^2}\right]$

3. Find $L^{-1}\left[\frac{s^2}{(s^2+1)(s^2+4)}\right]$

4. Find $L^{-1}\left[\frac{s}{(s+2)^4}\right]$

5. Find the inverse Laplace transforms of the following functions

(i) $\frac{1}{s^3(s-3)}$ (ii) $\frac{1}{s(s^2+a^2)}$ (iii) $\frac{1}{s^2(s^2+a^2)}$

(iv) $\frac{1}{s(s+2)^3}$ (v) $\frac{1}{s(s^2-4s+8)}$ (vi) $\frac{1}{s(s^2+4s+5)}$

6. Find $L^{-1}\left[\int_s^\infty \left(\frac{1}{s+a} - \frac{1}{s+b}\right) ds\right]$

7. Find $L^{-1}\left[\int_s^\infty \left(\frac{1}{s^2+a^2} - \frac{1}{s^2+b^2}\right) ds\right]$

8. Find $L^{-1}\left[\int_s^\infty \log\left(\frac{u+a}{u+b}\right) du\right]$

9. Find$L^{-1}\left[\int_s^\infty \tan^{-1}\frac{2}{s^2} ds\right]$.

ANSWERS

1. $2e^{-2t} - e^{-t}$

2. $\frac{1}{2w}[\sinh wt + wt\cosh wt]$

3. $\frac{1}{3}(\cos t - \cos 2t)$

4. $\frac{1}{6}t^2e^{-2t}(3-2t)$.

5.

(i) $\frac{1}{54}[2e^{3t} - 9t^2 - 6t - 2]$ (ii) $\frac{1}{a^2}(1 - \cos at)$

(iii) $\frac{1}{a^3}(at - \sin at)$ (iv) $\frac{1}{8}[1 - e^{-2t}(2t^2 + 2t + 1)]$

(v) $\frac{1}{8}[1 + e^{2t}(\sin 2t + \cos 2t)]$ (vi) $\frac{1}{5}[1 - e^{-2t}(2\sin t + \cos t)]$

6. $\frac{e^{-at}-e^{-bt}}{t}$

7. $\frac{1}{t}\left[\frac{1}{a}\sin at-\frac{1}{b}\sin bt\right]$

8. $\frac{e^{-bt}-e^{-at}}{t^2}$

9. $\frac{2\sin t\sinh t}{t^3}$.

5.12. Convolution Theorem

Definition: The *convolution* or *convolution integral* or *faltung* of two functions $f(t)$ and $g(t)$ is defined as $f(t)*g(t)=\int_0^t f(u)g(t-u)du.$

Convolution Theorem

Statement: If $f(t)$ and $g(t)$ are functions defined for $t\geq 0$, then

$L[f(t)*g(t)]=L[f(t)].L[g(t)]$

Proof: By the definition of Laplace transform,

$$L[f(t)*g(t)]=\int_0^\infty e^{-st}[f(t)*g(t)]dt$$

$$=\int_0^\infty e^{-st}\left[\int_0^t f(u)g(t-u)du\right]dt, \text{ by the definition of convolution}$$

$$=\int_0^\infty\int_0^t e^{-st}f(u)g(t-u)dudt \quad (1)$$

The region of integration for the double integral in (1) is bounded by the lines $u=0, u=t, t=0$ and $t=\infty$ and is shown in Figure 1.

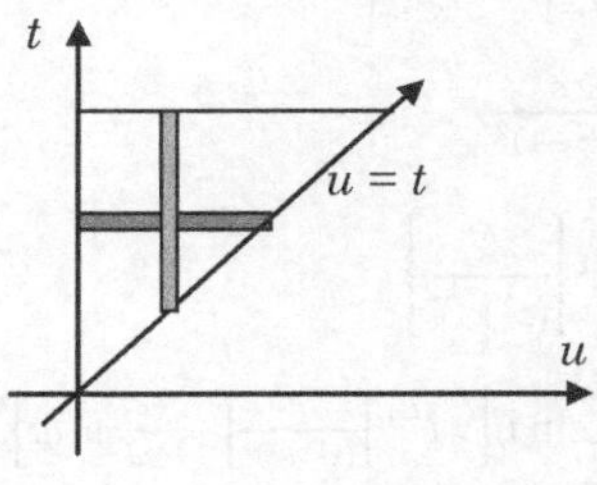

Figure 1

Changing the order of integration in (1), we get

$$L[f(t)*g(t)]=\int_0^\infty\int_u^\infty e^{-st}f(u)g(t-u)dtdu$$

$$=\int_0^\infty f(u)\left[\int_u^\infty e^{-st}g(t-u)dt\right]du \quad (2)$$

Put $t-u=v$ in the inner integral in (2).

Then $dt=dv$ and v varies from 0 to ∞.

$\therefore$(2) gives $L[f(t)*g(t)] = \int_0^\infty f(u)\int_0^\infty e^{-s(u+v)}g(v)dv\,du$

$$= \int_0^\infty e^{-su}f(u)du\int_0^\infty e^{-sv}g(v)dv$$

$$= \int_0^\infty e^{-st}f(t)dt.\int_0^\infty e^{-st}g(t)dt[\because u \text{ and } v \text{ are dummy variables}]$$

$$= L[f(t)].L[g(t)].$$

Corollary: $L^{-1}[F(s).G(s)] = L^{-1}[F(s)]*L^{-1}[G(s)]$.

Proof: By convolution theorem,

$$L[f(t)*g(t)] = L[f(t)].L[g(t)]$$

$$= F(s).G(s)$$

$$f(t)*g(t) = L^{-1}[F(s).G(s)]$$

i.e., $L^{-1}[F(s)]*L^{-1}[G(s)] = L^{-1}[F(s).G(s)]$.

Example 1: Using convolution theorem evaluate $\int_0^t \sin u \cos(t-u)\,du$.

Solution: ***[AU, Nov/Dec 2015]***

$L\left[\int_0^t \sin u\cos(t-u)\,du\right] = L[\sin t * \cos t]$, by definition of convolution

$= L[\sin t] \times L[\cos t]$, by convolution theorem

$$= \frac{1}{s^2+1} \times \frac{s}{s^2+1}$$

$$= \frac{s}{(s^2+1)^2}.$$

$$\therefore \int_0^t \sin u\cos(t-u)\,du = L^{-1}\left[\frac{s}{(s^2+1)^2}\right]$$

$$= \frac{t}{2}\ \sin t\left[\because L^{-1}\left[\frac{s}{(s^2+a^2)^2}\right] = \frac{t}{2a}\sin at\right].$$

Example 2: Use convolution theorem to find $L^{-1}\left[\frac{1}{s(s+a)}\right]$.

Solution: $L^{-1}\left[\frac{1}{s(s+a)}\right] = L^{-1}\left[\frac{1}{s}.\frac{1}{s+a}\right]$

$= L^{-1}\left[\frac{1}{s}\right]*L^{-1}\left[\frac{1}{s+a}\right]$, by convolution theorem

$$= 1*e^{-at}$$

$$= e^{-at}*1[\because f(t)*g(t) = g(t)*f(t)]$$

$= \int_0^t e^{-au}.1\,du,$ by the definition of convolution

$= \left[\frac{e^{-au}}{-a}\right]_0^t$

$= \left[\frac{e^{-at}-1}{-a}\right] = \frac{1}{a}[1-e^{-at}].$

Example 3: Find $L^{-1}\left[\frac{1}{s(s^2-a^2)}\right]$ using convolution theorem.

Solution: $L^{-1}\left[\frac{1}{s(s^2-a^2)}\right] = L^{-1}\left[\frac{1}{s}.\frac{1}{s^2-a^2}\right]$

$= L^{-1}\left[\frac{1}{s}\right] * L^{-1}\left[\frac{1}{s^2-a^2}\right]$, by convolution theorem

$= 1 * \frac{1}{a}\sinh at$

$= \frac{1}{a}\sinh at * 1 \;\; [\because f(t) * g(t) = g(t) * f(t)]$

$= \frac{1}{a}\int_0^t \sinh au.1\,du$ by the definition of convolution

$= \frac{1}{a}\left[\frac{\cosh\ au}{a}\right]_0^t = \frac{1}{a^2}(\cosh at - 1).$

Example 4: Find $L^{-1}\left[\frac{1}{s^2(s^2+1)}\right]$ using convolution theorem.

Solution: $L^{-1}\left[\frac{1}{s^2(s^2+1)}\right] = L^{-1}\left[\frac{1}{s^2}.\frac{1}{s^2+1}\right]$

$= L^{-1}\left[\frac{1}{s^2}\right] * L^{-1}\left[\frac{1}{s^2+1}\right]$

$= t * \sin t \; = \sin t * t$

$= \int_0^t \sin u\,(t-u)du = \int_0^t (t-u)\sin u\; du$

$= [(t-u)(-\cos u) - (-1)(-\sin u)]_0^t$

$= [(0-\sin t) - (-t-0)] = t - \sin t.$

Example 5: Find $L^{-1}\left[\frac{s^2}{(s^2+a^2)^2}\right]$ using convolution theorem. **[AU, *Nov/Dec 2017*]**

Solution: $L^{-1}\left[\frac{s^2}{(s^2+a^2)^2}\right] = L^{-1}\left[\frac{s}{s^2+a^2}\right] * L^{-1}\left[\frac{s}{s^2+a^2}\right]$

$= \cos at * \cos at$

$= \int_0^t \cos au \cos a(t-u)\,du$

$$= \int_0^t \left[\frac{\cos[(a-a)u+at]+\cos[(a+a)u-at]}{2}\right] du$$

$$= \int_0^t \left[\frac{\cos at+\cos[2au-at]}{2}\right] du$$

$$= \frac{1}{2}\left[cosat\,(u) + \frac{\sin[2au-at]}{2a}\right]_0^t$$

$$= \frac{1}{2}\left[cosat\,(t) + \left(\frac{\sin[2at-at]}{2a}\right) - 0 - \left(\frac{\sin[-at]}{2a}\right)\right]$$

$$= \frac{1}{2}\left[tcosat + \left(\frac{\sin at}{2a}\right) + \left(\frac{\sin at}{2a}\right)\right] [\because \sin(-x) = -\sin x]$$

$$= \frac{1}{2}\left[tcosat + 2\left(\frac{\sin at}{2a}\right)\right]$$

$$= \frac{1}{2}\left[tcosat + 2\left(\frac{\sin at}{2a}\right)\right].$$

$$= \frac{1}{2a}[\sin at + at\cos at]$$

Example 6: Find $L^{-1}\left[\frac{s}{(s^2+9)^2}\right]$ using convolution theorem.

Solution: $L^{-1}\left[\frac{s}{(s^2+9)^2}\right] = L^{-1}\left[\frac{1}{s^2+9}\cdot\frac{s}{s^2+9}\right]$

$$= L^{-1}\left[\frac{1}{s^2+9}\right] * L^{-1}\left[\frac{s}{s^2+9}\right] = \frac{1}{3}\sin 3t * \cos 3t$$

$$= \frac{1}{3}\int_0^t \sin 3u \cos 3(t-u)\ du$$

$$= \frac{1}{3}\int_0^t \left[\frac{\sin 3t+\sin(6u-3t)}{2}\right] du \left[\because \sin A\cos B = \frac{1}{2}[\sin(A+B)+\sin(A-B)]\right]$$

$$= \frac{1}{6}\left[\sin 3t\ .u - \frac{\cos(6u-3t)}{6}\right]_0^t$$

$$= \frac{1}{6}\left[\left(t\sin 3t - \frac{1}{6}\cos 3t\right) - \left(0 - \frac{\cos(-3t)}{6}\right)\right]$$

$$= \frac{1}{6}\left[t\sin 3t - \frac{1}{6}\cos 3t + \frac{1}{6}\cos 3t\right] = \frac{1}{6}\ t\sin 3t.$$

Example 7: Using convolution theorem, find $L^{-1}\left[\frac{s^2}{(s^2+a^2)(s^2+b^2)}\right]$.

Solution: ***[AU, May/June 2016, May 2014, Nov/Dec** 2014**]***

$$L^{-1}\left[\frac{s^2}{(s^2+a^2)(s^2+b^2)}\right] = L^{-1}\left[\frac{s}{s^2+a^2}\right] * L^{-1}\left[\frac{s}{s^2+b^2}\right]$$

$$= \cos at * \cos bt$$

$$= \int_0^t \cos au \cos b(t-u)\, du$$

$$= \int_0^t \left[\frac{[\cos\ [(a-b)u+bt]+cos\ [(a+b)u-bt]}{2}\right] du$$

$$= \frac{1}{2}\left[\frac{\sin[(a-b)u+bt]}{(a-b)} + \frac{\sin[(a+b)u-bt]}{(a+b)}\right]_0^t$$

$$= \frac{1}{2}\left[\left(\frac{\sin at}{a-b} + \frac{\sin at}{a+b}\right) - \left(\frac{\sin bt}{a-b} + \frac{\sin(-bt)}{a+b}\right)\right]$$

$$= \frac{1}{2}\left[\sin at\left(\frac{1}{a-b} + \frac{1}{a+b}\right) - \sin bt\left(\frac{1}{a-b} - \frac{1}{a+b}\right)\right] [\because\ \sin(-x) = -\sin x]$$

$$= \frac{1}{2}\left[\sin at\left(\frac{2a}{a^2-b^2}\right) - \sin bt\left(\frac{2b}{a^2-b^2}\right)\right]$$

$$= \frac{1}{a^2-b^2}(a \sin at - b\sin bt).$$

Example 8: Apply convolution theorem to evaluate $L^{-1}\left[\frac{s^2}{(s^2+4)(s^2+9)}\right]$.

Solution: **[*AU, April/May*2017]**

$$L^{-1}\left[\frac{s^2}{(s^2+4)(s^2+9)}\right] = L^{-1}\left[\frac{s}{s^2+4}\right] * L^{-1}\left[\frac{s}{s^2+9}\right]$$

$$= \cos 2t * \cos 3t$$

$$= \int_0^t \cos 2u \cos 3(t-u)\, du$$

$$= \int_0^t \left[\frac{\cos[(2-3)u+3t]+cos\ [(2+3)u-3t]}{2}\right] du$$

$$= \int_0^t \left[\frac{\cos[-u+3t]+cos\ [5u-3t]}{2}\right] du$$

$$= \int_0^t \left[\frac{\cos[3t-u]+cos\ [5u-3t]}{2}\right] du$$

$$= \frac{1}{2}\left[\frac{\sin[3t-u]}{-1} + \frac{\sin[5u-3t]}{5}\right]_0^t$$

$$= \frac{1}{2}\left[-\sin[3t-t] + \frac{\sin[5t-3t]}{5} + \sin(3t) - \frac{\sin(-3t)}{5}\right]$$

$$= \frac{1}{2}\left[-\sin 2t + \frac{\sin[2t]}{5} + sin3t + \frac{\sin 3t}{5}\right] [\because\ \sin(-x) = -\sin x]$$

$$= \frac{1}{2}\left[\frac{-4\sin[2t]}{5} + \frac{6\sin[3t]}{5}\right] = \left[\frac{-2\sin[2t]}{5} + \frac{3\sin[3t]}{5}\right]$$

Example 9: Find $L^{-1}\left[\frac{1}{(s^2+1)(s^2+4)}\right]$.

Solution: $L^{-1}\left[\frac{1}{(s^2+1)(s^2+4)}\right] = L^{-1}\left[\frac{1}{s^2+1}\right] * L^{-1}\left[\frac{1}{s^2+4}\right]$

$$= \sin t * \frac{1}{2}\sin 2t$$

$$= \frac{1}{2}\int_0^t \sin u \sin 2(t-u)\, du$$

$$= \frac{1}{2}\int_0^t \left[\frac{\cos(2t-3u) \quad -\cos(2t-u)}{2}\right] du$$

$$= \frac{1}{4}\left[\left(\frac{\sin(2t-3u)}{-3} - \frac{\sin(2t-u)}{-1}\right)\right]_0^t$$

$$= \frac{1}{4}\left[\left(-\frac{\sin(-t)}{3} + \sin t\right) - \left(\frac{-\sin 2t}{3} + \sin 2t\right)\right]$$

$$= \frac{1}{4}\left[\frac{1}{3}\sin t + \sin t + \frac{\sin 2t}{3} - \sin 2t\right] [\because \sin(-x) = -\sin x]$$

$$= \frac{1}{4}\left[\frac{4}{3}\sin t - \frac{2}{3}\sin 2t\right].$$

Example 10: Using convolution theorem find $L^{-1}\left[\frac{1}{(s+1)(s^2+1)}\right]$.

Solution: $L^{-1}\left[\frac{1}{(s+1)(s^2+1)}\right] = L^{-1}\left[\frac{1}{s+1}\right] + L^{-1}\left[\frac{1}{s^2+1}\right]$

$$= e^{-t} * \sin t = \sin t * e^{-t}$$

$$= \int_0^t \sin u\, e^{-(t-u)} du = e^{-t}\int_0^t e^u \sin u\, du$$

$$= e^{-t}\left[\frac{e^u}{1+1}(\sin u - \cos u)\right]_0^t$$

$$= \frac{e^{-t}}{2}[e^t(\sin t - \cos t) - (0-1)] = \frac{1}{2}(\sin t - \cos t) + \frac{1}{2}e^{-t}.$$

Example 11: Use convolution theorem to find $L^{-1}\left[\frac{s}{(s^2+2s+5)^2}\right]$

Solution: $L^{-1}\left[\frac{s}{(s^2+2s+5)^2}\right] = L^{-1}\left[\frac{s}{s^2+2s+5}\right] * L^{-1}\left[\frac{1}{s^2+2s+5}\right]$

$$= L^{-1}\left[\frac{s}{s^2+2s+1+4}\right] * L^{-1}\left[\frac{1}{s^2+2s+1+4}\right]$$

$$= L^{-1}\left[\frac{(s+1)-1}{(s+1)^2+4}\right] * L^{-1}\left[\frac{1}{(s+1)^2+4}\right]$$

$$= e^{-t}L^{-1}\left[\frac{s-1}{s^2+4}\right] * e^{-t}L^{-1}\left[\frac{1}{s^2+4}\right]$$

$$= e^{-t}\left\{L^{-1}\left[\frac{s}{s^2+4}\right] - L^{-1}\left[\frac{1}{s^2+4}\right]\right\} * e^{-t}L^{-1}\left[\frac{1}{s^2+4}\right]$$

$$= e^{-t}\left[\cos 2t - \frac{1}{2}\sin 2t\right] * e^{-t}\frac{1}{2}\sin 2t$$

$$= \int_0^t e^{-u}\left(\cos 2u - \frac{1}{2}\sin 2u\right)\frac{1}{2}e^{-(t-u)}\sin 2(t-u)du$$

$$= \frac{1}{2}e^{-t}\int_0^t\left(\cos 2u - \frac{1}{2}\sin 2u\right)\sin(2t-2u)\,du$$

$$= \frac{1}{2}e^{-t}\int_0^t\left[\cos 2u.\sin(2t-2u) - \frac{1}{2}\sin 2u\sin(2t-2u)\right]du$$

$$= \frac{1}{2}e^{-t}\int_0^t\left\{\frac{1}{2}[\sin 2t + \sin(2t-4u)] - \frac{1}{4}[\cos(2t-4u) - \cos 2t]\right\}du$$

$$= \frac{1}{2}e^{-t}\left\{\frac{1}{2}\left[\sin 2t.u + \frac{-\cos(2t-4u)}{-4}\right] - \frac{1}{4}\left[\frac{\sin(2t-4u)}{-4} - \cos 2t.u\right]\right\}_0^t$$

$$= \frac{1}{2}e^{-t}\left\{\frac{1}{2}(t\sin 2t - 0) + \frac{1}{4}(\cos 2t - \cos 2t) + \frac{1}{16}\left(-\sin 2t - \sin 2t + \frac{1}{4}(t\cos 2t - 0)\right)\right\}$$

$$= \frac{1}{2}e^{-t}\left\{\frac{t}{2}\sin 2t - \frac{1}{8}\sin 2t + \frac{1}{4}t\cos 2t\right\}.$$

Example 12: Find $L^{-1}\left[\frac{s^2+s}{(s^2+1)(s^2+2s+2)}\right]$, using convolution theorem.

Solution: $L^{-1}\left[\frac{s^2+s}{(s^2+1)(s^2+2s+2)}\right] = L^{-1}\left[\frac{s(s+1)}{(s^2+1)(s^2+2s+1+1)}\right]$

$$= L^{-1}\left[\frac{s}{s^2+1}\right] * L^{-1}\left[\frac{s+1}{(s+1)^2+1}\right]$$

$$= \cos t * e^{-t}L^{-1}\left[\frac{s}{s^2+1}\right] = \cos t * e^{-t}\cos t = e^{-t}\cos t * \cos t$$

$$= \int_0^t e^{-u}\cos u.\cos(t-u)\,du = \int_0^t e^{-u}\frac{1}{2}[\cos t + \cos(2u-t)]du$$

$$= \frac{1}{2}\left\{\cos t\int_0^t e^{-u}du + \int_0^t e^{-u}\cos(2u-t)du\right\}$$

$$= \frac{1}{2}\left\{\cos t\,[-e^{-u}]_0^t + \left[\frac{e^{-u}}{1+4}(-\cos(2u-t) + 2\sin(2u-t))\right]_0^t\right\}$$

$$= \frac{1}{2}\left\{\cos t\,(1-e^{-t}) + \frac{e^{-t}}{5}(-\cos t + 2\sin t) - \frac{1}{5}(-\cos t - 2\sin t)\right\}$$

$$= \frac{1}{2}\left\{\cos t - e^{-t}\cos t - \frac{1}{5}e^{-t}\cos t + \frac{2}{5}e^{-t}\sin t + \frac{1}{5}\cos t + \frac{2}{5}\sin t\right\}$$

$$= \frac{1}{2}\left[\frac{6}{5}\cos t + \frac{2}{5}\sin t + e^{-t}\left(\frac{2}{5}\sin t - \frac{6}{5}\cos t\right)\right]$$

$$= \frac{1}{5}(3\cos t + \sin t) + \frac{e^{-t}}{5}(\sin t - 3\cos t).$$

EXERCISES

Find the inverse Laplace transforms of the following functions using convolution theorem. (*i*) $\frac{1}{s(s-3)}$ (*ii*) $\frac{1}{s(s^2+a^2)}$ (*iii*) $\frac{1}{(s+a)(s+b)}$

(*iv*) $\frac{s}{(s^2+a^2)^2}$ (*v*) $\frac{16}{(s-2)(s+2)^2}$ (*vi*) $\frac{4}{(s^2+2s+5)^2}$

(*vii*) $\frac{1}{(s+1)(s^2+9)}$ (*viii*) $\frac{3}{(s+1)(s^2+4)}$ (*ix*) $\frac{s+2}{(s^2+4s+13)^2}$

ANSWERS

(*i*) $\frac{1}{3}(e^{3t}-1)$ (*ii*) $\frac{1}{a^2}(1-\cos at)$ (*iii*) $\frac{1}{a-b}(e^{-bt}-e^{-at})$

(*iv*) $\frac{1}{2a}\,t\sin at$ (*v*) $e^{2t}-e^{-2t}-4te^{-2t}$

(*vi*) $\frac{1}{4}e^{-t}(\sin 2t-2t\cos 2t)$ (*vii*) $\frac{1}{10}e^{-t}+\frac{1}{30}(\sin 3t+3\cos 3t)$

(*viii*) $\frac{3}{10}[2e^{-t}+\sin 2t-2\cos 2t]$ (*ix*) $\frac{1}{6}e^{-2t}\,t\sin 3t$.

5.13. Solutions of Ordinary Differential Equations

The Laplace transform technique is used to solve both differential and integral equations. In the usual procedure, to solve the ordinary differential equations first we find the general solution and then evaluate the arbitrary constants using the boundary conditions. But using the Laplace transform method we can find the solution directly.

To use Laplace transform technique, we can use the following formulae.

1. $L[y'] = s\,L[(y)] - y(0)$
2. $L[y''] = s^2L[y] - sy(0) - y'(0)$.
3. $L[y'''] = s^3L[y] - s^2y(0) - sy'(0) - y''(0)$.

Working Procedure

Step 1: Take Laplace transform on both sides of the given differential equation.

Step 2: Apply the given initial (boundary) conditions. This gives an algebraic equation in $\bar{y} = L[y(t)]$.

Step 3: Solve the algebraic equation to get $\bar{y}$ as a function of s.

Step 4: Apply inverse Laplace transform. It gives the required solution y or $y(t)$.

Example 1: Solve $\frac{d^2y}{dt^2}+5\frac{dy}{dt}+6y=2,\ y(0)=0,\ y'(0)=0.$ ***[AU, Nov/Dec 2017]***

Solution: The given equation is $y''+5y'+6y=2,\ y'(0)=0,\ y(0)=0.$

Apply L.T. on both sides, $L[y'']+5L[y']+6L[y]=L[2]$

$$[s^2L[y]-sy(0)-y'(0)]+5[sL[y]-y(0)]+6L[y]=\frac{2}{s}$$

$$s^2\bar{y}+5s\bar{y}+6\bar{y}=\frac{2}{s}$$

$$\bar{y}[s^2+5s+6]=\frac{2}{s}$$

$$\bar{y}=\frac{2}{s(s^2+5s+6)}=\frac{2}{s(s+2)(s+3)}$$

$$\therefore \quad y=L^{-1}\left[\frac{2}{s(s+2)(s+3)}\right] \quad (1)$$

By partial fractions, $\frac{2}{s(s+2)(s+3)}=\frac{A}{s}+\frac{B}{s+2}+\frac{C}{s+3}$

$\Rightarrow \quad 2=A(s+2)(s+3)+Bs(s+3)+Cs(s+2)$

Put $s=0$, we get $2=6A \quad \Rightarrow A=\frac{1}{3}.$

Put $s=-2$, we get $2=-2B \quad \Rightarrow B=-1.$

Put $s=-3$, we get $2=3C \quad \Rightarrow C=\frac{2}{3}.$

$\therefore$ (1) gives $y=L^{-1}\left[\frac{1/3}{s}+\frac{-1}{s+2}+\frac{2/3}{s+3}\right]$

$$=\frac{1}{3}-e^{-2t}+\frac{2}{3}e^{-3t}.$$

Example 2: Solve the differential equation $\frac{d^2y}{dt^2}-3\frac{dy}{dt}+2y=e^{3t}$ with $y(0)=1$ and $y'(0)=0$, using Laplace transform. ***[AU, Nov/Dec 2014]***

Solution: The given equation is $y''-3y'+2y=e^{3t}.$

Apply L.T. on both sides, $L[y'']-3L[y']+2L[y]=L[e^{3t}]$

$$[s^2L[y]-sy(0)-y'(0)]-3[sL[y]-y(0)]+2L[y]=\frac{1}{s-3}$$

$$s^2\bar{y}-s-3s\bar{y}+3+2\bar{y}=\frac{1}{s-3}$$

$$\bar{y}[s^2-3s+2]=\frac{1}{s-3}+s-3$$

$$\bar{y}(s-1)(s-2)=\frac{1+(s-3)^2}{(s-3)}$$

$$\overline{y} = \frac{1+(s-3)^2}{(s-1)(s-2)(s-3)}$$

$$\therefore \quad y = L^{-1}\left[\frac{1+(s-3)^2}{(s-1)(s-2)(s-3)}\right] \qquad (1)$$

By partial fractions, $\frac{1+(s-3)^2}{(s-1)(s-2)(s-3)} = \frac{A}{s-1} + \frac{B}{s-2} + \frac{C}{s-3}$ (2)

$$\Rightarrow 1+(s-3)^2 = A(s-2)(s-3) + B(s-1)(s-3) + C(s-1)(s-2)$$

Put $s = 1$, we get $5 = 2A \Rightarrow A = \frac{5}{2}$.

Put $s = 2$, we get $2 = -B \Rightarrow B = -2$.

Put $s = 3$, we get $1 = 2C \Rightarrow C = \frac{1}{2}$

$\therefore$ (1) gives $y = L^{-1}\left[\frac{5/2}{s-1} + \frac{-2}{s-2} + \frac{1/2}{s-3}\right]$

$$= \frac{5}{2}L^{-1}\left[\frac{1}{s-1}\right] - 2L^{-1}\left[\frac{1}{s-2}\right] + \frac{1}{2}L^{-1}\left[\frac{1}{s-3}\right]$$

$$= \frac{5}{2}e^{t} - 2e^{2t} + \frac{1}{2}e^{3t}.$$

Example 3: Using Laplace transform, solve the differential equation $\frac{d^2y}{dt^2} + 3\frac{dy}{dt} + 2y = e^{-t}$ with $y(0) = 1$ and $y'(0) = 0$. ***[AU, April/May 2017]***

Solution: The given equation is $y'' + 3y' + 2y = e^{-t}$.

Apply L.T. on both sides, $L[y''] + 3L[y'] + 2L[y] = L[e^{-t}]$

$$[s^2 L[y] - s\,y(0) - y'(0)] + 3[s\,L[y] - y(0)] + 2L[y] = \frac{1}{s+1}$$

$$s^2\overline{y} - s + 3s\overline{y} - 3 + 2\overline{y} = \frac{1}{s+1}$$

$$\overline{y}[s^2 + 3s + 2] = \frac{1}{s+1} + s + 3$$

$$\overline{y}(s+1)(s+2) = \frac{1+(s+3)(s+1)}{(s+1)}$$

$$\overline{y} = \frac{1+(s+3)(s+1)}{(s+1)(s+1)(s+2)} \qquad (1)$$

$$\therefore \quad y = L^{-1}\left[\frac{s^2+4s+4}{(s+1)^2(s+2)}\right] \qquad (2)$$

By partial fractions, $\frac{s^2+4s+4}{(s+1)^2(s+2)} = \frac{A}{s+1} + \frac{B}{(s+1)^2} + \frac{C}{s+2}$

$$\Rightarrow s^2 + 4s + 4 = A\,(s+1)(s+2) + B(s+2) + C(s+1)^2$$

Put $s = -1$, we get$1 - 4 + 4 = B \quad \Rightarrow B = 1.$

Put $s = -2$, we get$4 - 8 + 4 = C \quad \Rightarrow C = 0.$

Put $s = 0$, we get$4 = 2A + 2B + C \quad \Rightarrow A = 1$

$\therefore$ (1) gives $y = L^{-1}\left[\frac{1}{s+1} + \frac{1}{(s+1)^2} + \frac{0}{s-3}\right]$

$$= L^{-1}\left[\frac{1}{s+1}\right] - L^{-1}\left[\frac{1}{(s+1)^2}\right] = e^{-t} - te^{-t}.$$

Example 4: Solve using Laplace transform of $x'' - 2x' + x = e^t$ when $x(0) = 2$; $x'(0) = -1$. ***[AU, Nov/Dec 2015]***

Solution:

The given equation is $x'' - 2x' + x = e^t$.

Apply L.T. on both sides, $\quad L[x''] - 2L[x'] + L[x] = L[e^t]$

$$[s^2L[x] - sx(0) - x'(0)] - 2[sL[x] - x(0)] + L[x] = \frac{1}{s-1}$$

$$s^2\overline{x} - 2s + 1 - 2s\overline{x} + 4 + \overline{x} = \frac{1}{s-1}$$

$$\overline{x}[s^2 - 2s + 1] = \frac{1}{s-1} + 2s - 5$$

$$\overline{x}(s-1)^2 = \frac{1}{s-1} + 2s - 5$$

$$\overline{x} = \frac{1}{(s-1)^3} + \frac{2s}{(s-1)^2} - \frac{5}{(s-1)^2}$$

$\therefore$

$$x = L^{-1}\left[\frac{1}{(s-1)^3} + \frac{2s}{(s-1)^2} - \frac{5}{(s-1)^2}\right]$$

$$= L^{-1}\left[\frac{1}{(s-1)^3}\right] + 2L^{-1}\left[\frac{s}{(s-1)^2}\right] - 5L^{-1}\left[\frac{1}{(s-1)^2}\right]$$

$$= e^tL^{-1}\left[\frac{1}{s^3}\right] + 2L^{-1}\left[\frac{s-1+1}{(s-1)^2}\right] - 5e^tL^{-1}\left[\frac{1}{s^2}\right]$$

$$= e^t\frac{t^2}{2!} + 2L^{-1}\left[\frac{s-1}{(s-1)^2} + \frac{1}{(s-1)^2}\right] - 5e^tt$$

$$= e^t\frac{t^2}{2} + 2L^{-1}\left[\frac{1}{s-1}\right] + 2L^{-1}\left[\frac{1}{(s-1)^2}\right] - 5e^tt$$

$$= e^t\frac{t^2}{2} + 2e^t + 2e^tL^{-1}\left[\frac{1}{s^2}\right] - 5e^tt$$

$$= e^t\frac{t^2}{2} + 2e^t + 2e^tt - 5e^tt = e^t\frac{t^2}{2} + 2e^t - 3e^tt.$$

Example 5: Using Laplace transform, solve $\frac{d^2y}{dt^2} + 4y = \sin 2t$, given that $y(0) = 3$ and $y'(0) = 4$. ***[AU, May/June 2014]***

Solution: Given $y'' + 4y = \sin 2t\,,\; y(0) = 3, y'(0) = 4.$

Apply L.T. on both sides, $L[y''] + 4L[y] = L[\sin 2t]$

$$[s^2L[y] - sy(0) - y'(0)] + 4L[y] = \frac{2}{s^2+4}$$

$$[s^2L[y] - 3s - 4] + 4L[y] = \frac{2}{s^2+4}$$

$$L[y](s^2 + 4) - 3s - 4 = \frac{2}{s^2+4}$$

$$L[y](s^2 + 4) = \frac{2}{s^2+4} + 3s + 4$$

$$L[y] = \frac{2}{(s^2+4)^2} + \frac{3s}{s^2+4} + \frac{4}{s^2+4}$$

$$\therefore \quad y = 2L^{-1}\left[\frac{1}{(s^2+4)^2}\right] + 3L^{-1}\left[\frac{s}{s^2+4}\right] + 4L^{-1}\left[\frac{1}{s^2+4}\right]$$

$$= 2L^{-1}\left[\frac{1}{s^2+4}\frac{1}{s^2+4}\right] + 3\cos 2t + 2\sin 2t \qquad (1)$$

$$L^{-1}\left[\frac{1}{s^2+4}\frac{1}{s^2+4}\right] = L^{-1}\left[\frac{1}{s^2+4}\right] * L^{-1}\left[\frac{1}{s^2+4}\right]$$

$$= \frac{1}{2}\sin 2t * \frac{1}{2}\sin 2t$$

$$= \frac{1}{4}\int_0^t \sin 2u \sin 2(t-u)\,du$$

$$= \frac{1}{4}\int_0^t \left[\frac{\cos\,(4u-2t) - \cos\,2t}{2}\right] du$$

$$= \frac{1}{8}\left[\int_0^t \cos\,(4u-2t)du - \cos 2t\,\int_0^t du\right]$$

$$= \frac{1}{8}\left\{\left[\frac{1}{4}\sin(4u-2t)\right]_0^t - \cos 2t\,[u]_0^t\right\}$$

$$= \frac{1}{8}\left\{\frac{1}{4}[\sin 2t - \sin(-2t)] - \cos 2t\; t\right\}$$

$$= \frac{1}{8}\left\{\frac{1}{2}\sin 2t - t\cos 2t\right\}$$

$$\therefore \text{ (1) becomes, } y = \frac{1}{4}\left\{\frac{1}{2}\sin 2t - t\cos 2t\right\} + 3\cos 2t + 2\sin 2t$$

$$= \frac{17}{8}\sin 2t - \frac{t}{4}\cos 2t + 3\cos 2t.$$

Example 6: Solve $y'' + 4y' + 8y = \cos 2t$, given that $y = 2$ and $y' = 1$ when $t = 0$.

Solution: Given $y'' + 4y' + 8y = \cos 2t$, $y(0) = 2, y'(0) = 1$.

Apply L.T. on both sides, $L[y''] + 4L[y'] + 8L[y] = L[\cos 2t]$

$$[s^2L[y] - sy(0) - y'(0)] + 4[sL[y] - y(0)] + 8L[y] = \frac{s}{s^2+4}$$

$$[s^2L[y] - 2s - 1] + 4[sL[y] - 2] + 8L[y] = \frac{s}{s^2+4}$$

$$L[y](s^2 + 4s + 8) - 2s - 9 = \frac{s}{s^2+4}$$

$$L[y](s^2 + 4s + 8) = \frac{s}{s^2+4} + 2s + 9$$

$$L[y] = \frac{s}{(s^2+4)(s^2+4s+8)} + \frac{2s+9}{(s^2+4s+8)}$$

$$\therefore \quad y = L^{-1}\left[\frac{s}{(s^2+4)(s^2+4s+8)} + \frac{2s+9}{s^2+4s+8}\right] \qquad (1)$$

Consider $\frac{s}{(s^2+4)(s^2+4s+8)} = \frac{As+B}{s^2+4} + \frac{Cs+D}{s^2+4s+8}$

$\Rightarrow s = (As + B)(s^2 + 4s + 8) + (Cs + D)(s^2 + 4)$

Equating the coefficients, we get

$0 = A + C$

$0 = 4A + B + D$

$1 = 8A + 4B + 4C$

$0 = 8B + 4D$

Solving the above equations, we get

$A = \frac{1}{20}$, $B = \frac{1}{5}$, $C = \frac{-1}{20}$ and $D = \frac{-2}{5}$.

$$\therefore \text{ (1) gives, } \quad y = L^{-1}\left[\frac{\frac{1}{20}s + \frac{1}{5}}{s^2 + 4} + \frac{\frac{-1}{20}s - \frac{2}{5}}{s^2 + 4s + 8} + \frac{2s + 9}{s^2 + 4s + 8}\right]$$

$$= L^{-1}\left[\frac{\frac{1}{20}s+\frac{1}{5}}{s^2+4} + \frac{\frac{39}{20}s+\frac{43}{5}}{(s+2)^2+4}\right]$$

$$= \frac{1}{20}L^{-1}\left[\frac{s}{s^2+4}\right] + \frac{1}{5}L^{-1}\left[\frac{1}{s^2+4}\right] + L^{-1}\left[\frac{\frac{39}{20}(s+2) + \frac{47}{10}}{(s + 2)^2 + 4}\right]$$

$$= \frac{1}{20}\cos 2t + \frac{1}{5}.\frac{1}{2}\sin 2t + e^{-2t}L^{-1}\left[\frac{\frac{39}{20}s + \frac{47}{10}}{s^2 + 4}\right]$$

$$= \frac{1}{20}\cos 2t + \frac{1}{10}\sin 2t + e^{-2t}\left[\frac{39}{20}\cos 2t + \frac{47}{20}\sin 2t\right].$$

Example 7: Solve $(D^2 + D - 2)y = 3\cos 3t - 11\sin 3t,\ y(0) = 0$ and $y'(0) = 6$.

Solution: Given $y'' + y' - 2y = 3\cos 3t - 11\sin 3t$.

Apply L.T. on both sides, $L[y''] + L[y'] - 2L[y] = 3L[\cos 3t] - 11L[\sin 3t]$

$$[s^2L[y] - sy(0) - y'(0)] + [sL[y] - y(0)] - 2L[y] = \frac{3\,s}{s^2+9} - \frac{11*3}{s^2+9}$$

$$[s^2L[y] - 6] + sL[y] - 2L[y] = \frac{3s-33}{s^2+9}$$

$$L[y](s^2 + s - 2) = \frac{3s-33}{s^2+9} + 6$$

$$L[y](s-1)(s+2) = \frac{6s^2+3s+21}{s^2+9}$$

$$L[y] = \frac{6s^2+3s+21}{(s-1)(s+2)(s^2+9)}$$

$$\therefore \quad y = L^{-1}\left[\frac{6s^2+3s+21}{(s-1)(s+2)(s^2+9)}\right] \qquad (1)$$

By partial fractions, $\frac{6s^2+3s+21}{(s-1)(s+2)(s^2+9)} = \frac{A}{s-1} + \frac{B}{s+2} + \frac{Cs+D}{s^2+9}$

$$\Rightarrow 6s^2 + 3s + 21 = A(s+2)(s^2+9) + B(s-1)(s^2+9) + (Cs+D)(s-1)(s+2)$$

Putting $s = 1$, we get $30 = 30A \Rightarrow A = 1$.

Putting $s = -2$, we get $39 = -39B \Rightarrow B = -1$.

Equating constant terms, $21 = 18A - 9B - 2D$. Then $D = 3$.

Equating coefficient of s^3 on both sides,

$$0 = A + B + C \Rightarrow C = 0$$

$\therefore$ (1) gives, $y = L^{-1}\left[\frac{1}{s-1} + \frac{-1}{s+2} + \frac{0+6}{s^2+9}\right]$

$$= L^{-1}\left[\frac{1}{s-1}\right] - L^{-1}\left[\frac{-1}{s+2}\right] + L^{-1}\left[\frac{3}{s^2+9}\right]$$

$$= e^t - e^{-2t} + \sin 3t.$$

Example 8: Solve $y'' + 4y = 9t, y(0) = 0,\ y'(0) = 7$.

Solution: Apply L.T. on both sides,

$$L[y''] + 4L[y] = 9L[t]$$

$$s^2L[y] - s\,y(0) - y'(0) + 4L[y] = \frac{9}{s^2}$$

$$s^2L[y] - 7 + 4\,L[y] = \frac{9}{s^2}$$

$$L[y](s^2+4) = \frac{9}{s^2} + 7 = \frac{7s^2+9}{s^2}$$

$$L[y] = \frac{7(s^2+4)-19}{s^2(s^2+4)}$$

$$\therefore \quad y = L^{-1}\left[\frac{7(s^2+4)-19}{s^2(s^2+4)}\right]$$

$$= L^{-1}\left[\frac{7}{s^2} - \frac{19}{s^2(s^2+4)}\right]$$

$$= 7t - 19\int_0^t\int_0^t L^{-1}\left[\frac{1}{s^2+4}\right]dt\,dt$$

$$= 7t - \frac{19}{2}\int_0^t\int_0^t \sin 2t\,dt\,dt$$

$$= 7t - \frac{19}{2}\int_0^t\left[\frac{-\cos 2t}{2}\right]_0^t dt$$

$$= 7t - \frac{19}{4}\int_0^t(1-\cos 2t)dt$$

$$= 7t - \frac{19}{4}\left[t - \frac{\sin 2t}{2}\right]_0^t = 7t - \frac{19}{4}\left[t - \frac{\sin 2t}{2}\right]$$

$$= \frac{9}{4}\,t + \frac{19}{8}\sin 2t.$$

Example 9: Solve the differential equation $y'' - 3y' + 2y = 4t + e^{3t}$, where $y(0) = 1$ and $y'(0) = -1$, using Laplace transform. ***[AU, April/May 2015]***

Solution: Apply L.T. on both sides, $L[y''] - 3L[y'] + 2L[y] = 4L[t] + L[e^{3t}]$

$$\Rightarrow \quad [s^2L[y] - sy(0) - y'(0)] - 3[sL[y] - y(0)] + 2L[y] = \frac{4}{s^2} + \frac{1}{s-3}$$

$$\Rightarrow \quad [s^2L[y] - s + 1] - 3[sL[y] - 1] + 2L[y] = \frac{4}{s^2} + \frac{1}{s-3}$$

$$\Rightarrow \quad L[y](s^2 - 3s + 2) = \frac{4}{s^2} + \frac{1}{s-3} + s - 4$$

$$\Rightarrow \quad L[y](s-1)(s-2) = \frac{4(s-3) + s^2 + (s-4)\,s^2(s-3)}{s^2\,(s-3)}$$

$$\therefore \quad L[y] = \frac{4(s-3) + s^2 + (s-4)s^2\,(s-3)}{s^2(s-3)(s-1)(s-2)} \qquad (1)$$

By partial fractions,

$$\frac{4(s-3)+s^2+(s-4)s^2(s-3)}{(s-1)(s-2)(s-3)s^2} = \frac{A}{s-1} + \frac{B}{s-2} + \frac{C}{s-3} + \frac{D}{s} + \frac{E}{s^2}$$

$\Rightarrow 4(s-3) + s^2 + (s-4)s^2(s-3) = A(s-2)(s-3)s^2 + B(s-1)(s-3)s^2 +$

$C(s-1)(s-2)s^2 + D(s-1)(s-2)(s-3)s + E(s-1)(s-2)(s-3) \quad \ldots (2)$

Put $s = 1$ in (2) we get $A = \frac{-1}{2}$.

Put $s = 2$ in (2) we get $B = -2$.

Put $s = 3$ in (2) we get $C = \frac{1}{2}$.

Put $s = 0$ in (2) we get $E = 2$.

Equating coefficient of s^4, $1 = A + B + C + D \Rightarrow D = 3$

$\therefore$ (1) gives $L[y] = \frac{-1}{2(s-1)} - \frac{2}{s-2} + \frac{1}{2(s-3)} + \frac{3}{s} + \frac{2}{s^2}$

$$\therefore \quad y = L^{-1}\left[\frac{-1}{2(s-1)} - \frac{2}{s-2} + \frac{1}{2(s-3)} + \frac{3}{s} + \frac{2}{s^2}\right]$$

$$= \frac{-1}{2}L^{-1}\left[\frac{1}{s-1}\right] - 2L^{-1}\left[\frac{1}{s-2}\right] + \frac{1}{2}L^{-1}\left[\frac{1}{s-3}\right] + 3L^{-1}\left[\frac{1}{s}\right] + 2L^{-1}\left[\frac{1}{s^2}\right]$$

$$= \frac{-1}{2}e^t - 2e^{2t} + \frac{1}{2}e^{3t} + 3 + 2t.$$

Example 10: Using Laplace transforms technique solve $y'' + y' = t^2 + 2t$, given $y = 4, y' = -2$ when $t = 0$. ***[AU, Nov/Dec 2017, May/June 2016]***

Solution: Apply L.T. on both sides, $L[y''] + L[y'] = L[t^2] + L[2t]$

$$\Rightarrow \quad [s^2L[y] - sy(0) - y'(0)] + [sL[y] - y(0)] = \frac{2}{s^3} + \frac{2}{s^2}$$

$$\Rightarrow \quad [s^2L[y] - 4s + 2] + [sL[y] - 4] = \frac{2}{s^3} + \frac{2}{s^2}$$

$$\Rightarrow \quad L[y](s^2 + s) - 4s - 2 = \frac{2}{s^3} + \frac{2}{s^2}$$

$$\Rightarrow \quad L[y]\, s\,(s+1) = \frac{2}{s^3} + \frac{2}{s^2} + 4s + 2$$

$$\Rightarrow \quad L[y] = \frac{2 + 2s + 4s^4 + 2s^3}{s^3}$$

$$\therefore \quad L[y] = \frac{4s^4 + 2s^3 + 2s + 2}{s^4(s+1)}$$

$$y = L^{-1}\left[\frac{4s^4 + 2s^3 + 2s + 2}{s^4\,(s+1)}\right] \qquad (1)$$

By partial fractions,

$$\frac{4s^4+2s^3+2s+2}{s^4(s+1)} = \frac{A}{s} + \frac{B}{s^2} + \frac{C}{s^3} + \frac{D}{s^4} + \frac{E}{s+1}$$

$$\Rightarrow\ 4s^4 + 2s^3 + 2s + 2 = As^3(s+1) + Bs^2(s+1) + Cs(s+1) + D(s+1) + Es^4 \quad (2)$$

Put $s = 0$ in (2) we get $D = 2$.

Put $s = -1$ in (2) we get $E = 2$.

Equating coefficient of s^4, $4 = A + E \Rightarrow A = 2$

Equating coefficient of s^3, $2 = A + B \Rightarrow B = 0$

Equating coefficient of s^2, $0 = B + C \Rightarrow C = 0$

$\therefore$ (1) gives $y = L^{-1}\left[\frac{2}{s} + \frac{0}{s^2} + \frac{0}{s^3} + \frac{2}{s^4} + \frac{2}{s+1}\right]$

$$= 2L^{-1}\left[\frac{1}{s}\right] + 2L^{-1}\left[\frac{1}{s^4}\right] + 2L^{-1}\left[\frac{1}{s+1}\right]$$

$$= 2 + 2\frac{t^3}{6} + 2e^{-t} \quad = \frac{1}{3}t^3 + 2 + 2e^{-t}.$$

Example 11: Solve $y'' + 6y' + 9y = e^{-3t}t^2$, given $y = y' = 0$ at $t = 0$.

Solution:

Apply L.T. on both sides, $[y''] + 6L[y'] + 9L[y] = L[e^{-3t}t^2]$

$$[s^2L[y] - sy(0) - y'(0)] + 6[sL[y] - y(0)] + 9L[y] = L[t^2]_{s\to s+3}$$

$$s^2L[y] + 6sL[y] + 9L[y] = \left[\frac{2}{s^3}\right]_{s\to s+3}$$

$$L[y](s^2 + 6s + 9) = \frac{2}{(s+3)^3}$$

$$L[y](s+3)^2 = \frac{2}{(s+3)^3}$$

$$L[y] = \frac{2}{(s+3)^5}$$

$$\therefore \qquad y = L^{-1}\left[\frac{2}{(s+3)^5}\right] = e^{-3t}L^{-1}\left[\frac{2}{s^5}\right] = 2e^{-3t}\frac{t^4}{4!} = \frac{1}{12}\,t^4\,e^{-3t}.$$

Example 12: Solve $(D^2+1)y = t\cos 2t$, given $y = y' = 0$ when $t = 0$.

Solution: Apply L.T. on both sides, $L[y''] + L[y] = L[t\cos 2t]$

$$[s^2L[y] - sy(0) - y'(0)] + L[y] = \frac{-d}{ds}L[\cos 2t]$$

$$s^2L[y] + L[y] = \frac{-d}{ds}\left(\frac{s}{s^2+4}\right)$$

$$L[y](s^2+1) = \frac{s^2-4}{\left(s^2+4\right)^2}$$

$$L[y] = \frac{s^2-4}{\left(s^2+1\right)\left(s^2+4\right)^2}$$

$$\therefore \qquad y = L^{-1}\left[\frac{s^2-4}{(s^2+1)(s^2+4)^2}\right] \qquad (1)$$

By partial fractions,

$$\frac{s^2-4}{(s^2+1)(s^2+4)^2} = \frac{(u-4)}{(u+1)(u+4)^2} \quad [\text{putting } s^2 = u]$$

$$= \frac{A}{u+1} + \frac{B}{u+4} + \frac{C}{(u+4)^2} \qquad (2)$$

$$\Rightarrow \quad u-4 = A(u+4)^2 + B(u+1)(u+4) + C(u+1)$$

Put $u=-1$, we get $-5A = 9A \Rightarrow A = \frac{-5}{9}$.

Put $u=-4$, we get $-8 = -3C \Rightarrow C = \frac{8}{3}$.

Equating coefficient of u^2, $0 = A + B \Rightarrow B = \frac{5}{9}$.

$$\therefore \text{ (2) gives } \frac{s^2-4}{(s^2+1)(s^2+4)^2} = \frac{-5/9}{u+1} + \frac{5/9}{u+4} + \frac{8/3}{(u+4)^2}$$

$$= \frac{-5}{9}\frac{1}{s^2+1} + \frac{5}{9}\frac{1}{s^2+4} + \frac{8}{3}\frac{1}{(s^2+4)^2}.$$

Now (1) gives, $y = L^{-1}\left[\frac{-5}{9}\frac{1}{s^2+1} + \frac{5}{9}\frac{1}{s^2+4} + \frac{8}{3}\frac{1}{\left(s^2+4\right)^2}\right]$

$$= \frac{-5}{9}L^{-1}\left[\frac{1}{s^2+1}\right] + \frac{5}{9}L^{-1}\left[\frac{1}{s^2+4}\right] + \frac{8}{3}L^{-1}\left[\frac{1}{(s^2+4)^2}\right]$$

$$= \frac{-5}{9}\sin t + \frac{5}{9}.\frac{1}{2}\sin 2t + \frac{8}{3}L^{-1}\left[\frac{1}{(s^2+4)^2}\right] \qquad (3)$$

Now, $L^{-1}\left[\frac{1}{(s^2+4)^2}\right] = \frac{1}{8}L^{-1}\left[\frac{(s^2+4)-(s^2-4)}{(s^2+4)^2}\right]$

$$= \frac{1}{8}\left\{L^{-1}\left[\frac{1}{s^2+4}\right] - L^{-1}\left[\frac{s^2-4}{(s^2+4)^2}\right]\right\}$$

$$= \frac{1}{8}\left\{\frac{1}{2}\sin 2t - t\cos 2t\right\}$$

$\therefore$ (3) gives, $\quad y = \frac{-5}{9}\sin t + \frac{5}{18}\sin 2t + \frac{8}{3}.\frac{1}{8}\left[\frac{1}{2}\sin 2t - t\cos 2t\right]$

$$= \frac{-5}{9}\sin t + \sin 2t\left(\frac{5}{18}+\frac{1}{6}\right) - \frac{t}{3}\cos 2t$$

$$= \frac{-5}{9}\sin t + \frac{4}{9}\sin 2t - \frac{t}{3}\cos 2t.$$

Example 13: Solve $(D^2+9)y = \cos 2t, y(0) = 1$ and $y\left(\frac{\pi}{2}\right) = -1$.

Solution: Since y' is not given, it will be assumed as a constant, which will be evaluated using the condition $y\left(\frac{\pi}{2}\right) = -1$.

Given $y'' + 9y = \cos 2t$.

Apply L.T. on both sides,

$$[s^2L[y] - sy(0) - y'(0)] + 9L[y] = \frac{s}{s^2+4}$$

$$s^2L[y] - s - k + 9L[y] = \frac{s}{s^2+4} \quad \text{[assume that } y'(0) = k]$$

$$L[y](s^2+9) = \frac{s}{s^2+4} + s + k$$

$$L[y] = \frac{s}{(s^2+4)(s^2+9)} + \frac{s}{s^2+9} + \frac{k}{s^2+9}$$

$$= \frac{1}{5}\left[\frac{s}{s^2+4} - \frac{s}{s^2+9}\right] + \frac{s}{s^2+9} + \frac{k}{s^2+9}$$

$\therefore$ $$y = L^{-1}\left[\frac{1}{5}\frac{s}{s^2+4} + \frac{4}{5}\frac{s}{s^2+9} + \frac{k}{s^2+9}\right]$$

$$= \frac{1}{5}\cos 2t + \frac{4}{5}\cos 3t + \frac{k}{3}\sin 3t. \qquad (1)$$

Put $t = \frac{\pi}{2}$ in (1), we get $y\left(\frac{\pi}{2}\right) = \frac{1}{5}(-1) + \frac{4}{5}(0) + \frac{k}{3}(-1)$

i.e., $$-1 = \frac{-1}{5} - \frac{k}{3}\left[\because\ y\left(\frac{\pi}{2}\right) = -1\right]$$

$$\frac{k}{3} = \frac{4}{5}. \text{ i.e, } k = \frac{12}{5}.$$

$\therefore$ (1) gives, $\quad y = \frac{1}{5}\cos 2t + \frac{4}{5}\cos 3t + \frac{4}{5}\sin 3t.$

Example 14: Solve $y'' + 4y = f(t)$ with $y(0) = 0, y'(0) = 1$ and

$f(t) = \begin{cases} 1 & \text{for } 0 < t < 1 \\ 0 & \text{for } t > 1 \end{cases}.$

Solution: Apply L.T. on both sides.

$$[s^2L[y] - sy(0) - y'(0)] + 4L[y] = L[f(t)] = \int_0^\infty e^{-st} f(t)dt$$

$$s^2L[y] - 1 + 4L[y] = \int_0^1 e^{-st}\, dt$$

$$L[y](s^2 + 4) - 1 = \left[\frac{e^{-st}}{-s}\right]_0^1 = \frac{1 - e^{-s}}{s}$$

$$L[y](s^2 + 4) = \frac{1-e^{-s}}{s} + 1$$

$$L[y] = \frac{1-e^{-s}}{s(s^2+4)} + \frac{1}{s^2+4}$$

$$= \frac{1}{s^2+4} + \frac{1}{s(s^2+4)} - \frac{e^{-s}}{s(s^2+4)}$$

$$\therefore\ y = L^{-1}\left[\frac{1}{s^2+4}\right] + L^{-1}\left[\frac{1}{s(s^2+4)}\right] - L^{-1}\left[\frac{e^{-s}}{s(s^2+4)}\right]$$

$$= \frac{1}{2}\sin 2t + \int_0^t L^{-1}\left[\frac{1}{s^2+4}\right] dt - f(t-1)u_1(t), \text{where } f(t) = L^{-1}\left[\frac{1}{s(s^2+4)}\right]$$

$$= \frac{1}{2}\sin 2t + \int_0^t \frac{1}{2}\sin 2t\, dt - f(t-1)u_1(t)$$

$$= \frac{1}{2}\sin 2t + \frac{1}{2}\left[\frac{-\cos 2t}{2}\right]_0^t - f(t-1)u_1(t)$$

$$= \frac{1}{2}\sin 2t + \frac{1}{4}(1 - \cos 2t) - \frac{1}{4}(1 - \cos 2(t-1))u_1(t).$$

EXERCISES

1. Solve $y'' + 4y' + 5y = 0$ given $y(0) = 0, y'(0) = 1$.
2. Solve $y'' + 6y' + 8y = 2$ given $y(0) = 1, y'(0) = 1$.
3. Solve $\frac{d^2y}{dx^2} + 4\frac{dy}{dx} + 8y = 1$, given $y(0) = 0, y'(0) = 1$.
4. Solve $(D^2 - 4D + 3)y = e^{-t}$ given $y(0) = 1$ and $y'(0) = 0$.
5. Solve $y'' - 4y' + 8y = e^{2t}$, given $y(0) = 2$ and $y'(0) = -2$.
6. Solve $y'' + 4y' + 3y = 10 \sin t$ given $y = y' = 0$ when $t = 0$.

7. Solve $x'' + 4x' + 4x = \sin t$if $x' = 0$ and $x = 2$ when $t = 0$.
8. Solve $y'' + 5y' + 4y = t^2$, if $y'(0) = y(0) = 1$.
9. Solve $y'' - 8y' + 15y = 9te^{2t}, y(0) = 5, y'(0) = 10$.
10. Solve $y'' + 2y' + 5y = e^{-t}\sin t, y(0) = 0, y'(0) = 1$.
11. Solve$(D^2 + 4D + 13)y = e^{-t}\sin t, y = 0 = Dy$ at $t = 0$.
12. Solve $y'' + 9y = \sin t$, given $y(0) = 1, y\left(\frac{\pi}{2}\right) = -1$.
13. Solve $x'' + 9x = \cos 4t$, given $x(0) = 1, x\left(\frac{\pi}{2}\right) = -1$.

ANSWERS

1. $y = e^{-2t}\sin t$
2. $y = \frac{1}{4} + 2e^{-2t} - \frac{5}{4}e^{-4t}$
3. $y = \frac{1}{8}(1 - e^{-2x}\cos 2x + 3e^{-2x}\sin 2x)$
4. $y = \frac{5}{4}e^{t} + \frac{1}{8}e^{-t} - \frac{3}{8}e^{3t}$
5. $y = \frac{1}{4}e^{2t}(1 + 7\cos 2t - 12\sin 2t)$
6. $y = \frac{5}{2}e^{-t} - \frac{1}{2}e^{-3t} - 2\cos t + \sin t$
7. $x = \frac{1}{25}[54e^{-2t} + 105te^{-2t} - 4\cos t + 3\sin t]$
8. $y = \frac{21}{32} - \frac{5}{8}t + \frac{1}{4}t^2 + e^{-t} - \frac{21}{32}e^{-4t}$
9. $y = 4e^{2t} + 3te^{2t} + 3e^{3t} - 2e^{5t}$
10. $y = \frac{e^{-t}}{3}[\sin t + \sin 2t]$
11. $y = \frac{1}{85}\left[e^{-t}(-2\cos t + 9\sin t) + e^{-2t}\left(2\cos 3t - \frac{7}{3}\sin 3t\right)\right]$
12. $y = \frac{9}{8}\sin 3t + \frac{1}{8}\sin t + \cos 3t$
13. $x = \frac{8}{7}\cos 3t - \frac{1}{7}\cos 4t + \frac{3}{4}\sin 3t$.

5.14. Short Answer Questions

1. State the sufficiency conditions for the existence of Laplace transform.

Solution: [AU, April/May 2017, April/May 2015]

The existence conditions of Laplace transform are

(i) piecewise continuous in every finite interval

(ii) it is of exponential order.

2. Give two examples for a function for which Laplace transform does not exist.

Solution: The Laplace transforms of $\tan t$ and e^{t^2} do not exit.

3. Find the Laplace transform of $t^2 + \sin 2t$.

Solution: $L[t^2 + \sin 2t] = L[t^2] + L[\sin 2t] = \frac{2}{s^3} + \frac{2}{s^2+4}$.

4. Find $L[f(t)]$, if $f(t) = \begin{cases} 0, & \text{for } 0 < t < 1 \\ 1, & \text{for } 1 < t < 2 \\ 0, & \text{for } t > 2 \end{cases}$.

Solution: $L[f(t)] = \int_0^\infty e^{-st} f(t)\, dt$

$$= \int_1^2 e^{-st}\, dt = \left[\frac{e^{-st}}{-s}\right]_1^2$$

$$= \left[\frac{-e^{-2s}+e^{-s}}{s}\right] = \frac{1}{s}(e^{-s} - e^{-2s}).$$

5. Find $L[f(t)]$, if $f(t) = \begin{cases} e^{2t}, & \text{for } 0 < t < 1 \\ 0, & \text{for } t > 1 \end{cases}$.

Solution: $L[f(t)] = \int_0^\infty e^{-st} f(t)\, dt$

$$= \int_0^1 e^{-st}\, e^{2t} dt = \int_0^1 e^{(2-s)t}\, dt$$

$$= \left[\frac{e^{(2-s)\,t}}{2-s}\right]_0^1 = \frac{e^{2-s}-1}{2-s}.$$

6. Find $L[f(t)]$ if $f(t) = \begin{cases} e^{-t}, & 0 < t < 4 \\ 0, & t > 4 \end{cases}$ **[AU, *Nov/Dec 2017*]**

Solution: $L[f(t)] = \int_0^\infty e^{-st} f(t)\, dt$

$$= \int_0^4 e^{-st}\, e^{-t} dt = \int_0^4 e^{-(s+1)t}\, dt$$

$$= \left[\frac{e^{-(s+1)t}}{-(s+1)}\right]_0^4 = \frac{e^{-4(s+1)}-1}{-(s+1)} = \frac{1-e^{-4(s+1)}}{s+1}.$$

7. Find $\left[\sqrt{e^{3(t+2)}}\right]$.

Solution: $L\left[\sqrt{e^{3(t+2)}}\right] = L[\sqrt{e^{3t}}e^{3}]$

$$= e^3\, L\left[e^{\frac{3}{2}t}\right]$$

$$= e^3\, \frac{1}{s-\frac{3}{2}} = \frac{2e^3}{2s-3}.$$

8. Find $L[\sin^3 2t]$.

Solution: $L[sin^3 2t] = \frac{3}{4}L[\sin 2t] - \frac{1}{4}L[\sin 6t][\because \sin 3\theta = 3\sin\theta - 4\sin^3\theta]$

$$= \frac{3}{4}.\frac{2}{s^2+4} - \frac{1}{4}.\frac{6}{s^2+36}$$

$$= \frac{3}{2}.\frac{1}{s^2+4} - \frac{3}{2}.\frac{1}{s^2+36}.$$

9. Find $L[\sqrt{t}]$and $L\left[\frac{1}{\sqrt{\pi t}}\right]$.

Solution: (i) $L[\sqrt{t}] \;=\; L\left[t^{1/2}\right] = \frac{\Gamma\left(\frac{1}{2}+1\right)}{s^{\frac{1}{2}+1}}$

$$= \frac{\frac{1}{2}\Gamma\left(\frac{1}{2}\right)}{s^{\frac{3}{2}}} \quad [\because\ \Gamma(n+1) = n\,\Gamma(n)]$$

$$= \frac{\sqrt{\pi}}{2s^{3/2}}. \quad \left[\because\ \Gamma\left(\frac{1}{2}\right) = \sqrt{\pi}\right]$$

(ii) $L\left[\frac{1}{\sqrt{\pi t}}\right] = \frac{1}{\sqrt{\pi}}\ L\left[t^{-1/2}\right]$

$$= \frac{1}{\sqrt{\pi}}\frac{\Gamma\left(-\frac{1}{2}+1\right)}{s^{-\frac{1}{2}+1}} = \frac{1}{\sqrt{\pi}}\frac{\Gamma\left(\frac{1}{2}\right)}{s^{\frac{1}{2}}}$$

$$= \frac{1}{\sqrt{\pi}}\frac{\sqrt{\pi}}{s^{1/2}} = \ s^{-1/2}.$$

10. Find $L[\cos(at+b)]$.

Solution: $L[\cos(at+b)] \;=\; L[\cos at.\cos b - \sin at \sin b]$

$$= \cos b\, L[\cos at] - \sin b L[\sin at]$$

$$= \cos b \frac{s}{s^2+a^2} - \sin b \frac{a}{s^2+a^2}.$$

11. Find $L[\sin^2 3t]$.

Solution: $L[\sin^2 3t] \;=\; L\left[\frac{1-\cos 6t}{2}\right]\left[\because\ \sin^2 x = \frac{1-\cos 2x}{2}\right]$

$$= \frac{1}{2}(L\,[1] - L[\cos 6t]) \quad = \frac{1}{2}\left(\frac{1}{s} - \frac{s}{s^2+36}\right).$$

12. Find the Laplace transform of $\sin 3t \sin 5t$.

Solution: $L[\sin 5t \sin 3t] = L\left[\frac{\cos 2t - \cos 8t}{2}\right]$

$$= \frac{1}{2}(L[\cos 2t] - L[\cos 8t])$$

$$= \frac{1}{2}\left(\frac{s}{s^2+4} - \frac{s}{s^2+64}\right).$$

13. Find $L[e^{-at}\cos bt]$.

Solution: $L[e^{-at}\cos bt] = L[\cos bt]_{s\to s+a}$

$$= \left[\frac{s}{s^2+b^2}\right]_{s\to s+a} \;=\; \frac{s+a}{(s+a)^2+b^2}.$$

14. Find the Laplace transform of $e^{-t}\sin 2t$. **[AU, Nov/Dec 2014]**

Solution: $L[e^{-t}\sin 2t] = L[\sin 2t]_{s\to s+1}$

$$= \left[\frac{2}{s^2+2^2}\right]_{s\to s+1}$$

$$= \frac{2}{(s+1)^2+4} = \frac{2}{s^2+2s+5}.$$

15. Evaluate $L\,[e^{-2t}(t+\sin t)]$.

Solution: $L[e^{-2t}(t+\sin t)] = L[t+\sin t]_{s\to s+2}$

$$= \left[\frac{1}{s^2}+\frac{1}{s^2+1}\right]_{s\to s+2}$$

$$= \frac{1}{(s+2)^2}+\frac{1}{(s+2)^2+1} = \frac{1}{(s+2)^2}+\frac{1}{s^2+2s+5}.$$

16. Find the Laplace transform of t^3e^{2t} .

Solution: $L[t^3e^{2t}] = L[e^{2t}t^3] = L[t^3]_{s\to s-2}$

$$= \left[\frac{3!}{s^4}\right]_{s\to s-2} = \frac{6}{(s-2)^4}\,.$$

17. Find the Laplace transform of $t\cosh at$.

Solution: $L[t\ \cosh at] = \frac{-d}{ds}L[\cosh at]$

$$= \frac{-d}{ds}\left[\frac{s}{s^2-a^2}\right] = -\left[\frac{(s^2-a^2)-\ s.2s}{(s^2-a^2)^2}\right]$$

$$= -\left[\frac{-s^2-a^2}{(s^2-a^2)^2}\right] = \frac{s^2+a^2}{(s^2-a^2)^2}.$$

18. Find $L\left[\frac{\sin at}{t}\right]$.

Solution: $L\left[\frac{\sin at}{t}\right] = \int_s^\infty L[\sin at]ds \left[\because L\left[\frac{f(t)}{t}\right] = \int_s^\infty L[f(t)]\,ds\right]$

$$= \int_s^\infty \frac{a}{s^2+a^2}\,ds = a\int_s^\infty \frac{1}{s^2+a^2}\,ds$$

$$= a\left[\frac{1}{a}\tan^{-1}\frac{s}{a}\right]_s^\infty = \tan^{-1}\infty - \tan\frac{s}{a}$$

$$= \frac{\pi}{2} - \tan^{-1}\frac{s}{a} = \cot^{-1}\frac{s}{a} = \tan^{-1}\frac{a}{s}.$$

19. Find$L\left[\frac{\sin t}{t}\right]$. **[AU, May/June 2014, Nov/Dec 2017]**

Solution: $L\left[\frac{\sin t}{t}\right] = \int_s^\infty L[\sin t]ds \left[\because L\left[\frac{f(t)}{t}\right] = \int_s^\infty L[f(t)]\, ds\right]$

$$= \int_s^\infty \frac{1}{s^2+1^2}\, ds = \int_s^\infty \frac{1}{s^2+1}\, ds$$

$$= [\tan^{-1} s]_s^\infty = \tan^{-1}\infty - \tan s$$

$$= \frac{\pi}{2} - \tan^{-1} s = \cot^{-1} s = \tan^{-1}\frac{1}{s}.$$

20. Evaluate $L\left[\frac{1-e^t}{t}\right]$.

Solution: $L\left[\frac{1-e^t}{t}\right] = \int_s^\infty L\,[1-e^t]\, ds$

$$= \int_s^\infty \left[\frac{1}{s} - \frac{1}{s-1}\right] ds = [\log s - \log(s-1)]_s^\infty$$

$$= \left[\log\frac{s}{s-1}\right] = \left[\log\left(\frac{1}{1-1/s}\right)\right]$$

$$= \log 1 - \log\left(\frac{1}{1-\frac{1}{s}}\right)$$

$$= 0 - \log\left(\frac{s}{s-1}\right) \quad = \log\frac{s-1}{s}.$$

21. Find the Laplace transform of $\frac{e^{-t} - e^{-3t}}{t}$.

Solution: $L\left[\frac{e^{-t} - e^{-3t}}{t}\right] = \int_s^\infty L[e^{-t} - e^{-3t}]\, ds$

$$= \int_s^\infty \left[\frac{1}{s+1} - \frac{1}{s+3}\right] ds$$

$$= [\log(s+1) - \log(s+3)]_s^\infty$$

$$= \left[\log\left(\frac{s+1}{s+3}\right)\right]_s^\infty = \left[\log\left(\frac{1+1/s}{1+3/s}\right)\right]_s^\infty$$

$$= \log 1 - \log\frac{1+{}^1\!/_s}{1+{}^3\!/_s} = 0 - \log\frac{s+1}{s+3} = \log\frac{s+3}{s+1}.$$

22. Evaluate $L\left[\int_0^t \frac{\sin t}{t} dt\right]$.

Solution: $L\left[\int_0^t \frac{\sin t}{t} dt\right] = \frac{1}{s} L\left[\frac{\sin t}{t}\right]$

$$= \frac{1}{s}\int_s^\infty L[\sin t] ds$$

$$= \frac{1}{s}\int_s^\infty \frac{1}{s^2+1}\, ds = \frac{1}{s}[\tan^{-1} s]_s^\infty$$

$$= \frac{1}{s}\left[\frac{\pi}{2} - \tan^{-1} s\right] = \frac{1}{s}\cos^{-1} \text{s}.$$

23. Evaluate $\int_0^\infty \frac{\sin t}{t} dt$.

Solution: $\int_0^\infty \frac{\sin t}{t} dt \quad = \int_0^\infty e^{-st} \frac{\sin t}{t} dt, s = 0$

$$= L\left[\frac{\sin t}{t}\right]_{s=0}$$

$$= \left[\int_s^\infty L[\sin t] ds\right]_{s=0} = \left[\int_s^\infty \frac{1}{s^2+1} ds\right]_{s=0}$$

$$= [\tan^{-1} s]_s^\infty \ ; \quad s = 0$$

$$= \left[\frac{\pi}{2} - \tan^{-1} s\right]_{s=0}$$

$$= \frac{\pi}{2} - \tan^{-1} 0 = \frac{\pi}{2}.$$

24. Evaluate $\int_0^\infty e^{-2t} t \sin t \, dt$ using Laplace transform. ***[AU, April/May 2015]***

Solution: $\int_0^\infty e^{-2t} \, t \sin t \, dt \; = \left[\int_0^\infty e^{-st} \, t \sin t \, dt\right]_{s=2}$

$$= L[t \sin t]_{s=2}$$

$$= \left[\frac{2s}{(s^2+1)^2}\right]_{s=2} \quad = \frac{4}{25}.$$

25. State the initial value theorem.

Solution: $\lim_{t\to 0} f(t) = \lim_{s\to\infty} s\,F(s)$.

26. State final value theorem on Laplace transform. ***[AU, Nov/Dec2015]***

Solution: $\lim_{t\to\infty} f(t) = \lim_{s\to 0} s\,F(s)$.

27. If $L[f(t)] = \frac{s+3}{(s+1)(s+2)}$ find $\lim_{t\to 0} f(t)$.

Solution: By initial value theorem,

$$\lim_{t\to 0} f(t) \; = \lim_{s\to\infty} s\,F(s).$$

$$= \lim_{s\to\infty} \frac{s(s+3)}{(s+1)(s+2)}$$

$$= \lim_{s\to\infty} \frac{(1+{}^3/_s)}{(1+{}^1/_s)(1+{}^2/_s)} \; = \; 1.$$

28. If $L[f(t)] = \frac{4}{(s-3)(s^2+4)}$, find $\lim_{t\to\infty} f(t)$.

Solution: By final value theorem,

$$\lim_{t\to\infty} f(t) = \lim_{s\to 0} s\,F(s).$$

$$= \lim_{s\to 0} \frac{4s}{(s-3)(s^2+4)} = 0.$$

29. Find $f(\infty)$, if $L[f(t)] = \frac{1}{s(s+\alpha)}$. **[AU, *Nov/Dec 2017*]**

Solution: We know, $\lim_{t\to\infty} f(t) = \lim_{s\to 0} s\,F(s)$.

Given, $L[f(t)] = F(s) = \frac{1}{s(s+\alpha)}$.

$$\therefore\ f(\infty) = \lim_{t\to\infty} f(t) = \lim_{s\to 0} s\frac{1}{s(s+\alpha)} = \lim_{s\to 0} \frac{1}{(s+\alpha)} = \frac{1}{\alpha}.$$

30. Verify the final value theorem in Laplace transform for the function$1-e^{-at}$.

Solution: Let $f(t) = 1-e^{-at}$. Then $[f(t)] = F(s) = \frac{1}{s} - \frac{1}{s+a}$.

Now $\lim_{t\to\infty} f(t) = 1$ and $\lim_{s\to 0} s\,F(s) = \lim_{s\to 0}\left(1 - \frac{s}{s+a}\right) = 1.$

$$\therefore\ \lim_{t\to\infty} f(t) = \lim_{s\to 0} s\,F(s).$$

Hence final value theorem is verified.

31. State convolution theorem on Laplace transforms. **[*AU, May/June 2016*]**

Solution: If $f(t)$and $g(t)$ are functions defined for $t \geq 0$, then

$$L[f(t) * g(t)] = L[f(t)].L[g(t)].$$

32. Find $L^{-1}\left[\frac{s^2+2s+3}{s^3}\right]$. **[*AU, April/May 2017*]**

Solution : $L^{-1}\left[\frac{s^2+2s+3}{s^3}\right] = L^{-1}\left[\frac{1}{s} + \frac{2}{s^2} + \frac{3}{s^3}\right]$

$$= 1 + 2t + \frac{3}{2}t^2.$$

33. Find $L^{-1}\left[\frac{1}{(s+1)^{3/2}}\right]$.

Solution: $L^{-1}\left[\frac{1}{(s+1)^{3/2}}\right] = e^{-t}L^{-1}\left[\frac{1}{s^{3/2}}\right]$

$$= e^{-t}\frac{t^{\frac{3}{2}-1}}{\Gamma\left(\frac{3}{2}\right)} = e^{-t}\frac{t^{1/2}}{\frac{1}{2}\Gamma\left(\frac{1}{2}\right)} = \frac{2e^{-t}t^{1/2}}{\sqrt{\pi}}.$$

34. Find $L^{-1}\left[\frac{1}{(2s-3)^4}\right]$.

Solution: $L^{-1}\left[\frac{1}{(2s-3)^4}\right] = \frac{1}{2^4}L^{-1}\left[\frac{1}{\left(s-\frac{3}{2}\right)^4}\right]$

$$= \frac{1}{16}e^{-\frac{3}{2}t}L^{-1}\left[\frac{1}{s^4}\right]$$

$$= \frac{1}{16}e^{-\frac{3}{2}t}\frac{t^3}{3!} = \frac{1}{96}e^{-\frac{3}{2}t}t^3.$$

35. Find $f(t)$ if the Laplace transform of $f(t)$ is $\frac{s}{(s-2)^2}$. ***[AU, Nov/Dec 2014]***

Solution: $L^{-1}\left[\frac{s}{(s-2)^2}\right] = L^{-1}\left[\frac{s-2+2}{(s-2)^2}\right]$

$$= L^{-1}\left[\frac{1}{s-2} + \frac{2}{(s-2)^2}\right]$$

$$= L^{-1}\left[\frac{1}{s-2}\right] + L^{-1}\left[\frac{2}{(s-2)^2}\right]$$

$$= e^{2t} + 2e^{2t}L^{-1}\left[\frac{1}{s^2}\right]$$

$$= e^{2t} + 2te^{2t}.$$

36. Find $L^{-1}\left[\frac{2s+3}{s^2-4s+13}\right]$.

Solution: $L^{-1}\left[\frac{2s+3}{s^2-4s+13}\right] = L^{-1}\left[\frac{2s+3}{(s-2)^2+9}\right]$

$$= L^{-1}\left[\frac{2(s-2)+7}{(s-2)^2+9}\right] = e^{2t}L^{-1}\left[\frac{2s+7}{s^2+9}\right]$$

$$= e^{2t}\left\{2L^{-1}\left[\frac{s}{s^2+9}\right] + 7L^{-1}\left[\frac{1}{s^2+9}\right]\right\}$$

$$= e^{2t}\left\{2\cos 3t + \frac{7}{3}\sin 3t\right\}.$$

37. Evaluate $L^{-1}\left[\frac{s+2}{s^2+4s+8}\right]$. ***[AU, Nov/ Dec 2015]***

Solution: $L^{-1}\left[\frac{s+2}{s^2+4s+8}\right] = L^{-1}\left[\frac{s+2}{(s+2)^2+4}\right]$

$$= e^{-2t}L^{-1}\left[\frac{s}{s^2+4}\right] = e^{-2t}\cos 2t.$$

38. Evaluate the inverse Laplace transform of $\frac{2s-1}{s^2+4s+5}$.

Solution: $L^{-1}\left[\frac{2s-1}{s^2+4s+5}\right] = L^{-1}\left[\frac{2(s+2)-5}{(s+2)^2+1}\right] = e^{-2t}L^{-1}\left[\frac{2s-5}{s^2+1}\right]$

$$= e^{-2t}\left\{2L^{-1}\left[\frac{s}{s^2+1}\right] - 5L^{-1}\left[\frac{1}{s^2+1}\right]\right\}$$

$$= e^{-2t}\{2\cos t - 5\sin t\}.$$

39. Evaluate $L^{-1}\left[\frac{s}{s^2+4s+5}\right]$. ***[AU, May/June 2016]***

Solution: $L^{-1}\left[\frac{s}{s^2+4s+5}\right] = L^{-1}\left[\frac{s+2-2}{(s+2)^2+1}\right]$

$$= L^{-1}\left[\frac{s+2}{(s+2)^2+1} + \frac{-2}{(s+2)^2+1}\right]$$

$$= L^{-1}\left[\frac{s+2}{(s+2)^2+1}\right] + L^{-1}\left[\frac{-2}{(s+2)^2+1}\right]$$

$$= e^{-2t}L^{-1}\left[\frac{s}{s^2+1}\right] - 2e^{-2t}L^{-1}\left[\frac{1}{s^2+1}\right]$$

$$= e^{-2t}\cos t - 2e^{-2t}\sin t.$$

40. Evaluate $L^{-1}\left[\frac{1}{s^2+6s+13}\right]$. ***[AU, May/June 2014]***

Solution: $L^{-1}\left[\frac{1}{s^2+6s+13}\right] = L^{-1}\left[\frac{1}{(s+3)^2+4}\right]$

$$= e^{-3t}L^{-1}\left[\frac{1}{s^2+4}\right]$$

$$= \frac{1}{2}e^{-3t}\sin 2t.$$

41. Evaluate $L^{-1}\left[\frac{1}{s^2+9s+13}\right]$. ***[AU, Nov/Dec 2017]***

Solution: $L^{-1}\left[\frac{1}{s^2+6s+13}\right] = L^{-1}\left[\frac{1}{(s+3)^2+4}\right]$

$$L^{-1}\left[\frac{2s}{(s+3)^2-2}\right] = L^{-1}\left[\frac{2(s+3)-6}{(s+3)^2-2}\right]$$

$$= e^{-3t}L^{-1}\left[\frac{1}{s^2+4}\right]$$

$$= e^{-3t}L^{-1}\left[\frac{1}{s^2+2^2}\right]$$

$$= e^{-3t}\frac{1}{2}\sin 2t$$

$$= \frac{e^{-3t}}{2}\sin 2t$$

42. Evaluate $L^{-1}\left[\frac{1}{s(s+2)}\right]$.

Solution: $L^{-1}\left[\frac{1}{s(s+2)}\right] = \int_0^t L^{-1}\left[\frac{1}{s+2}\right] dt$

$$= \int_0^t e^{-2t}\, dt = \left[\frac{e^{-2t}}{-2}\right]_0^t = \frac{1-e^{-2t}}{2}.$$

43. Find the inverse Laplace transform of $\frac{1}{s(s^2+a^2)}$.

Solution: $L^{-1}\left[\frac{1}{s(s^2+a^2)}\right] = \int_0^t L^{-1}\left[\frac{1}{s^2+a^2}\right] dt$

$$= \int_0^t \frac{1}{a}\sin at\; dt$$

$$= \frac{1}{a}\left[-\frac{\cos at}{a}\right]_0^t = \frac{1}{a^2}[1-\cos at].$$

44. Find $L^{-1}\left[\log\frac{s+1}{s-1}\right]$.

Solution: Let $F(s) = \log\frac{s+1}{s-1} = \log(s+1) - \log(s-1)$.

Then $F'(s) = \frac{1}{s+1} - \frac{1}{s-1}$.

We know that $L^{-1}[F(s)] = -\frac{1}{t}L^{-1}[F'(s)]$

i.e., $L^{-1}\left[\log\frac{s+1}{s-1}\right] = -\frac{1}{t}L^{-1}\left[\frac{1}{s+1} - \frac{1}{s-1}\right] = \frac{-1}{t}[e^{-t} - e^{t}]$

$$= \frac{1}{t}[e^{t} - e^{-t}] = \frac{2}{t}\sinh t.$$

45. Find $L^{-1}\left[\log\left(1+\frac{a}{s}\right)\right]$.

Solution: Let $F(s) = \log\left(1+\frac{a}{s}\right) = \log\left(\frac{s+a}{s}\right) = \log(s+a) - \log s$.

Then $F'(s) = \frac{1}{s+a} - \frac{1}{s}$.

We know that, $L^{-1}[F(s)] = \frac{-1}{t}L^{-1}[F'(s)]$

i.e., $L^{-1}\left[\log\left(1+\frac{a}{s}\right)\right] = \frac{-1}{t}L^{-1}\left[\frac{1}{s+a} - \frac{1}{s}\right] = \frac{-1}{t}[e^{-at} - 1] = \frac{1}{t}[1 - e^{-at}]$.

46. Find $L^{-1}[\cot^{-1}s]$.

Solution: Let $F(s) = \cot^{-1}s$. Then $F'(s) = \frac{-1}{1+s^2}$.

We know that $L^{-1}[F(s)] = \frac{-1}{t}L^{-1}[F'(s)]$

$$L^{-1}[\cot^{-1}s] = \frac{-1}{t}L^{-1}\left[\frac{-1}{1+s^2}\right] = \frac{1}{t}\sin t.$$

47. Find $L^{-1}\left[\tan^{-1}\frac{a}{s}\right]$.

Solution: Let $F(s) = \tan^{-1}\frac{a}{s}$.

$$\text{Then } F'(s) = \frac{1}{1+\frac{a^2}{s^2}} \times \left(\frac{-a}{s^2}\right) = \frac{s^2}{s^2+a^2}\left(\frac{-a}{s^2}\right) = \frac{-a}{s^2+a^2}.$$

We know that $L^{-1}[F(s)] = \frac{-1}{t}L^{-1}[F'(s)]$

$$L^{-1}\left[\tan^{-1}\frac{a}{s}\right] = \frac{-1}{t}L^{-1}\left[\frac{-a}{s^2+a^2}\right] = \frac{1}{t}\sin at.$$

ANNA UNIVERSITY QUESTIONS

Unit I – Matrices

Part A(2 Marks)

1. If the Eigenvalues of the matrix A of order $3X3$ are $2, 3$ and 1, then find the Eigen values of adjoint of A. **[Refer page 29] [AU, *January 2014, May/June 2016*]**
2. If λ is the Eigenvalue of the matrix A, then prove that λ^2 is the Eigen value of A^2. **[Refer page 28] [AU, *January 2014, May/June 2016*]**
3. Find the Eigenvalues of the inverse of the matrix $A = \begin{bmatrix} 2 & 1 & 0 \\ 0 & 3 & 4 \\ 0 & 0 & 4 \end{bmatrix}$. **[Refer page 29] [AU, *May/June 2014*]**
4. If $2, -1, -3$ are the Eigenvalues of the matrix A, then find the Eigen values of the matrix $A^2 - 2I$. **[Refer page 28] [AU, *May/June 2014*]**
5. If 2, 3 are the Eigenvalues of $\begin{bmatrix} 2 & 0 & 1 \\ 0 & 2 & 0 \\ b & 0 & 2 \end{bmatrix}$, then find the value of b. **[Refer page 27] [AU, *Nov/Dec 2014*]**
6. State Caylely-Hamilton theorem. **[Refer page 71] [AU, *Nov/Dec 2014*]**
7. Find the sum and product of all the Eigenvalues of $\begin{bmatrix} 8 & -6 & 2 \\ -6 & 7 & -4 \\ 2 & -4 & 3 \end{bmatrix}$. **[Refer page 26] [AU, *April/May 2015*]**
8. Give the nature of a quadratic form whose matrix is $\begin{bmatrix} -1 & 0 & 0 \\ 0 & -1 & 0 \\ 0 & 0 & -2 \end{bmatrix}$. **[Refer page 74] [AU, *April/May 2015*]**
9. Find the eigenvalues of $3A + 2I$, where $A = \begin{bmatrix} 5 & 4 \\ 0 & 2 \end{bmatrix}$. **[Refer page 28] [*AU, Dec 2015/Jan 2016*]**
10. What is the nature of the quadratic form $x^2 + y^2 + z^2$ in four variables? **[Refer page 74] [*AU, Dec 2015/Jan 2016*]**
11. If the sum of two eigenvalues and trace of a matrix A are equal, find the value of $|A|$. **[Refer page 70] [AU, *Nov/Dec 2016*]**
12. Write down the matrix corresponding to the quadratic form $2x_1^2 + 5x_2^2 + 4x_1x_2 + 2x_3x_1$.**[Refer page 73] [*AU, Nov/Dec 2016*]**

13. Two Eigenvalues of the matrix $A = \begin{bmatrix} 8 & -6 & 2 \\ -6 & 7 & -4 \\ 2 & -4 & 3 \end{bmatrix}$ are 3 and 0. What is the third eigenvalue? What is the product of the eigenvalues of A?

[Refer page 67] **[AU, *April/May 2017*]**

14. Find the constants a and b such that the matrix $\begin{bmatrix} a & 4 \\ 1 & b \end{bmatrix}$ has 3 and -2 as its eigenvalues. **[Refer page 70]** **[AU, *April/May 2017*]**

15. Find the sum and product of the eigenvalues of a $3X3$ matrix A whose characteristic equation is $\lambda^3 - 7\lambda^2 + 36 = 0$. **[Refer page 27]** **[AU, *Nov/Dec 2017*]**

16. If $\lambda(\neq 0)$ is an eigenvalue of a square matrix A, then show that λ^{-1} is an eigenvalue of A^{-1}. **[Refer page 28]** **[AU, *Nov/Dec 2017*]**

Part B(16 Marks)

1. Find the eigen values and eigenvectors of $\begin{bmatrix} 2 & 2 & 1 \\ 1 & 3 & 1 \\ 1 & 2 & 2 \end{bmatrix}$. **[Refer page 12]**

[*AU, January 2014, Nov/Dec 2015*] (8 marks)

2. Using Cayley-Hamilton theorem find A^{-1} and A^4, if $A = \begin{bmatrix} 1 & 2 & -2 \\ -1 & 3 & 0 \\ 0 & -2 & 1 \end{bmatrix}$. (8 marks)

[Refer page 33] **[*AU, January 2014, April/May 2015*]**

3. Reduce the quadratic form $6x^2 + 3y^2 + 3z^2 - 4xy - 2yz + 4xz$ into a canonical form by an orthogonal reduction. Hence find its index, signature, rank and nature of the quadratic form. **[Refer page 59]** **[*AU, January 2014, Nov/Dec 2014, April/May 2015, Nov/Dec 2016, May/June 2016, Nov/Dec 2017*]**

4. Verify Cayley-Hamilton theorem for the matrix $\begin{bmatrix} 1 & 0 & 3 \\ 2 & 1 & -1 \\ 1 & -1 & 1 \end{bmatrix}$ hence find its inverse.

[Refer page 34] **[*AU, May/June 2014, April/May 2015*]**

5. Find the eigen values and eigenvectors of $\begin{bmatrix} -2 & 2 & -3 \\ 2 & 1 & -6 \\ -1 & -2 & 0 \end{bmatrix}$. **[Refer page 14]**

[*AU, May/June 2014, Nov/Dec 2015*] (8 marks)

6. Reduce the quadratic form $x_1^2 + 5x_2^2 + x_3^2 + 2x_1x_2 + 2x_2x_3 + 6x_3x_1$ to the canonical form through orthogonal transformation and find its nature.

[Refer page 63] **[*AU, May/June 2014, April/May 2015*]**

7. Verify Cayley Hamilton theorem, find A^4 and A^{-1} for the matrix $A = \begin{bmatrix} 2 & -1 & 2 \\ -1 & 2 & -1 \\ 1 & -1 & 2 \end{bmatrix}$. **[Refer page 42]** ***[AU, Nov/Dec 2014, April/May 2015& 2017]***

8. Find the eigen values and eigenvectors of $\begin{bmatrix} 7 & -2 & 0 \\ -2 & 6 & -2 \\ 0 & -2 & 5 \end{bmatrix}$. **[Refer page 16]**

   [AU, Nov/Dec 2014] (8 marks)

9. Reduce the quadratic form $3x^2 + 5y^2 + 3z^2 - 2yz + 2zx - 2xy$ to its canonical form through an orthogonal transformation. **[Refer page 52]**

   [AU, Nov/Dec 2014, Nov/Dec 2015]

10. If $A = \begin{bmatrix} 3 & -1 & 1 \\ -1 & 5 & -1 \\ 1 & -1 & 3 \end{bmatrix}$, verify Cayley-Hamilton theorem and hence find A^{-1}.

    [Refer page 36] ***[AU, April/May 2015]***

11. Find the eigenvalues and eigenvectors of $\begin{bmatrix} 2 & 0 & 1 \\ 0 & 2 & 0 \\ 1 & 0 & 2 \end{bmatrix}$. **[Refer page 18]**

    [AU, Dec 2015/Jan 2016] (8 marks)

12. Verify Cayley Hamilton theorem for the matrix $A = \begin{bmatrix} 1 & 2 & -2 \\ 2 & 5 & -4 \\ 3 & 7 & -5 \end{bmatrix}$. Hence find A^{-1}.

    [Refer page 37] ***[AU, Dec 2015/Jan 2016]*** (8 marks)

13. Find the eigenvalues and eigenvectors of the matrix $\begin{bmatrix} 1 & 1 & 3 \\ 1 & 5 & 1 \\ 3 & 1 & 1 \end{bmatrix}$. **[Refer page 63]**

    [AU, May/Jun 2016] (8 marks)

14. Verify Cayley Hamilton theorem for $A = \begin{bmatrix} 1 & 2 & 3 \\ 2 & -1 & 4 \\ 3 & 17 & -1 \end{bmatrix}$. Hence using it find A^{-1}.

    [Refer page 39] ***[AU, May/Jun 2016]***

15. Find the eigenvalues and eigenvectors of $\begin{bmatrix} 11 & -4 & -7 \\ 7 & -2 & -5 \\ 10 & -4 & -6 \end{bmatrix}$. **[Refer page 22]**

    [AU, Nov/Dec 2016] (8 marks)

16. Using Cayley-Hamilton theorem find A^{-1} where $A = \begin{bmatrix} 7 & 2 & -2 \\ -6 & -1 & 2 \\ 6 & 2 & -1 \end{bmatrix}$.

    [Refer page 38] ***[AU, Dec 2016]***

17. Reduce the matrix $\begin{bmatrix} 10 & -2 & -5 \\ -2 & 2 & 3 \\ -5 & 3 & 5 \end{bmatrix}$ to diagonal form. **[Refer page 48]** **[*AU, April/May 2017*]** (8 marks)

18. The eigenvectors of a 3×3 real symmetric matrix A corresponding to eigenvalues 1,3 and 3 are $(1 \quad 0 \quad -1)^T, (1 \quad 0 \quad 1)^T$ and $(0 \quad 1 \quad 0)^T$ respectively. Find the matrix A by an orthogonal transformation. **[Refer page 49]** **[*AU, Nov/Dec 2017*]** (8 marks)

19. Show that $A = \begin{bmatrix} 2 & 1 & 1 \\ 0 & 1 & 0 \\ 1 & 1 & 2 \end{bmatrix}$ satisfies its own characteristic equation and hence find A^{-1}.

[Refer page 40] **[*AU, Nov/Dec 2017*]** (8 marks)

Unit II - Vector Calculus

Part A (2 Marks)

1. Find the unit normal vector to the surface $x^2 + y^2 = z$ at $(1, -2, 5)$. **[Refer page 179]** **[AU, *May/June 2014, Nov/Dec 2017*]**
2. Prove that $curl\,(grad\,\varphi) = 0$. **[Refer page 177] [AU, *May/June 2014*]**
3. Find $\text{curl}\,\vec{F}$ if $\vec{F} = yz\vec{\imath} + zx\vec{\jmath} + xy\vec{k}$. **[Refer page 98] [AU, *Nov/Dec 2014*]**
4. State Stoke's theorem. **[Refer page 151] [AU, *Nov/Dec 2014, May/June 2016*]**
5. In what direction from (3,1,-2) is the directional derivative of $\varphi = x^2y^2z^4$ maximum? Find also the magnitude of this maximum. **[Refer page 179]** **[AU, *April/May 2015*]**
6. Find a such that $(3x - 2y + z)\vec{\imath} + (4x + ay - z)\vec{\jmath} + (x - y + 2z)\vec{k}$ is solenoidal. **[Refer page 181] [AU, *April/May 2015*]**
7. Prove that $3x^2y\,\vec{\imath} + (yz - 3xy^2)\vec{\jmath} - \frac{z^2}{2}\vec{k}$ is a solenoidal vector. **[Refer page 180]** **[AU, *Nov/Dec 2015*]**
8. State Green's theorem. **[Refer page 133] [AU, *Nov/Dec 2015*]**
9. Evaluate $\nabla^2\, log\,r$. **[Refer page 182] [AU, *May/June 2016*]**
10. Find the unit normal to the surface $xy = z^2$ at $(1, 1, -1)$.**[Refer page]** **[AU, *Nov/Dec 2016*]**
11. Using Green's theorem, evaluate $\int_C (xdy - ydx)$, where C is the circle $x^2 + y^2 = 1$ in the xy – plane. **[Refer page 187] [AU, *Nov/Dec 2016*]**
12. Find the unit normal vector to the surface $x^3 + y^2 = z$ at $(1, 1, 2)$.**[Refer page 178]** **[AU, *April/May 2017*]**
13. Using Green's theorem in the plane, find the area of the circle $x^2 + y^2 = a^2$. **[Refer page 148] [AU, *April/May 2017*]**
14. State Gauss divergence theorem. **[Refer page 163] [AU, *Nov/Dec 2017*]**
15. If $\vec{F} = (x + 3y)\vec{\imath} + (y - 2z)\vec{\jmath} + (x + 2k)\vec{k}$ has divergence zero, find the unknown value of k.**[Refer page 180]** **[AU, *Nov/Dec 2017*]**
16. Evaluate the integral $\int_C \vec{F}\,.d\vec{r}$ if $\vec{F} = xy^2\vec{\imath} + (x^2 + y^2)\vec{\jmath}$ and C is the curve given by $y = x^2 - 4$ from (2,0) to (4,12). **[Refer page 185]** **[AU, *Nov/Dec 2017*]**

Part B(16 Marks and 8 marks)

1. Verify Gauss divergence theorem for $\vec{F} = x^2\vec{i} + y^2\vec{j} + z^2\vec{k}$ taken over the cube bounded by the planes $x = 0,\ y = 0, z = 0, x = 1, y = 1$ and $z = 1$. **[Refer page 163] [AU, *May/June 2014, Nov/Dec 2017*]**
2. Verify Stoke's theorem for $\vec{F} = (x^2 - y^2)\vec{i} + 2xy\vec{j}$ in the rectangular region of $z = 0$ plane bounded by the lines $x = 0, x = a, y = 0$ and $y = b$. **[Refer page 151] [AU, *May/June 2014, April/May 2017*]**
3. Find the angle between the normal's to the surfaces $x^2 = yz$ at the points $(1,1,1)$ and $(2,4,1)$ **[Refer page 86] [AU, *Nov/Dec 2014*]**
4. Verify Green's theorem in the plane for $\int_C (3x^2 - 8y^2)dx + (4y - 6xy)dy$ where C is the boundary of the region defined by $x = 0, y = 0,\ x + y = 1$. **[Refer page 134] [AU, *Nov/Dec 2014*]**
5. Show that $\vec{F} = (x^2 - y^2 + x)\vec{i} - (2xy + y)\vec{j}$ is irrotational and hence find its scalar potential. **[Refer page 107] [AU, *Nov/Dec 2014, Nov/Dec 2017*]**
6. Show that $\nabla^2 r^n = n(n+1)r^{n-2}$ where $r^2 = x^2 + y^2 + z^2$. Hence find the value of $\nabla^2\left(\frac{1}{r}\right)$. **[Refer page 103] [AU, *April/May 2015*]**
7. Using Green's theorem, evaluate $\int_C (y - \sin x)dx + \cos x\, dy$, where C is the triangle formed by $y = 0, x = \frac{\pi}{2}\ , y = \frac{2x}{\pi}$. **[Refer page 145] [AU, *April/May 2015*]**
8. Find the directional derivative of $4x^2z + xy^2z$ at $(1,-1,2)$ in the direction of $2\vec{i} - \vec{j} - 3\vec{k}$. **[Refer page 82] [AU, *Nov/Dec 2015*]**
9. Find $\nabla^2 r^2$ and hence prove that $\nabla^2 r^n = n(n+1)r^{n-2}$. **[Refer page 102] [AU, *Nov/Dec 2015*]**
10. If $\nabla\varphi = 2xyz^3\,\vec{i} + x^2z^3\vec{j} + 3x^2yz^2\vec{k}$ find $\varphi(x,y,z)$ given that $\varphi(1,-2,2) = 4$. **[Refer page 88] [AU, *May/June 2016*]** (8)
11. Using Green's theorem in a plane evaluate $\int_C [x^2(1+y)dx + (x^3 + y^3)dy]$, where C is the square bounded by $x = \pm 1,\ y = \pm 1$. **[Refer page 138] [AU, *May/June 2016*]** (8)
12. Find the values of $'a'$ and 'b', if the surfaces $ax^3 - by^2z = (a+3)x^2$ and $4x^2y - z^3 = 1$ may cut orthogonally at the point $(2,-1,-3)$. **[Refer page 87] [AU, *May/June 2016*]** (8)
13. Prove that $curl\,(curl\,\vec{F}) = grad\ div\,\vec{F} - \nabla^2\vec{F}$. **[Refer page98] [AU,*May/June 2016*]** (8)

14. Find the angle between the surfaces $x^2 + y^2 + z^2 = 9$ and $z = x^2 + y^2 - 3$ at the point $(2, -1, 2)$. **[Refer page 85]** **[AU, *Nov/Dec 2016, April/May 2017*]**

15. Prove that $\vec{f} = (y^2 \cos x + z^3)\vec{i} + (2y \sin x - 4)\vec{j} + 3xz^2\vec{k}$ is irrotational and find its scalar potential. **[Refer page 105]** **[AU, *Nov/Dec 2016*]**

16. Find the directional derivative of $\varphi = 4xz^2 + x^2yz$ at the point $(1, -2, 1)$ in the direction of $2\vec{i} + 3\vec{j} + 4\vec{k}$. **[Refer page]** **[AU, *Nov/Dec 2016*]**

17. Verify Gauss divergence theorem for $\vec{F} = (x^2 - yz)\vec{i} + (y^2 - zx)\vec{j} + (z^2 - xy)\vec{k}$ where S is the surface of the cube formed by the planes $x = 0,\ y = 0,\ z = 0,$ $x = 1,\ y = 1$ and $= 1$. **[Refer page 166]** **[AU, *Nov/Dec 2016*]**

18. Verify divergence theorem for $\vec{F} = 4xz\,\vec{i} - y^2\vec{j} + yz\vec{k}$ taken over the cube bounded by the planes $x = 0,\ x = a, y = 0, y = a, z = 0, z = a$. **[Refer page 169]** ***[AU, April/May 2017, Apr/May 2015, Nov/Dec 2015, Nov/Dec 2014]***

19. Find the constants a, b, c so that $\vec{f} = (x + 2y + az)\vec{i} + (bx - 3y - z)\vec{j} + (4x + cy + 2z)\vec{k}$ is irrotational. For those values of a, b, c find its scalar potential. **[Refer page 108]** ***[AU, April/May 2017]***

20. Apply Green's theorem to evaluate in the plane for $\int_C (xy - x^2)dx + x^2y\,dy$ along the closed curve C formed by $y = 0, x = 1$ and $y = x$. **[Refer page 140]** **[AU, *Nov/Dec '17*]**

21. Evaluate $\iint_S \vec{F}.\hat{n}\, d$ where $\vec{F} = z\,\vec{i} + x\,\vec{j} - 3y^2z\,\vec{k}$ and S is the surface of the cylinder $x^2 + y^2 = 16$ included in the first octant between $z = 0$ and $z = 5$. **[Refer page 129]** **[AU, *Nov/Dec 2017*]**

22. Find the value of n such that the vector $r^n\vec{r}$ is both solenoidal and irrotational. **[Refer page 111]** **[AU, *Nov/Dec 2017, May/June 2014*]**

23. Evaluate $\int_C (x^2 + xy)dx + (x^2 + y^2)dy$ where C is the square formed by the lines $x = \pm 1, y = \pm 1$. **[Refer page 139]** ***[AU, Nov/Dec 2017]***

Unit III – Analytical Functions

Part A(2 Marks)

1. Is the function $f(z) = \bar{z}$ analytic? **[Refer page 255] [AU, *May/June 2014*]**
2. Find the invariant points of $f(z) = z^2$. **[Refer page 261] [AU, *May/June 2014*]**
3. Verify $f(z) = z^3$ is analytic or not. **[Refer page 256] [AU, *Nov/Dec 2014*]**
4. Find the critical points of the transformation $w^2 = (z-\alpha)(z-\beta)$. **[Refer page 232] [AU, *Nov/Dec 2014, May/June 2016*]**
5. Show that $|z|^2$ is not analytic at any point. **[Refer page 196][AU, *April/May 2015*]**
6. Find the invariant points of the transformation $w = \frac{z-1}{z+1}$. **[Refer page 261] [AU, *April/May 2015*]**
7. Prove that $w = \sin 2z$ is an analytic function. **[Refer page 255] [AU, *Nov/Dec 2015*]**
8. Define conformal mapping. **[Refer page 258] [AU, *Nov/Dec 2015*]**
9. Give an example of a function where u and v are harmonic but $u + iv$ is not analytic. **[Refer page 257] [AU, *May/June 2016*]**
10. Find the value of m if $u = 2x^2 - my^2 + 3x$ is harmonic. **[Refer page] [AU, *Nov/Dec 2016*]**
11. Find the image of the circle $|z| = 3$ under the transformation $w = 2z$. **[Refer page] [AU, *Nov/Dec 2016*]**
12. The real part of an analytic function $f(z)$ is constant, prove that $f(z)$ is a constant function. **[Refer page 257] [AU, *April/May 2017*]**
13. Find the critical points of the transformation $= z^2 + \frac{1}{z^2}$. **[Refer page 259] [AU, *April/May 2017*]**
14. Examine whether $y + e^x \cos y$ is harmonic. **[Refer page 257] [AU, *Nov/Dec 2017*]**
15. Find the image of the line $x = 1$ under the transformation $w = z^2$. **[Refer page 262] [AU, *Nov/Dec 2017*]**
16. Is the function $f(z) = z^2$ is analytic? Justify the claim. **[Refer page 255] [*AU, Nov/Dec 2017*]**
17. If $z = i$ is the fixed point of the bilinear transformation $= \frac{1}{z+c}$, then find 'c'. **[Refer page 261] [*AU, Nov/Dec 2017*]**

Part B(8 Marks)

1. Prove that the real and imaginary parts of an analytic function are harmonic functions. **[Refer page 204] [AU, *May/June 2014*]**

2. Find the bilinear transformation that maps $1, i$ and -1 of the z – plane onto $0, 1$ and ∞ of the w – plane. **[Refer page 249] [AU, *May/June 2014, Nov/Dec 2017*]**

3. Show that $v = e^{-x}(x\cos y + y\sin y)$ is harmonic function. Hence find the conjugate harmonic function for v, when $f(z) = u + iv$ is an analytic function of z. **[Refer page 223] [AU, *May/June 2014, Nov/Dec 2017*]**

4. Find the image of $|z+1| = 1$ under the map $w = \frac{1}{z}$. **[Refer page 237] [AU, *May/June 2014, Nov/Dec 2017*]**

5. If $f(z) = u + iv$ is an analytic function, then prove that $\left(\frac{\partial^2}{\partial x^2} + \frac{\partial^2}{\partial y^2}\right) |f(z)|^2 = 4\left|f'(z)\right|^2$. **[Refer page 208] [AU, *Nov/Dec 2014*]**

6. Find the bilinear transformation which maps the points $z = 1,\ i,\ -1$ onto the planes $w = i, 0, -i$. Hence find the image of $|z| < 1$. **[Refer page 250] [AU, *Nov/Dec 2014 &'17*]**

7. If $u = x^2 - y^2$ and $v = \frac{-y}{x^2+y^2}$, prove that u and v satisfies Laplace equation but is not an analytic function(u and v are not harmonic conjugates). **[Refer page 213] [AU, *Nov/Dec 2014*]**

8. Given that $u = \frac{\sin 2x}{\cosh 2y - \cos 2x}$. Find the analytic function whose real part is u. **[Refer page 224] [AU, *Nov/Dec 2014*]**

9. Determine the analytic function $w = u + iv$ if $u = e^{2x}(x\cos 2y - y\sin 2y)$. **[Refer page 223] [AU, *April/May 2015*]**

10. Show that a harmonic function 'u' satisfies the formal differential equation $\frac{\partial^2 u}{\partial z \partial \bar{z}} = 0$ and hence prove that $log|f'(z)|$ is harmonic, where $f(z)$ is a regular function. **[Refer page 209] [AU, *April/May 2015, April/May 2017*]**

11. Find the image in the w-plane of the infinite strip $\frac{1}{4} \le y \le \frac{1}{2}$ under the transformation $w = \frac{1}{z}$. **[Refer page 237] [AU, *April/May 2015*]**

12. Find the bilinear transformation that maps the points $z = 0, -1, i$ into the points $w = i, 0, \infty$ respectively. **[Refer page 248] [AU, *April/May 2015*]**

13. If $f(z) = u + iv$ is an analytic function, then prove that $\left(\frac{\partial^2}{\partial x^2} + \frac{\partial^2}{\partial y^2}\right) |f(z)|^p = p^2 \left|f'(z)\right|^2 |f(z)|^{p-2}$. **[Refer page 210]** **[AU, *Nov/Dec 2015*]**

14. Show that the transformation $w = \frac{1}{z}$ transforms all circles and straight lines. In the $z-$plane into circles and/or straight lines in the $w-$plane. **[Refer page 241]** **[AU, *Nov/Dec 2015, Nov/Dec 2016, April/May2017*]**

15. Determine the analytical function $f(z) = u + iv$, given $u - v = \frac{\cos x + \sin x - e^{-y}}{2(\cos x - \cosh y)}$, and $f\left(\frac{\pi}{2}\right) = 0$. **[Refer page 227]** **[AU, *Nov/Dec 2015*]**

16. Find the bilinear transformation which maps the points $-i, 0, i$ onto the point $-1, i, 1$ respectively. Into what curve the $y-$axis is transformed under this transformation? **[Refer page 253]** **[AU, *Nov/Dec 2015*]**

17. If $f(z) = u + iv$ is an analytic function in $z = x + iy$ then prove that $\left(\frac{\partial^2}{\partial x^2} + \frac{\partial^2}{\partial y^2}\right) |u|^2 = 2\left|f'(z)\right|^2$ **[Refer page 213]** **[AU, *May/June 2016*]**

18. Prove that $w = \frac{z}{z+a}$ where $a \neq 0$ is analytic whereas $w = \frac{\bar{z}}{\bar{z}+a}$ is not analytic.
[Refer page 197] **[AU, *May/June 2016*]**

19. Can $v = \tan^{-1}(y/x)$ be the imaginary part of an analytic function? If so construct an analytic $f(z) = u + iv$, taking v as the imaginary part and hence find u.
[Refer page 220] **[AU, *May/June 2016*]**

20. Find the bilinear transformation that transforms the points $z = 1,\ i, -1$ of the $z-$plane into the points $w = 2, i, -2$ of the $w-$plane. **[Refer page 249]** **[AU, *May/June 2016*]**

21. If $f(z) = u(x,y) + iv(x,y)$ is an analytic function, show that the curves $u(x,y) = c_1$ and $v(x,y) = c_2$ cut orthogonally. **[Refer page 205]** **[*AU, Nov/Dec 2016*]**

22. Find the analytic function $f(z) = u + iv$ whose real part is $u = e^x(x\cos y - y\sin y)$. Find also the conjugate harmonic of u. **[Refer page 223]** **[*AU, Nov/Dec 2016*]**

23. Find the bilinear transformation which maps the points $z = 0, 1, -1$ onto the Points $w = -1, 0, \infty$. Find also the invariant points of the transformation. **[Refer page]** **[*AU, Nov/Dec 2016*]**

24. Find the bilinear transformation which maps the point $-1,\ 0,\ 1$ of the z-plane into the points $-1, -i,\ 1$ of the w-plane respectively. **[Refer page 252]** **[*AU, April/May 2017*]**

25. Find the analytic function $f(z) = u + iv$, given that $2u + 3v = e^x(\cos y - \sin y)$. **[Refer page 225] [*AU, April/May 2017*]**

26. An electrostatic field in the xy-plane is given by the potential function $\varphi = 3x^2y - y^3$, find the stream function. **[Refer page 222] [*AU, Nov/Dec 2017*]**

27. Prove that $f(z) = z^n$ is analytic for all values of n and find its derivatives. **[Refer page 199] [AU, *Nov/Dec 2017*]**

28. Find the image of $|z - 2i| = 2$ under the transformation $w = \frac{1}{z}$. **[Refer page 238] [AU, *Nov/Dec 2017*]**

29. Prove that the function $u(x, y) = \log\sqrt{(x^2 + y^2)}$ is harmonic and hence find its harmonic conjugate. **[Refer page 219] [AU, *Nov/Dec 2017*]**

Unit IV – Complex Integration

Part A(2 Marks)

1. Evaluate $\int_C \frac{z}{z-2} dz$, where C is (a)$|z| = 1$, (b) $|z| = 3$. **[Refer page 349]** **[AU, *May/June 2014*]**

2. State Cauchy's residue theorem. **[Refer page 356] [AU, *May/June 2014, Nov/Dec '17*]**

3. Find the residue of $f(z) = \frac{z^2}{(z+1)^2(z-2)}$ at $z = 2$. **[Refer page 356] [AU, *Nov/Dec 2014*]**

4. State Cauchy's integral theorem. **[Refer page 349]** **[AU, *April/May 2015, Nov/Dec 2015 & 2016*]**

5. Identify the type of singularity of functio n $sin\left(\frac{1}{1-z}\right)$. **[Refer page 354]** **[AU, *April/May 2015*]**

6. Find the residue of $ze^{\frac{-2}{z}}$ at $z = 0$. **[Refer page 355] [AU, *Nov/Dec 2015*]**

7. Expand $f(z) = \frac{1}{z^2}$ as a Taylor's series about the point $z = 2$. **[Refer page 353]** **[AU, *May/June 2016*]**

8. Evaluate the residue of $f(z) = \tan z$ at its singularities. **[Refer page 355]** **[AU, *May/June 2016, Nov/Dec 2016*]**

9. Evaluate $\int_C \frac{e^z}{z-2} dz$, where C is the unit circle with centre as origin. **[Refer page 349] [AU, *April/May 2017*]**

10. Determine the residue of $f(z) = \frac{z+1}{(z-1)(z+2)}$ at $z = 1$. **[Refer page 356]** **[AU, *April/May 2017*]**

11. Expand $\frac{z-1}{z+1}$ about $z = 1$. **[Refer page 353]** **[AU, *Nov/Dec 2017*]**

12. Find the singular points of $(z) = \frac{\sin z}{z}$. **[Refer page 354]** **[AU, *Nov/Dec 2017*]**

13. Evaluate $\int_C \frac{z}{z-1} dz$, C is the circle $|z-1| = 1$. **[Refer page 351]** **[*AU, Nov/Dec 2017*]**

Part B(8 Marks)

1. Obtain the Laurent's series expansion of $f(z)=\frac{z^2-1}{(z+2)(z+3)}$ in $2<|z|<3$.

[Refer page 284] [AU, *May/June 2014, April/May 2015, Nov/Dec 2017*]

2. Evaluate $\int_0^{2\pi}\frac{d\theta}{13+5\sin\theta}$. **[Refer page 317] [AU, *May/June 2014, Nov/Dec 2014 & '17*]**

3. Evaluate $\int_C\frac{z-1}{(z+1)^2(z-2)}dz$, where C is $|z-i|=2$, using Cauchy's integral formula. **[Refer page 275] [AU, *May/June 2014*]**

4. Evaluate by contour integration $\int_0^{\infty}\frac{dx}{(1+x^2)^2}$. **[Refer page 329] [AU, *May/June 2014, Nov/Dec 2017*]**

5. Using Cauchy's integral formula, evaluate $\int_C\frac{z+1}{z^2+2z+5}dz$, where C is the circle $|z+1+i|=2$. **[Refer page 272] [AU, *Nov/Dec 2014*]**

6. Find the Laurent's series of $f(z)=\frac{1}{(z-1)(z-2)}$, valid in the region (a) $|z|>2$ (b) $0<|z-1|<1$. **[Refer page 286] [AU, *Nov/Dec 2014, Nov/Dec 2017*]**

7. Evaluate $\int_0^{\infty}\frac{x^2dx}{(x^2+9)(x^2+4)^2}$ using contour integration. **[Refer page 336] [AU, *Nov/Dec 2014*]**

8. Evaluate $\int_C\frac{z^2}{(z-1)^2(z+2)}dz$, where C is $|z|=3$. **[Refer page 306] [AU, *April/May 2015*]**

9. Evaluate $\int_0^{\infty}\frac{dx}{(x^2+a^2)^2}, (a>0)$ using contour integration. **[Refer page 331] [AU, *April/May 2015*]**

10. Evaluate $\int_C\frac{\tan\left(\frac{z}{2}\right)}{(z-a)^2}dz$, where $-2<a<2$ and C is the boundary of the square whose sides lie along $x=\pm2$ and $y=\pm2$. **[Refer page 276] [AU, *Nov/Dec 2015*]**

11. Evaluate $\int_{-\infty}^{\infty}\frac{\cos x\,dx}{(x^2+a^2)(x^2+b^2)}$ using contour integration, given $a>b>0$.

[Refer page 343] [AU, *Nov/Dec 2015*]

12. Expand Laurent's series of $f(z)=\frac{z}{(z-1)(z-2)}$, in the region (a) $1<|z|<2$

(b) $|z-1|<1$. **[Refer page 289] [AU, *Nov/Dec 2015, May/June 2016*]**

13. Evaluate $\int_0^{2\pi}\frac{\cos 3\theta}{5-4\cos\theta}d\theta$. **[Refer page 322] [AU, *Nov/Dec 2015*]**

14. Evaluate using Cauchy's integral formula: $\int_C\frac{z+1}{(z-3)(z-1)}dz$, where C is the circle $|z|=2$. **[Refer page 275] [AU, *May/June 2016*]**

15. Evaluate $\int_0^{2\pi}\frac{d\theta}{13+12\cos\theta}$ by using contour integration. **[Refer page 318] [AU, *May/June 2016*]**

16. Evaluate $\int_0^{\infty}\frac{x\sin mx}{x^2+a^2}dx$ where $a>0, m>0$. **[Refer page 344] [AU, *May/June 2016*]**

17. Using Cauchy's integral formula, evaluate $\int_C\frac{z\,dz}{(z-1)^2(z+2)}$, where C is the circle $|z-1|=1$. **[Refer page] [AU, *Nov/Dec 2016*]**

18. Using contour integration evaluate $\int_0^{\infty}\frac{\cos mx}{x^2+a^2}dx$.**[Refer page 341][AU, *Nov/Dec '16*]**

19. Find the Laurent's series expansion of $f(z)=\frac{1}{z^2+5z+6}$ valid in the region $1<|z+1|<2$. **[Refer page] [AU, *Nov/Dec 2016*]**

20. Evaluate $\int_C\frac{z\,dz}{(z^2+1)^2}$, where C is the circle $|z-i|=1$, using Cauchy's residue theorem. **[Refer page] [AU, *Nov/Dec 2016*]**

21. Evaluate $\int_C\frac{z+1}{(z-1)(z-2)^2}dz$, where C is the circle $|z-i|=\frac{1}{2}$, using Cauchy's integral formula. **[Refer page 274] [*AU, April/May 2017*]**

22. Evaluate $\int_0^{2\pi}\frac{d\theta}{1-2x\cos\theta+x^2}$, if $0<x<1$ using contour integration. **[Refer page 319] [*AU, April/May 2017*]**

23. Find the Laurent's series expansion of $f(z)=\frac{1}{z^2+4z+3}$, valid in the regions $|z|<1$ and $0<|z+1|<2$. **[Refer page 288] [*AU, April/May 2017*]**

24. Evaluate $\int_0^{\infty}\frac{x^2dx}{(x^2+4)(x^2+9)}$ using contour integration. **[Refer page 335] [*AU, April/May 2017*]**

25.Evaluate $\int_C \frac{\cos \pi z}{z^2-1}\, dz$ around a rectangle with vertices at $2 \pm i,\ -2 \pm i$.

[Refer page 271] [AU, *Nov/Dec 2017*] [Refer page]

26.Evaluate $\int_0^{2\pi} \frac{1}{2+\cos\theta}\, d\theta$ using contour integration. **[Refer page 313]**

[AU, *Nov/Dec 2017*]

27.Use calculus of residues to find $\int_0^{\infty} \frac{1}{(x^2+a^2)(x^2+b^2)}\, dx$ where $a, b > 0$.

[Refer page 333] [AU, *Nov/Dec 2017*]

28.Evaluate $\int_C \frac{z+1}{z^2+2z+4}\, dz$, where C is $|z+1+i| = 2$. **[Refer page 273]**

[*AU, Nov/Dec* 2017]

Unit V – Laplace Transforms

Part A(2 Marks)

1. Find $L\left[\frac{\sin t}{t}\right]$. **[Refer page 444]** **[AU, *May/June 2014, Nov/Dec 2017*]**

2. Evaluate $L^{-1}\left[\frac{1}{s^2+6s+13}\right]$. **[Refer page 448]** **[AU, *May/June 2014*]**

3. Find the Laplace transform of $e^{-t}\sin 2t$. **[Refer page 443]** **[AU, *Nov/Dec 2014*]**

4. Find $f(t)$ if the Laplace transform of $f(t)$ is $\frac{s}{(s-2)^2}$.

   [Refer page 447] **[AU, *Nov/Dec 2014*]**

5. State the sufficiency conditions for the existence of Laplace transform. **[Refer page 440]** **[AU, *April/May 2015, Nov/Dec 2016, April/May 2017*]**

6. Evaluate $\int_0^\infty te^{-2t}\sin t\,dt$ using Laplace transform. **[Refer page 445]**

   [AU, *April/May 2015*]

7. State final value theorem on Laplace transform. **[Refer page 445]** **[AU, *Nov/Dec '15*]**

8. Evaluate $L^{-1}\left[\frac{s+2}{s^2+4s+8}\right]$. **[Refer page 447]** **[AU, *Nov/Dec 2015*]**

9. Evaluate $L^{-1}\left[\frac{s}{s^2+4s+5}\right]$. **[Refer page 448]** **[AU, *May/June 2016*]**

10. State convolution theorem on Laplace transforms. **[Refer page 446]**

    [AU, *May/June 2016*]

11. Find the inverse Laplace transform of $\frac{s}{(s+2)^2}$. **[Refer page 447]** **[AU, *Nov/Dec 2016*]**

12. Find the inverse Laplace transform of $\frac{s^2-3s+2}{s^3}$. **[Refer page 446]**

    [AU, *April/May'17*]

13. Find $L[f(t)]$ if $f(t)=\begin{cases} e^{-t}, & 0<t<4 \\ 0, & t>4 \end{cases}$. **[Refer page 441]** **[AU, *Nov/Dec 2017*]**

14. Find $f(\infty)$, if $L[f(t)]=\frac{1}{s(s+\alpha)}$. **[Refer page 446]** **[AU, *Nov/Dec 2017*]**

15. Evaluate $L^{-1}\left[\frac{1}{s^2+9s+13}\right]$. **[Refer page 448]** **[*AU, Nov/Dec 2017*]**

Part B(8 Marks)

1. Find the Laplace transform of $f(t)$, where $f(t) = \begin{cases} \sin\omega t, & \text{for } 0 < t < \frac{\pi}{\omega} \\ 0, & \text{for } \frac{\pi}{\omega} < t < \frac{2\pi}{\omega} \end{cases}$ and $f(t + 2\pi) = f(t)$. **[Refer page 390] [AU, *May/June 2014*]**

2. Using convolution theorem find the inverse Laplace transform of $\frac{s^2}{(s^2+a^2)(s^2+b^2)}$. **[Refer page 423] [AU, *May/June 2014 &'16, Nov/Dec 2014 &'17*]**

3. Find the Laplace transform of $f(t) = t\,e^{-3t}\cos 2t$. **[Refer page 373] [AU, *May/June 2014, April/May 2017*]**

4. Using Laplace transform, solve $\frac{d^2y}{dt^2} + 4y = \sin 2t$, given that $y(0) = 3$ and $y'(0) = 4$. **[Refer page 431] [AU, *May/June 2014*]**

5. Find the Laplace transform for the following (i) $t\,e^{-t}\cos t$ (ii) $\frac{1-\cos t}{t}$ **[Refer page 373,368] [AU, *Nov/Dec 2014*]**

6. Find the Laplace transform of "triangular wave" function $f(t)$ defined by $f(t) = \begin{cases} t, & if\ 0 \le t \le a \\ 2a - t, & if\ a \le t \le 2a \end{cases}$ and $f(t + 2a) = f(t)$ for all t. **[Refer page 387] [AU, *Nov/Dec 2014, April/May 2015, Nov/Dec 2017*]**

7. Solve the differential equation $\frac{d^2y}{dt^2} - 3\frac{dy}{dt} + 2y = e^{3t}$ with $y(0) = 1$ and $y'(0) = 0$, using Laplace transform. **[Refer page 428] [AU, *Nov/Dec 2014*]**

8. Find $L^{-1}\left[\frac{s}{(s^2+1)(s^2+4)}\right]$. **[Refer page 398] [AU, *April/May 2015*]**

9. Solve the differential equation $y'' - 3y' + 2y = 4t + e^{3t}$, where $y(0) = 1$ and $y'(0) = -1$ using Laplace transforms. **[Refer page 434] [AU, *April/May 2015*]**

10. Find $L\left[\frac{\cos at - \cos bt}{t}\right]$. **[Refer page 369] [AU, *April/May 2015*]**

11. Find the Laplace transform of $t^2 e^{-3t}\cos t$ and $\int_0^t \frac{\sin t}{t}\,dt$. **[Refer page 374,378] [AU, *Nov/Dec 2015*]**

12. Using convolution theorem evaluate $\int_0^t \sin u \cos(t - u)\,du$. **[Refer page 421] [AU, *Nov/Dec 2015*]**

13. Find the Laplace transform of $f(t)$, where $f(t) = \begin{cases} \frac{4E}{T}t - E, & 0 < t < \frac{T}{2} \\ 3E - \frac{4E}{T}t, & \frac{T}{2} < t < T \end{cases}$ and $f(t + T) = f(t)$ and E is a constant). **[Refer page 388] [AU, *Nov/Dec 2015*]**

14.Solve using Laplace transform of $x'' - 2x' + x = e^t$ when $x(0) = 2; x'(0) = -1$. **[Refer page 430] [AU, *Nov/Dec 2015*]**

15.Evaluate (1) $L[t^2 e^{-t} \cos t]$ (2) $L^{-1}\left[\frac{e^{-2s}}{(s^2+2s+1)^2}\right]$. **[Refer page 374, 408] [AU, *May/June 2016*]**

16.Find the Laplace transform of $f(t)$ is defined by $f(t) = \begin{cases} E & if\ 0 < t < \frac{a}{2} \\ -E & if \frac{a}{2} < t < a \end{cases}$ where $f(t + a) = f(t)$. **[Refer page 386] [AU, *May/June 2016 , Nov/Dec 2017*]**

17.Using Laplace transforms technique solve $y'' + y' = t^2 + 2t$, given $y = 4$, $y' = -2$ when $t = 0$. **[Refer page 435] [AU, *May/June 2016, Nov/Dec 2017*]**

18.Find the Laplace transform of the following functions:

(1) $\frac{e^{-t} \sin t}{t}$ (2) $t^2 \cos t$ **[Refer page 365] [AU, *Nov/Dec 2016*]**

19.Using Laplace transform, solve $(D^2 + 3D + 2)y = e^{-3t}$ given $y(0) = 1$ and $y'(0) = -1$. **[Refer page] [AU, *Nov/Dec 2016*]**

20.Using convolution theorem, find $L^{-1}\left[\frac{s}{(s^2+4)(s^2+9)}\right]$.**[Refer page] [AU, *Nov/Dec 2016*]**

21.Find the Laplace transform of the square wave function defined by

$f(t) = \begin{cases} k & if\ 0 < t < \frac{a}{2} \\ -k & if \frac{a}{2} < t < a \end{cases}$, $f(t + a) = f(t)$. **[Refer page 386] [AU, *Nov/Dec 2016*]**

22.Using Laplace transform, solve the differential equation $\frac{d^2y}{dt^2} + 3\frac{dy}{dt} + 2y = e^{-t}$ with $y(0) = 1$ and $y'(0) = 0$. **[Refer page 429] [*AU, April/May 2017*]**

23.Evaluate $\int_0^{\infty} \frac{e^t (\cos 2t - \cos 5t)}{t} dt$. **[Refer page 381] [*AU, April/May 2017*]**

24.Apply convolution theorem to evaluate $L^{-1}\left[\frac{s^2}{(s^2+4)(s^2+9)}\right]$. **[Refer page 424]**

[*AU, April/May* 2017]

25. Find $L^{-1}\left[\log \frac{s^2+4}{(s-2)^2}\right]$ **[Refer page 411] [AU, *April/May* 2017]**

26. Evaluate (i) $\int_0^{\infty} te^{-2t} \sin 3t\, dt$ using Laplace transform and

(ii) $L^{-1}\left[\cot^{-1}\left(\frac{2}{s+1}\right)\right]$ **[Refer page 380, 414] [AU, *Nov/Dec 2017*]**

27. Find $L^{-1}\left[\frac{s^2}{(s^2+a^2)^2}\right]$ using convolution theorem. **[Refer page 422][AU, *Nov/Dec 2017*]**

28.Find Laplace transform of $f(t) = t\, e^{-3t} \sin 3t$. **[Refer page 374] [*AU, Nov/Dec 2017*]**

29.Solve $\frac{d^2y}{dt^2} + 5\frac{dy}{dt} + 6y = 2$, $y(0) = 0$, $y'(0) = 0$. **[Refer page 428][*AU, Nov/Dec 2017*]**

Modal Question Paper

Reg.No.: | | | | | | | | | | | |

Question Paper Code: 80606

B.E./B.Tech. DEGREE EXAMINATION, Nov/Dec 2017

Second Semester

Civil Engineering

MA8251–MATHEMATICS-II

(Common to all Branches except Marine Engineering)

(Regulation 2017)

Time: Three hours Maximum: 100 marks

Answer ALL questions

PART A-(10 x 2 = 20 marks)

1. If $\lambda(\neq 0)$ is an eigenvalue of a square matrix A, then show that λ^{-1} is an eigenvalue of A^{-1}.
2. State Caylely-Hamilton theorem.
3. Find the unit normal to the surface $xy = z^2$ at $(1,1,-1)$.
4. Using Green's theorem, evaluate $\int_C (xdy - ydx)$, where C is the circle $x^2 + y^2 = 1$in the $xy-$ plane.
5. Find the value of m if $u = 2x^2 - my^2 + 3x$ is harmonic.
6. Find the image of the circle $|z| = 3$ under the transformation $w = 2z$.
7. State Cauchy's integral theorem .
8. Find the residue of $f(z) = \tan z$ at $z = \frac{\pi}{2}$.
9. State the sufficient conditions for the existence of Laplace transform .
10. Find the inverse Laplace transform of $\frac{s}{(s+2)^2}$.

PART B– (5 x 16 = 80 marks)

11. (a) (i)Find the eigen values and eigenvectors of $\begin{bmatrix} -2 & 2 & -3 \\ 2 & 1 & -6 \\ -1 & -2 & 0 \end{bmatrix}$. (8)

(ii) Using Cayley-Hamilton theorem find A^{-1} and A^4, if $A = \begin{bmatrix} 1 & 2 & -2 \\ -1 & 3 & 0 \\ 0 & -2 & 1 \end{bmatrix}$. (8)

Or

(b) Reduce the quadratic form $6x^2 + 3y^2 + 3z^2 - 4xy - 2yz + 4xz$ into a canonical form by an orthogonal reduction. Hence find its index, signature, rank and nature of the quadratic form. (16)

12. (a) (i) Find the angle between the surfaces $x^2 + y^2 + z^2 = 9$ and $z = x^2 + y^2 - 3$ at the point $(2, -1, 2)$. (8)

(ii) Prove that $\vec{f} = (y^2 \cos x + z^3)\vec{i} + (2y \sin x - 4)\vec{j} + 3xz^2\vec{k}$ is irrotational and find its scalar potential. (8)

Or

(b)(i) Find the directional derivative of $\varphi = 4xz^2 + x^2yz$ at the point $(1, -2, 1)$ in the direction of $2\vec{i} + 3\vec{j} + 4\vec{k}$. (4)

(ii) Verify Gauss divergence theorem for $\vec{F} = (x^2 - yz)\vec{i} + (y^2 - zx)\vec{j} + (z^2 - xy)\vec{k}$ where S is the surface of the cube formed by the planes $x = 0,\ y = 0,\ z = 0,\ x = 1,\ y = 1$ and $z = 1$. (12)

13. (a) (i) If $f(z) = u(x, y) + iv(x, y)$ is an analytic function, show that the curves $u(x, y) = c_1$ and $v(x, y) = c_2$ cut orthogonally. (8)

(ii) Find the analytic function $f(z) = u + iv$ whose real part is $u = e^x(x \cos y - y \sin y)$. Find also the conjugate harmonic of u. (8)

Or

(b) (i) Show that the transformation $w = \frac{1}{z}$ transforms in general, circles and straight lines into circles or straight lines. (8)

(ii) Find the bilinear transformation which maps the points $z = 0, 1, -1$ onto the Points $w = -1, 0, \infty$. Find also the invariant points of the transformation. (8)

14. (a) (i) Using Cauchy's integral formula, evaluate $\int_C \frac{z\,dz}{(z-1)^2(z+2)}$, where C is the circle $|z - 1| = 1$. (8)

(ii) Using contour integration evaluate $\int_0^\infty \frac{\cos mx}{x^2 + a^2}\,dx$. (8)

Or

(b) (i) Find the Laurent's series expansion of $f(z) = \frac{1}{z^2 + 5z + 6}$ valid in the region $1 < |z + 1| < 2$. (8)

(ii) Evaluate $\int_C \frac{z\,dz}{(z^2+1)^2}$, where C is the circle $|z-i| = 1$, using Cauchy's residue theorem. (8)

15. (a) (i) Find the Laplace transform of the following functions:

(1) $\frac{e^{-t}\sin t}{t}$ (2) $t^2 \cos t$ (8)

(ii) Using Laplace transform, solve $(D^2 + 3D + 2)y = e^{-3t}$ given $y(0) = 1$ and $y'(0) = -1$. (8)

Or

(b) (i) Using convolution theorem, find $L^{-1}\left[\frac{s}{(s^2+4)(s^2+9)}\right]$ (8)

(ii) Find the Laplace transform of the square wave function defined by

$$f(t) = \begin{cases} k & if\ 0 < t < \frac{a}{2} \\ -k & if\ \frac{a}{2} < t < a \end{cases}, \quad f(t+a) = f(t).$$ (8)

BIBLIOGRAPHY

[1] B.S. Grewal, Higher Engineering Mathematics, 41st ed., Khanna Publications, Delhi, 2011.

[2] T. Veerarajan, Engineering Mathematics 3rd ed., Tata McGraw Hill Publishing Company, New Delhi, 2012.

[3] N.P. Bali and Manish Goyal, A Text book of Engineering Mathematics, 8th ed., Laxmi Publications Pvt Ltd, 2011.

[4] H.K. Dass and Er. Rajnish Verma, Higher Engineering Mathematics, S. Chand Private Ltd., 2011.

[5] Glyn James, Advanced Modern Engineering Mathematics, 3rd ed., Pearson Education, 2012.

[6] Peter V. O'Neil, Advanced Engineering Mathematics, 7th ed., Cengage learning, (2012).

[7] B.V. Ramana, Higher Engineering Mathematics, Tata McGraw Hill Publishing Company, New Delhi, 2008.

[8] P. Sivarama Krishna Das and E. Rukmangadachari, Engineering Mathematics Volume II, 2nd ed., PEARSON Publishing 2011.